D1472527

Molecular Approaches to Neurobiology

This is a volume in
CELL BIOLOGY
A series of monographs

Editors: D. E. Buetow, I. L. Cameron, G. M. Padilla, and A. M. Zimmerman

A complete list of the books in this series appears at the end of the volume.

Molecular Approaches to Neurobiology

Edited by

IAN R. BROWN

Department of Zoology
Scarborough College
University of Toronto
West Hill, Ontario, Canada

1982

ACADEMIC PRESS

A Subsidiary of Harcourt Brace Jovanovich, Publishers
New York   London   Toronto   Sydney   San Francisco

ACAP 1982

ACADEMIC PRESS, INC.
111 Fifth Avenue, New York, New York 10003

United Kingdom Edition published by
ACADEMIC PRESS, INC. (LONDON) LTD.
24/28 Oval Road, London NW1 7DX

Library of Congress Cataloging in Publication Data
Main entry under title:

Molecular approaches to neurobiology.

   (Cell biology)
   Includes bibliographies and index.
   1. Neurobiology. 2. Molecular biology. I. Brown,
Ian R. II. Series. [DNLM: 1. Molecular biology.
2. Nervous system. WL 100 M718]
QP356.M56   599.01'88   81-17593
ISBN 0-12-137020-8   AACR2

PRINTED IN THE UNITED STATES OF AMERICA

82 83 84 85   9 8 7 6 5 4 3 2 1

CONTENTS

4 MOLECULAR CHARACTERIZATION OF SYNAPSES OF THE CENTRAL NERVOUS SYSTEM
James W. Gurd

5 AXONAL TRANSPORT OF MACROMOLECULES
Jan-Olof Karlsson

6 MECHANISTIC STUDIES ON THE CELLULAR EFFECTS OF NERVE GROWTH FACTOR
David E. Burstein and Lloyd A. Greene

11 MOLECULAR CORRELATES BETWEEN PITUARY HORMONES AND BEHAVIOR
J. Jolles, V. J. Aloyo, and W. H. Gispen

12 MACROMOLECULES AND BEHAVIOR
Adrian J. Dunn

13 ISOLATION AND CULTURE OF SPECIFIC BRAIN CELLS AND THEIR EXPERIMENTAL USE
H. H. Althaus and V. Neuhoff

LIST OF CONTRIBUTORS

Numbers in parentheses indicate the pages on which the authors' contributions begin.

V. J. Aloyo (285), Division of Molecular Neurobiology, Rudolf Magnus Institute for Pharmacology, and Laboratory for Physiological Chemistry, Medical Faculty, Institute of Molecular Biology, State University of Utrecht, Utrecht 3508 TB, The Netherlands

H. H. Althaus (341), Forschungsstelle Neurochemie, Max-Planck-Institut für Experimentelle Medizin, 3400 Göttingen, West Germany

Xandra O. Breakefield (1), Department of Human Genetics, Yale University School of Medicine, New Haven, Connecticut 06510

Ian R. Brown (41, 221), Department of Zoology, Scarborough College, University of Toronto, West Hill, Ontario M1C 1A4, Canada

David E. Burstein (159), Departments of Pathology and Pharmacology, New York University School of Medicine, New York, New York 10016

James W. Cosgrove (221), Department of Zoology, Scarborough College, University of Toronto, West Hill, Ontario M1C 1A4, Canada

Adrian J. Dunn (317), Department of Neuroscience, College of Medicine, University of Florida, Gainesville, Florida 32610

P. C. Emson (255), Department of Pharmacology, MRC Neurochemical Pharmacology Unit, Medical School, Cambridge CB2 2QH, England

Caleb E. Finch (71), Department of Physiology and Biophysics, School of Medicine, University of Southern California, Los Angeles, California 90033

W. H. Gispen (285), Division of Molecular Neurobiology, Rudolf Magnus Institute for Pharmacology, and Laboratory for Physiological Chemistry, Medical Faculty, Institute of Molecular Biology, State University of Utrecht, Utrecht 3508 TB, The Netherlands

Lloyd A. Greene (159), Departments of Pathology and Pharmacology, New York University School of Medicine, New York, New York 10016

Paul Greenwood (41), Department of Zoology, Scarborough College, University of Toronto, West Hill, Ontario M1C 1A4, Canada

James W. Gurd (99), Department of Biochemistry, Scarborough College, University of Toronto, West Hill, Ontario M1C 1A4, Canada

John J. Heikkila (221), Department of Zoology, Scarborough College, University of Toronto, West Hill, Ontario M1C 1A4, Canada

S. P. Hunt (255), Department of Pharmacology, MRC Neurochemical Pharmacology Unit, Medical School, Cambridge CB2 2QH, England

J. Jolles* (285), Division of Molecular Neurobiology, Rudolf Magnus Institute for Pharmacology, and Laboratory for Physiological Chemistry, Medical Faculty, Institute of Molecular Biology, State University of Utrecht, Utrecht 3508 TB, The Netherlands

Barry B. Kaplan (71), Department of Anatomy, Cornell University Medical College, New York, New York 10021

Jan-Olof Karlsson (131), Institute of Neurobiology, Medical Faculty, University of Göteborg, S-400 33 Göteborg, Sweden

P. Linser (179), Developmental Biology Laboratory, Cummings Life Sciences Center, University of Chicago, Chicago, Illinois 60637

Bruce S. McEwen (195), The Rockefeller University, New York, New York 10021

A. A. Moscona (179), Developmental Biology Laboratory, Cummings Life Sciences Center, University of Chicago, Chicago, Illinois 60637

V. Neuhoff (341), Forschungsstelle Neurochemie, Max-Planck-Institut für Experimentelle Medizin, 3400 Göttingen, West Germany

John E. Pintar (1), Department of Human Genetics, Yale University School of Medicine, New Haven, Connecticut 06510

Michael B. Rosenberg (1), Department of Human Genetics, Yale University School of Medicine, New Haven, Connecticut 06510

*Present address: Psychiatric University Clinic, Nicolaas Beetsstraat 24, Utrecht 3511 HG, The Netherlands.

PREFACE

The neurosciences have experienced marked growth over the past few years as evidenced by the rapid expansion of journals and societies devoted to brain research. This field is of importance not only to those who traditionally have studied the nervous system but to a broad spectrum of biomedical scientists since it represents a frontier research area in which new discoveries are frequently made. The disciplines of anatomy, physiology, and neurochemistry have contributed greatly to the development of the neurosciences. More recent is the contribution of cellular and molecular biology which has opened avenues of investigation into mechanisms regulating differentiated function in cells of the nervous system.

This book deals with current molecular approaches to the analysis of the nervous system. The scope of topics ranges from the subcellular level of chromatin structure and complexity of RNA synthesis to the role of hormones in cell differentiation and molecular correlates between neuropeptides and behavior. The author of each chapter presents an overview of his or her area of speciality, highlighting recent advances and, where appropriate, pointing out current problems which might be impeding progress. Each contribution includes current results from the author's laboratory and concludes with a viewpoint on future directions. The last chapter includes a comprehensive review of procedures for the isolation of specific brain cells and their experimental use.

This book should be useful both to those working in the field of neurochemistry and to those engaged in morphological and physiological approaches to the analysis of the nervous system. Molecular biologists and biochemists working with non-neural tissues may also be interested in this book which surveys the current state of knowledge in a number of areas of molecular neurobiology.

Ian R. Brown

1

CURRENT GENETIC APPROACHES TO THE MAMMALIAN NERVOUS SYSTEM

Xandra O. Breakefield, John E. Pintar, and Michael B. Rosenberg

I. INTRODUCTION

Genetics, as other scientific disciplines, is a set of inherently consistent concepts which can provide unique insights into complex, natural phenomena. This chapter will examine how genetic concepts can be combined with new techniques in molecular biology, cell culture, biochemistry, and developmental biology to expand our understanding of the nervous system. Knowledge of the number, nature, and position of genes controlling neural properties can elucidate the molecular basis of the expression, structure, function, and interaction of these properties. Further, by perturbing neural function at the level of the gene, the relationships between specific molecules and behavior can be established. Genetic

Molecular Approaches to Neurobiology

studies should contribute not only to an integrated, multidimensional view of the nervous system, but also to an understanding of the molecular etiology of inherited neurologic and psychiatric diseases. Here we will summarize genetic approaches currently available and their potential use in the study of proteins critical to neural function.

II. INHERITED VARIATIONS IN GENES AND GENE PRODUCTS

A. Identifying Genes and Their Products

1. Number and Nature of Genes Coding for Proteins

In analyzing a particular protein and its genetic determinants, it is first helpful to establish its structure. If a protein can be purified in sufficient quantity, this can be determined biochemically. If, however, the protein is difficult to purify or exists in more than one form, a genetic analysis becomes useful. A discrete set of genes is necessary for the expression of any protein. These include structural genes coding for the primary amino acid sequence of the subunit(s) and other enzymes involved in processing of mRNA, post-translational modification of the protein, and metabolism of associated molecules. In addition, regulatory genes may affect the expression of structural genes. Examples will be discussed in which families of genes, arising from gene duplication and evolutionary divergence, can give rise to a set of functionally related or functionally distinct proteins, and in which one gene can give rise to a number of different polypeptides.

These types of potential genetic diversity in neural proteins will be considered for the following proteins: tubulin, the acetylcholine receptor, monoamine oxidase (MAO), the insulin family of polypeptides [which includes nerve growth factor (NGF)], adrenocorticotropin (ACTH) β-lipotropin precursor polypeptide, somatostatin, and myelin basic protein. These studies serve to illustrate the extent to which variation in protein structure can result from differences in structural genes coding for these proteins, as well as in processing of mRNA precursors and post-translational modification of polypeptides.

a. Related Genes Code for Functionally Related Proteins. Multiple forms of tubulin, the major component of microtubules, have been identified by biochemical criteria. Two related forms of this protein, α- and β-tubulin have been identified (for review see Raff, 1979). As isolated from chick brain these forms do not differ in molecular weight, but can be distinguished on the basis of peptide maps and isoelectric points (Nelles and Bamburg, 1979). The structural similarity of these two tubulin proteins suggests that genes coding for them arose from a common precursor gene (Cleveland *et al.,* 1980). DNA sequences containing genes coding for α- and β-tubulins have been identified using cDNA probes for them prepared from mRNA of chick brain, where tubulin represents 20% of the total cellular protein. Digestion of chick DNA with several restriction endonucleases reveals the presence of four unique fragments, which hybridize with both the 3' and 5' ends of each of the probes. This provides strong evidence that at least four separate genes code for each form of tubulin. Comparable analysis of human and rodent DNA reveals about ten genes coding for each form. Thus, what appeared to be two related proteins by biochemical criteria has been resolved into a family of eight to twenty related proteins by genetic criteria.

The nicotinic acetylcholine receptor, which mediates membrane events in synaptic transmission, represents another protein critical to nerve function for which two forms, junctional and extrajunctional, have been identified. Although in rat muscle these types of receptors can be distinguished on the basis of their affinities for d-tubocurarine, isoelectric points, and immunologic properties, they have the same subunit composition, and the subunits derived from them yield indistinguishable peptide maps (Nathanson and Hall, 1979). It is still not clear whether subunits of these two receptors originate from separate gene loci or represent different post-translational modifications of the same protein. Messenger RNA isolated from the electric organ of *Torpedo* can be used to direct the *in vitro* synthesis of polypeptides that cross-react with antibodies prepared against the acetylcholine receptor. Some of these newly synthesized polypeptides have apparent molecular weights different from the subunits of the receptor (Mendez *et al.*, 1980). These data are consistent with the known post-translational modifications of these subunits (Rafferty *et al.*, 1980). Amino acid sequencing of the five subunits of the *Torpedo* receptor shows that two are identical and all are structurally related, suggesting that they arose from a common precursor gene. Analysis of the number and structure of genes coding for the mammalian acetylcholine receptor may provide insight into the relationship between junctional and extrajunctional receptors.

Monoamine oxidase, which degradatively deaminates biogenic amines throughout the body, also has two distinct functional types. The A type of MAO activity has a higher affinity for serotonin and norepinephrine and is inhibited by lower concentrations of clorgyline; while the B type has a higher affinity for phenylethylamine and benzylamine and is selectively sensitive to low concentrations of deprenyl (Murphy, 1978; Houslay *et al.*, 1976). The flavin containing polypeptides of MAO-A and -B, isolated from rat and human sources, are structurally different on the basis of apparent molecular weight and peptide maps (Callingham and Parkinson, 1979; Cawthon *et al.*, 1981; Brown *et al.*, 1980; Cawthon and Breakefield, 1979). It is still not clear whether these structural differences result from variations in the primary amino acid sequence of different polypeptides or from post-translational modifications of the same polypeptide. For enzymes such as MAO, which are difficult to purify and for which there is no readily available source of mRNA, techniques of somatic cell genetics provide a means to resolve this issue (see Section II,B,1).

b. Related Genes Code for Functionally Distinct Proteins. A number of genes, which presumably have originated from gene duplications and evolutionary divergence in DNA sequence, can give rise to polypeptides that have substantial homology, but differ dramatically in function. Many "families" of related peptide hormones which serve both neural and endocrine functions have been described (Dockray, 1979; Stewart and Channabasavaiah, 1979). One family, the insulin-related polypeptides, includes insulin, relaxin, the insulin-like growth factors (Blundell and Humbel, 1980), and NGF (Bradshaw *et al.*, 1974). Preliminary studies (L. Villa-Komaroff, personal communication) using a DNA probe for the human insulin gene show that human brain contains at least six distinct mRNA species with substantial homology to this probe; some of these messages may code for as yet uncharacterized neurotropic peptides. Insulin will be considered here because it has been studied in the greatest detail and it serves as a prototype for neuronally active peptides. Insulin is synthesized as a larger molecular weight precursor polypeptide, pre-proinsulin. The initial cleavage in processing involves loss of an amino acid signal sequence. The resulting molecule, proinsulin, then folds with the formation of two disulfide bonds. Two subsequent site-specific cleavages yield two peptide chains joined by the disulfide bonds (insulin) and a free peptide which is subsequently lost. Using cloned cDNA fragments that correspond to mRNA coding for pre-proinsulin, it has been established that most mammalian species contain only one gene for insulin. However, some rodents have two separate gene loci for this protein which differ only slightly in structure (Cordell *et al.*, 1979; Lomedico *et al.*, 1979). Both forms of insulin are synthesized in equal amounts and appear to be functionally equivalent.

The biologically active subunit of NGF, β-NGF, is also derived from a larger molecular weight precursor pro-β-NGF (Berger and Shooter, 1977), which in turn may be derived from an even larger MW precursor (Wiche, 1979). Genes for β-NGF and insulin are believed to have arisen from a common precursor gene (Bradshaw *et al.*, 1974). However, proinsulin and β-NGF polypeptides have only about 30% sequence homology indicating that a substantial degree of diversification has occurred since the putative gene duplication event. These differences in amino acid sequence suggest that these two polypeptides may differ in conformation, and, in fact β-NGF is not able to compete with insulin for its receptor (Blundell and Humbel, 1980).

c. One Gene Codes for Many Proteins. An interesting example of one gene transcript giving rise to a number of functionally distinct polypeptides is provided by the ACTH-β-lipotropin precursor molecule synthesized in the pituitary (Roberts *et al.*, 1978; Eipper and Mains, 1980; Herbert, 1981) and possibly in the brain (Krieger and Liotta, 1979). Different subsets of biologically active peptides are generated from similar and possibly identical precursor molecules by differential post-translational modifications mediated by site-specific proteolysis and glycosylation (Fig. 1; Roberts *et al.*, 1978). In the anterior lobe of the pituitary, ACTH and β-lipotropin are the prominent end-product peptides, whereas in the intermediate lobe, α-melanotropin, corticotropin-like intermediate lobe peptide, and β-endorphin are the major end-products. These differences in processing are apparently lobe specific throughout development (J. Pintar and R. Allen, personal communication). Although there appears to be only one bovine gene coding for the ACTH β-lipotropin precursor (Nakanishi *et al.*, 1980), there appear to be two rodent genes (J. Roberts, personal communication), and thus differential processing in this latter species may result from differential gene expression.

Another example where one gene produces two structurally different, but functionally equivalent, peptides is provided by human somatostatin, a growth hormone produced by the pituitary. Two forms differing in molecular weight can be identified; the smaller is identical to the larger except that it lacks an internal sequence of amino acids (Lewis *et al.*, 1980). Nucleotide sequencing of the gene for somatostatin reveals that the smaller form is generated when an "error" occurs during excision of an intron region of the precursor mRNA (Fiddes *et al.*, 1979). Such processing errors can be explained by regions near the splicing signals which are similar in sequence to them. The occurrence of several com-

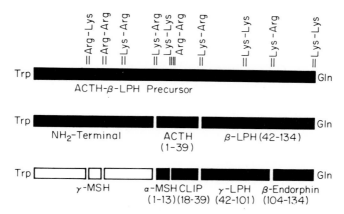

Fig. 1. The ACTH-β-lipotropin precursor polypeptide and different modes of processing. In the anterior lobe of the pituitary proteolytic cleavages and glycosylation events yield primarily ACTH and β-lipotropin (β-LPH) whereas in the intermediate lobe of the pituitary these events yield α-melanotropin, corticotropin-like intermediate lobe peptide (CLIP), γ-lipotropin (γ-LPH), and β-endorphin. (Reprinted from Herbert, 1981, with permission.)

mon modes of processing of the same precursor mRNA may also explain the four forms of myelin basic protein normally present in mice which are identical in sequence except for the presence or absence of two internal sequences (Barbarese *et al.,* 1977).

These findings show that the relationship between gene structure and protein structure is not always apparent, and that even functionally identical proteins are not necessarily the product of a single gene locus. Even in the case of monomeric proteins, structural diversity may result from differences in the sequence of separate gene loci or different alleles at the same locus, as well as in post-transcriptional processing of precursor mRNA and post-translational modification of polypeptides. A number of techniques are available to elucidate structural differences in proteins. The most complete involves amino acid sequencing, but not all proteins can be obtained in sufficient quantities for this procedure. Other techniques include electrophoretic separation of proteins on the basis of molecular weight, charge, and conformation; electrophoretic and chromato-graphic separation of peptides generated from proteins by site-specific proteases; and identification of antigenic differences between molecules using antisera or monoclonal antibodies. Although these latter techniques are not sensitive enough to resolve all possible differences in protein structure that could affect the functional capacity of the molecule, they have allowed resolution of some inherited variations in the structure of neural proteins (Sections II,A,2 and IV,B).

2. "Mutant" Proteins

Although all changes in DNA sequence can be classified as mutations, only a few of these mutations result in functionally defective gene products. A few examples will be discussed here in which disruption of function has resulted from mutations leading to production of structurally altered forms of tubulin, insulin, the neurophysins, myelin basic protein, and two components of the hormone-sensitive adenylate cyclase response.

Structurally altered forms of β-tubulin which interfere with microtubular assembly and confer resistance to antimitotic drugs have been described in *Drosophila* and in cultured Chinese hamster cells. In *Drosophila,* a unique form of β-tubulin specific to the testis has been demonstrated by two-dimensional gel electrophoresis (Kemphues *et al.,* 1979). A phenotypically dominant mutation that produces sterility in males is characterized by the production of equal amounts of a normal and an altered form of this testis-specific β-tubulin, which differs from the normal in apparent molecular weight and charge. The ultrastructure of the testes of heterozygous flies strongly suggests that coassembly of the mutant and wild-type tubulins leads to formation of structurally defective microtubules; meiotic events and formation of the sperm tail axoneme are disrupted.

A structurally altered form of β-tubulin has also been found in cultured Chinese hamster ovary cells following mutagenesis and selection for resistance to the antimitotic drugs colcemid and griseofulvin (Cabral *et al.*, 1980). The variant form of tubulin has a different isoelectric point from the normal. Mutant cells express both normal and variant forms of tubulin. *In vitro* translation of poly(A) mRNA isolated from these cells has revealed the presence of messages coding for both forms. Thus, the variant tubulin does not result from an alteration in post-translational modification, but must be due to a difference in gene structure or post-transcriptional processing of mRNA. Studies using cloned cDNA probes for β-tubulin have indicated that there may be up to ten copies of this gene in mammals (see Section II,A,1). It appears, then, that even a single mutation in one of the twenty allelic copies of this gene can alter the structure of microtubules sufficiently to change their sensitivity to antimitotic drugs. This phenomenon presumably results from the random insertion of altered tubulin subunits into microtubules composed predominantly of normal tubulin. The structural disorientation of a polymeric molecule by a minor component illustrates how phenotypically dominant mutations can occur even in multigene families.

Two interesting structurally altered forms of human insulin have been described which also show dominant phenotypic expression. A patient with diabetes mellitus produces insulin molecules which contain a single amino acid substitution (Tager *et al.*, 1979). This substitution does not affect the charge of the molecule, but results in a biologically inactive form of insulin that antagonizes the action of normal insulin. Another altered form of insulin has been described in which a single amino acid substitution in one of the two sites of proteolytic cleavage of the proinsulin molecule leads to formation of a larger, circulating proinsulin-like molecule (Gabay *et al.*, 1979). This altered form of the molecule is present in greater quantities than its normal counterpart, possibly due to a slower rate of degradation. However, this proinsulin-like molecule has some biologic activity and the condition asymptomatic.

The peptide hormones vasopressin and oxytocin are produced by separate neuronal populations in the posterior pituitary and are stored in granules along with their respective "carrier proteins," the neurophysins (Pickering and Jones, 1978). These neuropeptides and their carrier proteins appear to be synthesized from a common precursor molecule. Brattleboro rats (an inbred strain) are deficient in production of the antidiuretic hormone vasopressin, and have diabetes insipidus (Valtin *et al.*, 1974). These rats lack vasopressin, its associated neurophysin, and the normal precursor molecule for both peptides; while they retain oxytocin, its neurophysin, and their precursor (Brownstein *et al.*, 1980). Since inbred strains are homozygous at most alleles, this deficit presumably results from a mutation in both copies of a gene coding for this precursor molecule and/or for an enzyme involved in its processing.

Preliminary studies (J. H. Carson, personal communication) on the "shiverer" mutant in mice suggest that this recessive mutation leading to loss of myelin basic protein is responsible for the lack of myelin in the central nervous system of these animals. Although myelin basic protein is also absent in the peripheral nervous system, myelin is still present there and appears to be structurally normal. These mutants lack poly(A) mRNA coding for myelin basic protein, indicating either a change in the gene for myelin basic protein such that it is not transcribed, or an altered form of the message which is not processed normally and/or is degraded rapidly.

A series of "mutant" cell lines deficient in several aspects of hormone-responsive adenylate cyclase have been generated using lymphosarcoma cells that normally die in the presence of high levels of dibutyryl cyclic AMP (Coffino *et al.*, 1978) (also see Section III,A). In three of the mutants, structurally altered or missing proteins appear to be responsible for these deficits. The hormone-responsive adenylate cyclase system involves at least three membrane components: a receptor protein (e.g., a β-adrenergic or prostaglandins receptor), a "coupling" protein(s) (termed G/F factor or GTP binding protein), and the enzyme adenylate cyclase. The G/F factor can be labeled with ^{32}P-NAD, through ADP ribosylation mediated by cholera toxin, and resolved into two bands on SDS polyacrylamide gels of apparent MW 42,000 and 52,000 (Cassel and Pfeuffer, 1978). One class of mutants (AC⁻) able to grow in the presence of high levels of dibutyryl cyclic AMP has receptors and adenylate cyclase activity (with Mn^{2+}-ATP, but not Mg^{2+}-ATP as the substrate; Ross and Gilman, 1977), and lacks both protein components of the G/F factor (Johnson *et al.*, 1978). Another class of mutants (UNC⁻), selected for growth in the presence of a stable agonist of the β-adrenergic receptor and a phosphodiesterase inhibitor, has receptors and normal adenylate cyclase activity, but has an altered form of the G/F factor with the same apparent molecular weight 42,000 and 52,000 as the wild type, but an altered charge (Ross *et al.*, 1979; Schleifer *et al.*, 1980). These studies have helped to elucidate the role of the G/F factor in the hormone coupled adenylate cyclase response.

Variations in cyclic AMP dependent protein kinase have also been obtained by selecting lymphosarcoma cells able to grow in the presence of dibutyryl cyclic AMP (Hochman *et al.*, 1975; Steinberg *et al.*, 1977). One class of mutant cells required about ten times as much cyclic AMP, as wild-type cells, to dissociate protein kinase into its subunits and thus activate its catalytic activity (Hochman *et al.*, 1975; Nimmo and Cohen, 1977). Mixing experiments using subunits from both cell types revealed that the mutant regulatory subunit had an altered binding affinity for the catalytic subunit. By using affinity chromatography to enrich for the regulatory subunits and analyzing these polypeptides by two-dimensional gel electrophoresis, the mutant subunits were shown to have the same apparent molecular weight, but an altered net charge as compared to the wild type.

Although there are only a few current examples of inherited structural alterations in neural proteins that affect their function, still it is clear that by generating a set of mutations in different regions of the same peptide, one can elucidate the portions of the molecule critical to its binding to other subunits, substrates, and cellular components. Similarly, by obtaining mutants altered in the same function, it is possible to identify the number of different genes and proteins which contribute to that function.

3. DNA and RNA Structure

Since most of this volume deals with molecular biology, we will consider these techniques only briefly in this chapter. To fully understand the genetic control of a protein it is necessary to understand not only its structure and processing, but also its origins in RNA and DNA.

In tissues in which a protein of interest is being synthesized in substantial quantities, precursor polypeptides have been studied by isolating total poly(A) mRNA, using it to direct *in vitro* translation, and then immunoprecipitating and analyzing the immediate translation products. This approach has been used, for example, in studies of β-NGF from mouse submaxillary gland (Wiche, 1979), somatostatin from mouse and rat hypothalami (Joseph-Bravo *et al.*, 1980), and the acetylcholine receptor from the electric organ of *Torpedo* (Mendez *et al.*, 1980). In all three cases precursor molecules differing in apparent molecular weight from the final polypeptides were identified by SDS polyacrylamide gel electrophoresis. Because of potential cross-reactivity between proteins (e.g., Julliard *et al.*, 1980), it is important to establish the authenticity of the immunoprecipitated molecules. This can be done by comparing proteins precipitated with preimmune and immune sera, and by identifying protein structure by peptide mapping.

Total mRNA or partially purified mRNA (e.g., obtained by sucrose density gradient centrifugation or by immunoprecipitation of polyribosomes) can be used to generate cDNA probes. These probes can be used to study the structure of the mRNA. Sequencing of the probe can elucidate features of the untranslated portion of the mRNA molecule which may have a role in processing and turnover of mRNA, positioning of the mRNA and nascent polypeptides within the cell, and rates of protein translation. By hybridizing this probe with cellular RNA it is also possible to identify and characterize precursor forms of the mRNA.

DNA probes can also be used to identify specific gene sequences in the genome and to determine the number and structure of these genes, including intervening and flanking sequences. Probes can be generated not only from cDNA transcribed from mRNA, but also from synthetically synthesized nucleotide sequences which code for a unique amino acid sequence in the molecule, and from random fragments of nonrepetitive DNA sequences generated by restriction endonuclease digestion of total cellular DNA (Maniatis *et al.*, 1978).

Probes for neural proteins are just beginning to be obtained. The cloning of the cDNA sequences for brain tubulin and actin has recently been reported by Cleveland *et al.* (1980) and Ginzburg *et al.* (1980). Examples of the kinds of information that such probes can provide are illustrated by studies with probes for other types of proteins. Studies of globin genes have helped elucidate the evolutionary relatedness of the different globins, the position of these genes in the genome, the number of copies of each type of gene, and individual variations in gene number and structure in coding, noncoding, and flanking sequences (for review see Forget, 1979). For insulin genes, it has been shown that whereas two nonallelic loci of the gene occur in some rodents, most species including humans have only one locus (Cordell *et al.*, 1979; Lomedico *et al.*, 1979). Further, sequencing of genes for insulin in humans has revealed the presence of two allelic forms of the gene, which differ only in noncoding regions (Ullrich *et al.*, 1980). In addition to being used to analyze inherent variations in gene structure associated with normal and abnormal variations in protein structure and function, probes can also be specifically "mutated" *in vitro* and introduced into cells or organisms to create new variants (see Section III,A and B).

B. Locating Genes

1. Mapping the Chromosomal Location of Genes

The assignment of specific structural genes to individual chromosomes and to regions within chromosomes has been carried out extensively for human and rodent cells. Major advances in this field were gained by staining techniques that allow identification of individual chromosomes and specific regions within them (e.g., Francke and Oliver, 1978) and by somatic cell genetic techniques that allow introduction of chromosomes from donor cells into recipient cells (Giles and Ruddle, 1973). Donor chromosomes can be transferred into the recipient cell by fusion with the donor cell, with the subsequent loss of some donor chromosomes (Ringertz and Savage, 1976), or by fusion with donor cell vesicles ("microcells"), which contain only one or a few intact chromosomes (e.g., Fournier and Ruddle, 1977). "Clone panels" have been established in which a series of hybrid clones contain a small complementary set of donor chromosomes (e.g., Mellman *et al.*, 1979).

To map a gene it is necessary that the donor gene product is expressed in hybrids and can be distinguished from the recipient gene product, or that the presence of the donor gene itself can be established. Because it is difficult to construct appropriate hybrid cells that express differentiated neural properties, this latter approach may prove more useful in mapping neural genes. When the chromosome containing the gene of interest has been identified, one can use a series of hybrid lines which contain various portions of this chromosome to assign the gene to a particular region within the chromosome.

We have recently mapped the chromosomal location of a structural gene coding for human MAO using somatic cell hybridization techniques (Pintar *et al.*, 1981). Hybrids were formed between mouse neuroblastoma cells and normal human skin fibroblasts (Fig. 2). The neuroblastoma parent lacked hypoxanthine phosphoribosyltransferase (HPRT) activity and MAO activity (Breakefield *et al.*, 1976), both of which were present in the human parent. Following fusion, the cells were grown in medium containing hypoxanthine, aminopterin, and thymidine (HAT) (Littlefield, 1964), as well as ouabain (Mankovits *et al.*, 1974), to select against growth of the mouse and human parent cells, respectively. HAT medium also selects for hybrid cells which retain the human X chromosome since it bears the gene for HPRT (Pearson *et al.*, 1979). All hybrid lines selected in HAT medium expressed MAO activity (Table I); the A form, not the B form, of the enzyme predominated. Conversely, all hybrids except one cloned in medium containing 6-thioguanine (6TG), to select for cells which had lost HPRT activity and the human X chromosome, lost MAO activity. The

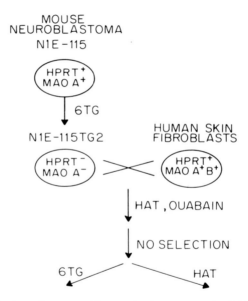

Fig. 2. Hybridization scheme used in mapping human gene for MAO. Variants of mouse neuroblastoma line N1E-115 were selected by growth in 6-thioguanine (6TG). One of these variants N1E-115TG2 lost both MAO-A and HPRT activities. This variant was fused to normal human skin fibroblasts from a male which expressed MAO-A and -B, as well as HPRT activities. Following fusion, cells were plated in medium containing HAT and ouabain. Survivors were plated at cloning density with no selection; colony clones were isolated, tested, and recloned in medium containing either 6TG or HAT. (Reprinted from Pintar *et al.*, 1981, with permission. Copyright 1981 by Williams & Wilkins, Baltimore.)

Table I.   Analysis of Neuroblastoma × Fibroblast Hybrids[a]

				Human		
Lines	Medium[b]	MAO[c]	HPRT[d]	X chromosome[e]	PGK[f]	G6PD[f]
NGM1E1	NS	+	+	+	+	+
NGM5F1	NS	+	+	+	+	+
NGM1E1HAT2B	HAT	+	+	+	+	+
NGM1D1HAT3A	HAT	+	+	+	+	+
NGM1E1TG1A	6TG	−	−	−	−	−
NGM5F1A2	6TG	−	−	−	−	−
NGM5F1D2	6TG	+	−	−	+	−

[a] Reprinted from Pintar *et al.*, 1981, with permission. Copyright 1981 by Williams & Wilkins, Baltimore.

[b] Hybrid lines were cloned and maintained in nonselective medium (NS), or in medium containing HAT or 6TG.

[c] MAO activity was measured using [³H]tryptamine as described (Costa *et al.*, 1980). Values below the level of detection (<1.0 pmoles/min/mg protein) are designated as negative.

[d] HPRT activity was measured using [¹⁴C]hypoxanthine as described (Pintar *et al.*, 1981). Values below the level of detection (<12 pmoles/min/mg protein) are designated as negative.

[e] The presence of the human X chromosome was determined by scanning metaphase chromosome spreads stained by the Giemsa banding technique (Francke and Oliver, 1978).

[f] Human forms of these enzymes were detected by electrophoresis on cellulose acetate thin layer sheets and staining for enzyme activity (Pintar *et al.*, 1981).

MAO-A in HAT selected hybrids was shown to be of human origin by limited proteolysis and peptide mapping of [³H]pargyline-labeled MAO molecules (Fig. 3). [Pargyline binds irreversibly to the flavin containing subunit of MAO (Chuang *et al.*, 1974)]. Thus the structural gene for the flavin-polypeptide of MAO-A is on the human X chromosome.

The one exceptional hybrid clone selected for loss of HPRT activity that continued to express human MAO-A has provided some information about the relative position of the MAO gene on the human X chromosome. These hybrid cells expressed the human form of phosphoglycerate kinase (PGK), but not the human form of glucose-6-phosphate dehydrogenase (G6PD) (Table I); genes for both of these enzymes are on the human X chromosome (Pearson *et al.*, 1979). Further, these cells contained no recognizable portion of the human X chromosome, but did contain several fragments of human chromosomal material, both free and attached to mouse chromosomes (Pintar *et al.*, 1981). This hybrid clone, then, appears to have retained the portion of the q arm of the human X chromosome that bears the genes for PGK and MAO.

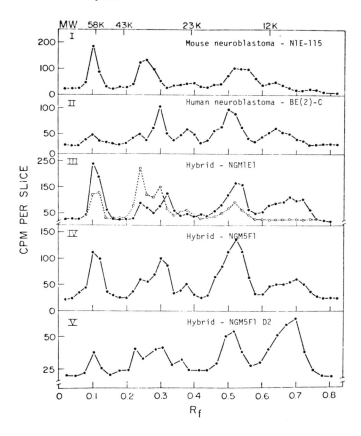

Fig. 3. Peptide maps of [³H]pargyline-labeled MAO-A from human, rodent, and hybrid cell lines. First, labeled proteins from crude mitochondrial preparations were resolved by SDS polyacrylamide gel electrophoresis, then the portion of the gel containing labeled MAO was cut out and placed in the well of another gel with *Staphylococcus aureus* V8 protease, and electrophoresis was carried out as described by Cawthon and Breakefield (1979). The distribution of peptides in the second gel was determined by slicing and counting. The migration relative to the bromphenol blue dye (R_f) is plotted on the lower abscissa; the migration of standard molecular weight markers on the upper abscissa. Radioactivity in each 2 mm slice is plotted on the ordinate. A representative map illustrating all peptides observed is shown for each line. The leftmost peak represents intact MAO molecules. (I) Mouse neuroblastoma line N1E-115 from which the parent line N1E-115TG2 was derived and which expresses MAO-A activity. (II) Human neuroblastoma line BE(2)-C (kindly provided by Dr. J. Biedler) which illustrates the peptide map also found for human MAO A from fibroblasts and placental tissue (unpublished data). (III–V) Hybrid lines. (Reprinted from Pintar *et al.*, 1981, with permission. Copyright 1981 by Williams & Wilkins, Baltimore.)

It is also possible to map the chromosomal location of genes by digesting total cellular DNA from hybrids with specific restriction endonucleases and identifying fragment(s) containing the gene(s) of interest by hybridization to labeled DNA probes. Since the position of restriction endonuclease sites can vary among species, the same gene from different species may be identified by the length of the DNA fragment(s) in which it is found. The advantage of this approach is that the gene need not be expressed to be mapped. Thus the human insulin gene was mapped by digestion of DNA from hybrids lines in a clone panel constructed from fusions between Chinese hamster ovary cells and human fibroblasts (Owerbach *et al.*, 1980). Only hybrids that contained human chromosome 11 had the human DNA fragment containing the insulin gene.

Radiolabeled DNA probes homologous to specific genes of interest can also be used to determine the chromosome location of these genes by using *in situ* hybridization techniques (Henderson *et al.*, 1978). Metaphase chromosome spreads are incubated under conditions that denature double stranded helices, a single stranded ^{125}I-labeled DNA probe is hybridized to these denatured chromosomes, and autoradiography is carried out to identify the position of the probe. This procedure works best for genes present in a number of tandem copies (e.g., genes for histones and ribosomal RNA genes, or amplified genes) and is usually not sensitive enough for unique copy sequences.

As DNA probes become available for neural genes, mapping can proceed rapidly using these techniques of somatic cell and molecular genetics. Establishing the chromosomal location of genes coding for proteins critical to nerve function should provide insight into their evolutionary relationships and their regulated expression during development, as well as aid in the construction of detailed linkage maps.

2. Linkage Analysis of Gene Loci

Assigning a gene locus to a particular chromosomal region is a relatively gross method of determining its relationship to nearby genes, since one microscopically distinct band on a chromosome may contain up to 100 gene loci. Linkage analysis can provide more detailed information about the position of genes relative to each other. Several approaches will be discussed here, including classical techniques which use recombination frequencies and protein polymorphisms, as well as newer methods which use DNA restriction endonuclease site polymorphisms and gene transfer.

Classical methods of linkage analysis of mammalian genes involve animal breeding and examination of human pedigrees to determine the frequency of co-inheritance of specific traits. The more frequently two traits are inherited together, the less recombination has occurred between the genes coding for them during meiosis, and the closer the genes are to each other on a linkage map. Although information derived from linkage analysis can be applied to a physical map of the chromosomes, most chromosomes contain more than one linkage group [unlinked genes are transmitted randomly to offspring], and the distances on a linkage map are not proportional to physical distances on metaphase chromosomes. In the past, linkage analysis has depended on the demonstration of phenotypic diversity (polymorphisms) in specific gene products. Traits with common polymorphic forms in the population provided the basis for the construction of linkage maps, and rare traits were positioned relative to these common polymorphic markers. In mice, where extensive breeding analysis has been carried out, out of 69 genes known to affect neural function, 48 have been positioned on the genome (Searle, 1979; 1980; Sidman *et al.*, 1965).

In humans classical linkage mapping has not yet provided any information about the chromosomal location of neuronally important genes, and has offered only limited insight into whether certain inherited neurologic and psychiatric diseases of unknown etiology are the result of variations in the structure of one or more gene loci. Limitations in these analyses result from the small size of family pedigrees and the relatively small number of common polymorphic traits in humans. (The 30 commonly used traits can provide linkage information for only about one-third of the genome.) Although some workers have reported that a gene responsible for a sex-linked form of bipolar (manic-depressive) illness is linked to genes on the X chromosome (Mendlewicz *et al.*, 1979), other workers have disputed these claims (Leckman *et al.*, 1979). Linkage analyses of schizophrenia (Elston *et al.*, 1973), Huntington disease (Beckman *et al.*, 1974; Hodge *et al.*, 1980), and unipolar depressive illness (Tanna *et al.*, 1976), to name a few, have not revealed the number or location of responsible gene loci.

Against this background of frustration in attempts to use classical linkage analysis to identify human genes affecting neural function, recent advances in molecular genetics have provided a single technique that can generate a large number of common polymorphic markers which can be used to establish linkage over the entire genome. In this approach polymorphisms in the sequence of the DNA itself are utilized as linkage markers (Botstein *et al.*, 1980). Probes for unique copy sequences in the genome are generated by cloning either random sequences from the total genome (see Maniatis *et al.*, 1978) or specific sequences homologous to mRNA's or proteins of interest. Total DNA from several individuals is digested with site-specific restriction endonucleases and fragments are separated on the basis of molecular weight in agarose gels. If the position of at least one of the restriction endonuclease sites on either side or within the sequence in the genome homologous to the probe is different among chromosomes from the same or different individuals, the probe will hybridize with fragments of different molecular weight (Fig. 4). Two or more hybridizing bands will be identified either when an individual is heterozygous for a given restriction endonuclease site polymorphism and/or when the site lies within the region of the genome homologous to the probe. A number of laboratories are actively engaged

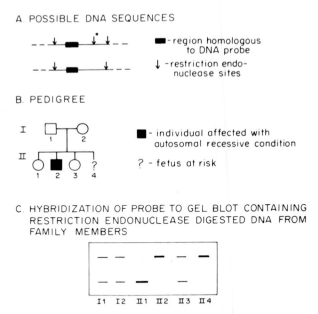

Fig. 4. Use of restriction endonuclease fragment length polymorphisms in prenatal diagnosis. (A) Restriction endonuclease sites for a specific endonuclease in a portion of DNA which is homologous to the probe used. The polymorphism (*) is the presence or absence of one of these sites. (B) In this pedigree one of the children (II2) is affected with an autosomal recessive disorder. Amniotic fluid cells are removed during a subsequent pregnancy (II4). (C) DNA is isolated from family members of the pedigree and digested with the endonuclease. Fragments are separated on the basis of molecular weight in agarose gels and the position of the fragment of DNA containing the sequence homologous to the probe is determined. (The gene causing this disorder is contained within this same fragment of DNA.) In this family the mutant gene cosegregates with lack of the polymorphic restriction endonuclease site. Thus the fetus (II4) has two copies of the mutant gene and will be affected. (Adapted from Fig. 1, Botstein, D., White, R. L., Skolnick, M., and Davis, D. W. "Construction of a genetic linkage map in man using restriction fragment length polymorphisms," *Am. J. Hum. Genet.* **32,** 314–331. Copyright 1980 by American Society of Human Genetics, Chicago, Illinois.)

in identifying probes that will reveal common polymorphism in DNA sequences in the human population. For example, a number of polymorphisms adjacent to the globin gene have been identified; one is tightly linked to a variant form of the β-globin gene responsible for sickle cell anemia (Kan and Dozy, 1978), another to the variant form of this gene causing β-thalassemia (Little *et al.*, 1980). At least eight polymorphic "alleles" in the control population have been characterized in sequences surrounding a random sequence probe (Wyman and White, 1980). It will take approximately 150 probes that reveal common polymorphisms in DNA sequences and are spaced evenly along the DNA to establish a linkage map for the entire human genome (Botstein *et al.*, 1980). By pedigree analysis the linkage relationship of other DNA probes and phenotypic traits can then be established.

This restriction endonuclease site linkage map can be used to establish the number and location of the gene(s) responsible for inherited neurologic and psychiatric diseases. Within a family, restriction endonuclease site polymorphisms which are tightly linked to the responsible gene can be used to detect the presence of the gene in heterozygous carriers, presymptomatic individuals, and affected fetuses. The advantages of this approach are apparent in a condition such as Huntington disease, which is inherited as an autosomal dominant mutation with late onset and where neither the biochemical lesion nor the chromosomal position of the defective gene has been established. Although it is not technically feasible at this time, eventually it should be possible to isolate the defective gene by "walking" closer and closer to it through adjacent restriction endonuclease site polymorphisms, and to identify the nature of the gene by *in vitro* transcription, translation, and characterization of the gene product. Antibodies could be raised against this protein and be used to determine its cellular location and to block its function in nervous tissue.

DNA mediated transfer of genes can also be used in linkage analysis (Ruddle, 1980; Pellicer *et al.*, 1980). It is possible to introduce foreign DNA into cultured cells, eggs or early embryos by endocytosis or microinjection directly into them. DNA can be added in the naked state (Wigler *et al.*, 1977) or in chromosomes (McBride and Ozer, 1973). Most of the DNA introduced into cells is degraded and lost, but a small portion is able to survive as self-replicating entities or through incorporation into the host genome (usually at random sites). These foreign genes may be expressed in the host cells and may be detected using homologous DNA probes and specific restriction endonucleases. In the case of cultured cells, it is convenient to have the foreign DNA contain a gene coding for a product that confers a selective advantage on the cells, e.g., drug resistance or nutritional auxotrophy. This allows the selection of the small number of cells in the population (usually about 1 in 10^5-10^6) to which genetic information has been successfully transferred. One can then establish whether other genes were cotransferred. The frequency of cotransfer of two genes is a measure of their proximity to each other in the genome. As selective systems for neural properties are developed, this approach should have more applications.

III. GENERATING GENETIC VARIANTS

A. In Cell Culture

Continuous cell lines, including neuroblastoma, glioma, and pheochromocytoma cells, are available which express a large number of neural properties in culture. In addition, other types of cell lines express some properties important in neural function, e.g., lymphosarcoma cells possess β-adrenergic receptors (Bourne *et al.*, 1975), L-cells produce NGF (Pantazis *et al.*, 1977), and hepatoma cells have MAO-A and -B activities (Hawkins and Breakefield, 1978). Such lines can be used to generate ''mutant'' or variant cells which are altered in these properties. Mutants can be used to understand the structure–function relationship of a particular protein by obtaining a series of different mutations in the gene locus or loci coding for this protein. Both mutants and variants can be used to dissect the functional components of a response by establishing the number of complementary lesions that can restore the response.

Before a mutational analysis of a particular protein or response is attempted with continuous cell lines, it is important to understand certain complicating aspects of the system. These involve variations in gene dosage and gene position, both of which can affect gene expression. Most continuous cell lines contain an aberrant number of chromosomes and rearrangements of chromosomal material, which occur through chromosome loss, nondisjunction, translocations, inversion, unequal sister chromatid exchange, etc. Thus some portions of the genome that are not critical to growth in culture may be lost, and other portions may be replicated and/or placed in new positions. Because these changes in chromosomal composition can occur continually during growth in culture, depending on the karyotypic stability of a line, even cells within a recently cloned population can be genetically different. When isolating variants, it is important to determine whether the variation in gene expression reflects a change in gene number and position, or a mutation in gene structure. One convenient way to test these alternatives is to determine whether the frequency of variant cells is increased by treatment with mutagens which cause single base substitutions in the DNA.

The number of copies of the gene(s) of interest present in cells can also affect the ability to isolate variants. A gene locus is usually thought to be present in one copy on each of two homologous chromosomes. As noted above, however, genes coding for some proteins, e.g., tubulin, are normally present in multiple copies within the genome. This situation is complicated by the presence of varying numbers of chromosomes and the redistribution of chromosomal material in some cell lines. If a mutation in one copy of a gene acts in a dominant manner, i.e., results in a phenotype which is distinct from the parent, then the mutant cells can be detected in the population irrespective of how many normal copies of the gene are also present. If a mutation is recessive, however, in order for the variant phenotype to be expressed, a cell must either have the mutation in all gene copies or have only one copy of the locus, the affected one (Breakefield *et al.*, 1979). The latter situation is frequently true for loci on the X chromosome, since cells usually have only one active X chromosome.

Many selective schemes have been developed that impart growth or survival advantages to variant cells in culture. Toxic drugs, altered culture conditions, and nutritional requirements have been used frequently. Negative selection protocols can be used to obtain cells that have lost certain neural properties. Examples include selection of variant cells resistant to stable β-adrenergic agonists to obtain lymphosarcoma cells with alterations in β-adrenergic receptors (Johnson *et al.,* 1979); to β-NGF, a protein that normally causes pheochromocytoma cells to cease proliferation and send out axons (Greene and Tischler, 1976), to obtain cells with an altered response to this neurotropic protein (Bothwell *et al.,* 1980); and to veratridine to obtain neuroblastoma cells with altered voltage-dependent Na^+ channels (West and Catterall, 1979). Examples of positive schemes include selection for neuroblastoma cells which remain attached to the substratum during exposure to the Ca^{2+}-specific chelator EGTA to obtain cells with altered cell surface properties (Culp, 1980); and for neuroblastoma cells able to grow in the absence of tyrosine, as only neural cells with tyrosine hydroxylase or tryptophan hydroxylase activities can synthesize this essential amino acid, to obtain lines able to produce norepinephrine or serotonin (Breakefield and Nirenberg, 1974). Positive schemes are more likely to select for variants in gene expression or gene dosage than in gene structure, since changes in gene structure usually compromise the functional capacity of the gene product. However, it is possible to turn a positive selective scheme into a negative one by including agents, such as BUdR, which kill only dividing cells (Puck and Kao, 1967). Variants which lack the property under selection survive by not growing and can be propagated later under nonselective conditions.

Variant cells can also be obtained by screening all cells in a population for quantitative variations in a particular trait by the use of a fluorescent cell sorter or by replica plating techniques. For example, cell surface proteins can be labeled with fluorescent antibodies or probes and individual cells can be sorted on the basis of fluorescent intensity (e.g., Dorman *et al.,* 1978). This approach should prove useful in obtaining cells with alterations in the number and structure of receptors and neural-specific antigens. Fluorescent moieties can also be attached to molecules that are taken up by living cells (Kaufman *et al.,* 1978; Henderson *et al.,* 1980). For replica plating, cells are plated at cloning density and then overlaid with filter paper; cells grow on the surface of the plate and up into the paper (Esko and Raetz, 1978). The paper, containing a replica pattern of the colonies on the dish, can be removed and tested, while viable cells remain attached to the plate. Screening of the replicas can be carried out with radiolabeled and fluorescent compounds or antibodies. We have used this technique to identify cells with and without MAO activity by virtue of the binding of [^{3}H]pargyline to active MAO molecules (unpublished data).

A newly revealed aspect of a quantitative selection for a trait is that variant lines can be obtained that have an amplified number of copies of the gene whose product is being detected. Cells express progressively more of a particular property through an increase in the amount of gene product which, in turn, is mediated by repetitive gene duplications. Only two examples of this phenomenon have been described to date, resistance to methotrexate (Alt *et al.*, 1978) and *N*-(phosphonacetyl)-L-aspartate (Wahl *et al.*, 1979), and neither is pertinent to the nervous system per se. However, as selective schemes for neural properties increase, this approach should prove extremely useful in obtaining cells with an increased amount of mRNA coding for specific neural proteins, as in many cases the amount of such mRNA is too low for isolation and DNA cloning protocols.

B. In Animals

Two approaches have been developed for introducing known mutations into animals. In the first, mutant donor cells contribute to the cellular organization of the animal and contain a complete complement of genetic information differing in many ways from that of the host cells. These donor cells can either be pluripotent teratocarcinoma cells, which participate in the development of chimeric animals, or "normal" cell types, which continue to divide in adult animals. In the second approach, mutant genes are transferred directly into embryonic cells, and in this case a relatively small amount of new genetic information is introduced into the existing genetic complement of the host cells.

Chimeric mice, formed by combining "embryonic" cells from individuals of different genotypes, have been used extensively to investigate the cellular site of action of mutant lesions (for review see Martin, 1980; Mullen and Herrup, 1979). This can be done either by combining cells from eight-cell embryos and allowing them to enter the blastocyst stage together *in vitro,* or by injecting embryonic cells from one donor into the blastocoel of another. In both cases, the chimeric embryos are introduced into a pseudopregnant foster mother and allowed to develop *in utero.* The potential usefulness of this technique in analyzing inherited diseases was expanded greatly by the finding that teratocarcinoma cells could serve as the donor cell type (Brinster, 1974). These cells can be grown as tumors or in culture, and retain a total pluripotency with respect to differentiation. They can contribute to the formation of every tissue type, including functional germ cells (Mintz and Illmensee, 1975).

Mutant teratocarcinoma cells can be obtained by any of the procedures used for selecting variant cells in culture, and can also serve as host cells in gene transfer experiments, thus providing a wide range of alterations in gene structure. Their limited potential to date for creating animal diseases of known genetic etiology has probably resulted from their aberrant karyotypes. Teratocarcinoma cells, as many tumor cell types, frequently contain an aneuploid chromosomal complement, and continued growth in culture can result in an increased number of chromosomal anomalies. It is essential that the teratocarcinoma cells be as close to euploid as possible so that they will not be at a disadvantage to normal cells during embryonic growth and can participate in gametogenesis (Martin, 1980).

In an attempt to create an animal model for a human neurologic disease, Mintz and co-workers selected teratocarcinoma cells in culture for loss of hypoxanthine phosphoribosyltransferase (HPRT) activity, the enzyme deficiency in the Lesch–Nyhan syndrome (Fig. 5; Dewey *et al.*, 1977). These mutant cells participated in the formation of a number of chimeric mice which had tissues, including the brain and gonads, that contained a large proportion of these cells. However, these chimeric mice did not give rise to progeny bearing the HPRT deficiency, and did not show any neurologic abnormalities. This latter finding is not surprising, since cells with HPRT activity can compensate for the lack of this activity in adjacent cells (Subak-Sharpe *et al.*, 1969). The ability to construct new mutant mouse strains awaits the creation of a chimeric mouse that can produce gametes containing the genome of the mutant teratocarcinoma cells.

Another genetic approach that can be used to alter the phenotype of an animal is to introduce a donor cell type which is capable of division and/or coexistence in the host, and which exerts a dominant phenotypic effect. Although this technique is limited in its application, it has the potential to correct some genetic

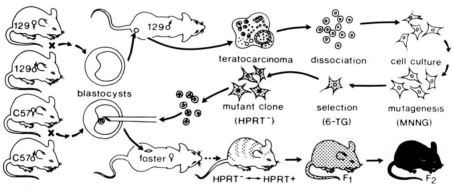

Fig. 5. Construction of new mouse mutants by using mutant teratocarcinoma cells to form chimeric mice. The plan of the experiment, spanning almost a decade, is diagrammed, starting at the upper left. A blastocyst from a mating of 129-strain mice was grafted under the testis capsule of a syngeneic host. The graft formed a malignant teratocarcinoma. After dissociation, tumor cells were explanted and the stem cells were established as an *in vitro* culture line. Following exposure to the mutagen *N*-methyl-*N*′-nitro-*N*-nitrosoguanidine (MNNG), HPRT⁻ cells (stippled) were selected for their resistance to 6-thioguanine (6TG). Cells from a resistant clone were then microinjected into the cavity of genetically marked blastocysts (e.g., of the C57BL/6 strain). The mice were born. Some were mosaics comprising HPRT⁻ cells derived from the mutant teratocarcinoma lineage, along with blastocyst-derived cells, in their coats (striped) and/or internal somatic tissues. Tissue-specific effects of the deficiency are analyzable in the mosaics. If mutant cells are of X/O (as in this case) or X/X sex chromosome type and contribute to the germ line of females, affected "Lesch–Nyhan" males would be obtained in the F₁ generation; if the mutant cells are X/Y, in mosaic (non-Lesch–Nyhan) males, affected males would occur in the F₂. (Reprinted from Dewey *et al.*, 1977.)

defects in individuals and hence has therapeutic applications. An example of this approach was the transfer of increased drug resistance to mice (Cline *et al.,* 1980). Genes for methotrexate resistance were introduced into cultured mouse bone marrow cells, which were identifiable by a distinctive chromosomal marker. These cells, and an equal number of cells lacking the resistance genes and the marker, were injected into mice that had been irradiated to destroy endogenous marrow cells. The animals were then treated with methotrexate. Thirty days later a sample of marrow cells was karyotyped and the majority of cells were found to contain the marker chromosome. In animals that were not treated with methotrexate, these "marked" cells did not predominate, indicating that they had a selective advantage *in vivo* only in the presence of the drug. In theory, any type of cells which is selectable *in vivo,* e.g., can supply necessary gene product or eliminate toxic metabolites, can be used as a vehicle to establish other nonselectable traits in the host. For example, drug-resistant cells containing normal β-globin genes could be introduced into individuals with sickle cell anemia or β-thalassemia. If compensation for inherited deficiencies could be mediated through serum, it is possible that in some cases non-neuronal cell types could be used to correct neuronal defects.

The preceding techniques produce animals with new genetic information in some subset of their cells. In the future, it may also be possible to produce animals in which all cells bear a specific change in their genotype. The groundwork for this approach has been carried out in mice by Ruddle and co-workers (Gordon *et al.,* 1980). Cloned plasmid DNA, containing inserted thymidine kinase genes from herpes simplex virus and the origin of replication from SV40 virus, was microinjected into pronuclei of fertilized oocytes, which were then implanted in foster mothers. DNA was extracted from newborn mice and screened for sequence homology to the injected material. Of 78 mice evaluated in one experimental series, two were found to contain these sequences at a level equivalent to one copy per cell. Thus, the injected DNA is capable of being replicated and distributed to daughter cells. However, not all of the inserted sequences were detected in either mouse and rearrangements in sequences were noted. No attempts were made to detect expression of the viral thymidine kinase gene. If genes introduced in this way can be expressed, then this approach can be used to alter both the genotype and phenotype of these animals and their progeny. Using DNA technology it is also possible to mutate genes at specific sites by introducing base substitutions or deletions and thus effect known changes in genotype (Shortle and Nathans, 1978; Shenk *et al.,* 1976).

Methods of cell culture and molecular biology allow direct manipulation of the cellular genome and incorporation of specifically altered genetic information into all or a substantial fraction of cells in living animals. It is possible, then, to construct animal models with known genetic alterations and to examine the effects of these alterations on the development, physiology, anatomy, and biochemistry of the nervous system. Such models will help us to unravel the complex interrelationships between the primary, secondary, and tertiary aspects of inherited diseased states. The use of these new approaches depends on the development of selective techniques for neural properties in cultured cells and the production of cloned DNA probes for neural genes.

IV. ASSESSING THE ROLE OF HEREDITY IN BEHAVIOR AND DISEASE

A. Populations and Individuals

Many examples could be given of attempts to determine how genetic contributions affect individual behavior and disease. In a few examples discussed here, variations in neural function have been assessed in mice and humans using genetic, biochemical, and anatomic analyses. These studies illustrate methods that neuroscientists are using to evaluate how variations in gene structure can alter the structure and function of the nervous system.

Reis and co-workers (1981) have carried out extensive analyses on the molecular and cellular basis of inherited differences in tyrosine hydroxylase levels in two inbred strains of mice, and on the pharmacologic and behavioral correlates of these differences. The specific activity of tyrosine hydroxylase is 20–50% lower in the brains of strain CBA/J mice than in strain BALB/cJ mice (Ciaranello *et al.*, 1974). This difference does not result from a change in the kinetic properties of the enzyme, but rather reflects differences in the number of enzyme molecules. The greatest differential between strains is found in areas of the brain containing dopaminergic nerve terminals or cell bodies, e.g., the substantia nigra. This differential reflects the fact that mice of strain CBA/J have fewer dopaminergic neurons (Baker *et al.*, 1980). The number of tyrosine hydroxylase molecules per dopaminergic neuron is actually the same in these two strains. Further, anatomic studies reveal parallel differences in areas of the brain which receive and send innervation to the substantia nigra. These two mouse strains also show differences in drug responsiveness and spontaneous behavior. Animals of strain BALB/cJ with greater dopaminergic innervation show greater locomotion and stereotypic behavior in response to a given dose of amphetamine, a drug that increases release of dopamine from presynaptic terminals. Strain BALB/cJ animals also have a higher level of spontaneous exploration and locomotory activity, behaviors also thought to be modulated in part through dopaminergic pathways in the brain.

Because changes in one area of the nervous system can affect development in other areas, it is difficult in cases such as this to establish the primary and secondary effects of critical gene differences between these strains. Further, although each inbred strain is homozygous at essentially all of its gene loci, the two strains differ at so many loci that it is impossible to identify the number or nature of the genetic differences that contribute to their phenotypic differences. Studies with isogenic strains of grasshoppers have shown that the number of identifiable neurons in the nervous system can be inherited (Goodman, 1979). Presumably inherited variation in neuronal number could account for these strain differences in dopaminergic neurons.

New approaches to mouse genetics involving recombinant and congenic strains have been developed to help identify the gene loci responsible for complex neural traits (Bailey, 1981). Inbred strains, such as those used above, are established by over 20 generations of brother–sister matings. The resulting individuals are homozygous at over 95% of gene loci, and differ from other inbred strains at about 30% of gene loci. A similiar breeding protocol is followed to generate recombinant strains, except that the original parents are chosen from two unrelated, highly inbred strains, and independent brother–sister matings are carried out from the offspring of the initial cross (Fig. 6). After 20 generations a set of inbred recombinant strains is established from the offspring of the original parents. Each new strain is homozygous at virtually all gene loci, but some alleles are contributed by one parent and some by the other, and the combination

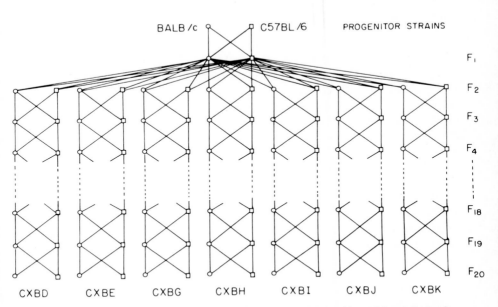

Fig. 6. Scheme for constructing the CSB set of seven recombinant-inbred strains from the progenitor strains, BALB/cBy and C57BL/6By. The symbol for any established RI set should bear the capital letter X (not a times sign). (Reprinted from Bailey, 1981.)

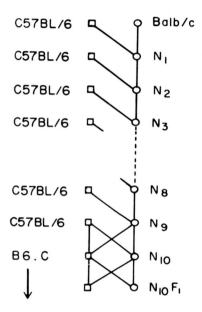

Fig. 7. Scheme for constructing a B6.C congenic strain from the C57BL/6By (B6) background strain and the BALB/cBy (C) differential-source strain. In each generation up to the tenth, a female carried of the differential gene is selected and backcrossed to a male of the background strain. In the tenth generation, both a female and a male carried are selected and intercrossed to establish by selection a homozygous true-breeding strain. (Reprinted from Bailey, 1981.)

of parental alleles is different for each recombinant strain. To assess a possible cause and effect relationship between any two properties, their coexpression can be evaluated in a series of recombinant strains. If the two properties are always inherited together, there is a high probability that they represent either different effects of the same allele or the separate affects of tightly linked alleles.

Another inbreeding technique involves construction of congenic strains in which a gene responsible for a specific neural trait can be introduced into the genetic background of an inbred strain (Bailey, 1981). The "foreign" strain carrying the gene of interest is crossed to an inbred strain and in each subsequent generation a carrier of the gene is identified and backcrossed to the inbred strain (Fig. 7). After ten backcrosses, mice containing the gene of interest have about 1% of the genome of the foreign strain, but are otherwise genetically identical to the original inbred strain. Comparison of animals that differ in only a small portion of their genome can allow resolution of the neural effects specific to the gene(s) of interest.

Establishing genetic contributions to human behavior has been a difficult process. Two approaches will be discussed here: (1) population genetic analysis of the complex inheritance patterns of psychiatric states, and (2) biochemical genetic studies of the inheritance patterns of neurochemical markers and the correlation of these markers with neurologic and psychiatric parameters. Heritability of a trait is the proportion of variation in the population due to genetic variation among individuals (Matthysse and Kidd, 1976). Because of the large contribution of the physical and cultural environment to behavior, and because of the non-Mendelian pattern of inheritance of most behavioral traits, it has been necessary to combine strategies of psychiatry and population genetics to elucidate inherited aspects of human behavior. Strategies have included diagnostic comparisons of affected and unaffected family members, monozygotic and dizygotic twins, and siblings raised by affected and unaffected parents. In addition, a set of Research Diagnostic Criteria has been developed which serve to subdivide psychic states into discrete symptoms and standardize diagnoses. Such studies have revealed clearly that certain forms of schizophrenia and bipolar depressive illness (for review see Gershon *et al.,* 1980) have a substantial genetic contribution. Determining the type and number of genes involved, however, is complicated by problems of variable age of onset, incomplete penetrance, etiologic heterogeneity, ascertainment bias, small family size, and environmental contributions. These complications can be weighted mathematically, and the probability that certain modes of genetic transmission operate for specific traits can be evaluated. For example, this approach can be used to determine whether certain psychiatric illnesses result from inherited variations at a single major gene locus or at many gene loci with additive effects (Matthysse and Kidd, 1976). In cases in which a single locus appears to be responsible, linkage analysis in pedigrees can be used to determine the position of this gene in the genome (see Section II,B,2). Linkage information can have immediate applications in genetic counseling. Further, linkage analysis using restriction endonuclease site polymorphisms should eventually allow isolation of the responsible gene and identification of its product.

The effects of inherited differences in neural properties have also been evaluated by measuring certain neurochemical marker traits and correlating variations in these traits with individual differences in other neurological parameters, such as drug sensitivity and psychiatric states. In humans, sampling of tissues from living individuals is limited to those tissues which can be obtained with minimal risk, e.g., blood and skin biopsies. Only a limited number of properties critical to nerve function are expressed in these non-neuronal tissues. Further, because a number of separate, but evolutionarily related, gene loci may code for similiar proteins, and because different loci may be active in different tissues, it is important to establish not only whether inherited variations in a protein exist in a given tissue, but whether these variations occur in other tissues of the same individual. The most extensive analysis of inherited variations in neurochemical properties in humans has been carried out for two enzymes important in catecholamine metabolism, catechol-O-methyltransferase (COMT) and dopamine β-hydroxylase (Weinshilboum, 1979). (Only the former will be discussed here.) Catechol-O-methyltransferase is a degradative enzyme which transfers methyl groups to the catechol moieties of amine neurotransmitters, drugs, and catechol-estrogens. Levels of enzyme activity measured in erythrocyte lysates correlate with levels in several other tissues from the same individual (Weinshilboum, 1978). Measurement of activity levels in over 800 normal individuals revealed that 25% of individuals have distinctly low levels of activity (Weinshilboum and Raymond, 1977). Pedigree analysis showed that the low activity trait is inherited as a single locus, autosomal recessive allele. Since COMT is a monomeric enzyme (White and Wu, 1975; Borchardt and Cheng, 1978), and since lysates from individuals with low activity also have a thermosensitive form of the enzyme, as compared to those with high activity (Scanlon *et al.*, 1979), it is reasonable to conclude that the locus controlling levels of activity is either the structural gene for COMT or a gene regulating a posttranslational event which affects activity and thermal stability. Although low COMT activity has not been associated with any altered behavioral states in humans, inherited levels of activity do affect the rate of degradation of certain catechol drugs, such as propranolol and L-Dopa, and thus affect the dosage:efficacy ratio in therapeutic treatment.

Monoamine oxidase is also involved in the degradation of amine transmitters (see Section II,A,1), and inherited variations in levels of activity are thought to affect human neurophysiology (Murphy, 1978). The evaluation of this phenomenon is complicated by the presence of two forms of the enzyme, MAO-A and MAO-B, which differ in substrate selectivity, drug sensitivity, molecular structure, and tissue distribution, and appear to be under separate genetic control (Cawthon *et al.*, 1981). The most extensive analysis has been carried out on MAO-B in human platelets. The correlation of activity levels in related and unrelated individuals indicates that approximately 25% of the variation in activity is genetically determined and that this variation could be under the control of a single major gene locus (Gershon *et al.*, 1980). The molecular basis of inherited differences in activity is not known, and it is not clear whether comparable differences exist for MAO-B in other tissues of the same individual. It does appear that individuals with low platelet MAO-B activity have an abnormally high incidence of psychopathology (Donnelly *et al.*, 1979), including bipolar depressive illness (Gershon *et al.*, 1980), alcoholism (Sullivan *et al.*, 1979), and schizophrenia (Wyatt *et al.*, 1975). Presumably, individuals with low MAO-B activity are more susceptible to the psychically detrimental effects of environmental stresses or other inherited neurologic traits.

Inhibition of MAO-A activity in humans appears to correlate with mood elevation (Lipper *et al.*, 1979). The high concordance of MAO-A activity measured in cultured skin fibroblasts from monozygotic twins suggests that these levels too are under genetic control (Breakefield *et al.*, 1980). In small sample populations, significantly different levels of MAO-A activity have been found in fibroblasts from patients with the Lesch–Nyhan syndrome (Costa *et al.*, 1980), as compared to age and sex matched controls, but not from patients with the Gilles de la Tourette syndrome (Giller *et al.*, 1980), schizophrenia (Groshong *et al.*, 1978), or bipolar depressive illness (Breakefield *et al.*, 1980). Because activity levels of an enzyme can be affected by many factors, including the structure and amount of the enzyme, and the presence of activating or inhibiting factors, it is important to understand the biochemical basis of these inherited variations in MAO-A and -B activities (Breakefield and Edelstein, 1980).

B. Cells and Molecules

Ultimately, genetic analyses of behavior can help to explain how variations in gene structure affect cellular metabolism and communication among cells, as well as the anatomy and physiology of the nervous system. First, it is necessary to demonstrate the existence of inherited variations in the structure of genes and proteins critical to nerve function. We will focus here on approaches that are applicable to humans as well as experimental animals. Although numerous animal models are available for inherited neurologic diseases and behavioral traits, in no case, to our knowledge, has the primary molecular lesion been identified conclusively. In animals full access to the nervous system at various stages of development can provide many levels of analysis (discussed in Section IV,A). In humans, one of the major difficulties in analysis is the inaccessibility and unavailability of appropriate neural tissue for study. Two approaches, however, may help to overcome this barrier: (1) peripheral tissue samples that contain proteins and metabolites of interest can be studied directly, e.g., blood, urine and cerebrospinal fluid; and (2) small numbers of living cells can be removed and propagated in culture, e.g., glia, lymphocytes, and fibroblasts from the brain, blood, and skin, respectively. Many proteins important in neural function are expressed in these other cell types. For example, cultured skin fibroblasts have β-adrenergic receptors and hormone-responsive adenylate cyclase (Haslam and Goldstein, 1974), glutamate decarboxylase Gray and Dana, 1979), a high affinity transport system for choline (Riker *et al.*, 1981), β-NGF (Schwartz and Breakefield, 1980) and voltage-dependent Na^+ channels (Munson *et al.*, 1979), as well as MAO-A and -B, COMT and phenol sulfotransferase (Roth *et al.*, 1976; Groshong *et al.*, 1978; Crooks *et al.*, 1978). Further, since these cells presumably contain the full complement of genetic information, specific DNA sequences can be examined directly using probes for neural genes. As more is learned about factors that affect the expression of genes, it may also be possible to activate "silent" neural genes in these non-neural cell types and analyze the gene products directly (for discussion see Breakefield and Pintar, 1981).

Familial dysautonomia is a neurologic disease inherited in an autosomal fashion and characterized by developmental defects in sympathetic and sensory nerve function (Riley, 1974). Similiarities in the neuropathology of ganglia from these patients and from newborn mice treated with antiserum to β-NGF led to the hypothesis of altered β-NGF in this disease (Pearson et al., 1975, 1978). Early studies on circulating β-NGF in serum from these patients suggested that they produced more of an immunoreactive molecule with reduced biologic activity, as compared to controls (Siggers et al., 1976). Although the tissue source(s) of β-NGF in humans in vivo is not known, cultured human skin fibroblasts produce a β-NGF-life protein (Schwartz and Breakefield, 1980). A comparison of biologic activity and immunoreactivity of this protein in fibroblasts indicated that patients produce a form of "β-NGF" which has approximately one-tenth the biological activity of its normal counterpart (Table II). Further, in contrast to control cells, production of β-NGF in patient cells is not affected by increases in intracellular cyclic AMP levels (Schwartz and Breakefield, 1980). These studies suggest that a defect in the structure of β-NGF or its precursor polypeptide may be the primary biochemical lesion in familial dysautonomia.

Table II. Biologic Activity of β-NGF in Fibroblasts[a]

Cell type	Age	Sex	Biologic activity[b] (B.U./ng immunoreactive β-NGF)
Control			
HF9	22	M	2.0
HF18	22	F	2.2, 5.2
HF8	22	M	2.9 ± 0.8 (3)
GM498	3	M	3.2 ± 0.9 (3)
GM23	31	F	3.4 ± 0.1 (4)
86	15	F	4.0, 6.4
Rid Mor	15	M	6.2
Familial dysautonomia			
HF65	1	F	0.09, 0.14
HF63	6	F	0.30
HF55	19	M	0.37 ± 0.04 (3)
HF56	17	M	0.46, 0.39
HF54	24	F	0.46 ± 0.15 (5)
GM850	26	M	0.33, 1.2

[a] Adapted from Schwartz and Breakefield, 1980.

[b] Biological activity was determined as the ratio of biological units (B.U.) to ng immunoreactive β-NGF. Biological units were determined by the dose dependent ability of fibroblast extracts to elicit neurite outgrowth from 9-day chick dorsal root ganglia in culture (outgrowth was inhibited by antiserum prepared against mouse β-NGF). Immunoreactive β-NGF was determined by radioimmunoassay using antiserum prepared against mouse β-NGF, ^{125}I-labeled mouse β-NGF, unlabeled mouse β-NGF, and fibroblast extracts. (For further details see Schwartz and Breakefield, 1980.)

Human erythrocytes have been used to evaluate whether alterations in the G/F component of hormone-responsive adenylate cyclase occur in patients with pseudohypoparathyroidism (Farfel *et al.*, 1980; Levine *et al.*, 1980). This is an autosomal recessive condition characterized by neuromuscular irritability, hypocalcemia, and calcifications of soft tissues. Patients show resistance to a number of hormones, including parathyroid hormone, which utilize cyclic AMP as a second messenger. This resistance appears to be explained, at least in some patients, by lack of the G/F component which couples hormone receptors to adenylate cyclase molecules.

The most complete analysis of inherited variations in structure involves gene sequencing, but this approach requires probes for specific neural proteins. Although these probes are available for some neural proteins, e.g., the acetylcholine receptor, ACTH, and somatostatin, and soon should be available for a number of neuropeptides, they may be difficult to obtain for many neural proteins. Another approach is to analyze variations in the structure of the proteins themselves. Ultimately, this approach will lead to amino acid sequencing of part of all of the molecule, which, in turn, will help in synthesizing DNA probes for the responsible genes. Meanwhile, we need to compare the structure of neural proteins in as detailed a way as possible using the small amounts of proteins available. In our laboratory, we are attempting to resolve possible structural differences between β-NGF produced by fibroblasts from patients with familial dysautonomia and controls (Grossman *et al.*, 1980). This study is complicated because only trace amounts of putative β-NGF are present in these cells, and because essentially nothing is known about the human form of this protein. Antiserum prepared against purified mouse β-NGF is used to immunoprecipitate cross-reacting material from human fibroblast extracts. Immunoreactive human β-NGF is then further purified by SDS polyacrylamide gel electrophoresis and identified by its comigration with purified mouse β-NGF. Human β-NGF in the gel is labeled at tyrosine residues with radioactive iodine, and cleaved with site-specific proteases. Peptide fragments are separated in two dimensions using electrophoresis and chromatography. Peptide maps made with putative human β-NGF from normal and dysautonomic fibroblasts can be compared with each other and with authentic mouse β-NGF, to detect differences in the number and location of peptide fragments. To date, in analyses that cover 20–50% of the putative β-NGF molecules, we have found no differences in proteins from controls and dysautonomic patients (both of which share fragments with authentic β-NGF). However, by labeling different amino acid residues and treating with different site-specific cleavage agents, it should be possible to achieve resolution of the full extent of the protein molecules.

Another way to analyze structural variations in small amounts of neural proteins is to separate all proteins by two-dimensional gel electrophoresis on the basis of molecular weight and net charge, and then to identify the protein of interest by consecutive incubations with antiserum and [^{125}I]protein A, which binds to IgG (Kessler, 1975). The protein can be labeled directly in the gel (Adair *et al.*, 1978) or after transfer to a solid support system, e.g., nitrocellulose sheets (Towbin *et al.*, 1979) or diazobenzyloxymethyl paper (Renart *et al.*, 1979). By successively probing the transferred proteins with monoclonal antibodies that have different specificities, additional differences in protein structure can be revealed.

By using antibodies which can be prepared against proteins purified from either human or nonhuman sources to enrich for small amounts of neural proteins and to identify antigenic differences at discrete sites in these proteins, it is possible to begin to establish the presence of inherited variations in the structure of neural proteins.

V. CONCLUSION

Genetic techniques are currently available for the characterization of the number, structure, and genomic location of genes coding for neural proteins. It is also possible to selectively alter the genotype of cells and to use these cells to construct animals with known genetic lesions. Further, naturally occurring variations in DNA and protein structure can be assessed and the affect of these variations on neural function can be evaluated. After a long period of time in which it was difficult to imagine how to study inherited variations in the structure and function of the mammalian nervous system, the necessary technology is now available.

REFERENCES

Adair, W. S., Jurivich, D., and Goodenough, U. W. (1978). Localization of cellular antigens in sodium dodecyl sulfate-polyacrylamide gels. *J. Cell Biol.* **79**, 281–285.

Alt, F. W., Kellems, R. E., Bertino, J. R., and Schimke, R. T. (1978). Selective multiplication of dihydrofolate reductase genes in methotrexate-resistant variants of cultured murine cells. *J. Biol. Chem.* **253**, 1357–1370.

Bailey, D. W. (1981). Strategic uses of recombinant inbred and congenic strains in behavior genetics research. *In* "Genetic Research Strategies in Psychobiology and Psychiatry" (E. S. Gershon, S. Matthysse, X. O. Breakefield, and R. D. Ciaranello, eds.), pp. 189–198. Boxwood Press, Pacific Grove, California.

Baker, H., Joh, T. H., and Reis, D. J. (1980). Genetic control of number of midbrain dopaminergic neurons in inbred strains of mice: Relationship to size and neuronal density of the striatum. *Proc. Natl. Acad. Sci. U.S.A.* **77**, 4369–4373.

Barbarese, E., Braun, P. E., and Carson, J. H. (1977). Identification of prelarge and presmall basic proteins in mouse myelin and their structural relationship to large and small basic proteins. *Proc. Natl. Acad. Sci. U.S.A.* **74**, 3360–3364.

Beckman, L., Cedergren, B., Mattsson, B., and J.-O. Ottoson (1974). Association and linkage studies of Huntington's chorea in relation to fifteen genetic markers. *Hereditas* **77**, 73–80.

Berger, E. A., and Shooter, E. M. (1977). Evidence for pro-β-nerve growth factor, a biosynthetic precursor to β-nerve growth factor. *Proc. Natl. Acad. Sci. U.S.A.* **74,** 3647–3651.

Blundell, T. L., and Humbel, R. E. (1980). Hormone families: Pancreatic hormones and homologous growth factors. *Nature (London)* **287,** 781–787.

Borchardt, R. T., and Cheng, C. F. (1978). Purification and characterization of rat heart and brain catechol methyltransferase. *Biochim. Biophys. Acta* **522,** 49–62.

Botstein, D., White, R. L., Skolnick, M., and Davis, D. W. (1980). Construction of a genetic linkage map in man using restriction fragment length polymorphisms. *Am. J. Hum. Genet.* **32,** 314–331.

Bothwell, M. A., Schechter, A. L., and Vaughn, K. M. (1980). Clonal variants of PC12 pheochromocytoma cells with altered responses to nerve growth factor. *Cell* **21,** 857–866.

Bourne, H. R., Coffino, P., Melmon, K. L., Tomkins, G. M., and Weinstein, Y. (1975). Genetic analysis of cyclic AMP in a mammalian cell. *Adv. Cyclic Nucleotide Res.* **5,** 771–786.

Bradshaw, R. A., Hogue-Angeletti, R. A., and Frazier, W. A. (1974). Nerve growth factor and insulin: Evidence of similiarities in structure, function and mechanism of action. *Recent Prog. Horm. Res.* **30,** 575–596.

Breakefield, X. O., and Edelstein, S. B. (1980). Inherited levels of A and B types of monoamine oxidase activity. *Schizophr. Bull.* **6,** 282–287.

Breakefield, X. O., and Nirenberg, M. W. (1974). Selection for neuroblastoma cells that synthesize certain transmitters. *Proc. Natl. Acad. Sci. U.S.A.* **71,** 2530–2533.

Breakefield, X. O., and Pintar, J. E. (1981). The use of cell culture to analyze the genetic basis of human neurologic disease. In ''Genetic Research Strategies in Psychobiology and Psychiatry'' (G. S. Gershon, S. Matthysse, X. O. Breakefield, and R. D. Ciaranello, eds.), pp. 407–414. Boxwood Press, Pacific Grove, California.

Breakefield, X. O., Castiglione, C., and Edelstein, S. (1976). Monoamine oxidase activity decreased in cells lacking hypoxanthine phosphoribosyltransferase activity. *Science* **192,** 1018–1020.

Breakefield, X. O., Edelstein, S. B., and Costa, M. R. C. (1979). Genetic analysis of neurotransmitter metabolism in cell culture: Studies on the Lesch–Nyhan syndrome. In ''Neurogenetics: Genetic Approaches to the Nervous System'' (X. O. Breakefield, ed.), pp. 197–234. Elsevier/North-Holland, New York.

Breakefield, X. O., Giller, E. L., Jr., Nurnberger, J. I., Castiglione, C. M., Buchsbaum, M. S., and Gershon, E. S. (1980). Monoamine oxidase type A in fibroblasts from patients with bipolar depressive illness. *Psychiatry Res.* **2,** 307–314.

Brinster, R. L. (1974). The effect of cells transferred into the mouse blastocyst on subsequent development. *J. Exp. Med.* **140,** 1049–1056.

Brown, G. K., Powell, J. F., and Craig, I. W. (1980). Molecular weight differences between human platelet and placental monoamine oxidase. *Biochem. Pharmacol.* **29,** 2595–2603.

Brownstein, M. J., Russell, J. T., and Gainer, H. (1980). Synthesis, transport and release of posterior pituitary hormones. *Science* **207,** 373–378.

Cabral, F., Sobel, M. E., and Gottesman, M. M. (1980). CHO mutants resistant to colchicine, colcemid, or griseofulvin have altered β-tubulin. *Cell* **20,** 29–36.

Callingham, B. A., and Parkinson, D. (1979). ³H-pargyline binding to rat liver mitochondrial MAO. In ''Monoamine Oxidase: Structure, Function and Altered Function'' (T. P. Singer, R. W. Von Korff, and D. L. Murphy, eds.), pp. 81–86. Academic Press, New York.

Cassel, D., and Pfeuffer, T. (1978). Mechanism of cholera toxin action: Covalent modification of the guanine nucleotide-binding protein of the adenylate cyclase system. *Proc. Natl. Acad. Sci. U.S.A.* **75,** 2669–2673.

Cawthon, R. M., and Breakefield, X. O. (1979). Differences in A and B forms of monoamine oxidase revealed by limited proteolysis and peptide mapping. *Nature (London)* **281,** 692–694.

Cawthon, R. M., Pintar, J. E., Haseltine, F. P., and Breakefield, X. O. (1981). Differences in the structure of A and B forms of human monoamine oxidase. *J. Neurochem.* **37,** 363–372.

Chuang, H. Y. K., Patek, D. R., and Hellerman, L. (1974). Mitochondrial monoamine oxidase activity. Inactivation by pargyline. Adduct formation. *J. Biol. Chem.* **249**, 2381-2384.

Ciaranello, R. D., Hoffman, H. J., Shuri, J. G., and Axelrod, J. (1974). Genetic regulation of the catecholamine biosynthetic enyzmes. II. Inheritance of tyrosine hydroxylase, dopamine β-hydroxylase and phenylethanolamine N-methyltransferase. *J. Biol. Chem.* **249**, 4528-4536.

Cleveland, D. W., Lopata, M. A., MacDonald, R. J., Cowan, N. J., Rutter, W. J., and Kirschner, M. W. (1980). Number and evolutionary conservation of α- and β-tubulin and cytoplasmic β- and α-actin genes using specific cloned cDNA probes *Cell* **20**, 95-105.

Cline, M. J., Stang, H., Mercola, K., Morse, L., Ruprecht, R., Brown, J., and Satser, W. (1980). Gene transfer in intact animals. *Nature (London)* **284**, 422-425.

Coffino, P., Bourne, H. R., Insel, P., Melmon, K. L., Johnson, G., and Vigue, J. (1978). Studies on cyclic-AMP action using mutant tissue culture cells. *In Vitro* **14**, 140-145.

Cordell, B., Bell, G., Tischer, E., DeNoto, F. M., Ullrich, A., Picter, R., Rutter, W. J., and Goodman, H. M. (1979). Isolation and characterization of a cloned rat insulin gene. *Cell* **18**, 533-543.

Costa, M. R. C., Edelstein, S. B., Castiglione, C. M., Chao, H., and Breakefield, X. O. (1980). Properties of monoamine oxidase in control and Lesch-Nyhan fibroblasts. *Biochem. Genet.* **18**, 577-580.

Crooks, P. A., Breakefield, X. O., Sulens, C. H., Castiglione, C. M., and Coward, J. K. (1978). Extensive conjugation of dopamine metabolites in cultured human skin fibroblasts and rat hepatoma cells. *Biochem. J.* **176**, 187-196.

Culp, L. A. (1980). Behavioral variants of rat neuroblastoma cells. *Nature (London)* **286**, 77-79.

Dewey, M. J., Martin, D. W., Jr., Martin, G. R., and Mintz, B. (1977). Mosaic mice with teratocarcinoma-derived mutant cells deficient in hypoxanthine phosphoribosyltransferase *Proc. Natl. Acad. Sci. U.S.A.* **74**, 5564-5568.

Dockray, G. J. (1979). Evolutionary relationships of the gut hormones. *Fed. Proc., Fed. Am. Soc. Exp. Biol.* **38**, 2295-2301.

Donnelly, E. F., Murphy, D. L., Waldman, I. N., Buchsbaum, M. S., and Coursey, R. D. (1979). Psychological characteristics corresponding to low versus high platelet monoamine oxidase activity. *Biol. Psychiatry* **14**, 375-383.

Dorman, B. P., Shimizu, N., and Ruddle, F. H. (1978). Genetic analysis of the human cell surface: Antigenic marker for the human X chromosome in human-mouse hybrids. *Proc. Natl. Acad. Sci. U.S.A.* **75**, 2363-2367.

Eipper, B. A., and Mains, R. E. (1980). Structure and biosynthesis of proadrenocorticotropin/endorphin and related peptides. *Endocr. Rev.* **1**, 1-27.

Elston, R. C., Kringlen, E., and Namboodire, F. K. (1973). Possible linkage analysis between certain blood groups and schizophrenic or other psychoses. *Behav. Genet.* **3**, 101-106.

Esko, J. D., and Raetz, C. R. H. (1978). Replica plating an *in situ* enzymatic assay of animal cell colonies established on filter paper. *Proc. Natl. Acad. Sci. U.S.A.* **75**, 1190-1193.

Farfel, Z., Brickman, A. S., Kaslow, H. R., Brothers, V. M., and Bourne, H. R. (1980). Defect of receptor-cyclase coupling in pseudohypoparathyroidism. *N. Engl. J. Med.* **303**, 237-242.

Fiddes, J. C., Seeburg, P. H., DeNoto, F. M., Hallewell, R. A., Baxter, J. D., and Goodman, H. M. (1979). Structure of genes for human growth hormone and chorionic somatomammotropin. *Proc. Natl. Acad. Sci. U.S.A.* **76**, 4294-4298.

Forget, B. G. (1979). Molecular genetics of human hemoglobin synthesis. *Ann. Intern. Med.* **91**, 605-616.

Fournier, R. E. K., and Ruddle, F. H. (1977). Microcell-mediated transfer of murine chromosomes into mouse, Chinese hamster, and human somatic cells. *Proc. Natl. Acad. Sci. U.S.A.* **74**, 319-323.

Francke, U., and Oliver, N. (1978). Quantitative analysis of high resolution trypsin-Giemsa bands on human prometaphase chromosomes. *Hum. Genet.* **45**, 137-165.

Gabay, K. H., Berglustal, R. M., Walff, J., Mako, M. E., and Rubenstein, A. H. (1979). Familial hyperproinsulinemia: Partial characterization of circulating proinsulin-like material. *Proc. Natl. Acad. Sci. U.S.A.* **76**, 2881-2885.

Gershon, E. S., Goldin, L. R., Lake, C. R., Murphy, D. L., and Guroff, J. J. (1980). Genetics of plasma dopamine β-hydroxylase, erythrocyte catechol-*O*-methyltransferase and platelet monoamine oxidase in pedigrees of patients with affective disorders. *In* "Enzymes and Neurotransmitters in Mental Disease" (E. Usdin, P. Sourkes, and M. B. H. Youdim, eds.). Wiley, New York.

Giles, R. E., and Ruddle, F. H. (1973). Production and characterization of proliferating somatic cell hybrids. *In* "Methods and Applications of Tissue Culture" (P. Kruse and M. K. Patterson, eds.), pp. 475-500. Academic Press, New York.

Giller, E. L., Young, J. G., Breakefield, X. O., Carbonari, C., Braverman, M., and Cohen, D. J. (1980). Monoamine oxidase and catechol-O-methyltransferase activities in cultured fibroblasts and blood cells from children with autism and the Gilles de la Tourette Syndrome. *Psychiatry Res.* **2**, 187-197.

Ginzburg, I., de Baetselier, A., Walker, M. D., Behar, L., Lehrach, H., Frishauf, A. M., and Littauer, U. Z. (1980). Brain tubulin and actin cDNA sequences isolation of recombinant plasmids. *Nucleic Acids Res.* **8**, 3553-3564.

Goodman, C. S. (1979). Isogenic grasshoppers: Genetic variability and development of identified neurons. *In* "Neurogenetics: Genetic Approaches to the Nervous System" (X. O. Breakefield, ed.), pp. 101-152. Elsevier/North-Holland, New York.

Gordon, J. W., Scangos, G. A., Plotkin, D. J., Barbosa, J. A., and Ruddle, F. H. (1980). Genetic transformation of mouse embryos by microinjection of purified DNA. *Proc. Natl. Acad. Sci. U.S.A.* **77**, 7380-7384.

Gray, P. N., and Dana, S. L. (1979). GABA synthesis by cultured fibroblast obtained from persons with Huntington's disease. *J. Neurochem.* **33**, 985-992.

Greene, L. A., and Tischler, A. S. (1976). Establishment of a noradrenergic clonal line of rat adrenal pheochromocytoma cells which respond to nerve growth factor. *Proc. Natl. Acad. Sci. U.S.A.* **73**, 2424-2428.

Groshong, R., Baldessarini, R. J., Gibson, A., Lipinski, J. G., Axelrod, D., and Pope, A. (1978). Activities of types A and B MAO and catechol-*O*-methyltransferase in blood cells and skin fibroblasts of normal and chronic schizophrenic subjects. *Arch. Gen. Psychiatry* **35**, 1198-1204.

Grossman, M. H., Schwartz, J. P., and Breakefield, X. O. (1980). Comparison of β-nerve growth factor produced by cultured fibroblasts from controls and patients with familial dysautonomia. *Soc. Neurosci. Symp.* Abstr. 248.7.

Haslam, R. J., and Goldstein, S. (1974). Adenosine 3':5'-cyclic monophosphate in young and senescent human fibroblasts during growth and stationary phase *in vitro*. *Biochem. J.* **144**, 253-263.

Hawkins, M., Jr., and Breakefield, X. O. (1978). Monoamine oxidase A and B in cultured cells. *J. Neurochem.* **30**, 1391-1397.

Henderson, A. S., Yu, M. T., and Atwood, K. C. (1978). The localization of mouse globin genes: A test of the effectiveness of hybridization in situ. *Cytogenet. Cell Gent.* **21**, 231-240.

Henderson, G. B., Russell, A., and Whiteley, J. M. (1980). A fluorescent derivative of methotrexate as an intracellular marker for dihydrofolate reductase in L1210 cells. *Arch. Biochem. Biophys.* **202**, 29-34.

Herbert, E. (1981). Discovery of pro-opiomelanocortin, a cellular polyprotein. *Trends Biomed. Sci.*, in press.

Hochman, J., Insel, P. A., Bourne, H. R., Coffino, P., and Tompkins, G. M. (1975). A structural gene mutation affecting the regulatory subunit of cyclic AMP-dependent protein kinase in mouse lymphoma cells. *Proc. Natl. Acad. Sci. U.S.A.* **72**, 5051-5055.

Hodge, S. E., Spence, M. A., Crandall, B. F., Sparkes, R. S., Sparkes, M. C., Crist, M., and Tideman, S. (1980). Huntington's disease: Linkage analysis with age-of-onset corrections. *Am. J. Hum. Genet.* **5**, 247–254.

Houslay, M. D., Tipton, K. F., and Youdim, M. B. H. (1976). Multiple forms of monoamine oxidase: Fact and artifact. *Life Sci.* **19**, 467–478.

Johnson, G. L., Kaslow, H. R., and Bourne, H. R. (1978). Genetic evidence that cholera toxin substrates are regulatory components of adenylate cyclase. *J. Biol. Chem.* **253**, 7120–7123.

Johnson, G. L., Bourne, H. R., Gleason, M. K., Coffino, P., Insel, P. A., and Melmon, K. L. (1979). Isolation and characterization of S49 lymphoma cells deficient in β-adrenergic receptors: Relation of receptor number to activation of adenylate cyclase. *Mol. Pharmacol.* **15**, 16–27.

Joseph-Bravo, P., Charli, J. L., Sherman, T., Boyer, H., Bolivar, F., and McKelvy, F. (1980). Identification of a putative hypothalamic mRNA coding for somatostatin and of its product in cell free translocation. *Biochem. Biophys. Res. Commun.* **94**, 1004–1012.

Julliard, J. H., Shibaski, T., Ling, N., and Guillemin, R. (1980). High molecular weight immunoreactive β-endorphin in extracts of human placenta is a fragment of immunoglobulin G. *Science* **208**, 183–185.

Kan, Y. W., and Dozy, A. M. (1978). Polymorphisms of DNA sequence adjacent to human β-globin structural gene: Relationship to sickle mutation. *Proc. Natl. Acad. Sci. U.S.A.* **75**, 5631–5635.

Kaufman, R. J., Bertino, J. R., and Schimke, R. T. (1978). Quantitation of dihydrofolate reductase in individual parental and methotrexate-resistant murine cells. Use of a fluorescent cell sorter. *J. Biol. Chem.* **253**, 5852–5860.

Kemphues, K. J., Raff, R. A., Kaufman, T. C., and Raff, E. C. (1979). Mutation in a structural gene for a β-tubulin specific to testis in *Drosophilia melanogaster. Proc. Natl. Acad. Sci. U.S.A.* **76**, 3991–3995.

Kessler, S. W. (1975). Rapid isolation of antigens from cells with a staphylococcal protein A-antibody adsorbent: Parameters of the interaction of antibody-antigen complexes with protein A. *J. Immunol.* **115**, 1617–1624.

Krieger, D. T., and Liotta, A. S. (1979). Pituitary hormones in brain: Where, how and why? *Science* **25**, 366–372.

Leckman, J. F., Gershon, E. S., McGinniss, M. H., Targum, S. D., and Dibble, E. D. (1979). New data do not suggest linkage between the Xg blood group and bipolar illness. *Arch. Gen. Psychiatry* **36**, 1435–1441.

Levine, M. A., Downs, R. W., Jr., Singer, M., Marx, S. J., Aurbach, G. D., and Spiegel, A. M. (1980). Deficient activity of guanine nucleotide regulatory protein in erythrocytes from patients with pseudohypothyroidism. *Biochem. Biophys. Res. Commun.* **94**, 1319–1324.

Lewis, U. J., Bonewald, L. F., and Lewis, L. J. (1980). The 20,000-dalton variant of human growth hormone: Location of the amino acid deletion *Biochem. Biophys. Res. Commun.* **92**, 511–516.

Lipper, S., Murphy, D. L., Slater, S., and Bushsbaum, M. S. (1979). Comparative behavioral effects of clorgyline and pargyline in man: A preliminary evaluation. *Psychopharmacogenetics* **62**, 123–128.

Little, P. F. R., Annison, G., Dailing, S., Williamson, R., Cababa, L., and Modell, B. (1980). Model for antenatal diagnosis of β-thalassemia and other monogenic disorders by molecular analysis of linked DNA polymorphisms. *Nature (London)* **285**, 144–147.

Littlefield, J. W. (1964). Selection of hybrids from matings of fibroblasts in vitro and their presumed recombinants. *Science* **145**, 709–711.

Lomedico, P., Rosenthal, N., Efstratiadis, A., Gilbert, W., Kolodner, R., and Tizard, R. (1979). The structure and evolution of the two nonallelic rat preproinsulin genes. *Cell* **18**, 545–558.

McBride, O. W., and Ozer, H. L. (1973). Transfer of genetic information by purified metaphase chromosomes. *Proc. Natl. Acad. Sci. U.S.A.* **70**, 1258–1262.

Maniatis, T., Hardison, R., Lowy, E., Laner, J., O'Connell, C., Quon, D., Sim, G. K., and

Efstradiatis, A. (1978). The isolation of structural genes from libraries of eucaryotic DNA. *Cell* **15**, 687–701.

Mankovits, R., Buchwald, M., and Baker, R. M. (1974). Isolation of ouabain-resistant human diploid fibroblasts. *Cell* **3**, 221–226.

Martin, G. R. (1980). Teratocarcinomas and mammalian embryogenesis. *Science* **209**, 768–776.

Matthysse, S. W., and Kidd, K. K. (1976). Estimating the genetic contribution to schizophrenia. *Am. J. Psychiatry* **133**, 185–191.

Mellman, I. S., Lin, P.-F., Ruddle, F. H., and Rosenberg, L. E. (1979). Genetic control of cobalamin binding in normal and mutant cells. Assignment of the gene for 5-methyltetrahydrofolate:L-homocysteine S-methyltransferase to human chromosome 1. *Proc. Natl. Acad. Sci. U.S.A.* **76**, 405–409.

Mendez, B., Valenzuela, P., Martial, J. A., and Baxter, J. D. (1980). Cell-free synthesis of acetylcholine receptor polypeptides. *Science* **209**, 695–697.

Mendlewicz, J., Linkowski, P., Guroff, J. J., and Van Praag, H. M. (1979). Color blindness linkage to bipolar manic-depressive illness. *Arch. Gen. Psychiatry* **36**, 1442–1447.

Mintz, B., and Illmensee, K. (1975). Normal genetically mosaic mice produced from malignant teratocarcinoma cells. *Proc. Natl. Acad. Sci. U.S.A.* **72**, 3585–3589.

Mullen, R. J., and Herrup, H. (1979). Chimeric analysis of mouse cerebellar mutants. *In* "Neurogenetics: Genetic Approaches to the Nervous System" (X. O. Breakefield, ed.), pp. 173–196. Elsevier/North-Holland, New York.

Munson, R., Westermark, B., and Glaser, L. (1979). Tetrodotoxin-sensitive sodium channels in normal human fibroblasts and normal glia-like cells. *Proc. Natl. Acad. Sci. U.S.A.* **76**, 6425–6429.

Murphy, D. L. (1978). Substrate-selective monoamine oxidases—Inhibitor, species and functional differences. *Biochem. Pharmacol.* **27**, 1889–1893.

Nakanishi, S., Teranishi, Y., Noda, M., Notake, M., Watanabe, Y., Kakidani, H., Jingami, H., and Numa, S. (1980). The protein-coding sequence of the bovine ACTH-β-LPH precursor gene is split near the signal peptide region. *Nature (London)* **287**, 752–755.

Nathanson, N. M., and Hall, Z. W. (1979). Subunit structure and peptide mapping of junctional and extrajunctional acetylcholine receptors from rat muscle. *Biochemistry* **18**, 3392–3401.

Nelles, L. P., and Bamburg, J. R. (1979). Comparative peptide mapping and isoelectric focusing of isolated subunits from chick embryo brain tubulin. *J. Neurochem.* **32**, 477–489.

Nimmo, H., and Cohen, P. (1977). Hormonal control of protein phosphorylation. *Adv. Cyclic Nucleotide Res.* **8**, 145–266.

Owerbach, D., Bell, G. I., Rutter, W. J. and Shows, T. B. (1980). The insulin-gene is located on chromosome 11 in humans. *Nature (London)* **286**, 82–85.

Pantazis, N. J., Blanchard, M. H., Arnason, B. G. W., and Young, M. (1977). Molecular properties of the nerve growth factor secreted by L cells. *Proc. Natl. Acad. Sci. U.S.A.* **74**, 1492–1496.

Pearson, J., Axelrod, F., and Dancis, J. (1975). Current concepts of dysautonomia: Neuropathological defects. *Ann. N.Y. Acad. Sci.* **228**, 288–300.

Pearson, J., Pytel, B. A., Grover-Johnson, N., Axelrod, F., and Dancis, J. (1978). Quantitative studies of dorsal root ganglia and neuropathologic observations on spinal cords in familial dysautonomia. *J. Neurol. Sci.* **35**, 77–92.

Pearson, P. L., Roderick, T. H., Davisson, M. T., Garver, J. J., Warburton, D., Lalley, P. A., and O'Brien, S. J. (1979). Report of the committee on comparative mapping. *Cytogenet. Cell Genet.* **25**, 82–95.

Pellicer, A., Robins, D., Wold, B., Sweet, R., Jackson, J., Lowry, I., Roberts, J. M., Sim, G. K., Silverstein, S., and Axel, R. (1980). Altering genotype and phenotype by DNA-mediated gene transfer. *Science* **209**, 1414–1422.

Pickering, B. T., and Jones, C. W. (1978). The neurophysins. *In* "Hormonal Proteins and Peptides" (C. H. Li, ed.), Vol. 5, pp. 103–158. Academic Press, New York.

Pintar, J. E., Barbosa, J., Francke, U., Castiglione, C. M., Hawkins, M., Jr., and Breakefield, X. O. (1981). Gene for monoamine oxidase type A assigned to the human X chromosome, *J. Neurosci.* **1**, 166–175.

Puck, T. T., and Kao, F.-T. (1967). Genetics of somatic mammalian cells. V. Treatment with 5-bromodeoxyuridine and visible eight for isolation of nutritionally deficient mutants. *Proc. Natl. Acad. Sci. U.S.A.* **58**, 1227–1234.

Raff, E. C. (1979). The control of microtubule assembly *in vivo. Int. Rev. Cytol.* **59**, 1–96.

Raftery, M. A., Hunkapiller, M. W., Strader, C. D., and Hood, L. E. (1980). Acetylcholine receptor: Complex of homologous subunits. *Science* **208**, 1454–1457.

Reis, D. J., Baker, H., and Fink, J. S. (1981). A genetic control of the number of dopamine neurons in mouse brain: Its relationship to brain morphology, chemistry and behavior. *In* "Genetic Research Strategies in Psychobiology and Psychiatry" (E. S. Gershon, S. Matthysse, X. O. Breakefield and R. D. Ciaranello), pp. 215–230. Boxwood Press, Pacific Grove, California.

Renart, J., Reiser, J., and Stark, G. R. (1979). Transfer of proteins from gels to diazobenzyloxymethyl-paper and detection with antisera: A method for studying antibody specificity and antigen structure. *Proc. Natl. Acad. Sci. U.S.A.* **76**, 3116–20.

Riker, D. K., Roth, R. H., and Breakefield, X. O. (1981). High-affinity ^{3}H-choline accumulation in cultured human skin fibroblasts. *J. Neurochem.* **36**, 746–752.

Riley, C. M. (1974). Familial dysautonomia—Clinical and pathological aspects. *Ann. N.Y. Acad. Sci.* **228**, 283–287.

Ringertz, N., and Savage, R. E. (1976). "Cell Hybrids." Academic Press, New York.

Roberts, J. L., Phillips, M., Rosa, P. A., and Herbert, E. (1978). Steps involved in processing of common precursor forms of adrenocorticotropin and endorphin in cultures of mouse pituitary cells. *Biochemistry* **17**, 3609–3618.

Ross, E. M., and Gilman, A. G. (1977). Resolution of some components of adenylate cyclase necessary for catalytic activity. *J. Biol. Chem.* **252**, 6966–6969.

Ross, E. M., Howlett, A. C., Sternweiss, P. C., and Gilman, A. C. (1979). Genetic and biochemical resolution of some components of hormone-sensitive adenylate cyclase. *In* "Neurogenetics: Genetic Approaches to the Nervous System" (X. O. Breakefield, ed.), pp. 235–255. Elsevier/North-Holland, New York.

Roth, J. A., Breakefield, X. O., and Castiglione, C. M. (1976). Monoamine oxidase and catechol-*O*-methyltransferase activities in cultured human skin fibroblasts. *Life Sci.* **19**, 1705–1710.

Ruddle, F. H. (1980). Gene transfer in eukaryotes. *In* "Transfer of Cell Constitutents into Eukaryotic Cells" (J. E. Celis, A. Graessmann, and A. Loyster, eds.). Plenum Press, New York.

Scanlon, P. D., Raymond, F. A., and Weinshilboum, R. M. (1979). Catechol-*O*-methyltransferase: Thermolabile enzyme in erythrocytes of subjects homozygous for allele for low activity. *Science* **203**, 63–65.

Schleifer, L. S., Garrison, J. C., Sternweiss, P. C., Northup, J. K., and Gilman, A. G. (1980). The regulatory component of adenylate cyclase from uncoupled S49 lymphoma cells differ in charge from the wild-type protein, *J. Biol. Chem.* **255**, 2641–2644.

Schwartz, J. P., and Breakefield, X. O. (1980). Altered nerve growth factor in fibroblasts from patients with familial dysautonomia. *Proc. Natl. Acad. Sci. U.S.A.* **77**, 1154–1158.

Searle, A. G. (1979). Linkage map of the mouse. *Mouse News Lett.* **16**, 19.

Searle, A. G. (1980). Mouse gene list. *Mouse News Lett.* **62**, 3–30.

Shenk, T. E., Carbon, J., and Berg, P. (1976). Construction and analysis of viable deletion mutants of simian virus 40. *J. Virol.* **18**, 664–671.

Shortle, D., and Nathans, D. (1978). Local mutagenesis: A method for generating viral mutants with base substitutions in preselected regions of the viral genome. *Proc. Natl. Acad. Sci. U.S.A.* **75**, 2170-2174.

Sidman, R. L., Green, M. C., and Appel, S. H. (1965). "Catalog of the Neurological Mutants of the Mouse." Harvard Univ. Press, Cambridge, Massachusetts.

Siggers, D. C., Rogers, J. G., Boyer, S. H., Margolet, L., Dorkin, H., Banerjee, S. P., and Shooter, E. M. (1976). Increased nerve-growth-factor beta-chain cross-reacting material in familial dysautonomia. *N. Engl. J. Med.* **295**, 629-634.

Steinberg, R. A., O'Farrell, P. H., Friednich, V., and Coffino, P. (1977). Mutations causing charge alterations in regulatory subunits of the cAMP-dependent protein kinase of cultured S49 lymphoma cells. *Cell* **10**, 381-391.

Stewart, J. M., and Channabasavaiah, K. (1979). Evolutionary aspects of some neuropeptides. *Fed. Proc. Fed. Am. Soc. Exp. Biol.* **38**, 2302-2308.

Subak-Sharpe, H., Bürk, R. R., and Pitts, J. D. (1969). Metabolic cooperation between biochemically marked mammalian cells in tissue culture. *J. Cell Sci.* **4**, 353-367.

Sullivan, J. L., Cavenar, J. O., Maltbie, A. A., Lister, P., and Zung, W. W. K. (1979). Familial biochemical and clinical correlates of alcoholics with low platelet monoamine oxidase activity. *Biol. Psych.* **14**, 385-394.

Tager, H., Given, B., Baldwin, D., Mako, M., Markese, J., and Rubenstein, A. (1979). A structurally abnormal insulin causing human diabetes. *Nature (London)* **281**, 122-125.

Tanna, V. L., Winokur, G., Elston, R. C., and Rodney, C. P. (1976). A linkage study of pure depressive disease: The use of the sib-pair method. *Biol. Psychiatry* **11**, 767-771.

Towbin, H., Staehelin, T., and Gordon, J. (1979). Electrophoretic transfer of proteins from polyacrylamide gels to nitrocellulose sheets: Procedure and some applications. *Proc. Natl. Acad. Sci. U.S.A.* **76**, 4350-4354.

Ullrich, A., Dull, T. J., Gray, A., Brosius, J., and Sures, I. (1980). Genetic variation in the human insulin gene. *Science* **209**, 612-615.

Valtin, H., Stewart, J., and Sokol, H. W. (1974). Genetic control of the production of posterior pituitary principles. *In* "Handbook of Physiology, Sect. 7: Endocrinology" (E. Knobil and W. H. Sawyer, eds.), Vol. IV, Part I, pp. 131-171. Am. Physiol. Soc., Washington, D.C.

Wahl, G. M., Padgett, R. A., and Stark, G. R. (1979). Gene amplification causes overproduction of the first three enzymes of UMP synthesis in *N*-(phosphonacetyl)-L-aspartate-resistant hamster cells. *J. Biol. Chem.* **254**, 8679-8689.

Weinshilboum, R. M. (1978). Human erythrocyte catechol-*O*-methyltransferase; correlation with lung and kidney activity. *Life Sci.* **22**, 625-630.

Weinshilboum, R. M. (1979). Catecholamine biochemical genetics in human populations. *In* "Neurogenetics: Genetic Approaches to the Nervous System" (X. O. Breakefield, ed.), pp. 257-282 Elsevier/North-Holland, New York.

Weinshilboum, R. M., and Raymond, F. A. (1977). Inheritance of low erythrocyte catechol-*O*-methyltransferase activity in man. *Am. J. Hum. Genet.* **29**, 125-135.

West, G. J., and Catterall, W. A. (1979). Selection of variant neuroblastoma clones with missing or altered sodium channels. *Proc. Natl. Acad. Sci. U.S.A.* **76**, 4136-4140.

White, H. L., and Wu, J. C. (1975). Properties of catechol-*O*-methyltransferases from brain and liver of rat and human. *Biochem. J.* **145**, 135-143.

Wiche, G. (1979). *In vitro* synthesis of nerve growth factor related glioma C6 cell polypeptides. *Biochem. Biophys. Res. Commun.* **89**, 620-626.

Wigler, M., Silverstein, S., Lee, L., Pellicer, A., Cheng, Y., and Axel, R. (1977). Transfer of purified herpes virus thymidine kinase gene to cultured mouse cells. *Cell* **11**, 223-232.

Wyatt, R. J., Belmaker, R., and Murphy, D. (1975). Low platelet monoamine oxidase and vulnerability to schizophrenia. *In* "Modern Problems in Pharmacopsychiatry" (J. Mendlewicz, ed.), Vol. 10, pp. 38–56. Karger, Basel, Switzerland.

Wyman, A. R., and White, R. (1980). A locus in human DNA which has restriction fragment length polymorphism. *Am. Soc. Hum. Genet. Ann. Meet. 31st,* p. 10A.

CHROMOSOMAL COMPONENTS IN BRAIN CELLS

Ian R. Brown and Paul Greenwood

I. INTRODUCTION

The process of cell differentiation in higher organisms involves the activation or repression of different sets of genes in different tissues. In a particular cell type only a small portion of the total set of genes is active at any particular stage of development. Proteins associated with DNA in the chromatin complex are thought to play an important role in the regulation of patterns of gene activity. To advance our understanding of mechanisms of gene regulation in neural tissue,

Molecular Approaches to Neurobiology

knowledge must be obtained on the interaction of DNA and chromosomal proteins in brain cells. In this chapter information on brain chromosomal components is reviewed with particular reference to recent progress on the organization of chromatin in neuronal and glial cells.

II. ORGANIZATION OF CHROMATIN

A. The Nucleosome Structure

Chromatin in eukaryotic organisms is organized in a series of repeating nucleoprotein particles which are termed nucleosomes (for reviews see Kornberg, 1977; Felsenfeld, 1978; Horgen and Silver, 1978; Sonnenbichler, 1979; Lilley and Pardon, 1979; McGhee and Felsenfeld, 1980). Each nucleosome consists of a "core" particle plus a "linker" region. The core particle is an evolutionarily conserved structure composed of double-stranded DNA associated with an octamer of four basic chromosomal proteins, histones H2A, H2B, H3, and H4 (Kornberg, 1974; Kornberg and Thomas, 1974). A fifth histone, H1, is associated with DNA in the linker region of the nucleosome (Varshavsky *et al.*, 1976; Whitlock and Simpson, 1976; Noll and Kornberg, 1977) and may be involved in the assembly of nucleosomes into the chromatin fiber (Cole *et al.*, 1977; Griffith and Christiansen, 1977; Renz *et al.*, 1977) and the condensation of interphase chromatin into chromosomes (Gurley *et al.*, 1978). The repeating nucleosome structure gives chromatin the "beads on a string" appearance which has been observed in the electron microscope, particularly in chromatin depleted of histone H1 (Olins and Olins, 1974; Oudet *et al.*, 1975; Woodcock *et al.*, 1976). Chromatin in the cerebral hemispheres of the brain has been shown to be organized in a nucleosome structure (Brown *et al.*, 1977). Using nucleic acid hybridization techniques this study demonstrated that nucleosomes were distributed throughout the repeated and nonrepeated sequences of the genome in the rabbit brain. No class of DNA sequences including transcriptively active sequences were found to be excluded from nucleosomes in brain nuclei.

B. A Short DNA Repeat Length in Cortical Neurons

The DNA content of the repeating nucleosome units can be analyzed by digesting nuclei with micrococcal nuclease which preferentially digests DNA in the linker regions releasing mononucleosomes and higher multiples. DNA fragments isolated from the released nucleosomes can be resolved on polyacrylamide–agarose gels as a series of bands representing monomer DNA and multiples of it (Noll, 1974a,b). This approach has been utilized to determine DNA repeat length which represents the average length of DNA associated with the nucleosome core plus linker region. Analysis of a wide range of eukaryotic cells has demonstrated that a DNA fragment of length 140–145 base pairs (bp) is associated with the nucleosome core particle and that variations in DNA repeat length appear to be due to differences in the length of DNA in the linker region (Kornberg, 1977). Most tissues in higher eukaryotes exhibit a DNA repeat length of approximately 200 bp while values 10–15 bp lower have been reported in actively dividing cells grown in tissue culture (Compton *et al.*, 1976; Kornberg, 1977).

Studies on nuclei isolated from total brain tissue suggested that the nucleosome structure of brain chromatin was basically similar to that found in other higher eukaryotic cell types i.e., nuclei isolated from total mouse brain demonstrated a DNA repeat length of 201–206 bp and no age-related changes in this value were noted between 1 and 28 months (Gaubatz *et al.*, 1979). Selection of a specific region of the brain, namely the cerebral cortex, and fractionation of nuclei from this region into "neuronal" and "nonastrocytic glial" enriched populations revealed that rabbit cortical neurons exhibit an unusually short DNA repeat length of 160–162 bp while the value in glial nuclei is 200 bp (Thomas and Thompson, 1977; Brown, 1977, 1978). Ermini and Kuenzle (1978) have also reported that a short DNA repeat length is present in cortical neurons of the rat brain.

The determination of the short DNA repeat length in cortical neurons was accomplished by digestion of the nuclei with micrococcal nuclease and the analysis of the resultant DNA fragments on polyacrylamide–agarose slab gels (Fig. 1). The length of DNA fragments in each of the bands was determined by parallel electrophoresis of phage T7[+] DNA fragments of known length obtained by digestion with *Hin*d II restriction endonuclease (Fig. 1A). As shown in Fig. 1B, DNA bands from cortical neurons were out of phase and contained shorter fragments compared to corresponding kidney DNA bands as multiples of increasing size were compared. DNA bands from cortical glial nuclei, however, were in phase with the kidney bands (Fig. 1C). DNA repeat lengths were determined by plotting DNA fragment length against band number (Fig. 2). An unusually short DNA repeat length of 162 bp was apparent for cortical neurons while glial nuclei from the same brain region demonstrated a more typical DNA repeat length of 200 bp as did kidney nuclei. Since all cell types demonstrated a DNA fragment of 140 bp in the nucleosome core particle, the short DNA repeat length appears to be associated with a decrease in the length of DNA in the linker region from 60 to approximately 20 bp. The short DNA repeat length, which is present in neurons of the cerebral cortex, but not in cerebellar neurons (Thomas and Thompson, 1977), appears to be unique in higher eukaryotic cells (for review see Kornberg, 1977).

C. Developmental Changes in DNA Repeat Length

The short DNA repeat length in cortical neurons was not apparent at fetal or newborn stages of development in the mouse and rabbit (Brown, 1978). In the

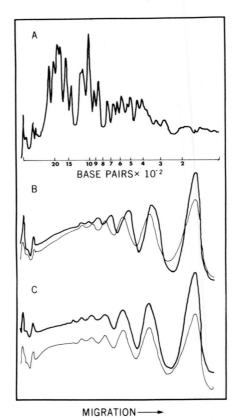

Fig. 1. Polyacrylamide gel electrophoresis of DNA fragments after micrococcal nuclease digestion of nuclei isolated from young adult rabbits. Brain and kidney nuclei were purified through dense sucrose according to Brown (1975). Total cerebral cortex nuclei were then subfractionated into "neuronal" and "nonastrocytic glial" populations as described by Thompson (1973). Digestion of nuclei with micrococcal nuclease and isolation of DNA fragments was as previously described (Brown, 1978). Electrophoresis of DNA samples was carried out on 2.5% polyacrylamide–0.5% agarose slab gels according to Loening (1967) as modified by Thomas and Furber (1976). Following staining with 20 μg/ml ethidium bromide for 15 min, the gels were photographed under short wavelength ultraviolet light. Negatives were scanned at 560 nm for calibration of migration distances. (A) marker DNA-*Hind* II digest of phage T7[+] DNA; (B) neuronal DNA (—), kidney DNA (—); (C) glial DNA (—), kidney DNA (—). (From Brown, 1978.)

development of the rabbit brain the conversion to the short DNA repeat occurred between $2\frac{1}{2}$ and $3\frac{1}{2}$ days after birth while in the mouse it took place between 4 and 7 days postnatally. This change in chromatin organization during early postnatal development of the brain was confirmed in experiments involving the isolation of cortical neuronal perikaryon, i.e., neuronal cell bodies with axons sheared off (Brown, 1978). A decrease in DNA repeat length has also been reported during early development of the rat brain (Ermini and Kuenzle, 1978). In rat cortical neurons the DNA repeat length was 195 bp 2 days prior to birth while at 7 days after birth the value decreased to 174 bp. Our studies on rat support these findings however, we have observed that the DNA repeat length in cortical neurons of the 8-day-old rat is 159 bp and that the postnatal change in chromatin structure in this mammal is a gradual process with onset between days 1 and 2 after birth (Table I).

Evidence obtained from three different mammals has suggested that chromatin in cortical neurons converts to a short DNA repeat length in the first few days after birth. To determine whether this alteration in chromatin structure was associated with a particular stage of neuronal cell differentiation or was induced as a result of the event of birth, we have selected the guinea pig brain for analysis. The length of the gestation period in this mammal is twice as long as in rabbits, mice, or rats, and as a result the guinea pig brain is much more advanced in development at birth (Himwich, 1962). In preliminary studies on the guinea

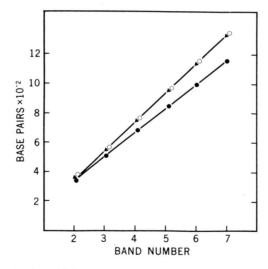

Fig. 2.   Determination of DNA repeat length in neuronal (●), glial (○), and kidney (▲) nuclei isolated from young adult rabbits. The average length of the DNA fragments in each of the bands shown in Fig. 1B and C was determined by parallel electrophoresis of phage T7[+] DNA fragments of known length which had been generated by digestion with Hind II restriction endonuclease (Fig. 1A). A calibration curve was obtained by plotting the known lengths of these phage DNA fragments (Ludwig and Summers, 1975) against their relative mobilities. DNA repeat lengths were determined as described by Thomas and Furber (1976) and Noll and Kornberg (1977). (From Brown, 1978.)

Table I. Developmental Change in DNA Repeat Length in Rat Cortical Neurons[a]

	Repeat length values in base pairs				
Days after birth:	1	2	5	6¾	8½
Neuronal:	195	185	174	176	159
Kidney:	197	190	201	197	199

[a] Kidney nuclei purified according to Brown (1975) and nuclei purified from rat neuronal perikaryon isolated according to Poduslo and McKhann (1977) were digested with micrococcal nuclease (Brown, 1978). DNA fragments were extracted and subsequently electrophoresed on polyacrylamide-agarose slab gels as in Fig. 1. DNA repeat lengths were determined as in Figs. 1 and 2. (P. D. Greenwood and I. R. Brown, unpublished results.)

pig we have observed that the shift to the short DNA repeat length occurs prior to birth at a stage of maturation of cortical neuronal cells which is equivalent to the postnatal stage at which neuronal chromatin in the rat, mouse, and rabbit brain shifted to the short DNA repeat length. The timing of the shift to the short DNA repeat length in cortical neurons therefore seems to be correlated to a stage of cellular differentiation of cortical neurons and not to the event of birth.

D. Significance of the Short DNA Repeat Length

Thomas and Thompson (1977) have suggested a possible correlation between the short DNA repeat length in cortical neurons and high transcriptional activity, although they advise caution in this interpretation. They have found that isolated cortical neuronal nuclei are more active in RNA synthesis measured *in vitro* compared to glial, kidney, or cerebellar neuronal nuclei. Mizobe *et al.* (1974) have also reported that neuronal nuclei have a higher transcriptional activity *in vitro* than glial nuclei. A correlation of this nature is also apparent in several types of fungal cells where a high percentage of the genome is expressed and short DNA repeats of 154–170 bp have been reported (Horgen and Silver, 1978; Silver, 1979). Similarly in chicken and sea urchin, transcriptionally inactive tissues such as erythrocytes and sperm have a longer repeat length relative to transcriptionally active tissues such as liver and gastrula (Morris, 1976b; Spadafora *et al.*, 1976; Weintraub, 1978). It should be pointed out, however, that little correlation has been found between DNA repeat length and the functional state of cells in culture (Compton *et al.*, 1976). Also, in ciliated protozoans, the transcriptionally inactive micronucleus has a shorter DNA repeat length relative to the transcriptionally active macronucleus (Gorovsky *et al.*, 1977; Lipps and Morris, 1977).

The short DNA repeat length in cortical neurons may be correlated with chromatin uncoiling and the amount of neuronal euchromatin. In electron microscopic studies, isolated cortical neuronal nuclei exhibited more dispersed chromatin compared to glial nuclei (Thompson, 1973; Sinha *et al.*, 1978; Stoykova *et al.*, 1979). This is also true of yeast nuclei (Gordon, 1977) which exhibit a short DNA repeat length (Lohr *et al.*, 1977). In several systems it has been reported that chromatin organized in a short DNA repeat is more sensitive to nuclease digestion compared to chromatin exhibiting a long DNA repeat (Todd and Garrard, 1977; Zongza and Mathias, 1979; Arceci and Gross, 1980).

In recent studies we have observed that chromatin in total cerebral hemisphere and kidney nuclei exhibited similar digestion kinetics with DNase I when nuclei were isolated $1-1\frac{1}{2}$ days after birth in the rat; however, at 8 days postnatally (i.e., after the conversion to the short DNA repeat length) chromatin in total cerebral hemisphere nuclei was found to be more digestable. This difference in sensitivity to DNase I was accentuated in neuronal nuclei which were subfractionated from total cerebral hemisphere nuclei (Fig. 3). It appears that the shift to a short

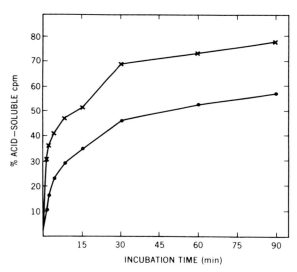

Fig. 3. DNase I digestion kinetics of neuronal (×) and kidney (●) nuclei. Nuclei were isolated from the cerebral hemispheres or kidneys of 8-day-old rats as described in Fig. 1 except that solutions A and B contained 1 mM phenylmethylsulfonylfluoride (PMSF) and Triton X-100 was absent in the second homogenate. Neuronal and kidney DNA was labeled by intraperitoneal injection of pregnant females 3 times/day with 250–300 μCi [³H]thymidine starting 7–10 days prior to birth. Nuclei were adjusted to 1 mg/ml DNA in reticulocyte standard buffer (RSB: 3 mM MgCl₂, 10 mM NaCl, 10 mM Tris HCl pH 7.4) and digested with 5 μg/ml DNase I (Sigma). The reaction was terminated by addition of EDTA to 5 mM and acid-insoluble material was precipitated by the addition of 10% trichloroacetic acid containing 100 μg/ml thymidine. Radioactivity in acid-soluble and acid-insoluble fractions was then determined. (From P. D. Greenwood and I. R. Brown, unpublished results.)

DNA repeat rendered cortical neuronal nuclei more sensitive to DNase I digestion whereas no change was observed in kidney nuclei which exhibited the typical 200 bp DNA repeat length throughout development. Since DNase I is known to preferentially digest extended chromatin (Burkholder and Weaver, 1975; Jalouzot *et al.*, 1980) and transcriptionally active chromatin (Weintraub and Groudine, 1976; Levy *et al.*, 1977), the postnatal shift to a short DNA repeat length may be associated with the uncoiling of neuronal chromatin and increased template activity. In preliminary experiments we have noted an increase in RNA template activity of cerebral hemisphere nuclei isolated subsequent to the conversion to the short DNA repeat length.

The timing of the developmental shift to the short DNA repeat length in cortical neurons seems to be correlated to a stage of cellular differentiation and not induced by the event of birth. It is unknown, however, whether the developmental change in chromatin structure is related to the process of axonal and dendritic outgrowth which is extensive in cortical neurons but is not as extensive in cerebellar neurons which do not exhibit the unusually short DNA repeat length (Thomas and Thompson, 1977). The significance of the short DNA repeat length remains unknown at present. Although correlations have been made, no firm connection between the existence of a short DNA repeat length and a particular cellular phenomenon has been demonstrated. Studies dealing with the developmental onset of this conformational change in neuronal chromatin, and particular cellular processes that may be altered as a result, should lead to a better understanding of the functional significance of the short DNA repeat length.

E. DNA Synthesis in Cortical Neurons

Neurons in the cortex of the mammalian forebrain do not divide after birth and it has been thought that these cells exhibit a diploid (2C) DNA content (Novakova *et al.*, 1970). More recently, it has been proposed that cortical neurons in rat, mouse and rabbit possess an elevated nuclear DNA content of approximately 3.5C as determined by cytofluorometry, ultraviolet absorption, and the diphenylamine reaction (Bregnard *et al.*, 1975, 1977, 1979; Kuenzle *et al.*, 1978). Staining with mithramycin, however, which binds to double-stranded but not single-stranded DNA demonstrated DNA values of only 2C (Kuenzle, 1979). Other reports have demonstrated increased amounts of DNA in neurons from the motor cortex of cats (Herman and Lapman, 1969) and 4C levels of DNA in all cerebellar and olifactory bulbar neurons except Purkinje cells (Swartz and Bhatnagar, 1981). The series of studies by Bregnard and Kuenzle has suggested that the extra DNA is synthesized predominantly in the first 7 days after birth as determined by postnatal injection of [³H]thymidine and that this DNA is qualitatively indistinguishable from other somatic DNA. Other studies, however, have not detected postnatal DNA synthesis in cortical neurons during the first week after birth following injection of [³H]thymidine into neonates or incubation of labelled precursor with isolated perikaryon (Altman, 1969; Burdman, 1972; Stambolova *et al.*, 1979; Brown, 1980).

Studies involving the incorporation of labeled precursors into the DNA of brain nuclei have demonstrated that incorporation occurred predominantly into glial nuclei in both young and adult animals; however, labeling of neuroblasts, spongioblasts and microneurons was also noted (Altman, 1972; Burdman, 1972; Stambolova *et al.*, 1973, 1979; Guiffrida *et al.*, 1976). Incorporation of labeled precursors into neuronal nuclei has often been attributed to contamination by glial nuclei (Burdman, 1972; Vilenchik and Tretjak, 1977; Stambolova *et al.*, 1979). Investigations which have observed DNA synthesis in mature neurons have ascribed it to one of two phenomena. Wintzerith *et al.* (1977) suggested that the incorporation of labeled DNA precursors into adult neuronal DNA is due to nucleolar DNA synthesis. Other investigations suggest that neuronal DNA synthesis involves DNA repair (Stambolova *et al.*, 1973; Wintzerith *et al.*, 1977; Vilenchik and Tretjak, 1977). Inoue *et al.* (1976) reported that neuronal nuclei synthesized DNA five times more actively than glial nuclei *in vitro*. In a later study this group suggested that the high level of neuronal DNA synthesis is associated with a DNA repair process rather than replicative DNA synthesis (Inoue *et al.*, 1979). This observation is supported by a related study which demonstrates that neurons are capable of DNA excision–repair activity (McCombe *et al.*, 1976).

III. BRAIN HISTONES

A. Total Histones

Histones are basic chromosomal proteins which are associated with nuclear DNA in the somatic cells of all higher eukaryotes (for review see Olson and Busch, 1974; Elgin and Weintraub, 1975; Isenberg, 1979). These proteins contain high concentrations of lysine and arginine, low amounts of cysteine, and no tryptophan. Following acid extraction from nuclei or chromatin, histones can be resolved by electrophoretic and chromatographic techniques into the following species: H1, H2A, H2B, H3, and H4.

Early studies suggested that brain histones displayed no tissue-specific features and little if any species-specific variations (Neidle and Waelsch, 1964; Piha *et al.*, 1966; Martenson *et al.*, 1969). Subsequent studies which achieved an increased resolution of histones have tended to confirm these earlier observations (Shaw and Huang, 1970; Duerre and Gaitonde, 1971; Uyemura *et al.*, 1974; Gaubatz *et al.*, 1979; Unger-Ullmann and Modak, 1979; Russanova *et al.*, 1980). In addition, no changes in the electrophoretic properties of total brain histones have been observed in different brain regions or at various stages of brain development (Shaw and Huang, 1970; Biessmann and Rajewsky, 1975; Kelly and Luttges, 1976; Gaubatz *et al.*, 1979).

Restriction endonucleases have been employed to separate chromatin fractions from mouse brain into components containing only satellite DNA or main band DNA (Mazrimas *et al.*, 1979). Analysis of histones isolated from these two chromatin fractions demonstrated that all five histones were present and that protein degradation was minimal. Endogenous levels of nucleases and proteases are an important factor in studies involving the analysis of chromatin. Fortunately, isolated brain nuclei contain very low levels of endogenous nuclease activity and minimal endogenous protease activity compared to other cell types (Balhorn *et al.*, 1978; Mazrimas *et al.*, 1979).

B. Histones of Neuronal and Glial Cells

Most of the studies on brain histones have focused on brain regions (i.e., cerebral hemispheres, cerebellum, etc.), rather than on specific cell types which are present within these regions. Recent observations have indicated the necessity of distinguishing neuronal and glial cells. Cortical neuronal nuclei exhibit a short DNA repeat length of 160 bp which arises in the first few days after birth in the mouse, rat, and rabbit, whereas cortical glial cell nuclei have the more typical DNA repeat of 200 bp (Thomas and Thompson, 1977; Brown, 1978; Ermini and Kuenzle, 1978). Since differences in histones, H1 in particular, may be correlated with variations in DNA repeat length (Morris, 1976a; Noll, 1976; Thomas and Thompson, 1977) an analysis of neuronal and glial histones is necessary.

Two studies dealing with neuronal and glial histones have revealed no qualitative differences; however, the electrophoretic resolution that was achieved was relatively poor (Albrecht and Hemminki, 1976; Schmitt and Matthies, 1979a). In our laboratory (Greenwood *et al.*, 1981a) we have analyzed cortical neuronal and glial histones on SDS polyacrylamide gels which resolve proteins on the basis of molecular weight. As shown in Fig. 4 neuronal (lane A) and glial histones (lane C) associated with both the core particle (H2A, H2B, H3 and H4) and the linker region (H1) of the nucleosome exhibited similar electrophoretic mobilities on SDS gels. Similar results were obtained using acid/urea polyacrylamide gels which separate protein on the basis of charge (Greenwood *et al.*, 1981a). Quantitative differences were noted however when neuronal histone H1 was compared to glial or kidney H1. As shown in Table II neuronal nuclei, which exhibited a short DNA repeat length of 160 bp, contained significantly less histone H1 per mg DNA compared to glial or kidney nuclei which exhibited the typical 200 bp DNA repeat length (Greenwood *et al.*, 1981a). It is interesting to note that several groups have found that chromatin enriched in transcriptionally active sequences is deficient in histone H1 (Pederson and Bhorjee, 1975; Peder-

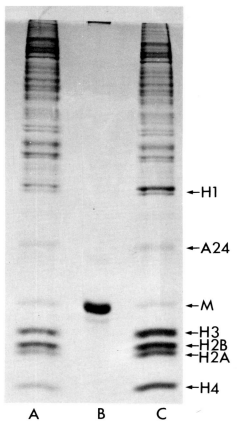

Fig. 4. Analysis of total histone by SDS–polyacrylamide slab gel electrophoresis. Neuronal and glial nuclei were isolated from the cerebral cortex of young adult rabbit brain as described in Fig. 1 except that solutions A and B contained 1 mM PMSF. Purified nuclei were resuspended in an equal volume of 0.5 N HCl–8 M urea containing 5 mg protamine per ml nuclei. An additional volume of 8 M urea–0.25 N HCl was then added and total histone was extracted on ice for 30–40 min. Histone was precipitated with acetone as described previously (Silver, 1979), dried in a nitrogen stream, dissolved in Laemmli (1970) sample buffer and electrophoresed on 15% SDS–polyacrylamide slab gels prepared according to Studier (1973) using the discontinuous buffer system of Laemmli (1970). Gels were stained with Coomassie Blue. M, Myelin basic protein; A24, histone H2A–ubiquitin conjugate. Gel lane A, neuronal histone; B, purified myelin basic protein; C, glial histone. (From Greenwood et al., 1981a.)

Table II.   Quantitation of Histone H1 per mg DNA[a]

| | μg histone H1/mg DNA | | |
Method	Kidney	Glial	Neuronal
Densitometric scanning	130.8 ± 15.5	115.5 ± 17.4	48.6 ± 5.7
Stained protein elution	127.9 ± 16.1	101.3 ± 26.8	42.0 ± 1.0

[a] Histone H1 was selectively extracted with 5% perchloric acid from rabbit kidney, glial and neuronal nuclei (isolated as in Fig. 4) and quantitated on a μg H1 per mg DNA basis by a densitometric scanning method (Goodwin *et al.*, 1977a) and a procedure involving elution of stained histone H1 from SDS polyacrylamide gels (Levy-Wilson *et al.*, 1980). In both methods, SDS polyacrylamide gels stained with Coomassie Blue were used since this method has previously been employed to quantitate histone H1 and total histone (Lawson and Cole, 1979; Djondjurov *et al.*, 1979). DNA was quantitated by the method of Dische and Schwarz (1937). (From Greenwood *et al.*, 1981a.)

son, 1978; Djondjurov *et al.*, 1979). The low concentration of histone H1 which we have observed in neuronal nuclei may be correlated with the elevated levels of transcriptional activity in these brain nuclei relative to glial nuclei (Thomas and Thompson, 1977).

Histone H1 also appears to be involved in the condensation of chromation (Cole *et al.*, 1977; Griffith and Christiansen, 1977; Renz *et al.*, 1977). Electron microscopic analysis of brain nuclei has revealed that neuronal nuclei exhibit less condensed chromatin compared to glial nuclei (Thompson, 1973; Sinha *et al.*, 1978; Stoykova *et al.*, 1979). The reduced amount of histone H1 in neuronal nuclei relative to glial nuclei may be related to the observed differences in chromatin condensation.

Our studies have demonstrated that two other proteins are extracted with histones from brain nuclei (Greenwood *et al.*, 1981a). Myelin basic protein was found to be present in extracts of total histones from both neuronal and glial nuclei (Fig. 4) as has previously been reported by Gaubatz *et al.* (1979) in total brain chromatin. A24, a protein composed of histone H2A linked to the nonhistone protein ubiquitin (Goldknopf *et al.*, 1975; Goldknopf and Busch, 1977), is also present in extracts of total histone prepared from neuronal and glial nuclei (Fig. 4). This protein which appears to be associated with the nucleosome core particle, has not been previously reported in brain tissue (Isenberg, 1979).

In a developmental analysis of neuronal histones, Bregnard *et al.* (1979) reported that the ratio of histones H1°:H1 increased at the time of birth and that this may be related to the postnatal arrest of neuronal cell division. Changes in the histones associated with nucleosome core particle were not apparent. Bregnard *et al.* (1979) have suggested that in cortical neurons, postnatal histone synthesis accompanied their observed postnatal synthesis of neuronal DNA. In our laboratory we have not detected postnatal synthesis of DNA; however, studies with isolated neuronal perikaryon have demonstrated a transient period of synthesis of core and linker histones which initiates just prior to the conversion to the short DNA repeat length (Brown, 1980).

As shown in Fig. 5A rabbit neuronal perikaryon isolated 18 h after birth did not incorporate labeled lysine and arginine into histones; however, cells isolated 30 and 42 h after birth showed labeling of histone H1, core histones (H3, H2B, H2A, H4), and A24. Labeling of all histones was apparent at a progressively decreasing level in neurons isolated 66 and 90 h after birth and by 114 h no incorporation into histones was observed (Fig. 5B). At no stage was incorporation of labeled thymidine into DNA observed.

A correlation may exist between the transient ability of neuronal perikaryon to incorporate labeled amino acids into histone and a possible requirement for additional histone for the conversion to the short DNA repeat length which occurs between 60 and 84 h after birth in the rabbit. The number of nucleosomes per unit length of DNA may increase when the linker DNA length is reduced from 60 to 20 bp at the time of the chromatin conversion in cortical neurons. This may suggest a need for the synthesis of additional histone to package DNA into the short DNA repeat length.

C. Modification of Brain Histones

The modification of histones by post-translational modification, i.e., acetylation or phosphorylation, may be a prerequisite to changes in gene activity (Allfrey, 1971). Initial reports on brain histone acetylation suggested that histone acetylase activity increased with neural maturation, that acetylation was higher in adult brain compared to liver, and that the arginine-rich and slightly lysine-rich brain histones were most heavily acetylated (Bondy *et al.*, 1970). More recent studies indicate that acetylation of brain histones may decrease with age (Thakur *et al.*, 1978; Das and Kanungo, 1979; Kanungo and Thakur, 1979), and that acetylation involved mostly lysine residues (Schmitt and Matthies, 1979b). A stimulation in the acetylation of specific brain histones has been reported following intravenous administration of LSD (Brown and Liew, 1975).

Modification of brain histones by phosphorylation has been demonstrated *in vitro* in brain slices and in isolated nuclei (OH'Hara and Yanagihara, 1977; Yanagihara *et al.*, 1978) and has been reported to decrease with age (Das and Kanungo, 1979). A third type of post-translational modification of histones, i.e., methylation, has been observed (Duerre and Lee, 1974). During development, levels of histone methylase activity appeared to drop (Miyake, 1975). Histone methylase activity may be higher in neuronal nuclei compared to glial nuclei (Lee and Loh, 1977).

With few exceptions research to date on the modification of brain histones has

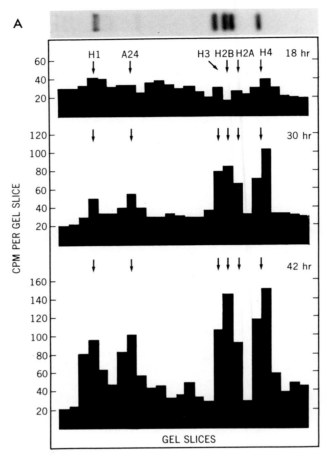

Fig. 5. Incorporation of [³H]lysine and [³H]arginine into histones by isolated neuronal perikaryon. Neuronal perikaryon were isolated from rabbit cerebral hemispheres as described by Poduslo and McKhann (1977) and subsequently incubated with [³H]lysine and [³H]arginine as previously described (Brown, 1980). Incorporation of radioactivity into specific histone species was determined by electrophoresis on SDS–polyacrylamide slab gels as in Fig. 4 with purified rabbit kidney histone as marker. The insert at the top of the figure is a gel profile prior to slicing and determination of radioactivity. Incorporation of [³H]lysine and [³H]arginine into histones prepared from neuronal perikaryon isolated 18, 30, and 42 h after birth (A), or 66, 90, and 114 h after birth (B) (From Brown, 1980.)

tended to focus on histones isolated from total brain regions. Since divergent chromatin structures exist in neuronal and glial nuclei it has become apparent that analysis of histone modification in these two types of brain nuclei is desirable. For example, acetylation of histones has been reported to be higher in

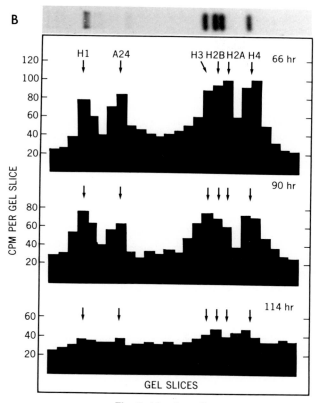

Fig. 5. (*Continued*)

neuronal versus glial nuclei (Sarkander and Dulce, 1978). Since acetylation of proteins in neuronal and glial chromatin has been suggested to induce an increase in the total number of initiation sites available for RNA polymerase (Sarkander and Dulce, 1978), modification of chromosomal proteins by this process may be involved in the control of diversity of transcription in brain cells.

IV. NONHISTONE CHROMOSOMAL PROTEINS (NHCP) OF THE BRAIN

A. General Characteristics and Modification

Nonhistone chromosomal proteins (NHCP) may play structural, regulatory, and enzymatic roles in chromatin (for reviews see Olson and Busch, 1974; Patel,

1974; Stein *et al.,* 1974; Elgin and Stumph, 1975; Elgin and Weintraub, 1975). These proteins demonstrate a wide range of molecular weights and often exhibit more tissue specific variations compared to histones. Efforts to isolate and characterize NHCP have been hampered by their tendency to aggregate with the histones; however, solubilization in detergents or salts may reduce this problem. NHCP are included at times in fractions termed total nuclear proteins, total chromatin proteins, and acidic nuclear proteins.

NHCP of the brain have demonstrated qualitative tissue specific differences when compared electrophoretically to NHCP from other tissues (Wu *et al.,* 1973; MacGillivray and Rickwood, 1974; Uyemura *et al.,* 1974; Choie *et al.,* 1977). In fact the brain NHCP pattern has been reported to be the most heterogeneous of the tissues analyzed and includes an exceptional number of high molecular weight proteins (Wu *et al.,* 1973). A comparison of NHCP associated with nuclei isolated from cerebral hemispheres, cerebellum, and remaining brain regions revealed similar patterns (Fujitani and Holoubek, 1974); however, a more recent study suggested that differences were present (Fujii *et al.,* 1980). During development, brain NHCP has been reported to show no change (Olpe *et al.,* 1972), quantitative changes (Choie *et al.,* 1977), and minor changes (Kelly and Luttges, 1976), yet Biessmann and Rajewsky (1975) report both qualitative and quantitative variations. A 0.35 *M* NaCl-extractable subnuclear fraction of brain NHCP has been reported to be preferentially associated with dispersed chromatin and absent from condensed chromatin, suggesting that this group of proteins is associated with active chromatin (Cocchia and Michetti, 1980).

The analysis of NHCP in brain would be clarified by the employment of more uniform isolation procedures, the analysis of specific cell types rather than total brain regions, and the improved resolution of proteins species by two-dimensional gel electrophoresis. Studies focusing on neuronal and glial nuclei have suggested that qualitative and quantitative differences exist in the NHCP pattern of these two nuclear types (Tashiro *et al.,* 1974; Tsitilou *et al.,* 1979; Heizmann *et al.,* 1980). The study by Heizmann *et al.* (1980), which employed two-dimensional gel electrophoresis, revealed pronounced developmental changes in NHCP in cortical and cerebellar neurons which appeared to correlate with the arrest of cell division and beginning of terminal differentiation. In particular the appearance of two proteins of molecular weight 35,000 and 38,000 coincided with the final loss of proliferative capacity (i.e., at day 0 in cortical neurons and day 22 in cerebellar neurons).

Acidic nuclear proteins have been selectively extracted from nuclei or chromatin with salt or phenol. Davis *et al.* (1972) identified 31 different brain acidic nuclear proteins and suggested that they played a role in gene derepression. This class of brain protein has been reported to exhibit tissue specific (Teng *et al.,*

1971; Uyemura *et al.*, 1974) and developmental variations (Szijan and Burdman, 1973) and differences between neuronal and glial nuclei have also been noted (Dravid and Burdman, 1968). A large percentage of the brain acidic nuclear proteins may be acidic lipoproteins (Lu and Koenig, 1973).

Relatively little work has been done on the post-translational modification of brain NHCP fractions. Total chromatin proteins (including both histones and NHCP) and nuclear acidic proteins appear to be highly phosphorylated in brain compared to liver (OH'Hara and Yanagihara, 1977; Yanagihara *et al.*, 1978). Differences in the levels of *in vitro* phosphorylation of various classes of brain nuclear proteins have been reported following cerebral anoxia and ischemia (Yanagihara, 1980). Acetylation of brain NHCP has been reported and this activity appears to decrease during development (Thakur *et al.*, 1978).

B. High Mobility Group (HMG) Proteins

A group of specific NHCP designated as the high mobility group (HMG) proteins have recently come under intense investigation in nonbrain tissues (for reviews see Goodwin and Johns, 1977; Goodwin *et al.*, 1978). Originally isolated as impurities of histone H1 fractions, the HMG proteins are prepared from nuclei or chromatin by 0.35 *M* NaCl extraction followed by precipitation with specific concentrations of trichloroacetic acid or alternatively by 5% PCA extraction followed by further fractionation and acetone precipitation (Goodwin and Johns, 1977). Electrophoresis on polyacrylamide gels, CM Sephadex chromatography, and amino acid analyses have revealed four major HMG proteins designated as HMG's 1, 2, 14, and 17. Like the histones, HMG proteins contain a high percentage of basic amino acids; however, unlike histones, they also contain a high percentage of acidic amino acids. A difficulty encountered in working with HMG proteins is their low concentration in nuclei, i.e., 1–5% of the yield of histones.

HMG proteins 1 and 2 have been reported to be associated with the linker region of the nucleosome (Goodwin *et al.*, 1978) and HMG 14 and 17 with core particles (Mathew *et al.*, 1979). Transcriptionally active chromatin is particularly sensitive to digestion with the nuclease DNase I (Weintraub and Groudine, 1976; Levy *et al.*, 1977). Association of HMG proteins 14 and 17 with nucleosome core particles appears to confer sensitivity to digestion with DNase I thus these proteins may be preferentially associated with transcriptionally active chromatin (Weisbrod and Weintraub, 1979).

Two studies have identified the presence of HMG proteins 1, 2, 14, and 17 in total brain nuclei (Levy and Dixon, 1978; Gordon *et al.*, 1980). Studies in our laboratory (Greenwood *et al.*, 1981b), have demonstrated the presence of putative HMG proteins 1, 2, and 17 in rabbit cerebral hemisphere nuclei (Fig. 6, lane

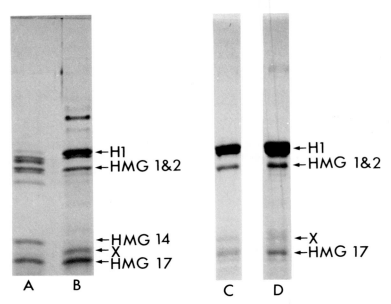

Fig. 6. Analysis of putative brain HMG proteins on SDS polyacrylamide gels. Total cerebral hemisphere nuclei were isolated from rabbits as in Fig. 1. Neuronal and glial nuclei were subfractionated as previously described (Thompson, 1973). Putative HMG proteins were extracted from total cerebral hemisphere nuclei as described by Goodwin and Johns (1977) and Goodwin *et al.* (1978). Purified neuronal and glial nuclei were quick-frozen in 5% perchloric acid and stored at −20°C. Nuclear fractions from 40 rabbit brains were thawed, pooled, and putative neuronal and glial HMG proteins extracted according to Goodwin *et al.* (1977b). Samples of HMG proteins were analyzed on SDS gels prepared as in Fig. 4. Gel lane A, purified calf thymus HMG proteins (Teng *et al.*, 1979); B, HMG proteins from total cerebral hemisphere nuclei; C, neuronal HMG proteins; D, glial HMG proteins. (From Greenwood *et al.*, 1981b.)

B). Protein X may represent a form of HMG 14 since this protein, isolated from some cell types, is known to exhibit variations in electrophoretic mobility compared to calf thymus HMG 14 (Reeves and Candido, 1980).

Since HMG proteins 1 and 2 have been reported to be associated with the linker region of the nucleosome (Goodwin *et al.*, 1978), and since the short DNA repeat length in cortical neurons is associated with a reduced length of DNA in the linker region of the nucleosome (Brown, 1978), it was of interest to compare HMG proteins in neuronal and glial nuclei which were subfractionated from cerebral hemisphere nuclei. As shown in Fig. 6, lanes C and D, the differences in DNA repeat lengths in neuronal and glial nuclei are not associated with qualitative differences in the pattern of putative HMG proteins as analyzed on SDS polyacrylamide gels (Greenwood *et al.*, 1981b).

V. DNA POLYMERASE

DNA polymerases in eukaryotic cells have been the subject of intense investigation (for reviews see Weissbach, 1977; Sheinin *et al.*, 1978). Three types of DNA polymerase designated α, β, and γ have been resolved. α is involved in nuclear replicative DNA synthesis, β in nuclear DNA repair, and γ in mitochondrial DNA replication. Initial studies on brain tissue revealed very little DNA polymerase activity in the adult tissue (Chiu and Sung, 1971); however, in a subsequent report the same authors identified two DNA polymerases A and B (corresponding to α and β) in the developing rat brain (Chiu and Sung, 1971), the A form probably associated with cell proliferation (Chiu and Sung, 1972). In the adult brain DNA polymerase B exhibited increased activity relative to A (Shimada and Terayama, 1972) and occurred in a particulate form (Chiu and Sung, 1972).

Early studies on DNA polymerase in brain utilized total brain tissue which was not fractionated into neuronal and glial cells. Recently, developmental changes in α, β, and γ DNA polymerase activity has been examined in purified cortical neurons (Hubscher *et al.*, 1977a). The level of activity of DNA polymerase α correlated with the *in vivo* rate of cell division and therefore to the rate of replicative DNA synthesis. In the rat as birth is approached and the rate of neuronal cell division stops, DNA polymerase α activity drops sharply and virtually disappears by 2 weeks postnatally. The developmental decrease in DNA polymerase α in neurons has been confirmed by the demonstration of a loss, as brain maturation proceeds, of binding in cell lysates of diadenosine tetraphosphate, a specific ligand of DNA polymerase α (Grummt *et al.*, 1979). In contrast, DNA polymerase β and γ activities are present at all stages of development and their levels seem independent of the rate of cell division (Hubscher *et al.*, 1977a).

Both DNA polymerase β and γ have been purified from brain tissue (Hubscher *et al.*, 1977a; Waser *et al.*, 1979). Neuronal nuclei of 60-day-old rats lack DNA polymerase α and contain mainly polymerase β (99.2%). Irradiation of these brain nuclei with UV light induced a 10-fold increase in incorporation of [^{3}H]dTTP into nuclear DNA, suggesting DNA polymerase β is involved in repair of damaged DNA (Hubscher *et al.*, 1978, 1979; Waser *et al.*, 1979). DNA polymerase γ has been shown to be present in neuronal nuclei and in mitochondria isolated from synaptosomes (Hubscher *et al.*, 1977b). This DNA polymerase is capable of supporting the replication of mitochondrial DNA in isolated brain synaptosomes (Hubscher *et al.*, 1977b, 1978). The novel use of mitochondria situated in synaptosomes for studies on DNA polymerases permit the purification of a mitochondrial fraction which is free of nuclear contamination since the synaptic ends of neurons are anatomically separated from the nuclei.

Bregnard *et al.*, (1975, 1977, 1979) and Keunzle *et al.*, (1978) have reported that cortical neurons in the adult mouse, rat and rabbit exhibit an elevated DNA content of 3.5C and that the "extra DNA" is synthesized in the first 2–3 weeks after birth. The activity profiles of DNA polymerase α, β, and γ were therefore examined in rat cortical neurons during development; however, none of the three DNA polymerase activities demonstrated any significant rise in activity concomitant with the reported increase in DNA content of cortical neurons from 2C to 3.5C (Kuenzle *et al.*, 1978).

VI. RNA POLYMERASE

Multiple forms of DNA-dependent RNA polymerase have been isolated from a variety of eukaryotic tissues (for review see Chambon, 1975). RNA polymerase I (or A) localized in the nucleolus is resistant to α-amanitin and is engaged in ribosomal RNA synthesis. RNA polymerase II (or B) localized in the nucleoplasm is sensitive to α-amanitin and is involved in the synthesis of precursors to messenger RNA while RNA polymerase III (or C), also in the nucleoplasm, shows intermediate sensitivity to α-amanitin and is involved in the synthesis of transfer and 5 S RNA.

Studies on RNA polymerase in brain have frequently involved the assay of RNA polymerase activity in isolated nuclei, rather than purification of the enzyme (Kato and Kurokawa, 1970; Austoker *et al.*, 1972; Banks and Johnson, 1972, 1973; Burdman *et al.*, 1973; Mizobe *et al.*, 1974; Guiffrida *et al.*, 1975; Banks-Schlegel and Johnson, 1975). In general, neuronal nuclei display higher levels of RNA polymerase activity than glial cell nuclei, and regional and developmental differences are apparent.

Using differential sensitivity to α-amanitin, two distinct RNA polymerase activities (I and II) have been observed in isolated neuronal and glial cell nuclei (Thompson, 1973). RNA polymerase from neuronal and glial nuclei appear to be similar in transcriptional efficiency (Sarkander and Dulce, 1978). The increased number of RNA initiation sites which was observed in isolated neuronal chromatin compared to glial chromatin appears to be related to properties of the chromosomal proteins rather than differences in RNA polymerases (Sarkander and Dulce, 1978).

Singh and Sung (1972) have purified RNA polymerase from total bovine brain nuclei using DEAE-cellulose chromatography and separated two forms corresponding to RNA polymerase I and II. The purification approach was extended by Yamamoto and Takahashi (1978) using total rat brain nuclei. The isolation and characterization of six forms of brain RNA polymerases (i.e., RNA polymerase I, two forms of RNA polymerase II, and three forms of RNA polymerase III) was accomplished using DEAE-Sephadex chromatography and sensitivities to α-amanitin. Brain RNA polymerase II was subsequently purified to homogeneity and its subunit composition examined (Yamamoto and Takahashi, 1980). The purification of RNA polymerase II will now permit the template specificity of brain chromatin to be examined using homologous enzyme rather than *E. coli* RNA polymerase since the *E. coli* enzyme may result in infidelity of transcription *in vitro* (Tsai *et al.*, 1976).

VII. CONCLUDING REMARKS

Brain is a tissue that is heterogeneous in cell types. The various cell types may exhibit divergent synthetic activities when a particular stage of neural development is examined. Studies on chromosomal components in brain cells highlight the desirability of carrying out molecular studies on specific cell types isolated from specific brain regions (see Chapter 13 by Althaus and Neuhoff for review of isolation procedures). While the analysis of total brain nuclei suggested that the organization of nucleosomes in brain chromatin was similar to that found in other higher eukaryotic cells, isolation of neurons from the cerebral cortex has indicated that this cell type converts to a unique chromatin conformation during the first few days after birth in the mouse, rat, and rabbit. Present studies have focused on the mechanism of the conversion to the short DNA repeat length and whether chromosomal proteins are altered. It remains to be determined whether the alteration in chromatin conformation is a prerequisite to developmental changes in cortical neurons.

ACKNOWLEDGMENT

Studies mentioned in this chapter were supported by grants to I.R.B. from the Natural Sciences and Engineering Research Council of Canada. The authors express thanks to Prof. Julie Silver for helpful comments on the manuscript.

REFERENCES

Albrecht, J., and Hemminki, K. (1976). Chromatin proteins of large and small brain nuclei. *J. Neurochem.* **26,** 1297–1299.

Allfrey, V. G. (1971). Functional and metabolic aspects of DNA-associated proteins. *In* "Histones and Nucleohistones" (D. M. P. Phillips, ed.), pp. 241–294. Plenum, New York.

Altman, J. (1969). DNA metabolism and cell proliferation. *In* "Handbook of Neurochemistry" (A. Lajtha, ed.), Vol. 2, pp. 137–182. Plenum, New York.

Altman, J. (1972). Autoradiographic examination of behaviorally induced changes in the protein and nucleic acid metabolism of the brain. *In* "Macromolecules and Behavior" (J. Gaito, ed.), 2nd ed., pp. 305–334. Meredith, New York.

Arceci, R. J., and Gross, P. R. (1980). Histone variants and chromatin structure during sea urchin development. *Dev. Biol.* **80,** 186–209.

Austoker, J., Cox, D., and Mathias, A. P. (1972). Fractionation of nuclei from brain by zonal centrifugation and a study of the ribonucleic acid polymerase activity in the various classes of nuclei. *Biochem. J.* **129,** 1139–1155.

Balhorn, R., Weston, S., Mazrimas, J. A., and Young, T. (1978). Endogenous nuclease and chromatin protease activities in rodent tissues. *J. Cell Biol.* **79,** 126a.

Banks, S. P., and Johnson, T. C. (1972). Maturation-dependent events related to DNA-dependent RNA synthesis in intact mouse brain nuclei. *Brain Res.* **41,** 155–169.

Banks, S. P., and Johnson, T. C. (1973). Synthesis of RNA—Polyadenylic acid by isolated brain nuclei. *Science* **181,** 1065–1066.

Banks-Schlegel, S. P., and Johnson, T. C. (1975). RNA metabolism in isolated mouse brain nuclei during early postnatal development. *J. Neurochem.* **24,** 947–952.

Biessmann, H., and Rajewsky, M. F. (1975). Nuclear protein patterns in developing and adult brain and in ethylnitrosourea-induced neuroectodermal tumors of the rat. *J. Neurochem.* **24,** 387–393.

Bondy, J. C., Roberts, S., and Morelos, B. S. (1970). Histone-acetylating enzyme of brain. *Biochem. J.* **119,** 665–672.

Bregnard, A., Knusel, A., and Kuenzle, C. C. (1975). Are all the neuronal nuclei polyploid? *Histochemistry* **43,** 59–61.

Bregnard, A., Kuenzle, C. C., and Ruch, F. (1977). Cytophotometric and autoradiographic evidence for postnatal DNA synthesis in neurons of the rat cerebral cortex. *Exp. Cell Res.* **107,** 151–157.

Bregnard, A., Ruch, F., Lutz, H., and Kuenzle, C. C. (1979). Histones and DNA increase synchronously in neurons during early postnatal development of the rat forebrain cortex. *Histochemistry* **61,** 271–279.

Brown, I. R. (1975). RNA synthesis in isolated brain nuclei after administration of *d*-lysergic acid diethylamide (LSD) *in vivo. Proc. Natl. Acad. Sci. U.S.A.* **72,** 837–839.

Brown, I. R. (1977). Analysis of gene activity in the mammalian brain. *In* ''Mechanisms, Regulation and Special Functions of Protein Synthesis in the Brain'' (S. Roberts, A. Lajtha, and W. Gispen, eds.), pp. 29–46. Elsevier/North Holland Biomed. Press, Amsterdam.

Brown, I. R. (1978). Postnatal appearance of a short DNA repeat length in neurons of the cerebral cortex. *Biochem. Biophys. Res. Commun.* **84,** 285–292.

Brown, I. R. (1980). Histone synthesis in isolated neuronal perikaryon relative to the postnatal appearance of a short DNA repeat length. *Dev. Biol.* **80,** 248–252.

Brown, I. R., and Liew, C. C. (1975). Lysergic acid diethylamide: effect on histone acetylation in rabbit brain. *Science* **188,** 1122–1123.

Brown, I., Heikkila, J., Silver, J., and Straus, N. (1977). Organization and transcriptional activity of brain chromatin subunits. *Biochim. Biophys. Acta* **477,** 288–294.

Burdman, J. A. (1972). The relationship between DNA snythesis and the synthesis of nuclear proteins in rat brain during development. *J. Neurochem.* **19,** 1459–1469.

Burdman, J. A., Szijan, I., Franzoni, L., and Garcia Argiz, C. A. (1973). The relationship between DNA synthesis and nuclear proteins in the cerebrum and cerebellum of 8-day-old rats. *J. Neurochem.* **20,** 1719–1725.

Burkholder, G. D., and Weaver, M. G. (1975). Differential accessibility of DNA in extended and condensed chromatin to pancreatic DNase I. *Exp. Cell Res.* **92,** 518–522.

Chambon, P. (1975). Eukaryotic nuclear RNA polymerases. *Annu. Rev. Biochem.* **44,** 613–638.

Chiu, J. F., and Sung, S. C. (1970). DNA nucleotidyltransferase activity of the developing rat brain. *Biochim. Biophys. Acta* **209,** 34–42.

Chiu, J. F., and Sung, S. C. (1971). Separation and properties of DNA polymerase from developing rat brain. *Biochim. Biophys. Acta* **246,** 44–50.

Chiu, J. F., and Sung, S. C. (1972). Particulate form of DNA polymerase in rat brain. *Nature (London), New Biol.* **239,** 176–178.

Choie, D. D., Friedberg, E. C., VanderBerg, S. R., and Herman, M. M. (1977). Non-histone chromosomal proteins in mouse brain at different stages of development and in a transplantable mouse teratoma. *J. Neurochem.* **29,** 811–817.

Cocchia, D., and Michetti, F. (1980). Localization of the loosely bound nuclear proteins of the rat brain: an immunocytochemical and ultrastructural study. *J. Histochem. Cytochem.* **28,** 552–556.

Cole, R. D., Lawson, G. M., and Hsiang, M. W. (1977). H1 histone and the condensation of chromatin and DNA. *Cold Spring Harbor Symp. Quant. Biol.* **42,** 253–263.

Compton, J. L., Bellard, M., and Chambon, P. (1976). Biochemical evidence of variability in the DNA repeat length in the chromatin of higher eukaryotes. *Proc. Natl. Acad. Sci. U.S.A.* **73,** 4382–4386.

Das, R., and Kanungo, M. S. (1979). Effects of polyamines on *in vitro* phosphorylation and acetylation of histones of the cerebral cortex of rats of various ages. *Biochem. Biophys. Res. Commun.* **90,** 708–714.

Davis, R. H., Copenhaver, J. H., and Carver, M. J. (1972). Characterization of acidic proteins in cell nuclei from rat brain by high-resolution acrylamide gel electrophoresis. *J. Neurochem.* **19,** 473–477.

Dische, Z., and Schwarz, K. (1937). Microchemical methods for determining various pentoses in the presence of one another and of hexoses. *Mikrochim. Acta.* **2,** 13–19.

Djondjurov, L., Ivanova, E., and Tsanev, R. (1979). Two chromatin fractions with different metabolic properties of nonhistone proteins and of newly synthesized RNA. *Eur. J. Biochem.* **97,** 133–139.

Dravid, A. R., and Burdman, J. A. (1968). Acidic proteins in rat brain nuclei: disc electrophoresis. *J. Neurochem.* **15,** 25–30.

Duerre, J. A., and Gaitonde, M. K. (1971). Isolation and electrophoretic identification of histones of rat brain and liver. *J. Neurochem.* **18,** 1921–1929.

Duerre, J. A., and Lee, C. T. (1974). *In vivo* methylation and turnover of rat brain histones. *J. Neurochem.* **23,** 541–547.

Elgin, S. C. R., and Stumph, W. E. (1975). Chemistry of the nonhistone chromosomal proteins. *In* "The Structure and Function of Chromatin" *Ciba Found. Symp.*, pp. 113–130. Associated Sci. Publ., Amsterdam.

Elgin, S. C. R., and Weintraub, H. (1975). Chromosomal proteins and chromatin structure. *Annu. Rev. Biochem.* **44,** 725–774.

Ermini, M., and Kuenzle, C. C. (1978). The chromatin repeat length of cortical neurons shortens during early postnatal development. *FEBS Lett.* **90,** 167–172.

Felsenfeld, G. (1978). Chromatin. *Nature (London)* **271,** 115–122.

Fujii, K., Yonemasu, Y., Kitamura, K., and Tsubota, Y. (1980). A genetic aspect of chromatin proteins of human brain tumors. *J. Neurol.* **223,** 23–34.

Fujitani, H., and Holoubek, V. (1974). Nonhistone nuclear proteins of rat brain. *J. Neurochem.* **23,** 1215–1224.

Gaubatz, J., Ellis, M., and Chalkley, R. (1979). The structural organization of mouse chromatin as a function of age. *Fed. Proc., Fed. Am. Soc. Exp. Biol.* **38,** 1973–1978.

Goldknopf, I. L., and Busch, H. (1977). Isopeptide linkage between nonhistone and histone 2A polypeptides of chromosomal conjugate-protein A24. *Proc. Natl. Acad. Sci. U.S.A.* **74,** 864–868.

Goldknopf, I. L., Taylor, C. W., Baum, R. M., Yeoman, L. C., Olson, M. O. J., Prestayko, A. W., and Busch, H. (1975). Isolation and characterization of protein A24, a "histone-like" nonhistone chromosomal protein. *J. Biol. Chem.* **250,** 7182–7187.

Goodwin, G. H., and Johns, E. W. (1977). Part D. Fractionation and characterization of nonhistone chromosomal proteins. I. The isolation and purification of the high mobility group (HMG) nonhistone chromosomal proteins. *Methods Cell Biol.* **16,** 257–267.

Goodwin, G. H., Nicolas, R. H., and Johns, E. W. (1977a). A quantitative analysis of histone H1 in rabbit thymus nuclei. *Biochem. J.* **167,** 485–488.

Goodwin, G. H., Woodhead, L., and Johns, E. W. (1977b). The presence of high mobility group non-histone chromatin proteins in isolated nucleosomes. *FEBS Lett.* **73**, 85–88.

Goodwin, G. H., Walker, J. M., and Johns, E. W. (1978). The high mobility group (HMG) nonhistone chromosomal proteins. *In* "The Cell Nucleus" (H. Busch, ed.), Vol. 6, pp. 181–219. Academic Press, New York.

Gordon, C. N. (1977). Chromatin behavior during the mitotic cell cycle of *Saccharomyces cerevisiae. J. Cell Sci.* **24**, 81–93.

Gordon, J. S., Rosenfeld, B. I., Kaufman, R., and Williams, D. L. (1980). Evidence for a quantitative tissue-specific distribution of the high mobility group chromosomal proteins. *Biochemistry* **19**, 4395–4402.

Gorovsky, M. A., Glover, C., Johmann, A., Kreevert, J. B., Mathis, O. J., and Samuelson, M. (1977). Histones and chromatin structure in *Tetrahymena* macro- and micronuclei. *Cold Spring Harbor Symp. Quant. Biol.* **42**, 493–505.

Greenwood, P. D., Silver, J. C., and Brown, I. R. (1981a). Analysis of histones associated with neuronal and glial nuclei exhibiting divergent DNA repeat lengths. *J. Neurochem.* **37**, 498–505.

Greenwood, P., Silver, J. C., and Brown, I. R. (1981b). Analysis of putative high mobility group (HMG) proteins in neuronal and glial nuclei from rabbit brain. *Neurochem. Res.* **6**, 673–679.

Griffith, J. D., and Christiansen, G. (1977). The multifunctional role of histone H1, probed with the SV40 minichromosome. *Cold Spring Harbor Symp. Quant. Biol.* **42**, 215–226.

Grummt, F., Waltl, G., Jontzen, H. M., Hamprecht, K., Hubscher, U., and Kuenzle, C. C. (1979). Diadenosine '5', 5'''-P^1, P^4-tetraphosphate, a ligand of the 57-kilodalton subunit of DNA polymerase α. *Proc. Natl. Acad. Sci. U.S.A.* **76**, 6081–6085.

Guiffrida, A. M., Cox, D., and Mathias, A. P. (1975). RNA polymerase activity in various classes of nuclei from different regions of rat brain during postnatal development. *J. Neurochem.* **24**, 749–755.

Guiffrida, A. M., Hamberger, A., and Serra, I. (1976). Biosynthesis of DNA and RNA in neuronal and glial cells from various regions of developing rat brain. *J. Neurosci. Res.* **2**, 203–215.

Gurley, L. R., Walters, R. A., Barham, S. S., and Deaven, L. L. (1978). Heterochromatin and histone phosphorylation. *Exp. Cell Res.* **111**, 373–383.

Heizmann, C. W., Arnold, E. M., and Kuenzle, C. C. (1980). Fluctuations of non-histone chromosomal proteins in differentiating brain cortex and cerebellar neurons. *J. Biol. Chem.* **255**, 11504–11511.

Herman, C. J., and Lapham, L. W. (1969). Neuronal polyploidy and nuclear volumes in the cat central nervous system. *Brain Res.* **15**, 35–48.

Himwich, W. A. (1962). Biochemical and neurophysiological development of the brain in the neonatal period. *Int. Rev. Neurobiol.* **4**, 117–158.

Horgen, P. A., and Silver, J. C. (1978). Chromatin structure in eukaryotic microbes. *Annu. Rev. Microbiol.* **32**, 249–284.

Hubscher, U., Kuenzle, C. C., and Spadari, S. (1977a). Variation of DNA polymerases α, β and γ during perinatal growth and differentiation. *Nucleic Acids Res.* **4**, 2917–2929.

Hubscher, U., Kuenzle, C. C., and Spadari, S. (1977b). Identity of DNA polymerase γ from synaptosomal mitochondria and rat brain nuclei. *Eur. J. Biochem.* **81**, 249–258.

Hubscher, U., Kuenzle, C. C., Limacher, W., Scherrer, P., and Spadari, S. (1978). Functions of DNA polymerases α, β and γ in neurons during development. *Cold Spring Harbor Symp. Quant. Biol.* **43**, 625–629.

Hubscher, U., Kuenzle, C. C., and Spadari, S. (1979). Functional roles of DNA polymerase β and γ. *Proc. Natl. Acad. Sci. U.S.A.* **76**, 2316–2320.

Inoue, N., Suzuki, O., and Kato, T. (1976). DNA synthesis in neuronal, glial and liver nuclei isolated from the adult guinea pig. *J. Neurochem.* **27**, 113–119.

Inoue, N., Ono, T., and Kato, T. (1979). Ligation and synthesis of chromatin deoxyribonucleic acid *in vitro* in neuronal, glial and liver nuclei isolated from the adult guinea pig. *Biochem. J.* **180**, 471–480.

Isenberg, I. (1979). Histones. *Annu. Rev. Biochem.* **48**, 159–191.

Jalouzot, R., Briane, D., Ohlenbusch, H. H., Wilhelm, M. L., and Wilhelm, F. X. (1980). Kinetics of nuclease digestion of *Physarum polycephalum* nuclei at different stages of the cell cycle. *Eur. J. Biochem.* **104**, 423–431.

Kanungo, M. S., and Thakur, M. K. (1979). Modulation of acetylation of histones and transcription of chromatin by butyric acid and 17β-estradiol in the brain of rats of various ages. *Biochem. Biophys. Res. Commun.* **87**, 266–271.

Kato, T., and Kurokawa, M. (1970). Studies on ribonucleic acid and homopolyribonucleotide formation in neuronal, glial and liver nuclei. *Biochem. J.* **116**, 599–609.

Kelly, P. T., and Luttges, M. W. (1976). Mouse brain protein composition during postnatal development: An electrophoretic analysis. *J. Neurochem.* **27**, 1163–1172.

Kornberg, R. D. (1974). Chromatin structure: A repeating unit of histones and DNA. *Science* **184**, 868–871.

Kornberg, R. D. (1977). Structure of chromatin. *Annu. Rev. Biochem.* **46**, 931–954.

Kornberg, R. D., and Thomas, J. O. (1974). Chromatin structure: Oligomers of the histones. *Science* **184**, 865–868.

Kuenzle, C. C. (1979). Postnatal DNA synthesis in neurons of the rat forebrain cortex. *Hoppe-Seyler's Z. Physiol. Chem.* **360**, 312.

Kuenzle, C. C., Bregnard, A., Hubscher, U., and Ruch, F. (1978). Extra DNA in forebrain cortical neurons. *Exp. Cell Res.* **113**, 151–160.

Laemmli, U. K. (1970). Cleavage of structural proteins during the assembly of the head of bacteriophage T4. *Nature (London)* **227**, 680–685.

Lawson, G. M., and Cole, R. D. (1979). Selective displacement of histone H1 from whole HeLa nuclei: Effect of chromatin structure in situ as probed by micrococcal nuclease. *Biochemistry* **18**, 2160–2166.

Lee, N. M., and Loh, H. H. (1977). Phosphorylation and methylation of chromatin proteins from mouse brain nuclei. *J. Neurochem.* **29**, 547–550.

Levy, W. B., and Dixon, G. H. (1978). A study of the localization of high mobility group proteins in chromatin. *Can. J. Biochem.* **56**, 480–491.

Levy, W. B., Wong, N. C. W., and Dixon, G. H. (1977). Selective association of the trout-specific H6 protein with chromatin regions susceptible to DNase I and DNase II: Possible location of HMG-T in the spacer region between core nucleosomes. *Proc. Natl. Acad. Sci. U.S.A.* **74**, 2810–2814.

Levy-Wilson, B., Kuehl, L., and Dixon, G. H. (1980). The release of high mobility group protein H6 and protamine gene sequences upon selective DNase I degradation of trout testis chromatin. *Nucleic Acids Res.* **8**, 2859–2869.

Lilley, D. M. J., and Pardon, J. F. (1979). Structure and function of chromatin. *Annu. Rev. Genet.* **13**, 197–233.

Lipps, H. J., and Morris, N. R. (1977). Chromatin structure in the nuclei of the ciliate *Stylonychia mytilus*. *Biochem. Biophys. Res. Commun.* **74**, 230–234.

Loening, U. E. (1967). The fractionation of high-molecular-weight ribonucleic acid by polyacrylamide gel electrophoresis. *Biochem. J.* **102**, 251–257.

Lohr, D., Corden, J., Tatchell, R. T., Kovacic, R. T., and Van Holde, K. E. (1977). Comparative subunit structure of HeLa, yeast and chicken erythrocyte chromatin. *Proc. Natl. Acad. Sci. U.S.A.* **74**, 79–83.

Lu, C. Y., and Koenig, H. (1973). Isolation of acidic lipoproteins from brain chromatin their relation to the acidic nonhistone proteins. *FEBS Lett.* **34**, 48–54.

Ludwig, R. A., and Summers, W. C. (1975). A restriction fragment analysis of the T7 left-early region. *Virology* **68**, 360–373.

MacGillivray, A. J., and Rickwood, D. (1974). The heterogeneity of mouse chromatin nonhistone proteins as evidenced by two-dimensional polyacrylamide gel electrophoresis and ion-exchange chromatography. *Eur. J. Biochem.* **41**, 181–190.

Martenson, R. E., Deibler, G. E., and Kies, M. W. (1969). Extraction of rat myelin basic protein free of other basic proteins of whole central nervous system tissue. *J. Biol. Chem.* **244**, 4268–4272.

Mathew, C. G. P., Goodwin, G. H., Johns, E. W. (1979). Studies on the association of the high mobility group non-histone chromatin proteins with isolated nucleosomes. *Nucleic Acids Res.* **6**, 167–179.

Mazrimas, J. A., Balhorn, R., and Hatch, F. T. (1979). Separation of satellite DNA chromatin and main band DNA chromatin from mouse brain. *Nucleic Acids Res.* **7**, 935–946.

McCombe, P., Lavin, M., and Kidson, C. (1976). Control of DNA repair linked to neuroblastoma differentiation. *Int. J. Rad. Biol. Relat. Stud. Phys. Chem. Med.* **29**, 523–531.

McGhee, J. D., and Felsenfeld, G. (1980). Nucleosome structure. *Annu. Rev. Biochem.*, **49**, 1115–1156.

Miyake, M. (1975). Methylases of myelin basic protein and histone in rat brain. *J. Neurochem.* **24**, 909–915.

Mizobe, R., Tashiro, T., and Kurokawa, M. (1974). Characterisation of RNA synthesis *in vitro* in neurone-rich, oligodendroglial and liver nuclei. *Eur. J. Biochem.* **48**, 25–33.

Morris, N. R. (1976a). Nucleosome structure in *Aspergillus nidulans*. *Cell* **8**, 357–363.

Morris, N. R. (1976b). A comparison of the structure of chicken erythrocyte and chicken liver chromatin. *Cell* **9**, 627–632.

Neidle, A., and Waelsch, H. (1964). Histones: Species and tissue specificity. *Science* **145**, 1059–1061.

Noll, M. (1974a). Subunit structure of chromatin. *Nature (London)* **251**, 249–251.

Noll, M. (1974b). Internal structure of the chromatin subunit. *Nucleic Acids Res.* **1**, 1573–1578.

Noll, M. (1976). Differences and similarities in chromatin structure of *Neurospora crassa* and higher eucaryotes. *Cell* **8**, 349–355.

Noll, M., and Kornberg, R. D. (1977). Action of micrococcal nuclease on chromatin and the location of histone H1. *J. Mol. Biol.* **109**, 393–404.

Novakova, V., Sandritter, W., and Schlueter, G. (1970). DNA content of neurons in rat central nervous system. *Exp. Cell Res.* **60**, 454–456.

OH'Hara, I., and Yanagihara, T. (1977). Nuclear chromatin proteins from rabbit cerebrum, cerebellum and liver: synthesis and phosphorylation. *J. Neurochem.* **29**, 1065–1073.

Olins, A. L., and Olins, D. E. (1974). Spheroid chromatin units (*v* bodies). *Science* **183**, 330–332.

Olpe, H. R., von Hahn, H. P., and Honegger, C. G. (1972). The non-histone protein pattern of rat brain during ontogenesis. *Experentia* **29**, 665–666.

Olson, M. O. J., and Busch, H. (1974). Nuclear proteins. *In* "The Cell Nucleus" (H. Busch, ed.), Vol. 3, pp. 211–268. Academic Press, New York.

Oudet, P., Gross-Bellard, M., and Chambon, P. (1975). Electron microscopic and biochemical evidence that chromatin structure is a repeating unit. *Cell* **4**, 281–300.

Patel, G. L. (1974). Isolation of the nuclear acidic proteins, their fractionation, and some general characteristics. *In* "Acidic Proteins of the Nucleus" (I. L. Cameron and J. R. Jeter, eds.), pp. 29–57. Academic Press, New York.

Pederson, T. (1978). Chromatin structure and gene transcription: nucleosomes permit a new synthesis. *Int. Rev. Cytol.* **55**, 1–22.

Pederson, T., and Bhorjee, J. S. (1975). A special class of non-histone protein tightly complexed with template-inactive DNA in chromatin. *Biochemistry* **14**, 3238–3242.

Piha, R. S., Cuenod, M. and Waelsch, H. (1966). Metabolism of histones of brain and liver. *J. Biol. Chem.* **241**, 2397–2404.

Poduslo, S. E., and McKhann, G. M. (1977). Maintenance of neurons isolated in bulk from rat brain: Incorporation of radiolabeled substrates. *Brain Res.* **132**, 107–120.

Reeves, R., and Candido, E. P. M. (1980). Partial inhibition of histone deacetylase in active chromatin by HMG 14 and HMG 17. *Nucleic Acids Res.* **8**, 1947–1963.

Renz, M., Nehls, P., and Hozier, J. (1977). Histone H1 involvement in the structure of the chromosome fiber. *Cold Spring Harbor Symp. Quant. Biol.* **42**, 245–252.

Russanova, V., Venkov, C., and Tsanev, R. (1980). A comparison of histone variants in different rat tissues. *Cell Diff.* **9**, 339–350.

Sarkander, H. I., and Dulce, H. J. (1978). Studies on the regulation of RNA synthesis in neuronal and glial nuclei isolated from rat brain. *Exp. Brain Res.* **31**, 317–327.

Schmitt, M., and Matthies, H. (1979a). Biochemische Untersuchungen an Histonen des Zentralnervensystems. II. Azetylierung von Histonen aus Neuronen und gliades Rattenhirns *in vitro*. *Acta Biol. Med. Germ.* **38**, 677–682.

Schmitt, M., and Matthies, H. (1979b). Biochemische Untersuchungen an Histonen des Zentralnervensystems. I. Charakterisierung und partielle Identifikation der Markierung von Histonen des RattheRattenhims nach intraventrikularer Gabe von 1-[^{14}C]-azetat. *Acta Biol. Med. Ger.* **38**, 673–676.

Shaw, L. M. J., and Huang, R. C. C. (1970). A description of two procedures which avoid the use of extreme pH conditions for the resolution of components isolated from chromatins prepared from pig cerebellar and pituitary nuclei. *Biochemistry* **9**, 4530–4542.

Sheinin, R., Humbert, J., and Pearlman, R. E. (1978). Some aspects of eukaryotic DNA replication. *Annu. Rev. Biochem.* **47**, 277–316.

Shimada, H., and Terayama, H. (1972). DNA synthesis in isolated nuclei from the brains of rats at different post-partal stages and the infant rat brain cytosol factor stimulating the DNA synthesis in infant rat brain nuclei. *Biochim. Biophys. Acta* **287**, 415–426.

Silver, J. C. (1979). Chromatin organization in the oomycete *Achlya ambisexualis*. *Biochim. Biophys. Acta* **561**, 261–264.

Singh, V. K., and Sung, S. C. (1972). Studies on isolated brain nuclear DNA-dependent RNA polymerase. *Can. J. Biochem.* **50**, 299–304.

Sinha, A. K., Rose, S. P. R., Sinha, L., and Spears, D. (1978). Neuronal and neuropil fractions from developing rat brain. *J. Neurochem.* **30**, 1513–1524.

Sonnenbichler, J. (1979). Advances in chromatin research. *Naturwissenschaften* **66**, 244–250.

Spadafora, C., Bellard, M., Compton, J. L., and Chambon, P. (1976). The DNA repeat lengths in chromatins from sea urchin sperm and gastrula cells are markedly different. *FEBS Lett.* **69**, 281–285.

Stambolova, M. A., Cox, D., and Mathias, A. P. (1973). The activity of deoxyribonucleic acid polymerase and deoxyribonucleic acid synthesis in nuclei from brain fractionated by zonal centrifugation. *Biochem. J.* **136**, 685–695.

Stambolova, M. A., Angelova, A. A., and Tsanev, R. G. (1979). Metabolic stability of nonhistone chromosomal proteins in two different classes of developing rat brain nuclei. *Cell Differentiation* **8**, 195–202.

Stein, G. S., Spelsberg, T. C., and Kleinsmith, L. J. (1974). Nonhistone chromosomal proteins and gene regulation. *Science* **183**, 817–824.

Stoykova, A. S., Dabeva, M. D., Dimova, R. N., and Hadijolov, A. A. (1979). Ribosomal RNA precursors in neuronal and glial rat brain nuclei. *J. Neurochem.* **33**, 931.–937.

Studier, F. W. (1973). Analysis of bacteriophage T7 early RNAs and proteins on slab gels. *J. Mol. Biol.* **79**, 237–248.

Swartz, F. J., and Bhatnagar, K. P. (1981). Are CNS neurons polyploid? A critical analysis based upon cytophotometric study of the DNA content of cerebellar and olfactory bulbar neurons of the bat. *Brain Res.* **208**, 267–281.

Szijan, I., and Burdman, J. A. (1973). Nuclear proteins in brain of 7-day-old and adult rats. *J. Neurochem.* **21**, 1093–1097.

Tashiro, T., Mizobe, R., and Kurokawa, M. (1974). Characteristics of cerebral non-histone chromatin proteins as revealed by polyacrylamide gel electrophoresis. *FEBS Lett.* **38**, 121–124.

Teng, C. S., Teng, C. T., and Allfrey, V. G. (1971). Studies of nuclear acidic proteins. *J. Biol. Chem.* **246**, 3597–3609.

Teng, C. S., Andrews, G. K., and Teng, C. T. (1979). Studies on the high mobility group non-histone proteins from hen oviduct. *Biochem. J.* **181**, 585–591.

Thakur, M. K., Das, R., and Kanungo, M. S. (1978). Modulation of acetylation of chromosomal proteins of the brain of rats of various ages by epinephrine and estradiol. *Biochem. Biophys. Res. Commun.* **81**, 828–831.

Thomas, J. O., and Furber, V. (1976). Yeast chromatin structure. *FEBS Lett.* **66**, 274–280.

Thomas, J. O., and Thompson, R. J. (1977). Variation in chromatin structure in two cell types from the same tissue: A short DNA repeat length in cerebral cortex neurons. *Cell* **10**, 633–640.

Thompson, R. J. (1973). Studies on RNA synthesis in two populations of nuclei from the mammalian cerebral cortex. *J. Neurochem.* **21**, 19–40.

Todd, R. D., and Garrard, W. T. (1977). Two-dimensional electrophoretic analysis of polynucleosomes. *J. Biol. Chem.* **252**, 4729–4738.

Tsai, M., Towle, H. C., Harris, S. E., and O'Malley, B. W. (1976). Effect of estrogen on gene expression in chick oviduct. Comparative aspects of RNA chain initiation in chromatin using homologous versus *Escherichia coli* RNA polymerase. *J. Biol. Chem.* **251**, 1960–1968.

Tsitilou, S. G., Cox, D., Mathias, A. P., and Ridge, D. (1979). The characterization of the non-histone chromosomal proteins of the main classes of nuclei from rat brain fractionated by zonal centrifugation. *Biochem. J.* **177**, 331–346.

Unger-Ullmann, C., and Modak, S. P. (1979). Gel electrophoretic analysis of histones in lens epithelium, lens fiber, liver, brain, and erythrocytes of late chick embryos. *Differentiation* **12**, 135–144.

Uyemura, K., Nakayama, T., Kitamura, K., Yamanaka, T., and Hirano, S. (1974). Nuclear proteins of the guinea pig brain. *J. Neurochem.* **23**, 65–70.

Varshavsky, A. J., Bakayev, V. V. and Georgiev, G. P. (1976). Heterogeneity of chromatin subunits *in vitro* and location of histone H1. *Nucleic Acids Res.* **3**, 477–492.

Vilenchik, M. M., and Tretjak, T. M. (1977). Evidence for unscheduled DNA synthesis in rat brain. *J. Neurochem.* **29**, 1159–1161.

Waser, J., Hubscher, U., Kuenzle, C. C., and Spadari, S. (1979). DNA polymerase β from brain neurons is a repair enzyme. *Eur. J. Biochem.* **97**, 361–368.

Weintraub, H. (1978). The nucleosome repeat length increases during erythropoiesis in the chick. *Nucleic Acids Res.* **5**, 1179–1188.

Weintraub, H., and Groudine, M. (1976). Chromosomal subunits in active genes have an altered conformation. *Science* **193**, 848–856.

Weissbach, A. (1977). Eucaryotic DNA polymerases. *Annu. Rev. Biochem.* **46**, 25–49.

Weisbrod, S., and Weintraub, H. (1979). Isolation of a subclass of nuclear proteins responsible for conferring a DNase-I sensitive structure on globin chromatin. *Proc. Natl. Acad. Sci. U.S.A.* **76**, 630–634.

Whitlock, J. P., and Simpson, R. T. (1976). Removal of histone H1 exposes a fifty base pair DNA segment between nucleosomes. *Biochemistry* **15**, 3307–3314.

Wintzerith, M., Wittendorp, E., Rechenmann, R. V., and Mandel, P. (1977). Nuclear, nucleolar repair, or turnover of DNA in adult rat brain. *J. Neurosci. Res.* **3**, 217–230.

Woodcock, C. L. F., Safer, J. P., and Stanchfield, J. E. (1976). Structural repeating units in chromatin. I. Evidence for their general occurrence. *Exp. Cell Res.* **97**, 101–110.

Wu, F. C., Elgin, S. C. R., and Hood, L. E. (1973). Nonhistone chromosomal proteins of rat tissues: A comparative study by gel electrophoresis. *Biochemistry* **12**, 2792–2797.

Yamamoto, H., and Takahashi, Y. (1978). Isolation and characterization of DNA-dependent RNA polymerase A, B and C from rat brain nuclei. *J. Neurochem.* **31**, 449–456.

Yamamoto, H., and Takahashi, Y. (1980). Purification and subunit structure of DNA-dependent RNA polymerase BII from rat brain nuclei. *J. Neurochem.* **34**, 255–260.

Yanagihara, T. (1980). Phosphorylation of chromatin proteins in cerebral anoxia and ischemia. *J. Neurochem.* **35**, 1209–1215.

Yanagihara, T., OH'Hara, I., Arvidson, C., and Gintz, J. (1978). Phosphorylation of nuclear proteins from rabbit cerebrum, cerebellum and liver *in vitro*. *J. Neurochem.* **31**, 225–231.

Zongza, V., and Mathias, A. P. (1979). The variation with age of the structure of chromatin in three cell types from rat liver. *Biochem. J.* **179**, 291–298.

3

THE SEQUENCE COMPLEXITY OF BRAIN RIBONUCLEIC ACIDS

Barry B. Kaplan and Caleb E. Finch

I. INTRODUCTION

A remarkable new property of mammalian brain was recently established by the techniques of RNA–DNA hybridization, namely, that the number of *different* types of brain RNA sequences (their ''complexity'' as defined precisely below) is manyfold greater than that of other somatic tissues or organs. The high com-

Molecular Approaches to Neurobiology
Copyright © 1982 by Academic Press, Inc.
All rights of reproduction in any form reserved.
ISBN 0-12-137020-8

plexity of brain messenger RNA (mRNA) bears on many issues in neurobiology including the ultimate function of this large amount of genetic information in mammals and the degree to which it is involved in the development and maintenance of specialized function in different brain cells and regions. The high complexity of brain mRNA also predicts a vast number of different polypeptides (100,000–200,000), greater than 95% of which, because of their relative scarcity, have not been distinguished by standard protein analysis. In this review, we will describe the techniques of RNA–DNA hybridization used to estimate RNA sequence complexity, and then discuss the available data on the manner in which brain RNA complexity differs between regions, cell types, and during various stages of postnatal development and aging. Although we do not attempt to systematically review the molecular biological literature, references to these studies are given in instances where they bear directly on questions of neurobiological interest.

II. RNA–DNA HYBRIDIZATION ANALYSIS

A. The Basis for Estimating Nucleic Acid Complexity

More than 15 years ago, double-stranded DNA was found to reversibly denature: after separation into single strands at 95°–100°C, DNA reannealed to form duplex structures during incubation at 60°–70°C at physiologic ionic strengths (reviewed in Wetmur, 1976). The hybrids that reformed from denatured prokaryotic DNA were nearly identical to native DNA with regard to their thermal stability and fidelity of base pairing. In addition, RNA was found to form stable RNA–DNA hybrids with homologous DNA sequences (i.e., the DNA sequences that served as template for their synthesis by RNA polymerase). The ability to form RNA–DNA hybrids was subsequently employed to estimate the fraction of DNA in a cell which was used as template for transcription; or, more precisely, the fraction of radiolabeled DNA that could form RNA–DNA hybrids with RNA from a given cell under appropriate experimental conditions.

Greater difficulty was encountered in applying this approach to higher organisms until it was recognized that 25–50% of eukaryotic DNA consisted of multiple copies of closely related DNA sequences (Britton and Kohne, 1968; Davidson and Britten, 1973). The function of this repetitive DNA is still largely obscure; most mRNA's are transcribed from nonrepetitive DNA sequences or "single-copy" DNA (i.e., DNA sequences found in only one copy per haploid genome). Most early hybridization studies using total eukaryotic DNA cannot be interpreted easily because hybridization can occur between related, but not identical, repetitive DNA sequences. If, however, RNA–DNA hybridization is carried out with single-copy DNA then direct estimates of RNA complexity can be made. RNA hybridization to single-copy DNA can be used to estimate not only RNA complexity, but also the relative number of copies of the RNA sequences present in a cell or tissue, as described in Section III,B.

B. Saturation Hybridization

Two general approaches are used to determine the complexity of RNA populations. The first involves the hybridization of radiolabeled single-copy DNA to RNA. Under conditions of vast RNA excess (mRNA:DNA>100:1) the hybridization reaction follows pseudo-first-order kinetics, since the concentration of the hybridizing RNA remains essentially unchanged throughout the experiment. The reaction kinetics are generally expressed as follows:

$$\frac{D}{D_0} = e^{-K_h R_0 t} \tag{1}$$

where D/D_0 represents the fraction of DNA that is single-stranded at time t (in seconds), R_0 the initial RNA concentration (in moles nucleotide per liter), and K_h the observed rate constant for the formation of RNA–DNA hybrids. The hybridization rate constant, expressed as a function of $R_0 t$ ($R_0 \times t$) is inversely proportional to the base sequence complexity of the RNA population (Bishop, 1969; Birnstiel *et al.*, 1972). Thus, the larger the RNA complexity the slower the reaction and the longer it takes to reach completion. The hybridization rate constant of a single kinetic component can be approximated from the $R_0 t_{1/2}$, the value at which the reaction is half complete ($K_h = \ln 2/R_0 t_{1/2}$). Interpretation of kinetic data derived from RNA-driven hybridization reactions will be discussed briefly in Sections III,B and IV,B. Typically, mammalian nuclear RNA requires prolonged incubations (>3 days) to complete the hybridization reaction. The sequence complexity of the hybridizing RNA is determined directly from the fraction of tracer DNA annealed to form RNA–DNA hybrids at the termination of the reaction (Davidson and Hough, 1971; Gelderman *et al.*, 1971; Hahn and Laird, 1971). Most simply defined, the *RNA sequence complexity* is the equivalent amount of single-copy DNA homologous to the transcribed (hybridizing) RNA and is expressed as the percentage of single-copy DNA complexity or as the equivalent number of nucleotides. The complexity does not indicate the relative amounts of the different types of RNA sequences. The amount of DNA present as RNA–DNA hybrids is experimentally determined by chromatography on hydroxylapatite or by hybrid resistence to single-strand specific nuclease. The number of different sequences comprising the RNA population is estimated from the sequence complexity of the population by assuming an average size RNA molecule.

For example, a typical RNA-driven hybridization experiment with brain polyadenylated [poly(A+)] mRNA hybridizes to 4% of the single-copy DNA tracer at the termination of the reaction (saturation). Assuming that gene transcription is asymmetric (involves only one DNA strand), the experimental value represents 8% of the available DNA. Given that the complexity of the single-copy genome is about 1.9×10^9 nucleotides [a typical mammalian haploid genome measures 2.9×10^9 nucleotides (nt) of which 65% is single-copy DNA] then the mRNA complexity is 1.5×10^8 nt ($0.08 \times 1.9 \times 10^9$). Assuming the average size of the mRNA is 1400 nt (Bantle and Hahn, 1976), the sequence complexity of the RNA is the equivalent of 107,000 different poly(A+)mRNA sequences ($1.5 \times 10^8 \div 1.4 \times 10^3$ nt).

C. cDNA Hybridization

In this approach, a viral reverse transcriptase is used to synthesize a highly radioactive complementary DNA probe (cDNA) from a RNA template *in vitro* using radiolabeled deoxyribonucleotides as substrates. The kinetic analysis of the hybridization of cDNA can be used to estimate the RNA sequence complexity and in addition the relative concentration of the different frequency classes of RNA (Bishop *et al.*, 1974). To obtain the RNA sequence complexity by this method, the hybridization kinetics of cDNA to its template RNA is compared to the hybridization of an RNA standard of known complexity with its cDNA annealed under identical conditions. Under conditions of RNA excess, the cDNA will hybridize to template RNA with pseudo-first-order kinetics and the hybridization rate constant for the reaction will be inversely proportional to the sequence complexity of the RNA. In practice, the sequence complexity of an unknown RNA population is estimated according to the equation:

$$X = \frac{\text{OBS} \cdot R_0 t_{1/2}(B)\,(C)}{A} \tag{2}$$

where the $\text{OBS} \cdot R_0 t_{1/2}$ is the value of $R_0 t$ at which the hybridization reaction is half complete, A is the $R_0 t_{1/2}$ of the standard reaction, C is the sequence complexity of the RNA standard, and B is the fraction of the total reaction which the component represents (see Getz *et al.*, 1976). In effect, B is a correction factor used to adjust the $\text{OBS} \cdot R_0 t_{1/2}$ to a value expected if the kinetic component being studied represented 100% of the total hybridization reaction (see below).

According to ideal pseudo-first-order kinetics, the reaction of a single RNA species to DNA will reach completion within 1.5 log units of $R_0 t$. Significant deviation from ideal kinetics (e.g., hybridization over 3–5 $\log_{10} R_0 t$) indicates heterogeneity in the concentration of RNA sequences that are driving the reaction. Thus, increases in the range of the reaction are an important factor in the evaluation of RNA sequence complexity by this method. Since most cells do not contain all RNA sequences in equal concentrations, complex hybridization kinetics are usually observed and represent a composite of several separate reactions, each of which involves a group of RNA sequences present in a similar range of concentrations. Those RNA's present in highest concentration (abundant sequences) will tend to hybridize most rapidly, whereas sequences present in low concentrations (rare sequences) will react more slowly, i.e., at higher R_0t values. Analysis of hybridization data is usually accomplished by a computer resolution (iterative, least squares, best fit) of the various kinetic components and will generate for the investigator both the sequence complexity and relative abundance of the sequences of each kinetic component (Pearson *et al.*, 1977; Quinlan *et al.*, 1978).

The validity and precision of cDNA hybridization analysis has been discussed extensively (Birnie *et al.*, 1974; Ryffel and McCarthy, 1975; Savage *et al.*, 1978). In theory, the validity of the approach rests on the assumption that the efficiency of cDNA synthesis is independent of the RNA concentration, and faithfully represents both the diversity and relative abundance of all RNA sequences in the population. This, however, may not always be the case. For example, conalbumin mRNA is a relatively inefficient template for the reverse transcriptase compared to other mRNA's synthesized in the oviduct (Buell *et al.*, 1978). Silk fibroin mRNA is a second example of a relatively inefficient template (Lizardi and Brown, 1973). A second difficulty is that, in practice, most cDNA probes are short (300–600 nucleotides) relative to their RNA templates and are only partially reactive (70–85%). Since, it becomes difficult to convincingly demonstrate termination of the hybridization reaction, the kinetics of the final transition (the complex, slowly hybridizing RNA sequence class) may be in considerable error. Compounding this difficulty, the final transition undoubtedly represents a small fraction of the total hybridization reaction and the accuracy of determining the $OBS \cdot R_0t_{1/2}$ of this phase of the reaction is limited.

In addition to the above considerations, the final calcualtion of the diversity of a given abundance class assumes that the RNA's that comprise each of these classes are of the same average size. The observations of Meyuhas and Perry (1979) call into question the validity of this last assumption. In mouse L-cells the high complexity RNA class was comprised of relatively larger-sized poly(A+)mRNA, while the low complexity class contained smaller mRNA's. In addition, the large mRNA's acted as less efficient templates for reverse transcriptase in terms of both cDNA yield and size. This situation, if found generally applicable, could further distort the RNA frequency distribution as determined by cDNA hybridization. For example, hybridization of cDNA to L-cell poly(A+)mRNA would overestimate the number of RNAs present in the low complexity RNA class while seriously underestimating the diversity of slowly hybridizing, high complexity RNA. The above reservations not withstanding, reasonably good agreement in complexity estimates (ca. 70%) have been obtained with cDNA and saturation hybridization analysis in the relatively few instances where the reactivity of the cDNA probes were greater than 90% (Axel *et al.*, 1976; Hahn *et al.*, 1978; Savage *et al.*, 1978; Van Ness and Hahn, 1980).

III. COMPLEXITY OF GENE TRANSCRIPTION IN THE BRAIN

A. Total Nuclear RNA Sequence Complexity

Results of early hybridization analysis of brain RNA from mouse (Brown and Church, 1971; Hahn and Laird, 1971; Grouse *et al.*, 1972) and rabbit (Brown and Church, 1972) suggested that at least twice as much single-copy DNA was transcribed in the brain as in some other organs. For example, mouse brain nuclear RNA hybridized to approximately 10–12% of the single-copy DNA, whereas liver and kidney nuclear RNA was complementary to 4–5% (Hahn and Laird, 1971). Recent estimates of brain nuclear RNA complexity, obtained under more advantageous conditions, are even larger than these pioneering studies indicated (Table I). Assuming asymmetric transcription, 30–40% of the single-copy genome is transcribed in the brain, a value equivalent to 130,000–180,000 different heterogeneous nuclear RNA (hnRNA) sequences averaging 4500 nucleotides in length (Table I, footnotes *b* and *c*). Presumably, the early hybridization studies underestimated brain nuclear RNA complexity because of (a) loss of poly(A+)RNA and other sequences during RNA isolation, (b) suboptimal hybridization conditions, which frequently lead to a breakdown of RNA and DNA with premature termination of the reaction, and (c) diffilulties in reaching maximum hybridization values in the presence of contaminating ribosomal RNA (rRNA) which does not hybridize to single-copy DNA, and therefore slows (by dilution effects) hybridization of the RNA's of interest.

The extent of similarity (sequence homology) in the RNA sequences transcribed in various tissues was estimated by hybridizing single-copy DNA to mixtures of nuclear RNA from different organs. At saturation, hybridization of mixtures of brain and liver, or brain and kidney, RNA yielded values equal to that obtained for brain alone (Chikaraishi *et al.*, 1978). Theoretically, if brain and liver contained totally nonoverlapping populations of RNA then the complexity values would be additive, or equal to the sum of values obtained for each organ alone. Conversely, if all the RNA sequences transcribed in one organ appear in another the hybridization values would not summate but reflect the value of the organ with the highest complexity RNA. Although mixing experiments provide a relatively insensitive index of sequence homology, these data suggest that most nuclear RNA's present in liver and kidney also occur in brain. It follows that gene transcripts required for brain-specific function are present in addition to a RNA population that is apparently shared by several less complex organs.

Table I. The Base Sequence Complexity of Whole Brain RNA's

RNA source	Species	Fraction of single-copy DNA hybridized (%)[a]	Sequence complexity (nt)[b]	No. of diff. sequences[c]	References
Total cellular RNA[f]	Mouse[d]	18.8	7.3×10^8	163,000	Grouse et al. (1978a)
	Cat[d]	14.8	5.8×10^8	129,000	Grouse et al. (1979b)
	Rat[d]	16.4	6.4×10^8	143,000	Grouse et al. (1978b)
Nuclear RNA	Mouse[d]	17.2	6.5×10^8	144,000	Maxwell et al. (1978)
	Mouse[e]	21.2	8.1×10^8	179,000	Bantle and Hahn (1976)
	Rat[d]	15.6	5.9×10^8	131,000	Chikaraishi et al. (1978)
	Rat[d]	16.9	6.4×10^8	142,000	Grouse et al. (1978b)
Total poly(A+)RNA	Mouse[e]	13.3	5.0×10^8	111,000	Bantle and Hahn (1976)
	Rat[e]	12.5	4.8×10^8	107,000	Kaplan et al. (1978)
	Rat[e]	13.0	4.9×10^8	109,000	Colman et al. (1980)
Poly(A+)mRNA	Mouse[e]	3.8	1.4×10^7	100,000	Bantle and Hahn (1976)
	Rat[d]	3.2	1.2×10^7	86,000	Grouse et al. (1978b)
	Rat[e]	4.8	1.8×10^7	129,000	Chikaraishi et al. (1978)
	Rat[e]	4.4	1.7×10^7	121,000	Colman et al. (1980)
Poly(A−)mRNA	Rat[e]	4.6	1.7×10^7	121,000	Bernstein et al. (1980)
	Mouse[e]	3.6	1.4×10^7	100,000	Van Ness et al. (1979)
	Rat[d,e]	4.0	1.5×10^7	107,000	Chikaraishi (1979)

[a] Hybridization values given for each study are corrected for the reactivity of the ^{3}H-DNA tracer.

[b] Complexity estimates assume that DNA transcription is asymmetric and that the rodent single-copy DNA represents 65% of a haploid genome comprising 2.8×10^9 nucleotide pairs.

[c] Values obtained considering the number average length of brain hnRNA and mRNA are 4500 and 1400 nucleotides, respectively (Bantle and Hahn, 1976).

[d] Hybridization values estimated by S1 nuclease assay.

[e] Hybridization values determined by hydroxylapatite chromatography.

[f] A computer extrapolation to saturation from hybridization kinetic data.

B. Polyadenylated Nuclear RNA's

Recent studies show that nuclear RNA consists of two major populations of RNA, one containing a poly(adenylic acid) sequence approximately 200 nucleotides in length located at the 3′-OH terminus (DeLarco *et al.*, 1975; Mahony and Brown, 1975) and another RNA fraction that is nonadenylated. Polyadenylation is a post-transcriptional event, the significance of which remains unclear.

Poly(A+)hnRNA constitutes about 10–20% of the mass of nuclear RNA and has a sequence complexity of 5.0×10^8 nt (Table 1), a value representing 65–75% of the total nuclear RNA complexity (Bantle and Hahn, 1976; Kaplan *et al.*, 1978; Colman *et al.*, 1980). Because polysomal poly(A+)RNA consists of sequences transcribed in the nucleus, we will consider total tissue poly(A+)RNA complexity to be equivalent to that of nuclear poly(A+)RNA. Ample evidence indicates that much of the hnRNA serves as a precursor to cytoplasmic messengers on average 2–5 times smaller than the nuclear transcript (reviewed by Molloy and Puckett, 1976; Monahan, 1978; Tobin, 1979). Recently, Hahn *et al.* (1978) hybridized cDNA synthesized from poly(A+)RNA fragments cleaved from large poly(A+)hnRNA's to poly(A+)mRNA isolated from mouse brain polysomes. The cDNA hybridized almost to completion with the poly(A+)mRNA demonstrating that a large proportion of the sequences adjacent to the 3′-terminus of poly(A+)hnRNA (>8000 nt in length) appears in the poly(A+)mRNA population in brain.

The percentage of the nuclear RNA mass that drives the hybridization reaction has been estimated by comparing the observed first-order rate constant (K_{obs}) with that expected (K_{exp}) if all RNA sequences in the population participated equally in the reaction (Hough *et al.*, 1975). In the brain, approximately 2% of the nuclear RNA (Bantle and Hahn, 1976) and 4% of the poly(A+)hnRNA mass (Kaplan *et al.*, 1978) drives the hybridization reaction. The presence of ribosomal RNA precursors, abundant RNA, and RNA sequences transcribed from repetitive DNA add mass but little complexity (single-copy) to the nuclear RNA population and hence reduce the K_{obs}.

The small fraction of nuclear RNA driving the hybridization reaction has led to an interesting calculation concerning the distribution of the rare RNA class among the many cell types resident in brain. Once the sequence complexity and average size of the rare nuclear RNA's is obtained, the number of copies of these sequences per cell can be estimated. Given that the amount of total poly(A+)RNA per brain cell is 0.3 pg (Kaplan *et al.*, 1978) and that 4% of this RNA drives the hybridization reaction, the rare nuclear RNA sequences are present in about 0.03 copies per cell (Kaplan *et al.*, 1978). Similar results were obtained for total nuclear RNA by Chikaraishi *et al.*, (1978). Assuming that a cell must have at least one RNA copy, this value suggests that each rare RNA sequence is represented only once in every 30 cells. Therefore, rare RNA sequences could be restricted to particular cell types or cell populations in specific brain regions. In this instance, the RNA species could be present in many copies per cell or region, but at the whole brain level appear very rare. Alternatively, a fractional RNA copy number could reflect a high RNA turnover. For example, in sea urchins the rate of synthesis of the complex nuclear RNA is estimated to be one copy every 6.8 h, while their half-lives are approximately 10–20 min (Kleene and Humphreys, 1977). Therefore, a steady-state copy number less than one would be expected from the relative rates of synthesis and degradation.

Similar kinetic considerations indicate that most abundant brain nuclear RNAs are present in about 18 copies per cell (Grouse *et al.*, 1978b). In contrast to rare RNA's, the abundant nuclear RNA sequences may represent precursors to the "housekeeping" genes (Galau *et al.*, 1976) present in relatively high concentrations in most, if not all, brain cells.

Most recently, there is evidence to suggest that RNA driver mass estimates derived using single-copy DNA probes underestimates the fraction of RNA that drives the hybridization reaction (for discussion see Van Ness and Hahn, 1980). For example, kinetic data obtained with cDNA indicate that 40% of brain poly(A+)mRNA drives the reaction, in contrast to a value of 5–10% obtained with single-copy DNA. This discrepancy results, in part, from a retardation in the rate of hybridization caused by sequence discontinuity in the single-copy DNA tracer and RNA driver (W. E. Hahn, personal communication). The single-copy DNA probes used in these experiments are frequently a mixture of coding and noncoding (intervening) sequences. Although the mechanism whereby sequence discontinuity affects hybridization is unclear, data obtained with single-copy DNA could lead to underestimates in the driver fraction by as much as 4- to 8-fold.

IV. RNA SEQUENCE COMPLEXITY AT THE LEVEL OF TRANSLATION

A. Total Polysomal RNA

Since nuclear mRNA precursors undergo extensive post-transcriptional modification and processing prior to their transport to the cytoplasm, it does not necessarily follow that the large amount of genetic information transcribed in brain is finally expressed at the level of protein. The possibility exists that comparatively less of the total nuclear RNA sequence complexity is conserved in this tissue and that the diversity of brain mRNA is similar to that of liver and/or kidney. This question was addressed by Chikaraishi (1979), who compared the complexity of rat brain polysomal RNA with that of other organs. To eliminate possible hnRNA contamination of the polysomal RNA due to nuclear leakage or breakage, polysomes were isolated and RNA extracted from the mRNA released from polysomes by puromycin. In this approach, mRNA is released from the polysome complex as relatively small ribonucleoprotein particles (mRNP), which can be separated from larger hnRNP by velocity sedimentation in sucrose density gradients. At maximum hybridization, total brain polysomal RNA was complementary to 8% of the single-copy DNA, a finding consistent with the remarkable complexity manifest in brain nuclear RNA. Assuming asymmetric transcription, this value represents sufficient information to code for 190,000 different proteins averaging 45,000 daltons. By comparison, liver and kidney mRNA contained information to code for the equivalent of 57,000 and 39,000 different proteins, respectively. Similar findings were obtained in mouse brain (Van Ness *et al.*, 1979) with special care given to obtain polysomes free of hnRNA contamination (see below).

As in nuclear RNA, polysomal RNA contains both polyadenylated and nonadenylated sequences, whose properties are described below.

B. Polyadenylated mRNA

The base sequence complexity of poly(A+)mRNA is summarized in Table I. About half the complexity of polysomal mRNA's resides in the polyadenylated RNA fraction. Analysis of brain polysomal poly(A+)mRNA by DNA-excess hybridization indicates that the majority of brain mRNA is complementary to single-copy DNA, with a small kinetic component complementary to repeated DNA sequences (Heikkila and Brown, 1977). The hybridization kinetics of poly(A+)mRNA to single-copy DNA suggests that there are at least three RNA abundance classes (Bantle and Hahn, 1976; Grouse *et al.*, 1978b). The most abundant mRNA sequences found hybridize rapidly, and contain about 20% of the total poly(A+)mRNA sequence complexity. These abundant poly(A+)mRNA sequences are believed present in most cells and include the so-called "housekeeping" genes that are requisite for basic cell function. Given that these sequences are distributed uniformly in all cells, the calculations from hybridization kinetics suggest that there are about 6 copies per cell (Grouse *et al.*, 1978b; but see Section III,B). Such mRNA concentrations are sufficient to maintain known steady-state quantities of several histospecific proteins in liver (calculated by Galau *et al.*, 1977).

The slower kinetic fractions include the intermediate and rare DNA abundance classes comprising 60% and 20% of the total sequence complexity, respectively. As discussed above (Section III,B), the rare RNA sequences may represent brain cell-specific and/or region-specific gene transcripts and may appear as seldom as once in every 500 cells.

Comparison of liver and kidney poly(A+)mRNA populations by cDNA kinetic analysis indicates that they also contain at least three major frequency classes (Hastie and Bishop, 1976). Interestingly, there are significant tissue-specific differences in the complexity and composition of the most abundant class of messengers. For example, a liver cDNA probe enriched for rapidly hybridizing, *abundant* sequences cross-hybridized with brain poly(A+)mRNA with slow kinetics indicative of a *rare* RNA sequence class. Thus, gene transcripts present in liver in great abundance are also expressed in brain, but in greatly reduced amounts. Similar conclusions were reached in an independent cDNA analysis conducted by Young *et al.* (1976). These findings are also consistent with the hybridization data of Chikaraishi and associates (1978), who demonstrated that most nuclear RNA sequences present in liver were also transcribed in brain (see Section III,A). Taken together, these studies indicate that the establishment of organ-specific function involves both quantitative and qualitative alterations in the pattern of gene expression.

C. Nonadenylated mRNA

In the past, it was sometimes assumed that the poly(A+)mRNA class repre-sented the entire mRNA population, including the nonadenylated [poly(A−)] fraction. Although it first appeared that, except for histone mRNA, all eukaryote messengers were adenylated (Greenberg and Perry, 1972; Adesnick *et al.*, 1972), recent data show that poly(A−)mRNA's comprise as much as 30–50% of the mass of nascent mRNA (Milcarek *et al.*, 1974; Nemer *et al.*, 1974; Green-berg, 1976a). Poly(A−)mRNA, like its adenylated counterpart, has a AU-rich base composition, 5'-terminal cap structure (Surrey and Nemer, 1976; Faust *et al.*, 1976), is released from polyribosomes by EDTA and puromycin (Milcarek *et al.*, 1974), is associated with protein (Greenberg, 1976b), and serves as a template for protein synthesis when assayed in a heterologous cell-free transla-tion system (Kaufman *et al.*, 1977). The rates of metabolic turnover of these two classes of mRNA also appear similar. In cross-hybridization experiments in which DNA complementary to poly(A+)mRNA was reacted with excess polysomal poly(A−)RNA, little hybridization was observed at R_0t values suffi-cient to anneal abundant RNA sequences (Milcarek *et al.*, 1974). These results were taken as evidence that polysomal poly(A−)RNA did not arise from po-ly(A+)mRNA by nonadenylation, deadenylation, or random degradation. Re-cent hybridization data suggest that poly(A−)mRNA from mouse liver and cul-tured fibroblasts might represent as much as 30–40% of the total mRNA se-quence complexity and that the poly(A+) and poly(A−) RNA classes contained relatively few sequences in common (Grady *et al.*, 1978). These findings cannot be considered conclusive, however, as no data were provided to indicate that the poly(A−)RNA fraction was free of hnRNA contamination. Since nuclear RNA contains about four times the complexity of cytoplasmic RNA, small amounts of hnRNA contamination (by mass) could contribute disproportionately to the apparent complexity of the mRNA population.

Poly(A−)mRNA is also found in the brain (Chikaraishi, 1979; Van Ness *et al.*, 1979). Hybridization of rat brain polysomal poly(A−)RNA to single-copy DNA yielded complexity values equal to those obtained for poly(A+)mRNA (Chikaraishi, 1979; and Table I). Addition–hybridization experiments with polysomal poly(A−)RNA and poly(A+)mRNA clearly demonstrated little overlap in these RNA populations, i.e., that they contained few sequences in common. In support of the hybridization data, most translation products synthesized in a heterlogous cell-free system derived from rabbit reticulocytes were different with respect to poly(A+)mRNA vs. poly(A−)mRNA templates, as analyzed by two-dimensional gel electrophoresis (Chikaraishi, 1979). However, several proteins, the most abundant being actin, were present in the translation products of each messenger population. That the poly(A+)- and poly(A−)mRNA fractions code for qualitatively similar sets of abundant proteins was also observed in sea urchin embryos (Brandhorst *et al.*, 1979), HeLa (Milcarek, 1979) and mouse Friend cells (Minty and Gros, 1980).

Identical results were obtained by Van Ness *et al.* (1979) using a different analytic approach. In this study, polysomes were isolated by either (a) velocity centrifugation in sucrose gradients, or (b) centrifugation of formaldehyde-fixed polysomes in CsCl gradients and mRNA isolated from total polysomal RNA by benzoylated cellulose chromatography. The poly(A+)mRNA sequences were subsequently removed from the poly(A−)RNA fraction by affinity chromatography using oligo(dT)-cellulose. The poly(A−)mRNA fraction obtained contained neglible amounts of poly(A), the equivalent of one short adenosine tract (20 nt long) per 100 average size poly(A−)RNA molecules as judged by hybridization to ^{3}H-labeled poly(uridylic acid). The sequence homology of the two mRNA populations was estimated by hybridizing cDNA synthesized from poly(A−)RNA (after removal of DNA complementary to contaminating rRNA sequences) to excess poly(A+)mRNA. Results of this careful investigation demonstrate that polysomal poly(A−)RNA and poly(A+)RNA share few sequences in common.

It is pertinent that the proteins coded for by most brain mRNA (>95%) have not yet been resolved. Perhaps 1000–2000 different polypeptides have been detected by two-dimensional polyacrylamide gel electrophoresis, yet mRNA complexity measurements predict 100,000–200,000 different polypeptides. Since the great majority are probably present at very low concentrations per average cell it may be some time before their cellular distribution, subcelluar location, and function are known.

V. REGIONAL AND CELLULAR DISTRIBUTION OF GENE TRANSCRIPTS

A. Sequence Complexity in Major Brain Regions

The manner in which the gene transcripts are distributed in brain is knowledge fundamental to understanding the role played by the genome in the maintenance of brain structure and function. Conceivably, the large diversity of genetic information expressed in brain could result from a summation of gene subsets active in a wide variety of highly organized and functionally distinct regions. Given this situation, one might hope to identify genes whose transcription underlies region-specific structure and function.

To investigate this possibility, we compared the sequence complexity of nuclear (total) poly(A+)RNA of several brain regions (Kaplan *et al.*, 1978). The regions selected for study differed markedly in cell composition, neuronal/glial ratio, structure, and function. In general, a surprising degree of homology (>80%) was observed in the nuclear poly(A+)RNA sequences present in cerebral cortex, cerebellum, hypothalamus, and hippocampus (Fig. 1). Similar findings were reported in the early study of Hahn (1973), who hybridized large excesses of total nuclear RNA from monkey cerebrum and cerebellum to radiolabeled single-copy DNA and found no differences, albeit maximum hybridization values were not obtained. These data do not eliminate, however, the possibility that substantial region-specific nuclear transcripts exist, as histospecific sequences may be present in concentrations too low to effectively hybridize. Considering the resolution of the hybridization assay (±5–10%), we calculate that regions differ by less than 5000 nuclear poly(A+)RNA sequences averaging 4000 nt in length. Furthermore, there is no information available on possible regional differences in either the relative frequency of poly(A+)RNA sequences or the nonadenylated nuclear RNA population.

Although gene transcription in different brain regions appears similar, post-transcriptional mechanisms could regulate region-specific gene expression by controlling the sequences that ultimately appear in the cytoplasm. There is now good evidence to indicate that post-transcriptional mechanisms are involved in the regulation of organ-specific gene expression in sea urchin embryos (Wold *et al.*, 1978; Davidson and Britten, 1979; Shepherd and Nemer, 1980), frog embryos (Shepherd and Flickinger, 1979), and plants (Kamalay and Goldberg, 1980). At present, there is little information on the regional compartmentalization of brain mRNA. Preliminary data indicate that poly(A+)mRNA from rat cerebral cortex has a 10% greater complexity compared to that of cerebellum,

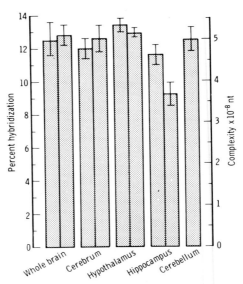

Fig. 1. Sequence complexity of nuclear (total) poly(A+)RNA from rat brain regions. RNA was reacted to single-copy DNA (0.6 M Na+, 70°C) to R_0t values $> 2 \times 10^4$ M^{-1} sec^{-1} and hybridization values determined by hydroxylapatite chromatography. Hybrid values ($\pm$SEM) are corrected for DNA duplex content and reactivity of the ^{3}H-DNA tracer. Each bar in the histogram represents the data from an independent RNA preparation. Animals used in this study were 90-day-old male Sprague–Dawley rats. Complexity estimates were made assuming that transcription is asymmetric and the complexity of the rat single-copy genome is 2.0 $\times$ 10^9 nt. (Adapted with permission from Kaplan *et al.*, 1978. Copyright 1978 American Chemical Society.)

whereas cerebral cortex mRNA complexity is indistinguishable from that of whole brain (Kaplan *et al.*, unpublished results). This modest regional difference in RNA–DNA hybridization values is equivalent to sufficient information to code for 5000–10,000 different proteins, averaging 50,000 daltons. Clearly, a great deal more information is required on both the poly(A+)- and poly(A−)mRNA populations before one can reach a meaningful conclusion concerning the regional specificity of gene expression in brain. Although the search for brain region-specific proteins has yielded relatively few clear examples (e.g., Margolis, 1972), it is likely that many more exist, but are not detected due to their low average concentrations.

B. Cell Lines of Neuroectodermal Origin

The striking complexity of brain mRNA is generally interpreted as reflecting the heterogeneity of cell types present in the tissue. Based on RNA complexity estimates and frequency distribution data, Grouse *et al.* (1978b) speculated that rat brain could contain as many as 500 different cell types each differing by 10–20 gene transcripts. The possibility arises, however, that neuroectodermal derivatives express more genes than most other somatic cell types. The data available on the subject are summarized in Table II. The neuroectodermal derivatives selected for study retain many of the morphologic and physiologic properties of neurons and glia *in vivo* and are commonly employed as convenient models for the study of differentiated cell function. Surprisingly, nuclear poly(A+)RNA from neuroblastoma and glioma cells contained 75% of the total complexity of brain nuclear RNA, while the poly(A+)mRNA contained about 60% of that from whole brain. Complexity estimates for other somatic cell lines obtained by saturation hybridization analysis are given in Table II. For additional comparison, rat liver nuclear poly(A+)RNA and poly(A+)mRNA complexities were also included. These results suggest that, as in the brain, cultured cells of neural origin express 1.5- to 3.0-fold more genetic information than most other somatic cell types. It is interesting that the ratio of nuclear/cytoplasmic poly(A+)RNA complexity is 3–4 over a 4-fold range of sequence complexity in these diverse cells. This ratio approximates the size differences in the nuclear and polysomal poly(A+)RNA sequence transcripts, and infers that the fraction of hnRNA destined for transport to the cytoplasm as mRNA is the same for a broad class of gene products.

Grouse *et al.* (1979a) using a cDNA probe to analyze the sequence complexity and frequency distribution of poly(A+)mRNA from rat C6 glioma found three major frequency classes, with a complexity equivalent of 33,000 different messengers. In contrast, some other mammalian cell lines express, on average, about 5000–15,000 different mRNA types as measured by cDNA analysis (Birnie *et al.*, 1974; Ryfell and McCarthy, 1975; Williams and Penman, 1975; Getz *et al.*, 1976; Affara *et al.*, 1977; Patterson and Bishop, 1977). Cells of neuroectodermal origin may thus have an unusually high mRNA complexity which is "constitutive" and maintained in diverse states of growth and differentiation.

Since the above results are derived from transformed cells, the findings should be interpreted cautiously. Although transformation of cultured cells by either chemical (Getz *et al.*, 1977; Moyzis *et al.*, 1980; Supowit and Rosen, 1980) or viral agents (Rolton *et al.*, 1977; Williams *et al.*, 1977) does not necessarily alter gene expression qualitatively, elucidation of the diversity of gene activity in nerve cells will ultimately require the isolation of mRNA from a number of specific brain cell types.

Neuroblastoma cell lines have frequently been used as models for the study of neuronal development. Under the appropriate culture conditions these cells will differentiate from round, immature neuroblasts to cells which exhibit many properties of mature neurons as assessed by morphologic, biochemical, and physiologic criteria. Using cDNA probes, Felsani *et al.* (1978) analyzed the

Table II. Sequence Complexity of Polyadenylated RNA from Cultured Mammalian Cells

Cell type	Species	Single-copy DNA hybridized (%)	Sequence complexity (nt)	References
A. Nuclear poly(A+)RNA				
1. Neuroectodermal				
Neuroblastoma (B104)	Rat	8.7	3.1×10^8	Kaplan et al. (1978)
Glioma (C6)	Rat	9.3	3.4×10^8	Kaplan et al. (1978)
2. Other				
Friend cells (M2)	Mouse	5.4	1.9×10^8	Kleiman et al. (1977)
Embryonal carcinoma (PCC3)	Mouse	2.5	0.9×10^8	Jacquet et al. (1978)
SHE[a]	Hamster	2.8	1.0×10^8	Moyzis et al. (1980)
Liver	Rat	6.0	2.2×10^8	Kaplan et al. (1978)
B. Cytoplasmic poly(A+)mRNA				
1. Neuroectodermal				
Neuroblastoma (B104)	Rat	2.7	8.6×10^7	B. Kaplan et al. (unpublished results)
Neuroblastoma (NS20)	Mouse	2.1	7.6×10^7	Schrier et al. (1978)
Glioma (C6)	Rat	2.7	9.7×10^7	B. Kaplan et al. (unpublished results)
2. Other				
Friend cells (M2)	Mouse	1.3	4.7×10^7	Kleiman et al. (1977)
Embryonal carcinoma	Mouse	0.5	1.8×10^7	Jacquet et al. (1978)
SHE[a]	Hamster	0.7	2.5×10^7	Moyzis et al. (1980)
Fibroblasts (PYAL/N)	Mouse	1.0	3.6×10^7	Grady et al. (1978)
Liver	Rat	1.8	6.5×10^7	Savage et al. (1978)

[a] Syrian hamster embryo cells. Complexity estimates calculated assuming hamster single-copy DNA is 4.16×10^9 nucleotides.

sequence complexity and frequency distribution of polysomal poly(A+)mRNA from a noradrenergic neuroblastoma cell line (NIE 115) at two different developmental states. The complexity of mRNA from neuroblasts (suspension grown) and differentiated cells (monolayer culture) were identical, corresponding to about 7000 average-size sequences of 1750 nucleotides. The RNA frequency distribution and size of the poly(A) segments associated with these RNA's were also identical. Determination of the sequence homology between these mRNA's using heterologous cross-hybridization experiments showed that neuroblasts contained all the sequences present in differentiated cells. In contrast, neuroblasts contained 100–150 mRNA's not expressed in cells exhibiting the properties of mature neurons.

Similar results were reported in a cholinergic neuroblastoma cell line (Grouse *et al.*, 1980). Both immature and differentiated cells expressed approximately 8000 different poly(A+)mRNA's with the undifferentiated cells containing about 200 mRNA's not present in mature cells. In contrast to the results of Felsani *et al.* (1978), however, these workers detected a mRNA population specific to the differentiated state. This discrepancy was attributed to either the use of a more sensitive stage-specific cDNA probe or improvements in the differentiation paradigm. Nonetheless, it bears emphasis that cDNA measurements tend to underestimate RNA complexity and the number of different mRNA's detected in these studies are at varience with estimates obtained by saturation hybridization analysis (see Table II) and some cDNA analysis (Grouse *et al.*, 1979a). Thus, observations derived from both of these cDNA studies may pertain to less than 20% of the total mRNA sequence complexity.

VI. GENE ACTIVITY DURING DEVELOPMENT AND AGING

A. Postnatal Development

In view of the importance of gene expression in cell differentiation and organ development, it was of obvious interest to examine brain RNA sequence complexity as a function of age. Results of early studies of small mammals suggested that the percentage of the genome transcribed increased rather dramatically during postnatal development (Brown and Church, 1972; Grouse *et al.*, 1972), and decreased with age after maturity (Cutler, 1975). Since most early hybridization studies seriously underestimated the complexity of gene transcription (see Section III,A), we reinvestigated the effects of age on RNA diversity using improved RNA extraction techniques and *in vitro* hybridization conditions to obtain more accurate estimates of RNA sequence complexity. These studies are summarized in Fig. 2, which shows the hybridization values of whole brain nuclear (total)

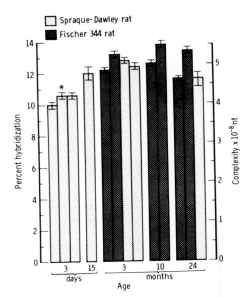

Fig. 2. Effect of age on brain nuclear (total) poly (A+)RNA sequence complexity. Hybridization values (±SEM) and complexity estimates were obtained as described in Fig. 1. *, *p* < 0.01.

poly (A+)RNA at several postpartum ages. A small but reproducible increase (ca.15%) was detected in nuclear poly (A+)RNA complexity during the first two weeks after birth. Marked increases in the sequence complexity of nonadenylated nuclear and messenger RNA were recently reported in mouse brain during early postnatal development (Hahn and Chaudari, 1981). Taken together, these studies confirm the early observations. The question of complexity changes during adult aging will be discussed below.

In contrast to the data obtained with whole brain, however, cerebellar nuclear poly(A+)RNA complexity decreases by 15–20% during postnatal development (Bernstein *et al.*, 1980). This decrement occurred during the second and third week postpartum, when cell acquisition and differentiation in rat cerebellum nears completion (Zagon and McLaughlin, 1979).

The above studies were recently extended to the mRNA populations of the cerebral cortex and cerebellum, two brain regions differing substantially in their patterns of postnatal development. Cerebral cortex poly(A+)mRNA complexity did not change during postnatal development. Furthermore, hybridization of mixtures of neonate and adult poly(A+)mRNA indicated that an extensive sequence homology is maintained in these RNA populations during postnatal development. Therefore, development in rat cerebral cortex may involve altered rates of production and processing of poly(A+)mRNA, which result in changes in the relative abundance of mRNA sequences rather than major qualitative changes in mRNA populations. A similar situation appears to characterize the resting and proliferating state of cells in tissue culture, where cells appear to regulate their poly(A+)mRNA content by controlling the efficiency with which nuclear poly(A+)RNA is processed to cytoplasmic messengers (Johnson *et al.*, 1975; Getz *et al.*, 1976).

Consistent with the findings obtained with nuclear poly(A+)RNA, postnatal development in the cerebellum entails a small but significant reduction (5–10%) in the sequence complexity of the poly(A+)mRNA population (Bernstein *et al.*, 1980). Combined, these data suggest that there may be region-specific differences in the developmental pattern of gene expression in brain. Additionally, the coordinate decrease in the sequence complexity of cerebellar nuclear poly(A+)RNA and poly(A+)mRNA also implies, but does not prove, that the developmental differences seen in this brain region are regulated at the level of gene transcription rather than post-transcriptional modification and processing.

Development of the nervous system clearly includes both regulation by intrinsic mechanisms which operate independently of external factors and environmental cues that can modify the developmental program in an important, but as yet ill-defined manner. That the rearing environment can markedly effect animal physiology and behavior with corresponding alterations in brain weight, thickness, neuronal/glial ratio, and neuronal size and ultrastructure was reported by many investigators (reviewed by Rosenzweig and Bennett, 1980). Whether experience can also alter the pattern of gene expression in brain is a question of fundamental importance to molecular neurobiology. Two studies suggest plasticity in brain genomic expression. In preliminary investigations, Uphouse and Bonner (1975), and Grouse *et al.* (1978b) reported increases in total RNA sequence complexity in rats reared in a complex environment compared to animals raised in a sensory and behaviorally deprived environment. Consistent with these findings is a report of Grouse *et al.* (1979b) on total brain RNA from several brain regions of cats reared normally or deprived of patterned visual stimulation. There was a relatively large decrease (25%) in the complexity of RNA obtained from visual cortex (area 17) of deprived animals, a loss equivalent to 1.5×10^4 average size hnRNA's. No differences were seen in other areas of cortex or subcortical structures. These data were interpreted to reflect either a failure to (a) maintain certain features of the system established at birth, or (b) continue a developmental process dependant on visual stimulation.

However, as these studies using *total* RNA (including rRNA) did not achieve the maximum hybridization of DNA observed in most recent studies (Table I), it is likely that only a portion of the rare RNA sequences was assayed. In addition, there is, as yet, no indication that environmental factors alter the brain mRNA diversity or frequency distribution. These criticisms notwithstanding, the above studies address important questions concerning gene expression in brain and will, no doubt, stimulate much needed work in the future.

B. Aging Studies

The effect of aging on RNA populations in the brain bears on some major issues of neurogerontology including the neuroendocrine theories of aging (reviewed by Finch, 1976, 1979) and the hypothesis that postmitotic cells are at particular risk for accumulating various types of genomic damage during aging (Strehler, 1977). In an extensive study of poly(A+)RNA from rat brain no age changes were found in the yield or complexity of nuclear poly(A+)RNA (Fig. 2) or poly(A+)mRNA (Fig. 3) in two rat strains, aged 2, 12, 24. and 32 months. In addition, mixtures of poly(A+)mRNA from different age groups gave complexity values that were indistinguishable from the complexity of each component, indicating that most poly(A+)mRNA sequences were common to all adult age groups. In accord with these findings is the study by Farquhar *et al.* (1979), who found no age-related differences in the sequence complexity of cerebral cortex hnRNA from female macaques.

The results in Fig. 3 also suggest that the sequence complexity of polysomal poly(A+)mRNA of Sprague–Dawley rats is somewhat greater than that of the Fischer strain. In view of the anatomic similarity of the Fischer and Sprague–Dawley rat brains, strain differences in the complexity of its single-copy genome or in the extent of transcription seem unlikely. Although the higher complexity of Sprague–Dawley poly(A+)mRNA could conceivably represent contamination by nuclear poly(A+)RNA sequences in this strain only, the reproducibility of the strain differences argues against it. Strain differences in the concentration of the rare poly(A+)RNA sequences could, however, result in slightly different complexity values. Interestingly, consistent differences (ca. 15%; $p < 0.005$) were also observed in the hybridization values of hnRNA obtained from cerebral cortex of two subspecies of macaques (Farquhar *et al.*, 1979). The unexpected possibility of variations in brain RNA sequence complexity within a species suggests a need to use highly inbred animal models reared under standard conditions. Regardless of possible strain differences, however, the invariance in RNA complexity expressed as a function of age is consistent within each strain.

Kinetic studies of the hybridization of total poly(A+)RNA to single-copy DNA revealed no significant differences in either the rate constants or complexity of the major hybridizing component (Fig. 4). The absence of age changes in hybridization kinetics implies that the rare RNA sequences driving this phase of the reaction are present in the same relative frequency. Although these data indicate that no major age-related changes occur in the poly(A+)RNA popula-

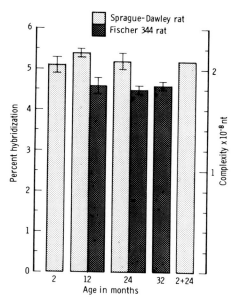

Fig. 3. Sequence complexity of polysomal poly(A+)RNA from aging rats. Hybridization values (±SD) and complexity estimates were obtained as described in Fig. 1. 2 + 24, hybridization reaction mixtures containing equal amounts of brain poly(A+)mRNA from 2-month- and 24-month-old male animals. (Data adapted from Colman *et al.*, 1980.)

tion, they do not rule out differences in (a) a minority class of RNA sequences, (b) poly(A−)RNA, or (c) a minority of brain cells. For example, two studies detected 10–20% loss of bulk RNA (principally rRNA) in the striatum (Chaconas and Finch, 1973; Shaskan, 1977), a brain region that contains about 2% of the brain's cells. Microspectrophotometric studies show 10–30% decreases of RNA in human cerebellar Purkinje cells (Mann *et al.*, 1978; Mann and Yates, 1979), as well as in rodent neurons (Bohn and Mitchell, 1977; Zs-Nagy *et al.*, 1977). Considerations of interassay variance set upper limits (±5%) on the extent of age-related changes which could not be detected with present techniques.

The possibility of genomic damage during aging in brain and other nondividing cells has been widely discussed (Strehler, 1977), but there is little evidence to support this premise as a general principle. The template activity of mouse brain chromatin does not change with age (Hill, 1976), nor do the levels of most brain proteins when considered on a regional basis (Cicero *et al.*, 1972; Gordon and Finch, 1974; Wilson *et al.*, 1978). Perhaps study of proteins from neurons of the type showing reduced RNA synthesis (see above) would reveal age-related alterations in protein metabolism.

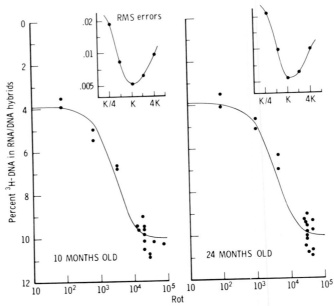

Fig. 4. Kinetics of hybridization of single-copy ³H-DNA to brain total poly(A+)RNA. Semilogarithmic plot of data for RNA–DNA hybrids. Data are uncorrected for DNA reactivity. Line drawn is a computer graphed, best least-squares fit to the data determined according to pseudo-first-order kinetics (Section II,B). The insert in each panel shows the root mean square error that results from varying the rate constant from its "best fit" value. Rate constants for 10-month- and 24-month-old data are 2.0×10^{-4} and 1.4×10^{-4} M^{-1} sec^{-1}, respectively. (Reproduced from Colman *et al.*, 1980.)

VII. CONCLUDING REMARKS

Recent applications of RNA-driven hybridizations with single-copy DNA se-quences or with complementary DNA indicate a remarkable number of different types of mRNA sequences in the adult brain. Sequence complexity estimates predict the existence of 100,000–200,000 different brain polypeptides. Rela-tively few ($< 5\%$) of this predicted number has been identified by standard gel-electrophoretic analysis. At the time of birth, the rodent brain has nearly achieved the adult inventory of polyadenylated nuclear and mRNA sequences; only modest changes are detectable in the whole brain or cerebellum, despite this region's cellular and morphologic immaturity at parturition. No subsequent changes in poly(A+)RNA complexity were observed throughout the remainder of the adult lifespan. In contrast, major developmental changes occur in the com-plexity of the nonadenylated nuclear and messenger RNA populations (Hahn and

Chaudhari, 1981). This is an important new finding as the age-related alterations in gene expression occur in an RNA population largely specific to the brain. At present, it is still unclear as to whether experiential or environmental cues effect gene expression in brain.

A key question concerns the regional distribution of brain mRNA and protein. Although only limited data are available, most brain regions appear to have RNA populations that are indistinguishable from each other or from whole brain. This observation is consistent with the paucity of reported brain region-specific proteins.

Studies of cultured cells of neuroectodermal origin suggest that as a class, neuroectodermal derivatives may have an unusually large number of different mRNA's. This appears as a constituitive property that is relatively uninfluenced by environment or state of differentiation. How this enormous amount of genetic information is used in the differentiation and maintenance of specific cell function(s) remains unknown.

Considering the complexity of brain RNA and the limits in the resolution of RNA–DNA hybridization analysis, progress in the future will depend on the availability of more specific DNA hybridization probes. These will no doubt come in the near future from the purification of brain-specific mRNA's and the isolation of the corresponding genes from an appropriate "shotgun-cloned" gene library. Additionally, the identification and isolation of a random population of DNA sequences transcribed specifically in brain would provide an invaluable tool for the study of the regional distribution and control of gene expression in brain. Thus, molecular neurobiology appears to be moving inexorably into the area of recombinant DNA research.

ACKNOWLEDGMENTS

We are grateful to our colleagues S. Bernstein, A. Gioio, E. Baetge (CUMC), H. Osterburg, B. Schachter, and P. Colman (USC), who actively participated in the studies conducted in our laboratories. We are also grateful to Dr. W. E. Hahn (University of Colorado, School of Medicine) for a critical evaluation of the manuscript. The neuroblastoma and glioma cell lines were obtained through the generosity of Dr. J. deVellis (UCLA) and were cultured in the laboratory of Dr. T. H. Joh (CUMC). This work was supported by NIH grants HD11392 (BBK) and AG00117 and AG00446 (CEF).

REFERENCES

Adesnik, M., Salditt, M., Thomas, W., and Darnell, J. E. (1972). Evidence that all messenger RNA molecules (except histone mRNA) contain poly(A) sequences and that the poly(A) has a nuclear function. *J. Mol. Biol.* **71**, 21–30.

Affara, N. A., Jacquet, M., Jakob, H., Jacob, F., and Gros, F. (1977). Comparison of polysomal polyadenylated RNA from embryonal carcimona and committed myogenic and erythropoietic cells lines. *Cell* **12**, 509–520.

Axel, R., Fiegelson, P., and Schutz, G. (1976). Analysis of the complexity and diversity of mRNA from chicken liver and oviduct. *Cell* **7**, 247–254.

Bantle, J. A., and Hahn, W. E. (1976). Complexity and characterization of polyadenylated RNA in mouse brain. *Cell* **8**, 139–150.

Bernstein, S. L., Gioio, A. E., and Kaplan, B. B. (1980). Sequence complexity of polyadenylated RNA in rat cerebellum during postnatal development. *Soc. Neurosci. Symp. Abstr.* **6**, 775.

Birnie, G. D., Macphail, E., Young, B. D. Getz, M. J., and Paul, J. (1974). The diversity of the messenger RNA population in growing Friend calls. *Cell Differ.* **3**, 221–237.

Birnstiel, M. L., Sells, B. H., and Purdom, I. F. (1972). Kinetic complexity of RNA molecules. *J. Mol. Biol.* **63**, 21–39.

Bishop, J. O. (1969). The effect of genetic complexity on the time-course of RNA-DNA hybridization. *Biochem. J.* **113**, 805–811.

Bishop, J. O., Morton, J. G., Rosbash, M., and Richardson, M. (1974). Three abundance classes in HeLa cell messenger RNA. *Nature (London)* **250**, 199–203.

Bohn, R. C., and Mitchell, R. B. (1977). Age-related cytochemical changes in rat Purkinje neuron nucleic acids. *Exp. Neurol.* **57**, 161–178.

Brandhorst, B. P., Verma, D. P. S., and Fromson, D. (1979). Polyadenylated and nonadenylated messenger RNA fractions from sea urchin embryos code for the same abundant proteins. *Dev. Biol.* **71**, 128–141.

Britten, R. J., and Kohne, D. E. (1968). Repeated sequences in DNA. *Science* **161**, 529–540.

Brown, I. R., and Church, R. B. (1971). RNA transcription from nonrepetitive DNA in the mouse. *Biochem. Biophys. Res. Commun.* **42**, 850–856.

Brown, I. R., and Church, R. B. (1972). Transcription of nonrepeated DNA during mouse and rabbit development. *Dev. Biol.* **29**, 73–84.

Buell, G. N., Wicken, M. P., Farhang, P., and Schimke, R. T. (1978). Synthesis of full length cDNAs from four partially purified oviduct mRNAs. *J. Biol. Chem.* **253**, 2471–2482.

Chaconas, G., and Finch, C. E. (1973). The effect of aging on RNA/DNA ratios in brain regions of the C57BL/6J male mouse. *J. Neurochem.* **21**, 1469–1473.

Chikaraishi, D. M. (1979). Complexity of cytoplasmic polyadenylated and nonadenylated rat brain RNAs. *Biochemistry* **18**, 3249–3256.

Chikaraishi, D. M., Deeb, S. S., and Sueoka, N. (1978). Sequence complexity of nuclear RNA in adult rat tissues. *Cell* **13**, 111–120.

Cicero, T. J., Ferrendelli, J. A., Suntzeff, V., and Moore, B. W. (1972). Regional changes in CNS levels of the S-100 and 14-2-3 proteins during development and aging of the mouse. *J. Neurochem.* **19**, 2119–2125.

Colman, P. D., Kaplan, B. B., Osterburg, H. H., and Finch, C. E. (1980). Brain poly(A)RNA during aging: Stability of yield and sequence complexity in two rat strains. *J. Neurochem.* **34**, 335–345.

Cutler, R. G. (1975). Transcription of unique and reiterated DNA sequences in mouse liver and brain tissues as a function of age. *Exp. Gerontol.* **10**, 37–60.

Davidson, E. H., and Britten, R. J. (1973). Organization, transcription, and regulation in the animal genome. *Q. Rev. Biol.* **48**, 565–613.

Davidson, E. H., and Britten, R. J. (1979). Regulation of gene expression: Possible role of repetitive sequences. *Science* **204**, 1052–1059.

Davidson, E. H., and Hough, B. (1971). Genetic information in oocyte RNA. *J. Mol. Biol.* **56**, 491–506.

DeLarco, J., Abramowitz, A., Bromwell, K., and Guroff, G. (1975). Polyadenylic acid-containing RNA from rat brain. *J. Neurochem.* **24**, 215–222.

Farquhar, M. N., Kosky, K. J., and Omenn, G. S. (1979). Gene Expression in Brain. *In,* "Aging in Nonhuman Primates" (D. M. Bowden, Ed.), pp. 71–79. Van Nostrand-Reinhold, New York.

Faust, M., Millward, S., Duchastel, A., and Fromson, D. (1976). Methylated constituents of poly(A−) and poly(A+) polyribosomal RNA of sea urchin embryos. *Cell* **9**, 597–604.

Felsani, A., Bethelot, F., Gros, F., and Croizat, B. (1978). Complexity of polysomal polyadenylated RNA in undifferentiated and differentiated neuroblastoma cells. *Eur. J. Biochem.* **92**, 569–577.

Finch, C. E. (1976). The regulation of physiological changes during mammalian aging. *Q. Rev. Biol.* **51**, 49–83.

Finch, C. E. (1979). Neuroendocrine mechanisms and aging. *Fed. Proc., Fed. Am. Soc. Exp. Biol.* **38**, 178–183.

Galau, G. A., Klein, W. H., Davis, M. M., Wold, B. J., Britten, R. J., and Davidson, E. H. (1976). Structural gene sets active in embryos and adult tissues of the sea urchin. *Cell* **7**, 487–505.

Galau, G. A., Klein, W. H., Britten, R. J., and Davidson, E. H. (1977). Significance of rare mRNA sequences in liver. *Arch. Biochem. Biophys.* **179**, 584–599.

Gelderman, A. H., Rake, A. V., and Britten, R. J. (1971). Transcription of nonrepeated DNA in neonatal and fetal mice. *Proc. Natl. Acad. Sci. U.S.A.* **68**, 172–176.

Getz, M. J., Elder, P. K., Benz, E. W., Stephens, R. E., and Moses, H. L. (1976). Effect of cell proliferation on levels and diversity of poly(A)-containing mRNA. *Cell* **7**, 255–265.

Getz, M. J., Reiman, H. M., Siegel, G. P., Quinlan, T. J., Proper, J., Elder, P. K., and Moses, H. L. (1977). Gene expression in chemically transformed mouse embryo cells: Selective enhancement of the expression of C type RNA tumor virus genes. *Cell* **11**, 909–921.

Gordon, S. M., and Finch, C. E. (1974). An electrophoretic study of protein synthesis in brain regions of senescent male mice. *Exp. Gerontol.* **9**, 269–273.

Grady, L. J., North, A. B., and Campbell, W. P. (1978). Complexity of poly(A+) and poly(A−) polysomal RNA in mouse liver and cultured mouse fibroblasts. *Nucleic Acids Res.* **5**, 697–712.

Greenberg, J. R. (1976a). Isolation of L-cell messenger RNA which lacks poly(adenylate). *Biochemistry* **15**, 3516–3522.

Greenberg, J. R. (1976b). Isolation of messenger ribonucleoproteins in cesium sulfate density gradients: Evidence that polyadenylated and non-adenylated messenger RNAs are associated with protein. *J. Mol. Biol.* **108**, 403–416.

Greenberg, J. R., and Perry, R. P. (1972). Relative occurrence of polyadenylic acid sequences in messenger and heterogeneous nuclear RNA of L cells as determined by poly(U)-hydroxylapatite chromatography. *J. Mol. Biol.* **72**, 91–98.

Grouse, L., Chilton, M. D., and McCarthy, B. J. (1972). Hybridization of RNA with unique sequences of mouse DNA. *Biochemistry* **11**, 798–805.

Grouse, L. D., Nelson, P. G., Omenn, G. S., Schrier, B. K. (1978a). Measurements of gene expression in tissues of normal and dystrophic mice. *Exp. Neurol.* **59**, 470–478.

Grouse, L. D., Schrier, B. K., Bennett, E. L., Rosenzweig, M. R., and Nelson, P. G. (1978b). Sequence diversity studies of rat brain RNA: Effects of environmental complexity on rat brain RNA diversity. *J. Neurochem.* **30**, 191–203.

Grouse, L. D., Lettendre, C., and Schrier, B. K. (1979a). Sequence complexity and frequency distribution of poly(A)-containing messenger RNA sequences from the glioma cell line C6. *J. Neurochem.* **33**, 583–585.

Grouse, L. D., Schrier, B. K., and Nelson, P. G. (1979b). Effect of visual experiences on gene expression during the development of stimulus specificity in cat brain. *Exp. Neurol.* **64**, 354–364.

Grouse, L. D., Schrier, B. K., Lettendre, C. H., Zubairi, M. Y., and Nelson, P. G. (1980). Neuroblastoma differentiation involves both the disappearance of old and the appearance of new poly(A+) messenger RNA sequences in polyribosomes. *J. Biol. Chem.* **255**, 3871–3877.

Hahn, W. E. (1973). Genetic transcription in the cerebrum and cerebellum of the primate brain. *Soc. Neurosci. Symp. Abstr.* **3**, 139.

Hahn, W. E., and Laird, C. D. (1971). Transcription of nonrepeated DNA in mouse brain. *Science* **173,** 158–161.

Hahn, W. E., and Chaudhari, N. (1981). Late expression of genes represented by poly(A−)mRNA in developing brain. *Am. Soc. Neurochem. Abstr.* **12,** 71.

Hahn, W. E., Van Ness, J., and Maxwell, I. H. (1978). Complex population of mRNA sequences in large polyadenylated nuclear RNA molecules. *Proc. Natl. Acad. Sci. U.S.A.* **75,** 5544–5574.

Hastie, N. D., and Bishop, J. O. (1976). The expression of three abundance classes of messenger RNA in mouse tissues. *Cell* **9,** 761–774.

Heikkila, J. J., and Brown, I. R. (1977). Analysis of rabbit brain poly(A+)mRNA by DNA excess hybridization. *Biochim. Biophys. Acta* **474,** 141–153.

Hill, B. T. (1976). Influence of age on chromatin transcription in murine tissues using heterologous and homologous RNA polymerases. *Gerontology* (Basel) **22,** 111–123.

Hough, B. R., Smith, M. J., Britten, R. J., and Davidson, E. H. (1975). Sequence complexity of heterogeneous nuclear RNA in sea urchin embryos. *Cell* **5,** 291–299.

Jacquet, M., Affara, N. A., Robert, B., Jakob, H., Jacob, F., and Gros, F. (1978). Complexity of nuclear and polysomal polyadenylated RNA in a pluripotent embryonal carcinoma cell line. *Biochemistry* **17,** 64–79.

Johnson, L. F., Williams, J. G., Abelson, H. T., Green, H., and Penman, S. (1975). Changes in RNA in relation to growth of the fibroblast. III. Post-transcriptional regulation of mRNA formation in resting and growing cells. *Cell* **4,** 69–75.

Kamalay, J. C., and Goldberg, R. B. (1980). Regulation of structural gene expression in tobacco. *Cell* **19,** 935–946.

Kaplan, B. B., Schachter, B. S., Osterburg, H. H., deVellis, J. S., and Finch, C. E. (1978). Sequence complexity of polyadenylated RNA obtained from rat brain regions and cultured cells of neural origin. *Biochemistry* **17,** 5516–5524.

Kaufman, Y., Milcarek, C., Berissi, H., and Penman, S. (1977). HeLa cell poly(A−)mRNA codes for a subset of poly(A+)mRNA-directed proteins with an actin as the major product. *Proc. Natl. Acad. Sci. U.S.A.* **74,** 4801–4805.

Kleene, K. C., and Humphreys, T. (1977). Similarity of hnRNA sequences in blastula and pluteus stage sea urchin embryos. *Cell* **12,** 143–155.

Kleiman, L., Birnie, G. D., Young, B. D., and Paul, J. (1977). Comparison of the base-sequence complexities of polysomal and nuclear RNAs in growing Friend erythroleukemia cells. *Biochemistry* **16,** 1218–1223.

Lizardi, P. M., and Brown, D. D. (1973). The length of fibroin structural gene sequences in *Bombyx mori. Cold Spring Harbor Symp. Quant. Biol.* **38,** 701–706.

Mahony, J. B., and Brown, I. R. (1975). Characterization of poly(A) sequences in brain RNA. *J. Neurochem.* **25,** 503–507.

Mann, D. M. A., and Yates, P. O. (1979). The effects of aging on the pigmented nerve cells of the human locus coeruleus and substantia nigra. *Acta Neuropathol.* **47,** 93–97.

Mann, D. M. A., Yates, P. O., and Stamp, J. E. (1978). The relationship between lipofuscin pigment and aging in the human nervous system. *J. Neurol. Sci.* **37,** 83–93.

Margolis, F. L. (1972). A brain protein unique to the olfactory bulb. *Proc. Natl. Acad. Sci. U.S.A.* **69,** 1221–1224.

Maxwell, I. H., Van Ness, J., and Hahn, W. E. (1978). Assay of DNA-RNA hybrids by S1 nuclease digestion and adsorption to DEAE-cellulose filters. *Nucleic Acids Res.* **5,** 2033–2038.

Meyuhas, O., and Perry, R. P. (1979). Relationship between size, stability and abundance of the messenger RNA of mouse L cells. *Cell* **16,** 139–148.

Milcarek, C. (1979). Hela cell cytoplasmic mRNA contains three classes of sequences: Predominantly poly(A)-free, predominently poly(A)-containing and biomorphic. *Eur. J. Biochem.* **102,** 467–476.

Milcarek, C., Price, R., and Penman, S. (1974). The metabolism of a poly(A)minus mRNA fraction in HeLa cells. *Cell* **3**, 1–10.

Minty, A. J., and Gros, F. (1980). Coding potential of non-polyadenylated messenger RNA in mouse Friend cells. *J. Mol. Biol.* **139**, 61–83.

Molloy, G., and Puckett, L. (1976). The metabolism of heterogeneous nuclear RNA and the formation of cytoplasmic messenger RNA in animal cells. *Prog. Biophys. Mol. Biol.* **31**, 1–38.

Monahan, J. J. (1978). The structure, origin and function(s) of RNA in the nuclei of eukaryotic cells. *Int. Rev. Cytol.* **8**, 229–290.

Moyzis, R. K., Grady, D. L., Li, D. W., Mirvis, S. E., and Ts'o, P. O. P. (1980). Extensive homology of nuclear RNA and polysomal poly(adenylic acid) messenger RNA between normal and neoplastically transformed cells. *Biochemistry* **19**, 821–837.

Nemer, M., Graham, M., and Dubroff, L. M. (1974). Co-existence of nonhistone messenger RNA species lacking and containing polyadenylic acid in sea urchin embryos. *J. Mol. Biol.* **89**, 435–454.

Paterson, B. M., and Biship, J. O. (1977). Changes in the mRNA population of chick myoblasts during myogenesis *in vitro. Cell* **12**, 751–765.

Pearson, W. R., Davidson, E. H., and Britten, R. J. (1977). A progran for least squares analysis of reassociation and hybridization data. *Nucleic Acids Res.* **4**, 1727–1737.

Quinlan, T. J., Beeker, G. W., Cox, R. F., Elder, P. K., Moses, H. L., and Getz, M. J. (1978). The concept of mRNA abundance classes: A critical reevaluation. *Nucleic Acids Res.* **5**, 1611–1625.

Rolton, H. A., Birnie, G. D., and Paul, J. (1977). The diversity and specificity of nuclear and polysomal poly(A+)RNA populations in normal and MSV-transformed cells. *Cell Differ.* **6**, 25–39.

Rosenzweig, M. R., and Bennett, E. L. (1980). How plastic is the nervous system? *In* "A Comprehensive Handbook of Behavioral Medicine" (B. Taylor and J. Ferguson, eds.). Spectrum Publications, Jamaica, New York.

Ryffel, G. U., and McCarthy, B. J. (1975). Complexity of cytoplasmic RNA in different mouse tissues measured by hybridization of polyadenylated RNA to complementary DNA. *Biochemistry* **14**, 1379–1384.

Savage, M. J., Sala-Trepat, J. M., and Bonner, J. (1978). Measurement of the complexity and diversity of poly(adenylic acid) containing messenger RNA from rat liver. *Biochemistry* **17**, 462–467.

Schrier, B. K., Zubairi, M. Y., Lettendre, C. H., and Grouse, L. D. (1978). Bromodeoxyuridine effects on the RNA sequence complexity and phenotype in a neuroblastoma clone. *Differentiation (Berlin)* **12**, 23–30.

Shaskan, E. G. (1977). Brain regional spermidine and spermine levels in relationship to RNA and DNA in aging rat brain. *J. Neurochem.* **28**, 509–516.

Shepherd, G. W., and Flickinger, R. (1979). Post-transcriptional control of messenger RNA diversity in frog embryos. *Biochim. Biophys. Acta* **563**, 413–421.

Shepherd, G. W., and Nemer, M. (1980). Developmental shifts in the frequency distribution of polysomal mRNA and their posttranscriptional regulation in the sea urchin embryo. *Proc. Natl. Acad. Sci. U.S.A.* **77**, 4653–4656.

Strehler, B. L. (1977). "Time, Cells and Aging." 2nd ed. Academic Press, New York.

Supowit, S. C., and Rosen, J. M. (1980). Gene expression in normal and neoplastic mammary tissue. *Biochemistry* **19**, 3452–3460.

Surrey, S., and Nemer, M. (1976). Methylated blocked 5′terminal sequences in sea urchin embryo messenger RNA classes containing and lacking poly(A). *Cell* **9**, 589–595.

Tobin, A. J. (1979). Evaluating the contribution of posttranscriptional processing to differentiated gene expression. *Dev. Biol.* **68**, 47–58.

Uphouse, L. L., and Bonner, J. (1975). Preliminary evidence for the effect of environmental complexity on hybridization of rat brain RNA and rat unique DNA. *Dev. Psychobiol.* **8,** 171–178.

Van Ness, J., and Hahn, W. E. (1980). Sequence complexity of cDNA transcribed from a diverse mRNA population. *Nucleic Acids Res.* **8,** 4259–4269.

Van Ness, J., Maxwell, I. H., and Hahn, W. E. (1979). Complex population of nonadenylated messenger RNA in mouse brain. *Cell* **18,** 1341–1349.

Vaughn, W. J., and Calvin, M. (1977). Electrophoretic analysis of brain proteins from young adult and aged mice. *Gerontology* (Basel) **23,** 110–126.

Wetmur, J. G. (1976). Hybridization and renaturation kinetics of nucleic acids. *Annu. Rev. Biophys. Bioengin.* **5,** 337–361.

Williams, J. G., and Penman, S. (1975). The messenger RNA sequences in growing and resting mouse fibroblasts. *Cell* **6,** 197–206.

Williams, J. G., Hoffman, R., and Penman, S. (1977). The extensive homology between mRNA sequences of normal and SV40-transformed human fibroblasts. *Cell* **11,** 901–907.

Wilson, D. E., Hall, M. E., and Stone, G. C. (1978). Test of some aging hypothesis using two-dimensional protein mapping. *Gerontology (Basel)* **24,** 426–433.

Wold, B. J., Klein, W. H., Hough-Evans, B. R., Britten, R. J., and Davidson, E. H. (1978). Sea urchin embryo mRNA sequences in the nuclear RNA of adult tissues. *Cell* **14,** 941–950.

Young, B. D., Birnie, G. D., and Paul, J. (1976). Complexity and specificity of polysomal poly(A+)RNA in mouse tissues. *Biochemistry* **15,** 2823–2829.

Zagon, I. S., and McLaughlin, P. J. (1979). Morphological identification and biochemical characterization of isolated brain cell nuclei from the developing rat cerebellum. *Brain Res.* **170,** 443–457.

Zs-Nagy, V., Bertoni-Freddari, C., Zs-Nagy, I., Pieri, C., and Giuli, C. (1977). Alterations in the numerical density of perichromatin granules in different tissues during aging and cell differentiation. *Gerontology (Basel)* **23,** 267–276.

MOLECULAR CHARACTERIZATION OF SYNAPSES OF THE CENTRAL NERVOUS SYSTEM

James W. Gurd

I. INTRODUCTION

Synapses of the central nervous system (CNS) consist of pre- and postsynaptic membranes, pre- and postsynaptic membrane specializations of varying degrees

Molecular Approaches to Neurobiology

of complexity, and material within the synaptic cleft. Until recently knowledge concerning the molecular composition of the synapse has been derived almost exclusively from cytochemical studies (for recent reviews see Pappas and Waxman, 1972; Pfenninger, 1973, 1979; Gray, 1975b; Jones, 1977; Elvin, 1976). While providing a considerable amount of information relating to synaptic structure, the molecular identity and characteristics of the components responsible for the cytochemical staining properties remain largely unknown. More recently immunocytochemical procedures have permitted the localization of several identified components to the synaptic structure (see, for example, Wood *et al.*, 1980; Matus, 1975; Toh *et al.*, 1976). The ultimate goal of the neurochemist, however, has remained the ability to dissect the synapse into its component molecules in order that these may be chemically and biochemically characterized and their interactions within the synaptic structure elucidated.

The past five years have witnessed the first steps towards the achievement of this goal. The development in several laboratories of procedures for the isolation of subcellular fractions enriched in subsynaptic structures, including synaptic junctional complexes (deRobertis *et al.*, 1967; Davis and Bloom, 1973; Cotman and Taylor, 1972; Thérien and Mushynski, 1976), postsynaptic densities (Cotman *et al.*, 1974; Cohen *et al.*, 1977), and postsynaptic membranes (deBlas and Mahler, 1978), has made possible for the first time the direct biochemical analysis of these synaptic organelles. Isolated synaptic junctional complexes are characterized, and indeed identified by, the presence of a prominent postsynaptic thickening and are derived primarily, if not exclusively from asymmetric, type I excitatory synapses (Cotman and Taylor, 1972; Davis and Bloom, 1973). The molecular characteristics of the synapse described in the following article therefore apply to this type of synaptic structure. Symmetrical synapses (Gray type II) have generally not been identified in isolated fractions either because they do not survive the isolation procedures or because of the absence of a characteristic morphologic marker such as the postsynaptic density (see Matus and Walters, 1976).

The present chapter reviews the current state of information relating specifically to the molecular composition of synaptic junctions and postsynaptic densities. The discussion has been limited to results obtained with isolated SJC and PSD* fractions and no attempt has been made to survey the vast literature relating to the composition of synaptic plasma membranes which has been frequently reviewed (Mahler, 1977, 1979; Morgan and Gombos, 1976; Barondes, 1974; Kornguth, 1974; Mahler *et al.*, 1975).

---

*Abbreviations: SJC, synaptic junctional complexes; PSD, postsynaptic density; SPM, synaptic plasma membranes; PSM, postsynaptic membranes.

II. ISOLATION AND CHARACTERIZATION OF SYNAPTIC JUNCTIONAL COMPLEXES AND POSTSYNAPTIC DENSITIES

A. Isolation Procedures

1. Synaptic Junctional Complexes

The isolation of synaptic junctional complexes and postsynaptic densities depends on the selective resistance of these nerve terminal organelles to detergent solubilization. deRobertis *et al.* (1967) and Fiszer and deRobertis (1967) initially observed that the insoluble residue obtained following the extraction of synaptic membranes with the neutral detergent Triton X-100 was enriched in synaptic junctional complexes. This finding was subsequently confirmed and extended by Davis and Bloom (1970, 1973) and by Cotman's laboratory (Cotman *et al.*, 1971; Cotman and Taylor, 1972). Optimal conditions for the preservation of synaptic junctions include the presence of Ca^{2+} during the preparation of the parent membrane fraction or during the detergent extraction step, and the use of a Triton X-100:protein ratio of 1 to 2:1 (wt/wt). Following detergent extraction synaptic junctions are separated from other insoluble materials by sedimentation on a discontinuous sucrose gradient (Cotman and Taylor, 1972; Davis and Bloom, 1973). The isolated synaptic junction fraction consists predominantly of postsynaptic structures (PSD plus the associated PSM) as well as some intact junctions and free membranes. The yield is between 0.1 and 0.2 mg of protein per gram wet weight of brain. The isolation procedure of Cotman and Taylor (1972) has been used for the isolation of synaptic junctions from adult and neonatal rat brain (Cotman and Taylor, 1972; deSilva *et al.*, 1979), from human brain (Rostas *et al.*, 1979), and from several regions of bovine brain (Rostas *et al.*, 1979) and, with minor modifications, from rabbit brain (Freedman *et al.*, 1980) and day-old chick brains (Webster and Klingman, 1979).

An alternate approach to the isolation of junctional complexes based on the homogenization of synaptic membranes in a biphasic system consisting of Freon 113 and an aqueous phase containing .02% Triton X-100 was developed by Thérien and Mushynski (1976). Although a greater percentage of intact junctions (pre- and postsynaptic membranes with associated dense projections and densities) are obtained using this procedure, the yield is an order of magnitude lower (0.012 mg/gm brain) than that obtained using Triton X-100 alone.

2. Postsynaptic Densities

Extraction of synaptic membranes with anionic detergents or higher detergent/protein ratios than those used for the preparation of SJC's effectively solubilizes all of the membrane bilayer leaving an insoluble residue which is highly enriched in postsynaptic densities. Detergents used for the preparation of PSD's include 3.9% sodium lauryl sarcosinate (Cotman *et al.*, 1974), 10% deoxycholate (Walters and Matus, 1975c), 3% Triton X-100 (Matus and Taff-Jones, 1978), 3% deoxycholate (Matus and Taff-Jones, 1978), and 0.5% Triton X-100 (Cohen *et al.*, 1977). Each of these procedures yields a protein enriched, insoluble residue that sediments through 1.2 to 1.5 *M* sucrose and which morphologically and cytochemically resembles the postsynaptic density observed in intact tissue sections. The yield as well as the morphology and composition of PSD fractions varies with the particular isolation procedure employed.

3. Additional Isolation Procedures

The biochemical analysis of synaptic structures has been primarily carried out using fractions obtained by the procedures described in Sections II,A,1 and 2 above. In addition to these methods a novel procedure for the isolation of synaptic junctions based on affinity chromatography on convanavalin A sepharose was described (Bittiger, 1976) but further biochemical characterization of this fraction has not been reported. More recently a preliminary report of an isolation procedure for the postsynaptic apparatus which does not entail the use of detergents was given (Ratner and Mahler, 1979). This procedure, which involves extraction with salt and EGTA followed by sonication of the extracted membranes and sucrose gradient centrifugation, represents an important methodological advance in so far as the effects of detergent on the composition, enzymatic activity and structural integrity of isolated SJC and PSD fractions are unknown. A postsynaptic membrane fraction, lacking postsynaptic densities and enriched in receptors for a variety of neurotransmitters, has also been described by Mahler's laboratory (deBlas and Mahler, 1978). Finally, although a specific isolation procedure for the presynaptic membrane has not been developed it may be noted that the synaptic membrane fraction previously described by Gurd *et al.* (1974) is derived from a synaptosome fraction which has been stripped of associated postsynaptic structures and may therefore be primarily derived from the presynaptic nerve terminal. A summary of isolation procedures for various sub-synaptic fractions is presented in Table I.

B. Ultrastructure of Isolated Synaptic Junctions and Postsynaptic Densities

In the absence of appropriate enzymatic or chemical markers, estimates of the purity of isolated synaptic junctional and postsynaptic density fractions have relied almost exclusively on morphologic characterization and the ability of isolated fractions to react with histochemical stains such as ethanolic phosphotungstic acid (Bloom and Aghajanian, 1966, 1968) and bismuth iodide-uranyl lead (Pfenninger, 1971a). Although some loss of staining specificity may

Table I. Summary of Isolation Procedures for Synaptic Junctional Complexes and Postsynaptic Densities

Fraction	Detergent	Yield[a]	Purity[b] (%)	Reference
SJC	0.2% Triton X-100	0.15–0.20	50	Davis and Bloom (1973)
SJC	0.5% Triton X-100	0.1–0.15	65[c]	Cotman and Taylor (1972)
SJC	0.5% Triton X-100/Freon 113	0.012	65[d]	Thérien and Mushynski (1976)
SJC[e]	0.2% Triton	0.24		Webster and Klingman (1979)
PSD	3% sarcosyl	0.125	73–93	Cotman et al. (1974)
PSD[f]	0.5% Triton X-100	0.1–0.15	>90	Cohen et al. (1977)
PSD	3% Triton X-100	0.68[g]	Not reported	Matus and Taff-Jones (1978)
PS lattice	3% deoxycholate	0.17	Not reported	Matus and Taff-Jones (1978)
SJC[h]	No detergent used	0.075 to 0.15	Not reported[i]	Bittiger (1976)
PS apparatus[j]	No detergent used	Not reported	Not reported[k]	Ratner and Mahler (1979)
PSM[l]	No detergent used	0.013	Not reported	deBlas and Mahler (1978)

[a] mg protein/gm wet weight of brain.
[b] Purity estimates on the basis of morphologic analysis of fractions.
[c] 10% of structures have pre- and postsynaptic membranes attached; 55% of structures consist of PSM and associated PSD.
[d] 62% of structures have recognizable pre- and postsynaptic structures attached to each other.
[e] Isolated from 1-day-old chick brain by a modification of the procedure of Cotman and Taylor (1972).
[f] Isolated from dog forebrain.
[g] Calculated from Matus and Taff-Jones (1978).
[h] Isolated by affinity chromatography on concanavalin A Sepharose.
[i] Enriched 4- and 7-fold, respectively, in pilocarpine and GABA binding sites.
[j] Isolated by salt and EGTA washes of synaptic membranes followed by sonication.
[k] Mitochondrial contamination estimated as 13%.
[l] Isolated by gradient centrifugation from the microsomal fraction.

occur following detergent extraction procedures (see Davis and Bloom, 1973) morphologic analyses have been interpreted as demonstrating a high enrichement of either SJC or PSD in the isolated fractions.

1. Synaptic Junctional Complexes

Synaptic junctional complexes isolated by the procedures described above resemble the asymmetric type I synapses described by Gray (1959) and are therefore most probably derived primarily from excitatory axodendritic synapses. The SJC fraction obtained following Triton extraction of membranes (Cotman and Taylor, 1972; Davis and Bloom, 1973) consists predominantly of postsynaptic membranes with associated postsynaptic densities (55% of structures) with fewer (10%) intact junctions containing presynaptic membranes and cleft material. Most of the presynaptic dense projections are lost during the isolation procedure (Matus *et al.*, 1975a; Cotman and Taylor, 1972). Estimates of purity based on morphologic analysis suggest that 50 to 65% of the structures are clearly derived from the synaptic junction. The external surface of the postsynaptic membrane is characterized by a series of knobs which extend 70 to 100 Å out from the membrane. These knobs may correspond to the particles observed on the PSM by freeze fracture techniques (Sandri *et al.*, 1972; Landis and Reese, 1974) and which Matus *et al.* (1975a) suggested may represent elements of the PSD which span the membrane. Structures that retain the presynaptic membrane in association with the postsynaptic apparatus exhibit fibers traversing the cleft region (Cotman and Taylor, 1972).

In contrast to the junctional fraction prepared using Triton X-100, the Triton X-100/Freon 113 procedure (Thérien and Mushynski, 1976) yields a fraction in which 62% of the structures contain both pre- and postsynaptic specializations. In these synaptic complexes the presynaptic specializations remain connected by a fibrous network to the postsynaptic structure even when the presynaptic membrane has been largely dissolved away, suggesting that they may be directly linked with the PSM and/or the PSD. The relationship between these transsynaptic fibers and the cleft substance present in isolated synaptic junctions (Cotman and Taylor, 1972) or synaptic membrane fractions (Matus *et al.*, 1975a) which have lost the presynaptic dense projections is not known, but they may be identical to the fibers that extend from the postsynaptic membrane, possibly linking it to the presynaptic structure as previously described by Kornguth (1974).

2. Postsynaptic Densities

Postsynaptic density fractions are generally of a higher degree of purity than SJC fractions and estimates of purity may be as high as 90% (Cohen *et al.*, 1977; Cotman *et al.*, 1974). Postsynaptic densities isolated following extraction of membranes with Triton X-100 (3% and 0.5%) or low concentrations (1.2%) of deoxycholate consist of arrays of 10 to 20 nm electron dense particles (Cohen *et al.*, 1977; Matus and Taff-Jones, 1978; Matus and Walters, 1975) which resemble the particles seen in unextracted synaptic membrane preparations and *in situ* (Matus *et al.*, 1975a; Sandri *et al.*, 1972; Gray, 1975b). Negative staining of the densities reveals the presence of filaments arising from and occasionally connecting individual densities (Cohen *et al.*, 1977). Extraction of synaptic membranes with 3% deoxycholate on the other hand yields a less electron dense preparation interpreted by Matus and Taff-Jones (1978) as consisting of a lattice-like array of 5 nm fibers enclosing ring-shaped spaces which, when filled with protein, correspond to the particles seen in the Triton preparations. The morphology of the postsynaptic "lattice" described by Matus and Taff-Jones, however, resembles in several respects the "loosened" PSD structures obtained following various chemical treatments of PSD's (Blomberg *et al.*, 1977). The latter authors conclude on the basis of an extensive analysis of the morphology and protein content of PSDs obtained following several different extraction procedures that the electron dense "particles" consist of cross-sectional views of 10–13 nm filaments and that the PSD consists of a tightly packed array of these filaments, perhaps held together by actin filaments (Blomberg *et al.*, 1977). Resolution of these views awaits additional characterization of the composition and structural organization of the PSD.

III. PROTEIN COMPOSITION OF ISOLATED SYNAPTIC JUNCTIONS AND POSTSYNAPTIC DENSITIES

A. General

Staining of synaptic structures by ethanolic phosphotungstic acid (Bloom and Aghajanian, 1968) and bismuth iodide-uranyl-lead (Pfenninger, 1971a), both of which stain primarily proteins, indicated the proteinaceous nature of synaptic structures. The selectivity of these stains for the synaptic region further suggested a uniqueness with respect to the composition of synaptic proteins and it was proposed that they may be highly enriched in basic amino acids (Bloom and Aghajanian, 1968; Pfenninger, 1971b). The high protein content of isolated postsynaptic densities (Banker *et al.*, 1974; Cohen *et al.*, 1977) is thus in general agreement with the cytochemical staining results although nonprotein components may be lost during the isolation procedure. Although isolated junctions and densities retain their selective ability to react with synaptic specific stains (Cotman *et al.*, 1974; Cohen *et al.*, 1977) they are not enriched in basic amino acids as suggested earlier and their amino acid composition does not differ significantly from that of synaptic membranes (Churchill *et al.*, 1976; Banker *et al.*, 1974).

The protein composition of synaptic junctions is heterogeneous. In spite of the extensive solubilization of membrane proteins (80% or more of the membrane protein is solubilized during extraction with Triton X-100) the protein composition of SJC's is qualitatively similar to that of the parent membrane fraction (Fig. 1; Kelly and Cotman, 1977; Gurd, 1980; Goodrum and Tanaka, 1978) although certain proteins and glycoproteins are relatively enriched in synaptic junctions (see below). No major differences in the polypeptide or glycoprotein composition of SJC from various species or brain regions were revealed by gel electrophoretic analysis (Rostas *et al.*, 1979).

The protein composition of postsynaptic densities is generally less complex than that of the synaptic junctional fraction but is dependent on the method of isolation. Thus the procedure of Cotman *et al.*, (1974), using 3% sarcosyl yields PSD's having a highly simplified complement of proteins with 1 major polypeptide (MW≃51,000) accounting for 50% or more of the total protein (Banker *et al.*, 1974; and Fig. 1). In contrast the protein composition of PSD's prepared using the milder extraction conditions of Cohen *et al.* (1977) is similar in overall complexity to that of the SJC and SPM fraction although an enrichment in the 51,000 component is also apparent (Fig. 1). The differences in protein content of PSD's prepared by these two procedures contrasts with the generally similar morphology of the two fractions and serves to emphasize the difficulty of relating composition to structure when using detergent-extracted subcellular components. The relatively simplified protein composition of the sarcosyl PSD's may reflect a "minimum" structure from which less tightly bound proteins have been removed and in this respect resembles the postsynaptic lattice of Matus and Taff-Jones (1978). In contrast, PSD's prepared according to the milder procedure of Cohen *et al.* (1977) may still retain a number of relatively loosely associated PSD proteins as well as membrane proteins which remain associated with the proteinaceous network of the density in a manner analogous to the retention of membrane proteins by the Triton extracted cytoskeletal network of cells (Yu *et al.*, 1973; Ben Ze'ev *et al.*, 1979).

B. Fibrous Proteins

The presence of filamentous proteins in the region of the synapse has stimulated considerable speculation concerning the possible role(s) of these proteins in synaptic function and several theories involving chemomechanical modulation of pre- and postsynaptic events during chemical transmission have been formulated (Bowler and Duncan, 1966; Duncan, 1965; LeBeaux and Willemot, 1975; Berl *et al.*, 1973; Blomberg *et al.*, 1977).

Studies of isolated synaptic fractions have demonstrated the association of tubulin, actin, and neurofilament-like proteins with synaptic organelles and have also identified an additional protein, the "PSD" protein, which may be unique to the postsynaptic apparatus.

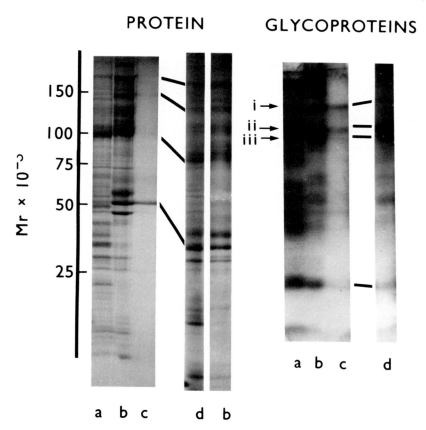

Fig. 1. Protein and glycoprotein composition of synaptic membranes, synaptic junctions and postsynaptic densities. Synaptic membranes (a) were isolated according to the procedure of deSilva *et al.* (1979). Synaptic junctions (b) were obtained using the method of Cotman and Taylor (1972). Postsynaptic densities were isolated using 3% sarcosyl (c) as described by Cotman *et al.* (1974) or with 0.5% Triton X-100 (d) according to Cohen *et al.* (1977). Samples were separated on 7–17% polyacrylamide gradient slab gels and stained with Coomassie blue for protein or reacted with [^{125}I]concanavalin A for the identification of glycoproteins. i, ii and iii indicate concanavalin A receptors with apparent molecular weights of 180,000, 130,000, and 110,000, respectively.

1. Actin

Morphologic studies have demonstrated the presence of 40–60 Å actin-like filaments in association with both the pre- and postsynaptic membranes and membrane specializations (LeBeaux and Willemot, 1975; Hanson and Hyden, 1974; Metuzals and Mushynski, 1974), and actin was located in the postsynaptic structure by immunocytochemical means (Toh *et al.*, 1976). Studies of isolated fractions have identified actin as a component of synaptic plasma membranes (Blitz and Fine, 1974; Berl *et al.*, 1973; Puszkin and Kochwa, 1974; Kornguth and Sunderland, 1975) and as one of the major proteins of synaptic junctional complexes and postsynaptic densities (Mushynski *et al.*, 1978; Kelly and Cotman, 1977; Cohen *et al.*, 1977; Blomberg *et al.*, 1977; Walters and Matus, 1975a; Matus and Taff-Jones, 1978). Synaptic junctional actin coelectrophoresed in two-dimensional gels with β- and γ-actin from nonmuscle tissues but not with α-actin (Kelly and Cotman, 1978). It did not contain the more acidic forms of actin described in synaptosomes (Marotta *et al.*, 1978) and which may in fact be artifacts of proteolysis (Lee *et al.*, 1979). In addition to actin, α-actinin (MW 100,000) and tropomyosin (MW 33,000 and 35,000) have been tentatively identified in PSD's on the basis of similar migration in one-dimensional gels. Myosin was not identified as a component of the PSD fraction in these studies (Blomberg *et al.*, 1977) but has recently been identified in rat and steer SJC fractions using antimyosin antisera (P. T. Kelly, personal communication).

2. Neurofilament Protein

The association of neurofilament like proteins with the synaptic junction and postsynaptic density is indicated by the reaction of these structures with antineurofilament antisera (Yen *et al.*, 1977; Berzins *et al.*, 1977). The reaction of the major 51,000 dalton PSD protein with the same antisera (Yen *et al.*, 1977) may reflect small amounts of neurofilament protein contaminating the major PSD protein band. Thus although small amounts of neurofilament may be associated with the PSD it is now apparent that it is not the major component as earlier suggested (Blomberg *et al.*, 1977).

3. Tubulin

Morphologic studies have recently succeeded in demonstrating microtubules in close association with synaptic vesicles and the presynaptic dense projections as well as underlying (and linked to?) the postsynaptic density (Gray, 1975a,b; Westrum and Gray, 1976; Bird, 1976). Consistent with these results tubulin, previously identified as a component of nerve terminal membranes (Feit *et al.*, 1971; Bhattacharyya and Wolf, 1975; Blitz and Fine, 1974; Mahler *et al.*, 1975), is also present as a prominent component of postsynaptic densities and synaptic-junctional complexes (Matus *et al.*, 1975a; Mushynski *et al.*, 1978; Kelly and Cotman, 1978; Matus and Taff-Jones, 1978; Feit *et al.*, 1977; Yen *et al.*, 1977). The earlier suggestion that tubulin was the major PSD protein (Walters and Matus, 1975b,c) appears to have resulted from the identification of tubulin in the PSD by immunocytochemical procedures (Matus *et al.*, 1975b) coupled with the incomplete electrophoretic resolution of proteins in the 50,000 and 55,000 dalton gel region resulting in peptide maps that were a composite of two or more proteins. Tubulin is more highly concentrated in synaptic junctions prepared according to Thérien and Mushynski (1976) in which it constitutes the major protein species (Mushynski *et al.*, 1978). The greater retention of presynaptic structures in this preparation suggests that tubulin may be preferentially localized in the presynaptic dense projections (Mushynski *et al.*, 1978).

4. "PSD Protein"

The major protein component of isolated PSD's and synaptic junctional complexes has a molecular weight of between 50,000 and 53,000 daltons (Cohen *et al.*, 1977; Banker *et al.*, 1974; Walters and Matus, 1975b; Thérien and Mushynski, 1976; Kelly and Cotman, 1978). Variously identified as tubulin (Matus and Walters, 1975b) and as neurofilament or neurofilament-like protein (Blomberg *et al.*, 1977; Berzins *et al.*, 1977) it is now apparent that the 51,000 dalton PSD protein is distinct from other known fibrous proteins, including actin, α- and β-tubulin, and neurofilament associated proteins (Kelly and Cotman, 1978), and may be unique to the postsynaptic density. Peptide mapping experiments indicated that the primary sequence of the PSD protein is similar in several different species and brain regions (Rostas *et al.*, 1979).

C. Glycoproteins

1. Histochemical Localization of Synaptic Carbohydrates

The potential structural diversity, and consequently informational content, of the oligosaccharide prosthetic groups of glycoproteins ideally suits these molecules for a variety of roles in synaptic function.

Histochemical procedures have demonstrated the pressence of high concentrations of carbohydrates in the region of the synaptic cleft (Rambourg and LeBlond, 1967; Pfenninger, 1973). More recently receptors for concanavalin A and *Ricinus communis* lectin were identified on postsynaptic membranes (Bittiger and Schnebli, 1974; Matus *et al.*, 1973; Cotman and Taylor, 1974; Kelly *et al.*, 1976) where they occur in close proximity to bristle-like (postsynaptic) membrane specializations which may represent structural units of the PSD which extend through the membrane (Matus and Walters, 1975). Lectin receptors are more highly concentrated in the junctional region than in the surrounding nonjunctional membranes. Lectin receptors overlying the postsynaptic density do not form clusters under the influence of added lectin as is the case for receptors in adjoining nonjunctional membranes (Cotman and Taylor, 1974; Matus and Walters, 1976; Kelly *et al.*, 1976). These observations suggest that a direct or indirect interaction of membrane components with the proteinaceous network of the underlying PSD may serve in restricting their ability to diffuse in the plane of the membrane (Kelly *et al.*, 1976; Cotman and Taylor, 1974; Gurd, 1977a). Lectin receptors are not detected on the cytoplasmic side of the postsynaptic membrane or in direct association with the PSD of asymmetrical type I synapses. These results are in general agreement with topographic studies which have indicated that synaptic glycoproteins are located on the external surface of synaptic membranes (Wang and Mahler, 1976; Chiu and Babitch, 1978) although the latter studies have considered primarily the presynaptic nerve terminal membrane. In contrast to their external localization on type I synapses, concanavalin A binding sites were observed on the cytoplasmic surface of the postsynaptic junctional membrane of type II (symmetric) synapses, suggesting a fundamental difference in the structural organization of these two broad classes of synapses (Matus and Walters, 1976).

2. Carbohydrate Composition of Synaptic Glycoproteins

The level of protein bound carbohydrates is similar in synaptic membranes, synaptic junctions, and postsynaptic densities although there are variations in the relative amounts of individual sugars associated with the different fractions (Churchill *et al.*, 1976; Webster and Klingman, 1980). Synaptic junctions and postsynaptic densities are relatively deficient in sialic acid while the latter fraction contains increased levels of mannose* (Churchill *et al.*, 1976; J. W. Gurd, unpublished observations). Of particular interest is the high level of glucose which is found to be associated with isolated SJC and PSD fractions (Churchill *et al.*, 1976). Although not commonly present as a component of glycoproteins, glucose has been consistently found in association with neuronal, and in particular, synaptic glycoproteins (Margolis *et al.*, 1975; Zanetta *et al.*, 1977). The particularly high levels of glucose found in synaptic junctions and postsynaptic densities may reflect the presence of glucose-containing glycoproteins in these fractions or the tight association of endogenous glucose polymers (Margolis *et al.*, 1976) with the synaptic apparatus. Some of the glucose may also be derived from tightly adhering sucrose originating in the isolation media (Churchill *et al.*, 1976).

3. Identification of Synaptic Glycoproteins

In accord with histochemical results demonstrating the presence of lectin receptors on the postsynaptic membrane, isolated synaptic junctions react with concanavalin A. Synaptic junctions and synaptic membranes react with similar amounts of concanavalin A (3.5 to 4.0 $\times$ 10^{-9} equivalents of concanavalin A bound/mg protein (Gurd, 1980). However, whereas the latter fraction contains

---

*In contrast to these results Margolis *et al.* (1975) found that the residue obtained following extraction of synaptic membranes with 0.2% Triton X-100 contained only 50% as much carbohydrate as unextracted membranes. The reasons for this apparent discrepancy in the results described above may be related to the small numbers of junctions present in the parent synaptic membrane fraction used by these authors (Gurd *et al.*, 1974) resulting in proportionally fewer synaptic structures in the triton residue.

10 to 12 concanavalin A reacting glycoproteins (Gurd, 1977b; Zanetta *et al.*, 1975) synaptic junctional concanavalin A receptor activity is associated primarily with 3 high molecular weight glycoproteins with apparent molecular weights of 180,000 (GP 180), 130,000 (GP 130), and 110,000 (GP 110) (Fig. 1; Gurd, 1977a,b; Kelly and Cotman, 1977). GP 180, GP 130, and GP 110 are also associated with isolated postsynaptic densities prepared using a variety of extraction conditions (Gurd, 1977a; Blomberg *et al.*, 1977), indicating that at least a portion of these glycoproteins is firmly anchored in the postsynaptic apparatus. Similar molecular weight classes of concanavalin A receptors are present in synaptic junctions and/or postsynaptic densities isolated from various brain regions (Rostas *et al.*, 1979) and from a variety of species, including rat (Gurd, 1977a; Kelly and Cotman, 1977), dog (Blomberg *et al.*, 1977), cow and human (Rostas *et al.*, 1979), and rabbit (Freedman *et al.*, 1980). GP 110, 130, and 180 are uniquely associated with the synaptic apparatus and are not detected in microsomes, axolemma, synaptic vesicles, myelin, or extrajunctional nerve-terminal membranes (Gurd, 1980). The reaction of concanavalin A with GP 110, 130, and 180 is sensitive to digestion with α-mannosidase and endoglycosidase H, indicating that the lectin reacts primarily, if not exclusively, with asparagine-linked, polymannose oligosaccharides. (Gurd, 1980; Gurd and Fu, 1981). The majority of concanavalin A receptor activity associated with these glycoproteins is accounted for by 2 classes of high-mannose oligosaccharides containing 5 and 9 mannose residues (Gurd and Fu, 1981). Synaptic junction and postsynaptic density glycoproteins also react with wheat germ agglutinin, *Ricinus communis* lectin, lentil lectin, and fucose binding protein (Gurd, 1977a, 1979, 1980) indicating the presence of a variety of oligosaccharide structures in addition to the polymannose chains responsible for concanavalin A binding. The wide regional and species distribution of these glycoproteins coupled with their unique localization in the synaptic apparatus strongly suggests a central role for them in synaptic function.

The antigen thy-1 is associated with synaptosomes and is also a component of isolated synaptic junctions (Acton *et al.*, 1978; Stohl and Gonatas, 1977). Thy-1 is a glycoprotein which has been purified from brain and lymphoid cells and its chemical properties described. Thy-1 from brain has an apparent molecular weight on polyacrylamide gels in the presence of sodium dodecyl sulfate of 25,000 and a corrected molecular weight as determined by sedimentation equilibrium of 17,500 (Barclay *et al.*, 1976). Although the amino acid composition of brain and thymus thy-1 are very similar, the oligosaccharide components of the two proteins are quite distinct, suggesting that its functional role in different tissues may be related to the carbohydrate prosthetic group(s) (Barclay *et al.*, 1976).

It may also be noted here that a number of receptor proteins, including the acetylcholine receptor (Salvaterra *et al.*, 1977), a glutamate binding protein

(Michaelis, 1975), and a catecholamine receptor (Lee, 1974), are also glycoproteins and are presumably located at the synaptic junction. However, with the possible exception of cholinergic receptor activity (deRobertis *et al.,* 1967) other glycoprotein receptors have not been identified in isolated junctional fractions. Table II provides a summary list of proteins that have been identified in isolated SJC and PSD fractions.

D. Identification of Synaptic Specific Proteins

The unique structural and functional properties of the synapse suggest that it may have a distinct molecular composition. Identification of such synaptic specific molecules has, however, proved to be elusive. Biochemical analyses of synaptic membrane fractions have failed to reveal any components unique to this neuronal membrane (see Morgan and Gombos, 1976). Although immunochemical approaches have demonstrated the enrichment of various antigens in synaptic membranes, the unambiguous identification of a synaptic membrane specific antigen has yet to be achieved (Herschman *et al.,* 1972; Jørgensen and Bock, 1974).

In contrast to the results obtained with synaptic membranes, analysis of isolated synaptic junction and postsynaptic density fractions have identified components which appear to be uniquely localized to these structures, namely, the "PSD" protein (Kelly and Cotman, 1978) and the concanavalin A receptors GP 180, GP 130, and GP 110 (Gurd, 1980). The tentative identification of these molecules as "synaptic specific" suffers from the difficulties inherent in subcellular fractionation studies. However, they provide the first examples of biochemically (as opposed to immunochemically) identified molecules whose location, and presumably function, appears to be specifically synaptic. Further biochemical and functional characterization of these components and confirmation of their subcellular distribution (e.g., through immunocytochemical procedures) will clarify their role in synaptic organization and activity.

Immunologic procedures have also indicated the presence of antigens specific to the postsynaptic apparatus (Matus, 1975; Orosz *et al.,* 1973, 1974). However, these studies have not been followed up by the molecular identification of the antigens in question and the nature of these apparently postsynaptic specific components remains unknown.

E. Stability and Structural Organization of the Synapse

Little is known concerning the specific molecular arrangements and interactions between the various protein components of the synapse. On the presynaptic side the dense projections, which are associated with tubulin and actin (Mushynski *et al.,* 1978), appear to be continuous with fibrous networks extend-

Table II. Identified Proteins of Synaptic Junctions and Postsynaptic Densities

Protein	Method(s) of identification[a]	Comments[a]
Tubulin (α and β bands)	Coelectrophoresis in two-dimensional gels[1,3] Immunocytochemistry[2] Peptide mapping[1,3]	May be concentrated in presynaptic specializations[3]
Actin	Coelectrophoresis in two-dimensional gels[1,3] Immunocytochemistry[4] Peptide mapping[1,3]	β and γ forms present, α form absent[1]
Neurofilament	Cross-reaction with anti-neurofilament antisera[5,6]	
Calmodulin	Coelectrophoresis in one-dimensional gels[7] Amino acid analysis[7] Immunocytochemistry[8]	Similar amino acid composition to total brain calmodulin, contains ε-N-trimethyl lysine
Proteins Ia and Ib	Solubility properties Phosphorylation with cAMP dependent protein kinase[9,10]	Substrate for cAMP stimulated protein kinase
Thy-1	Reaction with anti-thy-1 antisera[11]	
Myosin	Reaction with antimyosin sera and peptide mapping[12]	

[a] Superscript numbers refer to references: 1, Kelly and Cotman (1978); 2, Matus et al. (1975b); 3, Mushynski et al. (1978); 4, Toh et al. (1976); 5, Yen et al. (1977); 6, Berzins et al. (1977); 7, Grab et al. (1979); 8, Wood et al. (1980); 9, Ueda et al. (1979); 10, Kelly et al. (1979); 11, Acton et al. (1978); 12, P.T. Kelly, personal communication.

ing away from both surfaces of the membrane and in some instances providing a direct link with the postsynaptic apparatus (Thérien and Mushynski, 1976). A fibrous system connecting pre- and postsynaptic membranes is also characteristic of intact junctions present in the SJC fraction prepared according to Cotman and Taylor (1972). Isolated junctions are highly stable, withstanding dissociation by a variety of conditions including hypertonicity, mild proteolysis, extremes of pH, 1 M urea, EDTA, and EGTA (Cotman and Taylor, 1972; Matus and Walters, 1976). Although these results suggest that the synaptic contact is very strong the following should also be noted: (1) During the isolation procedure of Cotman and Taylor (1972) the majority of synapses are disrupted so that intact junctions represent less than 15% of the total structures in the isolated fraction. (2) The procedure of Thérien and Mushynski (1976), which yields a much greater percentage of intact junctions, also results in an almost 10-fold lower yield and may therefore select for strongly adhering synapses, the stability of which may be related to the presence of connecting fibers crossing the cleft region. (3) Matus and Walters (1976) observed that incubation in Krebs-Ringer solution at 37°C for 30 min was sufficient to detach many type I (but no type II) synapses in a synaptosome preparation. (4) Highly purified synaptic membrane fractions are associated with few if any intact synaptic complexes (Morgan *et al.*, 1971; Gurd *et al.*, 1974), suggesting that the physical stresses of the isolation procedures are sufficient to dissociate the majority of contacts. Taken together these results indicate that synaptic contacts which survive the isolation procedures currently employed may be atypically strongly adhering and not representative of the majority of synapses. The recent report that trypsin dissociates greater than 90% of the type I junctions present in a synaptosome fraction (Gordon-Weeks *et al.*, 1981), in contrast to isolated junctional structures which were resistent to protelytic dissection (Cotman and Taylor, 1972), is consistent with this interpretation.

The postsynaptic apparatus consists of a stable association between the postsynaptic density and the overlying membrane. Direct interaction between these structures is suggested by (1) the apparent lack of fluidity of the membrane immediately overlying the PSD (Kelly *et al.*, 1976; Matus and Walters, 1976); (2) the selective resistance of the postsynaptic membrane to solubilization by low concentrations of detergent as also reported for neuromuscular junctions (Rash and Ellisman, 1974); (3) the resistance of glycoproteins to detergent extraction under conditions that dissolve the PSM, suggesting that they are firmly anchored in the PSD and may span the PSM (Bittiger and Schnebli, 1974; Gurd, 1977a); and (4) the occurrence of many proteins common to both the PSD and synaptic membrane fractions (Blomberg *et al.*, 1977), suggesting that some membrane proteins remain associated with the PSD following detergent extraction. The interaction between the PSD and PSM may serve to segregate the latter from the surrounding fluid bilayer of nonjunctional membranes, thus preserving local concentrations of components required for synaptic transmission, such as receptor molecules.

The PSD itself consists of a basic network of fibrous proteins (actin, tubulin, "PSD" protein) which serves as a framework for the interaction of a wide variety of other proteins, originating in the membrane, the cytoplasm, and the density. The major component of this structure is the 51,000 dalton "PSD" protein. The specific nature of the interactions between the PSD protein, other fibrous proteins and other proteins associated with the density remain unknown although an involvement of disulfide bridges (Kelly and Cotman, 1976; Blomberg et al., 1977) and Ca^{2+} ions (Blomberg et al., 1977) has been suggested.

IV. FUNCTIONAL PROPERTIES OF SYNAPTIC JUNCTIONS AND POSTSYNAPTIC DENSITIES

A. Phosphorylation of Synaptic Proteins

The stimulatory effect of cAMP on the phosphorylation of synaptic membrane proteins by intrinsic protein kinase has led to the suggestion that cyclic nucleotides, and in particular cAMP, may play a role as a second messenger during neurotransmission. This hypothesis and the results supporting it have been reviewed in several recent publications (Bloom, 1975; Iversen, 1975; Von Hungen and Roberts, 1974; Daly, 1977; Beam and Greengard, 1976; Williams and Rodnight, 1977; Nathanson, 1977; Greengard, 1976, 1978; Berridge, 1979; Weller, 1980). More recent work has led to the identification and characterization of phosphoproteins and related enzymes associated with isolated synaptic junctions and postsynaptic densities.

1. The Phosphorylation of SJC and PSD Proteins

The phosphorylation of SJC and PSD proteins by intrinsic synaptic protein kinase has been demonstrated either by incubating the parent SPM fraction with $[\gamma^{-32}P]ATP$ and then isolating the appropriate subsynaptic component or by direct incubation of the isolated fractions with $[^{32}P]ATP$. Under either condition as many as 15 proteins may be labeled in synaptic junctions (deBlas et al., 1979; Kelly et al., 1979) or postsynaptic densities prepared with Triton X-100 (Ueda et al., 1979; deBlas et al., 1979; Ng and Matus, 1979). Postsynaptic densities prepared using 3% sarcosyl or 3% deoxycholate incorporate less ^{32}P into fewer protein species commensurate with the simplified protein composition of these fractions and the possible denaturing effects of the more rigorous extraction conditions (Ng and Matus, 1979; Kelly et al., 1979). The intrinsic enzyme preferentially phosphorylates endogenous synaptic junctional or postsynaptic density proteins as opposed to exogenous substrates or extrajunctional membrane proteins (Weller and Morgan, 1976a; Ng and Matus, 1979; Kelly *et al.*, 1979). This suggests that the specificity of the enzyme is such that the junctional and/or density proteins are preferred substrates or that constraints placed upon the enzyme, for example, by the low fluidity of the postsynaptic membrane or by direct interaction with the protein network of the PSD, limit its access to potential nonjunctional substrates.

2. The Role of cAMP

Protein kinase activity associated with isolated SJC's and PSD's is stimulated by cAMP (Carlin *et al.*, 1980; Kelly *et al.*, 1979; Ng and Matus, 1979; deBlas *et al.*, 1979; Ueda *et al.*, 1979; Weller and Morgan, 1976b; Dunkley *et al.*, 1976). The overall stimulation of ^{32}P incorporation into junctional proteins produced by cAMP varies from 10% (Weller and Morgan, 1976b) to 40% (Kelly *et al.*, 1979), with the phosphorylation of individual proteins increased by as much as 100% (Kelly *et al.*, 1979; Ng and Matus, 1979).

The stimulation of protein kinase activity by cAMP indicates that the association between regulatory and catalytic subunits at least partially survives the preparative procedure.* Of the two types of protein kinases† associated with synaptic membrane fractions (Uno *et al.*, 1977; Walter *et al.*, 1977, 1978), the type IIm enzyme predominates (Walter *et al.*, 1977, 1978). In general agreement with this, Ueda *et al.* (1979) estimated, on the basis of labeling with 8-N$_3$ [^{32}P]cAMP (Walter and Greengard, 1978), that the type IIm regulatory subunit was present in a nearly 2-fold excess over the type Im form in isolated PSD's. Kelly *et al.* (1979) on the other hand found 32% more of the type Im regulatory subunit in synaptic junctions using a similar approach. Whether or not these apparent differences in the distribution of the two types of protein kinases are due to the different procedures employed for the isolation of the PSD and SJC fractions or reflect a real difference in the subsynaptic distribution of the two enzyme forms remains to be clarified.

Termination of the effects produced by cAMP and protein kinase requires the action of cyclic nucleotide phosphodiesterase and protein phosphatase. Cytochemical procedures have localized the former emzyme to the region of the postsynaptic density (Florendo *et al.*, 1971) and it is active in isolated PSD

*The failure of cAMP to stimulate protein kinase activity of sarcosyl-PSD's (Kelly *et al.*, 1979) and Triton/Freon SJC's (Thérien and Mushynski, 1979a) may be due to dissociation of the regulatory and catalytic subunits during the more rigorous preparative procedures used in these instances. In the latter case the 2.6-fold enrichment in cAMP binding sites relative to SPM indicates concentration of the regulatory subunit in the SJC fraction (Thérien and Mushynski, 1979a).

†The regulatory subunits of the type Im and IIm enzymes have molecular weights of 47,000 and 52,000–55,000, respectively (Walter *et al.*, 1978). For further discussion of the different types of protein kinases the reader is referred to Walter and Greengard (1978) and Walter *et al.* (1978).

(Cotman *et al.*, 1974) and SJC (Thérien and Mushynski, 1979b) fractions. The preferential hydrolysis of cGMP rather than cAMP by postsynaptic phosphodiesterase (Ariano and Appleman, 1979) is consistent with the proposed involvement of the former nucleotide in modulating synaptic function (Berridge, 1979, and see below). Protein phosphatase was demonstrated in isoalted PSD's where it was enriched 2 to 3 times relative to the SPM (Ng and Matus, 1979) but was not detected in SJC's (Thérien and Mushynski, 1979a).

3. The Role of Ca^{2+}

The poshophorylation of synaptic membrane and PSD proteins is also stimulated by Ca^{2+} in the presence of exogenous calmodulin (Ca^{2+}-dependent regulator protein) (Schulman and Greengard, 1978a,b; deBlas *et al.*, 1979; Grab *et al.*, 1979; deLorenzo, 1980). The recent demonstration that calmodulin and a calmodulin-binding protein are associated with the postsynaptic density (Grab *et al.*, 1979; Wood *et al.*, 1980) indicates that fluctuations in postsynaptic Ca^{2+} levels in response to receptor binding may directly alter the phosphorylation of postsynaptic membrane and PSD proteins by its effect on a Ca^{2+}-sensitive protein kinase. Different effects of Ca^{2+} plus calmodulin on the phosphorylation of proteins associated with PSD's from cerebrum and cerebellum were reported (Carlin *et al.*, 1980). An alternate role for Ca^{2+} in the regulation of protein phosphorylation is suggested by the activating action of calmodulin on Ca^{2+}-dependent cyclic nucleotide phosphodiesterase (Cheung, 1970; Lin *et al.*, 1975; Teshina and Kakiuchi, 1974). Although a direct effect of calmodulin on the phosphodiesterase present in isolated PSD's has not been shown, a Ca^{2+}-sensitive phosphodiesterase activity is present in isolated synaptic junctions where it is located primarily in the postsynaptic density (Thérien and Mushynski, 1979b).

The proposal that Ca^{2+}-dependent brain phosphodiesterase preferentially hydrolyzes cGMP rather than cAMP (Kakiuchi, 1975) coupled with the preferential hydroylsis of cGMP by postsynaptic phosphodiesterase (Ariano and Appleman, 1979) and the activating effect of calmodulin on adenyl cyclase (Cheung *et al.*, 1975) indicates that changes in calcium levels may result in reciprocal changes in the cGMP and cAMP concentrations in the postsynaptic cell. cGMP and cAMP frequently have opposing effects on cellular metabolism and these proposed effects of calcium on postsynaptic cyclic nucleotide levels are consistent with the suggestion that the relative amounts of cGMP and cAMP determine the cellular response to applied stimuli. [For further discussion of the effects of Ca^{2+} and cyclic nucleotides on neuronal function see Berridge (1979).]

B. Modification of Synaptic Glycoproteins

Modification of synaptic glycoproteins through the local addition and/or removal of sugars has been a controversial issue (Barondes, 1974; Den *et al.*,

1975; Reith *et al.*, 1972; Raghupathy *et al.*, 1972). Although it is now clear that the majority, if not all, of synaptic glycoproteins is synthesized in the cell body and subsequently incorporated into synaptic membranes (Mahler, 1979; Barondes, 1974), the potential for some local alteration of oligosaccharide structures remains.

Studies of synaptic glycosyltransferases have in general been difficult to interpret because of potential contamination of the isolated SPM fractions by cytoplasmic membrane components (Raghupathy *et al.*, 1972). However, solubilization of these membranes by the detergents used for the preparation of SJC and PSD fractions considerably reduces this problem. Thus the recent finding of intrinsic galactosyltransferase activity in isolated synaptic junctions (Goodrum *et al.*, 1979) again raises the possibility of the local addition of sugar residues at the synaptic junction.

The removal of sialyl residues by intrinsic synaptic sialidase has been suggested as a possible regulatory mechanism of synaptic function (Rahman *et al.*, 1976; Schengrund and Nelson, 1974). Ganglioside and glycoprotein sialidase is a component of synaptic membranes (Schengrund and Rosenberg, 1970; Tettamanti *et al.*, 1972; Yohe and Rosenberg, 1977; Cruz and Gurd, 1978) and has recently been demonstrated in synaptic junctions and postsynaptic densities (T. Cruz and J. W. Gurd, unpublished observations). Short-term modulation of synaptic activity by variation in sialic acid concentration requires the activity of sialyltransferase in order to add back sialyl residues removed by the sialidase. The former enzyme was recently shown to be present in a purified synaptic membrane fraction (Preti *et al.*, 1980; Den *et al.*, 1975; but see Ng and Dain, 1980).

Recent results have demonstrated that intrinsic SJC protein kinase phosphorylates the three major junctional glycoproteins providing a further mechanism for the modification of synaptic glycoproteins (Gurd *et al.*, 1981; and Fig. 2).

Finally, it should be noted that the oligosaccharide moieties of synaptic glycoproteins and gangliosides are oriented toward the extracellular space. Modification of these structures *in situ* therefore requires the action of ectoenzymes present on the outer surface of the pre- or postsynaptic membrane, a localization that must be established before the functional significance of these enzyme activities can be properly assessed.

Table III summarizes functional activities that have been identified in PSD and SJC fractions.

V. DEVELOPMENT OF THE SYNAPSE

The formation of a synapse involves several steps, including (1) the growth and extension of the presynaptic axon; (2) recognition and formation of the initial contact between pre- and postsynaptic cells; (3) development of pre- and post-

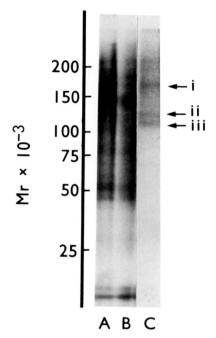

Fig. 2. Phosphorylation of synaptic junctional glycoproteins. Synaptic junctions were incubated in the presence of [³²P]ATP for 3 min, solublized in sodium dodecyl sulfate, and fractionated by affinity chromatography on concanavalin A Sepharose. Samples of the total junctions (A), the concanavalin A negative (B), and concanavalin A positive (C) fractions were separated on 7–17% polyacrylamide gradient slab gels and radioautograms prepared. i, ii, and iii indicate phosphorylated concanavalin A receptor proteins with apparent molecular weights of 180,000, 130,000, and 110,000, respectively.

synaptic specializations; and (4) stabilization and functional and structural maturation of the synapse (for reviews of synaptogenesis see Rees, 1978; Cotman and Banker, 1974; Pfenninger and Rees, 1976; Bloom, 1972; Pfenninger, 1979).

Prior to formation of the initial contact, putative synaptic junctions cannot be isolated since present isolation procedures depend on the preexistence of a synaptic contact that will withstand the physical stresses of homogenization and isolation of synaptic membranes. However, the development of the growth cone and the events accompanying the initial encounter between the growth cone and its target cells have been described morphologically both *in vivo* and in tissue culture systems (the reader is referred to the above reviews for details of these studies).

The development of synaptic membrane specializations during the latter stages of synaptogenesis have been described in a number of morphologic studies

Table III. Enzyme and Receptor Activities Associated with Synaptic Junction and Postsynaptic Density Fractions

| Enzyme | Fraction[a] | | Comments[a] |
	SJC	PSD	
Protein kinase	1, 2	3, 4, 9	
cAMP-stimulated protein kinases	1, 2, 5, 6	3, 4	PSD and SJC are enriched in type IIm and type Im regulatory subunits, respectively. Sarcosyl PSD PK was not stimulated by cAMP[2]
Ca^{2+}-stimulated protein kinase	—	7, 8	Ca^{2+} stimulation requires presence of exogenous calmodulin
Protein phosphatase	—	4	Protein phosphatase was not detected in SJC[9]
Cyclic nucleotide phospho-diesterase	10	11	SJC enzyme shows Ca^{2+} stimulation
Galactosyltransferase	12	—	Most active against endogenous substrates
Glycoprotein sialidase	13	—	
γ-Aminobutyric acid receptor	14, 15, 18	—	
Cholinergic receptors	16	—	
Kainic binding sites	17	—	

[a] Numbers refer to references: 1, deBlas *et al.* (1979); 2, Kelly *et al.* (1979); 3, Ueda *et al.* (1979); 4, Ng and Matus (1979); 5, Weller and Morgan (1976b); 6, Dunkley *et al.* (1976); 7, Grab *et al.* (1979); 8, deLorenzo (1980); 9, Thérien and Mushynski (1979a); 10, Thérien and Mushynski (1979b); 11, Cotman *et al.* (1974); 12, Goodrum *et al.* (1979); 13, T. Cruz and J. W. Gurd, unpublished results; 14, Giambalvo and Rosenberg (1976); 15, Bittiger (1976); 16, deRobertis *et al.* (1967); 17, Foster *et al.* (1981); 18, Matus *et al.* (1981).

(Aghajanian and Bloom, 1967; Adinolfi, 1972; Dyson and Jones, 1976; Pffeninger and Maylie-Pffeninger, 1979) and it is clearly desirable to relate these morphologic events to the molecular composition of synaptic organelles. To this end, synaptic junctional fractions have been isolated from immature rat brains using the procedure of Cotman and Taylor (1972) and their molecular and enzymatic composition compared to that of similar fractions obtained from older animals. Of several developmentally related changes in composition perhaps the most striking is the very low concentration of the "PSD protein" found in association with immature SJC (Fu *et al.*, 1981; Kelly and Cotman, 1981; de Silva *et al.*, 1979; Kelly *et al.*, 1979; Fig. 3). The marked increase in the concentration of this protein during development suggests that a fundamental change in the structural organization of the PSD accompanies maturation. In contrast to the major changes in the concentration of the PSD protein junctional preparations from immature brains are associated with a full complement of glycoproteins, although some variation in the absolute and relative amounts of individual glycoproteins accompany development (deSilva *et al.*, 1979; Fu *et al.*,

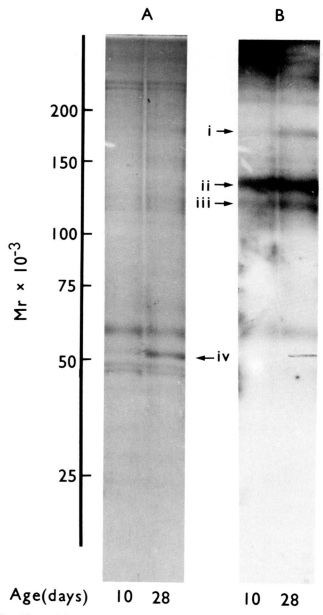

Fig. 3. Effect of development on synaptic junctional protein and glycoprotein composition. Synaptic junctions were isolated from the cortices of 10- and 28-day-old rats according to the procedure of Cotman and Taylor (1972) and separated by electrophoresis on 7–17% polyacrylamide gradient slab gels. Gels were stained with Coomassie blue for protein (A) or reacted with [^{125}I]concanavalin A and radioautograms prepared for the identification of glycoproteins (B). i, ii, iii, and iv indicate the positions of glycoproteins with apparent molecular weights of 180,000, 130,000, and 110,000, and the major "PSD protein" (MW 51,000), respectively.

1981; Kelly and Cotman, 1981; Fig. 3). The developmental appearance of the glycoprotein antigen Thy-1, a component of synaptic junctions (Acton *et al.*, 1978), also closely parallels the postnatal development of synaptic junctions (Zwerner *et al.*, 1977; Schachner and Hämmerling, 1974). Changes in the lectin receptor activity of the developing photoreceptor synapse have also been reported (McLaughlin and Wood, 1977; McLaughlin *et al.*, 1980).

In addition, synaptic junctional protein kinases and their substrates display a rapid maturation between postnatal days 7 and 15 which closely parallels the appearance of mature synaptic structures and functions (P. T. Kelly, personal communication). These results indicate that the maturational processes previously described at a morphologic level are paralleled by changes in the molecular composition, enzymatic activity, and structural organization of the developing synapse. The precise relationship between the biochemical and morphologic changes remains to be defined.

VI. CONCLUSIONS

The biochemical characterization of subsynaptic structures reviewed in this chapter clearly represents only the initial phases in the unraveling of the molecular basis of synaptic organization and function.

In spite of the considerable progress made over the past 5 years with respect to the characterization of synaptic components, the function and identity of the majority of synaptic proteins remains unknown. Similarly the principles behind the organization and interactions between the presumptive structural elements of the pre- and postsynaptic specializations and the various enzyme and receptor activities associated with them remain to be elucidated. We know little concerning the dynamics of pre- and postsynaptic functions. What effect(s) does neurotransmitter binding have on the molecular organization of the postsynaptic apparatus? How is this affected by phosphorylation reactions and/or the modification of oligosaccharide chains associated with synaptic glycoproteins? Glycoproteins have been implicated in various processes of synaptogenesis (Pfenninger, 1979) and changes in glycoprotein composition accompany synapse formation and development. Yet the role of individual glycoproteins in this process remains to be defined as does the functional significance of the developmental increase in the concentration of the major PSD protein. The whole area of the synthesis, assembly, and turnover of synaptic organelles remains fertile ground for future investigations. Finally, although some progress has been made with respect to the identification of synaptic specific molecules, questions relating to higher orders of specificities between different synaptic types are unanswered.

The significance of these questions lies not in their number but rather in the fact that we are now able to seek answers to them by directly probing the molecular composition of the synapse.

REFERENCES

*Acton, R. T., Addis, J., Carl, G. F., McClain, L. D., and Bridgers, W. (1978). Association of thy-1 differentiation alloantigen with synaptic complexes isolated from mouse brain. *Proc. Natl. Acad. Sci. U.S.A.* **75**, 3283–3287.

Adinolfi, A. M. (1972). The organization of paramembranous densities during post-natal maturation of synaptic junctions in the cerebral cortex. *Exp. Neurol.* **34**, 383–393.

Aghajanian, G. K., and Bloom, F. E. (1967). The formation of synaptic junctions in developing rat brain: A quantitative electron microscopic study. *Brain Res.* **6**, 716–727.

Ariano, M. A., and Appleman, M. M. (1979). Biochemical characterization of post-synaptically localized cyclic nucleotide phosphodiesterase. *Brain Res.* **177**, 301–309.

*Banker, G., Churchill, L., and Cotman, C. W. (1974). "Proteins of the postsynaptic density." *J. Cell Biol.* **63**, 456–465.

Barclay, A. N., Letarte-Muirhead, M., Williams, A. F., and Faulkes, R. A. (1976). Chemical characterization of the thy-1 glycoproteins from the membranes of rat thymocytes and brain. *Nature (London)* **263**, 563–567.

Barondes, S. H. (1974). Synaptic macromolecules; identification and metabolism. *Annu. Rev. Biochem.* **43**, 147–168.

Beam, K. G., and Greengard, P. (1976). Cyclic nucleotides, protein phosphorylation and synaptic function. *Cold Spring Harbor Symp. Quant. Biol.* **60**, 157–168.

Ben Ze'ev, A., Duerr, A., Solomon, F., and Penman, S. (1979). The outer boundary of the cytoskeleton: A lamina derived from plasma membrane proteins. *Cell* **17**, 859–865.

Berl, S., Puszkin, S., and Nicklas, W. J. (1973). Actomyosin-like protein in brain: Actomyosin-like protein may function in the release of transmitter material at synaptic endings. *Science* **179**, 441–446.

Berridge, M. J. (1979). Modulation of nervous activity by cyclic nucleotides and calcium. *In* "The Neurosciences Fourth Study Program" (F. O. Schmitt and F. G. Worden, eds.), pp. 873–889. MIT Press, Cambridge, Massachusetts.

*Berzins, K., Cohen, R. S., Grab, D., and Seikevitz, P. (1977). Immunological and biochemical analysis of some of the major protein components of post-synaptic densities. *Soc. Neurosci. Abstr.* **3**, 331.

Bhattacharyya, B., and Wolf, J. (1975). Membrane bound tubulin in brain and thyroid tissue. *J. Biol. Chem.* **250**, 7639–7646.

Bird, M. I. (1976). Microtubule–synaptic vesicle associations in cultured rat spinal cord neurons. *Cell Tissue Res.* **168**, 101–115.

*Bittiger, H. (1976). Separation of subcellular fractions from rat brain. *In* "Concanavalin A as a Tool" (H. Bittiger and H. P. Schnebli, eds.), pp. 436–445. Wiley, London.

Bittiger, H., and Schnebli, H. P. (1974). Binding of concanavalin A and Ricin to synaptic junctions of rat brain. *Nature (London)* **249**, 370–371.

Blitz, A. L., and Fine, R. E. (1974). Muscle-like contractile proteins and tubulin in synaptosomes. *Proc. Natl. Acad. Sci. U.S.A.* **71**, 4472–4476.

Blomberg, F., Cohen, R. S., and Siekevitz, P. (1977). The structure of postsynaptic densities isolated from dog cerebral cortex II. Characterization and arrangement of some of the major proteins within the structure. *J. Cell Biol.* **74**, 204–225.

Bloom, F. E. (1972). The formation of synaptic junctions in developing rat brain. *In* "Structure and Function of Synapses" (G. D. Pappas and D. P. Purpura, eds.), pp. 101–120. Raven, New York.

---

*References dealing with the characterization of isolated synaptic junctions or postsynaptic densities.

Bloom, F. E. (1975). Role of cyclic nucleotides in central synaptic functions. *Rev. Physiol. Biochem. Pharmacol.* **74**, 1–103.

Bloom, F. E., and Aghajanian, G. K. (1966). Cytochemistry of synapses: Selective staining for electron microscopy. *Science* **154**, 1575–1577.

Bloom, F. E., and Aghajanian, G. K. (1968). Fine structural and cytochemical analysis of the staining of synaptical junctions with phosphotungstic acid. *J. Ultrastruct. Res.* **22**, 361–375.

Bowler, K., and Duncan, C. K. (1966). Actomyosin-like protein from crayfish nerve: A possible molecular explanation of permeability changes during excitation. *Nature (London)* **211**, 642–643.

*Carlin, R. K., Grab, D. J., Cohen, R. S., and Seikevitz, P. (1980). Isolation and characterization of postsynaptic densities from various brain regions: enrichment of different types of postsynaptic densities. *J. Cell Biol.* **86**, 831–843.

Cheung, W. Y. (1970). Cyclic 3′5′-nucleotide phosphoditesterase: demonstration of an activator. *Biochem. Biophys. Res. Commun.* **38**, 533–538.

Cheung, W. Y., Bradham, L. S., Lynch, T. J., Lin, Y. M., and Tallant, E. A. (1975). Protein activator of cyclic 3′5′-nucleotide phosphodiesterase of bovine or rat brain also activates its adenyl cyclase. *Biochem. Biophys. Res. Commun.* **66**, 1055–1062.

Chiu, T. C., and Babitch, J. A. (1978). Topography of glycoproteins in the chick synaptosomal plasma membrane. *Biochim. Biophys. Acta* **510**, 112–123.

*Churchill, L., Cotman, C., Banker, G., Kelly, P., and Shannon, L. (1976). Carbohydrate composition of central nervous system synapses. Analysis of isolated synaptic junctional complexes and postsynaptic densities. *Biochim. Biophys. Acta* **448**, 57–72.

*Cohen, R. S., Blomberg, F., Berzins, K., and Seikevitz, P. (1977). The structure of postsynaptic densities isolated from dog cerebral cortex. Overall Morphology and Protein Composition. *J. Cell Biol.* **74**, 181–203.

Cotman, C. W., and Banker, G. A. (1974). The making of a synapse. *Rev. Neurosci.* **1**, 1–62.

*Cotman, C. W., and Taylor, D. (1972). Isolation and structural studies on synaptic complexes from rat brain. *J. Cell Biol.* **55**, 696–711.

Cotman, C. W., and Taylor, D. (1974). Localization and characterization of concanavalin A receptors in the synaptic cleft. *J. Cell Biol.* **62**, 236–242.

*Cotman, C. W., Levy, W., Banker, G., and Taylor, D. (1971). An ultrastructural and chemical analysis of the effect of Triton X-100 on synaptic plasma membranes. *Biochim. Biophys. Acta* **249**, 406–418.

*Cotman, C. W., Banker, G., Churchill, L., and Taylor, D. (1974). Isolation of post- synaptic densities from rat brain. *J. Cell Biol.* **63**, 441–455.

Cruz, T., and Gurd, J. W. (1978). Reaction of synaptic membrane sialoglycoproteins with intrinsic sialidase and wheat germ agglutinin. *J. Biol. Chem.* **253**, 7314–7318.

Daly, J. W. (1977). The formation, degradation and function of cyclic nucleotides in the central nervous system. *Int. Rev. Neurobiol.* **20**, 105–168.

*Davis, G. A., and Bloom, F. E. (1970). Proteins of synaptic junctional complexes. *J. Cell Biol.* **47**, 46a.

*Davis, G. A., and Bloom, F. E. (1973). Isolation of synaptic junctional complexes from rat brain. *Brain Res.* **62**, 135–153.

*deBlas, A., and Mahler, H. R. (1978). Studies on nicotinic acetylcholine receptors in mammalian brain. Characterization of a microsomal subfraction enriched in receptor function for different neurotransmitters. *J. Neurochem.* **30**, 563–577.

*deBlas, A. L., Wang, Y. J., Sorensen, R., and Mahler, H. R. (1979). Protein phosphorylation in synaptic membranes regulated by adenosine 3′-5′-monophosphate: Regional and subcellular distribution of the endogenous substrates. *J. Neurochem.* **33**, 647–659.

*deLorenzo, R. J. (1980). Post-synaptic densities: Calmodulin and calcium dependent phosphorylation. *Trans. Am. Soc. Neurochem.* **11**, 81.

Den, H. B., Kaufman, B., and Roseman, S. (1975). The sialic acids. XVIII. Subcellular distribution of seven glycosyl transferases in embryonic chick brain. *J. Biol. Chem.* **250**, 739–746.

*deRobertis, E., Azcurra, J. M., and Fiszer, S. (1967). Ultrastructure and cholinergic binding capacity of junctional complexes isolated from rat brain. *Brain Res.* **5**, 45–56.

*deSilva, N. S., and Gurd, J. W. (1979). Developmental alterations of rat brain synaptic membranes. Reaction of glycoproteins with plant lectins. *Brain Res.* **165**, 283–293.

Duncan, C. J. (1965). Cation-permeability control and depolarization in excitable cells. *J. Theor. Biol.* **8**, 403–418.

Dunkley, P. R., Holmes, H., and Rodnight, R. (1976). Phosphorylation of synaptic-membrane proteins from ox cerebral cortex *in vitro*. *Biochem. J.* **157**, 661–666.

Dyson, S. E., and Jones, D. G. (1976). The morphological categorization of developing synaptic connections. *Cell Tissue Res.* **167**, 363–371.

Elvin, L. G. (1976). The ultrastructure of neuronal contacts. *Prog. Neurobiol.* (Oxford) **8**, 45–79.

Feit, H., Dutton, G. R., Barondes, S. H., and Shelanski, M. (1971). Microtubule protein, identification in and transport to nerve endings. *J. Cell Biol.* **51**, 138–147.

*Feit, H., Kelly, P., and Cotman, C. W. (1977). Identification of a protein related to tubulin in the post-synaptic density. *Proc. Natl. Acad. Sci. U.S.A.* **74**, 1047–1051.

Fiszer, S., and deRobertis, E. (1967). Action of Triton X-100 on ultrastructure and membrane bound enzymes of isolated nerve endings from rat brain. *Brain Res.* **5**, 31–44.

Florendo, N. T., Barrnett, R. J., and Greengard, P. (1971). Cyclic 3′5′ nucleotide phosphodiesterase cytochemical localization in cerebral cortex. *Science* **173**, 745–748.

*Foster, A. C., Mena, E. E., Monaghan, D. T., and Cotman, C. W. (1981). Synaptic localization of kainic acid binding sites. *Nature (London)* **289**, 73–76.

*Freedman, M. S., Clark, B. D., Cruz, T. F., Gurd, J. W., and Brown, I. R. (1980). Selective effects of LSD and hypothermia on the synthesis of synaptic proteins and glycoproteins. *Brain Res.* **207**, 129–145.

*Fu, S. C., Cruz, T., and Gurd, J. W. (1981). Development of synaptic glycoproteins: Effect of postnatal age on the synthesis and concentration of synaptic membrane and synaptic junctional fucosyl- and sialyl-glycoproteins. *J. Neurochem.*, in press.

*Giambalvo, C. T., and Rosenberg, P. (1976). The effects of phospholipases and proteases on the binding of γ-amino butyric acid to junctional complexes of rat cerebellum. *Biochim. Biophys. Acta* **436**, 741–756.

*Goodrum, J. F., and Tanaka, R. (1978). Protein and glycoprotein composition of sub-synaptosomal fractions. *Neurochem. Res.* **3**, 599–617.

*Goodrum, J. F., Bosmann, H. B., and Tanaka, R. (1979). Glycoprotein galactosyltransferase activity in synaptic junctional complexes isolated from rat forebrain. *Neurochem. Res.* **4**, 331–337.

Gordon-Weeks, P. R., Jones, D. H., Gray, E. G., and Barron, J. (1981). Trypsin separates synaptic junctions to reveal pre- and post-synaptic concanavalin A receptors. *Brain Res.* **219**, 224–230.

*Grab, D. J., Berzins, K., Chen, R. S., and Seikevitz, P. (1979). Presence of calmodulin in post-synaptic densities isolated from canine cerebral cortex. *J. Biol. Chem.* **254**, 8690–8696.

Gray, E. G. (1959). Axosomatic and axodendritic synapses of the cerebral cortex: An electron microscopic study. *J. Anat.* **93**, 420–433.

Gray, E. G. (1975a). Presynaptic microtubules and their association with synaptic vesicles. *Proc. R. Soc. London, Ser. B.*, **190**, 369–372.

Gray, E. G. (1975b). Synaptic fine structure and nuclear, cytoplasmic and extracellular networks. *J. Neurocytol.* **4**, 315–339.

Greengard, P. (1976). Possible role for cyclic nucleotides and phosphorylated membrane proteins in post-synaptic actions of neurotransmitters. *Nature (London)* **260**, 101–108.

Greengard, P. (1978). Cyclic Nucleotides, Phosphorylated Proteins and Neuronal Functions.'' Raven, New York.

*Gurd, J. W. (1977a). Identification of lectin receptors associated with rat brain post-synaptic densities. *Brain Res.* **126,** 154–159.

*Gurd, J. W. (1977b). Synaptic plasma membrane glycoproteins. Molecular identification of lectin receptors. *Biochemistry* **16,** 369–374.

*Gurd, J. W. (1979). Molecular and biosynthetic heterogeneity of fucosyl glycoproteins associated with rat brain synaptic junctions. *Biochim. Biophys. Acta* **555,** 221–229.

*Gurd, J. W. (1980). Subcellular distribution and partial characterization of the major concanavalin A receptors associated with rat brain synaptic junctions. *Can. J. Biochem.* **58,** 941–1003.

*Gurd, J. W. and Fu, S. C. (1981). Concanavalin A receptors associated with rat brain synaptic junctions are high-mannose type oligosaccharides. Submitted.

Gurd, J. W., Jones, L., Mahler, H. R., and Moore, W. (1974). Isolation and partial characterization of rat brain synaptic plasma membranes. *J. Neurochem.* **22,** 281–290.

*Gurd, J. W., Bissoon, N., and Kelly, P. T. (1981). Synaptic junctional glycoproteins are phosphorylated by cyclic AMP dependent protein kinase. Submitted.

Hanson, H. A., and Hyden, H. (1974). A membrane associated network of protein filaments in nerve cells. *Neurobiology (Copenhagen)* **4,** 364–375.

Herschman, H. R., Cotman, C., and Matthews, D. A. (1972). Serological specificity of brain subcellular organelles. *J. Immunol.* **108,** 1362–1369.

Iversen, L. L. (1975). Dopamine receptors in the brain. *Science* **188,** 1084–1089.

Jones, D. G. (1977). Morphological features of central synapses, with emphasis on the presynaptic terminal. *Life Sci.* **21,** 477–491.

Jones, D. H., and Matus, A. I. (1974). Isolation of synaptic plasma membranes by combined flotation-sedimentation density gradient centrifugation. *Biochim. Biophys. Acta* **356,** 276–287.

Jørgensen, O. S., and Bock, E. (1974). Brain specific synaptosomal membrane proteins demonstrated by crossed immunoelectrophoresis. *J. Neurochem.* **23,** 879–880.

Kakiuchi, S. (1975). Ca^{2+} plus Mg^{2+}-dependent phosphodiesterase and its modulator from various tissues. *In* "Advances in Cyclic Nucleotide Research 5" (G. I. Drummond, P. Greengard, and G. A. Robison, eds.). Raven, New York.

*Kelly, P., and Cotman, C. W. (1976). Intermolecular disulfide bonds at central nervous system synapses. *Biochem. Biophys. Res. Commun.* **73,** 858–864.

*Kelly, P. T., and Cotman, C. W. (1977). Identification fo glycoproteins and proteins at synapses in the central nervous system. *J. Biol. Chem.* **252,** 786–793.

*Kelly, P. T., and Cotman, C. W. (1978). Synaptic Proteins: Characterization of tubulin and actin and identification of a distinct post-synaptic density polypeptide. *J. Cell Biol.* **79,** 173–183.

Kelly, P., Cotman, C. W., Gentry, C., and Nicolson, G. (1976). Distribution and mobility of lectin receptors on synaptic membranes of identified neurons in the central nervous system. *J. Cell Biol.* **71,** 487–496.

*Kelly, P. T., Cotman, C. W., and Largen, M. (1979). Cyclic AMP-stimulated protein kinases at brain synaptic junctions. *J. Biol. Chem.* **254,** 1564–1575.

*Kelly, P. T., and Cotman, C. W. (1981). Developmental changes in morphology and molecular composition of isolated synaptic junctional structures. *Brain Res.* **206,** 251–271.

Kornguth, S. E. (1974). The synapse: A perspective from *in situ* and *in vitro* studies. *In* "Reviews of Neuroscience" (S. Ehrenpreis and I. J. Kopin, eds.), Vol. 1, pp. 63–114. Raven, New York.

Kornguth, S. E., and Sunderland, E. (1975). Isolation and partial characterization of a tubulin-like protein from human and swine synaptosomal membranes. *Biochim. Biophys. Acta* **393,** 100–114.

Landis, D. M. D., and Reese, T. S. (1974). Differences in membrane structure between excitatory and inhibitory synapses in the cerebral cortex. *J. Comp. Neurol.* **155,** 93–125.

LeBeaux, Y. J., and Willemot, J. (1975). An ultrastructural study of the microfilaments in rat brain by means of E-PTA staining and heavy meromyosin labeling II. The Synapses. *Cell Tissue Res.* **160,** 37–68.

Lee, C. J. (1974). Catecholamine-Binding Protein in Mouse Brain: Isolation and Characterization. *Brain Res.* **81**, 497–509.

Lee, R. W. H., Mushynski, W. H., and Trifaro, J. M. (1979). Two forms of cytoplasmic actin in adrenal chromaffin cells. *Neurosci.* **4**, 843–852.

Lin, Y. M., Liu, Y. P., and Cheung, W. Y. (1975). Cyclic nucleotide phosphodiesterase: Ca^{2+} dependent formation of bovine brain enzyme-activator complex. *FEBS Lett.* **49**, 356–360.

Mahler, H. R. (1977). Proteins of the synaptic membrane. *Neurochem. Res.* **2**, 119–147.

Mahler, H. R. (1979). Glycoproteins of the synapse. *In* "Complex Carbohydrates of Nervous Tissue" (R. U. Margolis and R. K. Margolis, eds.), pp. 165–184. Plenum, New York.

Mahler, H. R., Gurd, J. W., and Wang, Y.-J. (1975). Molecular topography of the synapse. *In* "The Nervous System" (D. B. Tower, ed.), Vol. 1, pp. 455–466. Raven, New York.

Margolis, R. K., Margolis, R. U., Preti, C., and Lai, D. (1975). Distribution and metabolism of glycoproteins and glycosamino glycans in subcellular fractions of brain. *Biochemistry* **14**, 4797–4804.

Margolis, R. K., Preti, C., Lai, D., and Margolis, R. U. (1976). Developmental changes in brain glycoproteins. *Brain Res.* **112**, 363–369.

Marotta, C. A., Strocchi, P., and Gilbert, J. M. (1978). Microheterogeneity of brain cytoplasmic and synaptoplasmic actins. *J. Neurochem.* **30**, 1441–1451.

Matus, A. I. (1975). Immunohistochemical demonstration of antigen associated with the post-synaptic lattice. *J. Neurocytol.* **4**, 47–53.

*Matus, A. I., and Taff-Jones, D. H. (1978). Morphology and molecular composition of isolated post-synaptic junctional structures. *Proc. R. Soc. Lond. Ser. B.* **203**, 135–151.

*Matus, A. I., and Walters, B. B. (1975). Ultrastructure of the synaptic junctional lattice isolated from mammalian brain. *J. Neurocytol.* **4**, 369–375.

Matus, A. I., and Walters, B. B. (1976). Type I and Type II synaptic junctions: Differences in distribution of concanavalin A binding sites and stability of the junctional adhesions. *Brain Res.* **108**, 249–256.

Matus, A., DePetris, S., and Raff, M. C. (1973). Mobility of concanavalin A receptors in myelin and synaptic membranes. *Nature (London) New Biol.* **244**, 278–80.

Matus, A. I., Walters, B. B., and Jones, D. H. (1975a). Junctional ultrastructure in isoalted synaptic membranes. *J. Neurocytol.* **4**, 357–367.

Matus, A. I., Walters, B. B., and Mughal, S. (1975b). Immunohistochemical demonstration of tubulin associated with microtubules and synaptic junctions in mammalian brain. *J. Neurocytol.* **4**, 733–744.

*Matus, A. I., Rehling, G., and Wilkinson, D. (1981). γ-aminobutyric acid receptors in brain postsynaptic densities. *J. Neurobiol.* **12**, 67–73.

McLaughlin, B. J., and Wood, J. G. (1977). The localization of concanavalin A binding sites during photoreceptor synaptogenesis in the chick retina. *Brain Res.* **119**, 57–71.

McLaughlin, B. J., Wood, J. G., and Gurd, J. W. (1980). The localization of lectin binding sites during photoreceptor synaptogenesis in the chick retina. *Brain Res.* **191**, 345–357.

Metuzals, J., and Mushynski, W. E. (1974). Electron microscopic and experimental investigations of the neurafilamentous network in Dieter's neuron. *J. Cell Biol.* **61**, 701–722.

Michaelis, E. K. (1975). Partial purification and characterization of a glutamate-binding membrane protein from rat brain. *Biochem. Biophys. Res. Commun.* **65**, 1004–1012.

Morgan, I. G., and Gombos, G. (1976). Biochemical specificity of synaptic components. *In* "Neuronal Recognition" (S. H. Barondes, ed.), pp. 179–202. Plenum, New York.

Morgan, I. G., Wolfe, L. S., Mandel, P., and Gombos, G. (1971). Isolation of plasma membrane from rat brain. *Biochim. Biophys. Acta* **241**, 737–751.

*Mushynski, W. E., Glen, S., and Thérien, H. M. (1978). Actin-like and tubulin-like proteins in synaptic junctional complexes. *Can. J. Biochem.* **56**, 820–830.

Nathanson, J. A. (1977). Cyclic nucleotides and nervous system functions. *Physiol. Rev.* **57**, 157–256.

Ng, S. S., and Dain, J. A. (1980). Specificity and membrane properties of young rat brain sialyl transferases in cell surface glycolipids (C. Sweeley, ed.), pp. 345–358. Am. Chem. Soc., Washington.

*Ng, M., and Matus, A. (1979). Protein phosphorylation in isolated plasma membranes and post-synaptic junctional structures from bain synapses. *Neuroscience* **4**, 169–180.

Orosz, A., Hamori, J., Falus, A., Madarasz, E., Lakos, I., and Adam, G. (1973). Specific antibody fragments against the post-synaptic web. *Nature (London) New Biol.* **245**, 18–19.

Orosz, A., Madarasz, E., Falus, A., and Adam, G. (1974). Demonstration of detergent-soluble antigen specific for the synaptosomal membrane fraction isolated from the cat cerebral cortex. *Brain Res.* **76**, 119–131.

Pappas, G. D., and Waxman, S. G. (1972). Synaptic fine structure—Morphological correlates of chemical and electrotonic transmission. *In* "Structure and Function of Synapses" (G. D. Pappas and D. P. Purpura, eds.), pp. 1–43. Raven, New York.

Pfenninger, K. H. (1971a). The cytochemistry of synapitc densities. I. An analysis of the bismuth iodide impregnation method. *J. Ultrastruct. Res.* **34**, 103–122.

Pfenninger, K. H. (1971b). The cytochemistry of synaptic densities. II. Proteinaceous components and mechanism of synaptic connectivity. *J. Ultrastruct. Res.* **35**, 451–475.

Pfenninger, K. H. (1973). Synaptic morphology and cytochemistry. *Prog. Histochem. Cytochem.* **5**, 1–86.

Pfenninger, K. H. (1979). Synaptic membrane differentiation. *In* "The Neurosciences Fourth Study Program" (F. O. Schmitt and F. G. Warden, eds.), pp. 779–795. MIT Press, Cambridge, Massachusetts.

Pfenninger, K. H., and Maylie-Pfenninger, M. F. (1979). Surface glycoproteins in the differentiating neuron. *In* "Complex Carbohydrates of Nervous Tissue" (R. U. Margolis and R. K. Margolis, eds.), pp. 185–191. Plenum, New York.

Pfenninger, K. H., and Rees, R. P. (1976). From the growth cone to the synapse. *In* "Neuronal Recognition" (S. H. Barondes, ed.), pp. 131–178. Plenum, New York.

Puszkin, S., and Kochwa, S. (1974). Regulation of neurotransmitter release by a complex of actin with relaxing protein isolated from rat brain synaptosomes. *J. Biol. Chem.* **249**, 7711–7714.

Preti, A., Fiorilli, A., Lombardo, A., and Tettamanti, G. (1980). Occurrence of sialyltransferase activity in the synaptosomal membranes prepared from calf brain cortex. *J. Neurochem.* **35**, 281–296.

Raghupathy, E., Ko, G. K., and Peterson, N. A. (1972). Glycoprotein biosynthesis in the developing rat brain. III. Are glycoprotein glycosyl transferases present in synaptosomes. *Biochim. Biophys. Acta* **786**, 339–349.

Rahman, H., Rösner, H., and Breer, H. (1976). A functional model of sialo-glycomacromolecules in synaptic transmission and memory formation. *J. Theor. Biol.* **57**, 231–237.

Rambourg, A., and LeBlond, C. P. (1967). Electron microscopic observations on the carbohydrate-rich cell coat present at the surface of cells in the rat. *J. Cell Biol.* **32**, 27–53.

Rash, J. E., and Ellisman, M. H. (1974). Studies of Excitable Membranes. I. Macromolecular specializations of the neuromuscular junction and the non-junctional sarcolemma. *J. Cell Biol.* **63**, 567–586.

*Ratner, N., and Mahler, H. R. (1979). Isolation of post-synaptic membranes retaining structural and functional integrity. *J. Cell Biol.* **83**, 271a.

Rees, R. P. (1978). The morphology of interneuronal synaptogenesis: a review. *Fed. Proc., Fed. Am. Soc. Exp. Biol.* **37**, 2000–2009.

Reith, M., Morgan, I. G., Gombos, B., Breckenridge, W. C., and Vincendon, G. (1972). Synthesis of synaptic glycoproteins. I. The distribution of UDP-galactose: *N*-acetyl glucosamine galactosyl transferase and thiamine diphosphatase in adult rat brain subcellular fractions. *Neurobiology* **2**, 169–175.

*Rostas, J. A. P., Kelly, P. T., Pesin, R. H., and Cotman, C. W. (1979). Protein and glycoprotein composition of synaptic junctions prepared from discrete synaptic regions and different species. *Brain Res.* **168**, 151–167.

Salvaterra, P. M., Gurd, J. W., and Mahler, H. R. (1977). Interations of the nicotinic acetylcholine receptor from rat brain with lectins. *J. Neurochem.* **29**, 345–348.

Sandri, C., Akert, K., Livingston, R. B., and Moor, H. (1972). Particle aggregations at specialized sites in freeze-etched post-synaptic membranes. *Brain Res.* **41**, 1–16.

Schachner, M., and Hämmerling, U. (1974). The post-natal development of antigens on mouse brain cell surfaces. *Brain Res.* **73**, 362–371.

Schengrund, C.-L., and Nelson, J. T. (1974). Influence of cation concentration on the sialidase activity of neuronal synaptic membranes. *Biochem. Biophys. Res. Commun.* **63**, 217–223.

Schengrund, C.-L., and Rosenberg, A. (1970). Intracellular location and properties of bovine brain sialidase. *J. Biol. Chem.* **245**, 6196–6200.

Schulman, H., and Greengard, P. (1978a). Stimulation of brain protein phosphorylation by calcium and an endogenous heat stable protein. *Nature (London)* **271**, 478–479.

Schulman, H., and Greengard, P. (1978b). Ca^{2+} dependent protein phosphorylation systones in membranes from various tissues and its activation by calcium dependent regulator. *Proc. Natl. Acad. Sci. U.S.A.* **75**, 5432–5436.

Stohl, W., and Gonatas, N. K., (1977). Distribution of the thy-1 antigen in cellular and subcellular fractions of adult mouse brain. *J. Immunol.* **119**, 422–427.

Teshina, Y., and Kakiuchi, S. (1974). Mechanism of stimulation of Ca^{2+} and Mg^{2+} dependent phosphodiesterase from rat cerebral cortex by the modulator protein and Ca^{2+}. *Biochem. Biophys. Res. Commun.* **56**, 489–495.

Tettamanti, G., Morgan, I. G., Gombos, G., Vincendon, G., and Mandel, P. (1972). Subsynaptosomal localization of brain particulate neuraminidase. *Brain Res.* **47**, 515–518.

*Thérien, H. M., and Mushynski, W. E. (1976). Isolation of synaptic junctional complexes of high structural integrity from rat brain. *J. Cell Biol.* **71**, 807–822.

*Thérien, H. M., and Mushynski, W. E. (1979a). Distribution and properties of protein kinase and protein phosphatase activities in synaptosomal plasma membranes and synaptic junctions. *Biochim. Biophys. Acta* **585**, 188–200.

*Thérien, H. M., and Mushynski, W. E. (1979b). Characterization of the cyclic 3′5′-nucleotide phosphodiesterase activity associated with synaptosomal plasma membranes and synaptic junctions. *Biochim. Biophys. Acta* **585**, 201–209.

Toh, B. H., Gallichio, H. A., Jeffrey, P. L., Livett, B. G., Muller, H. K., Cauchi, M. N., and Clarke, F. M. (1976). Anti-Actin Stains Synapses. *Nature (London)* **264**, 648–650.

*Ueda, T., Greengard, P., Berzins, K., Cohen, R. S., Blomberg, F., Grab, D. J., and Siekevitz, P. (1979). Subcellular distribution in cerebral cortex of two proteins phosphorylated by a cAMP-dependent protein kinase. *J. Cell Biol.* **83**, 308–319.

Uno, I., Ueda, T., and Greengard, P. (1977). Andensine 3′5′-monophosphate regulated phospho-protein systems of neuronal membranes. II. Solubilization purification and some properties of an endogenous adenosine 3′5′-monophosphate-dependent protein kinase. *J. Biol. Chem.* **252**, 5164–5174.

von Hungen, K., and Roberts, S. (1974). Neurotransmitter-sensitive adenylate-cyclase systems in the brain. *In* "Review of Neuroscience 1," (S. Ehrenpreis and I. J. Kopin, eds.), pp. 231–281. Raven, New York.

Walter, U., and Greengard, P. (1978). Quantitative labeling of the regulatory subunit of type II cAMP dependent protein kinase from bovine heart by a photo-affinity analog. *J. Cyclic Nucleotide Res.* **4**, 437–444.

Walter, U., Uno, I., Liu, A. Y.-C., and Greengard, P. (1977). Identification, characterization and quantitative measurement of cyclic AMP receptor proteins in cytosol of various tissues using a photoaffinity ligand. *J. Biol. Chem.* **252**, 6494–6500.

Walter, U., Kanof, P., Schulman, H., and Greengard, P. (1978). Adenosine 3'5'-monophosphate receptor proteins in brain. *J. Biol. Chem.* **253**, 6275–6280.

*Walters, B. B., and Matus, A. I. (1975a). Ultrastructural organization in isolated synaptic densities. *J. Anat. (London)*, **119**, 415.

*Walters, B. B., and Matus, A. I. (1975b). Tubulin in post-synaptic junctional lattice. *Nature (London)* **257**, 496–498.

*Walters, B. B., and Matus, A. I. (1975c). Proteins of the synaptic junctions. *Biochem. Soc. Trans.* **3**, 109–112.

*Wang, Y. S., and Mahler, H. R. (1976). Topography of the synaptosomal plasma membrane. *J. Cell Biol.* **71**, 639–658.

*Webster, J. C., and Klingman, J. D. (1979). Isolation of a synaptic junction enriched fraction from the forebrain of day-old chicks. *Neurochem. Res.* **4**, 137–153.

*Webster, J. C., and Klingman, J. D. (1980). Synaptic junctional glycoconjugates from chick brain. Glycoprotein identification and carbohydrate composition. *Neurochem. Res.* **5**, 404–414.

Weller, M. (1980). "Protein Phosphorylation, The Nature Function and Metabolism of Proteins Which Contain Covalently Bound Phosphorous." Pion, London.

Weller, M., and Morgan, I. G. (1976a). Distribution of protein kinase in subcellular fractions of rat brain. *Biochim. Biophys. Acta* **436**, 675–685.

*Weller, M., and Morgan, I. G. (1976b). Localization in the synaptic junction of the cyclic AMP stimulated intrinsic protein kinase activity of synaptosomal plasma membranes. *Biochim. Biophys. Acta* **433**, 223–228.

Westrum, L. E., and Gray, E. G. (1976). Microtubules and membrane specializations. *Brain Res.* **105**, 547–550.

Williams, M., and Rodnight, R. (1977). Protein phosphorylation in nervous tissue: Possible Involvement in nervous tissue function and relationship to cyclic nucleotide metabolism. *Prog. Neurobiol. (Oxford)* **8**, 183–250.

Wood, J. G., Wallace, R. W., Whitaker, J. N., and Cheung, W. Y. (1980). Immunocytochemical localization of calmodulin and a heat labile calmodulin binding protein (CaM-BP$_{80}$) in basal ganglia of mouse brain. *J. Cell Biol.* **84**, 66–76.

*Yen, S. H., Liem, R. K. H., Kelly, P. T., Cotman, C. W., and Shelanski, M. L. (1977). Membrane linked proteins at CNS synapses. *Brain Res.* **132**, 172–175.

Yohe, H. C., and Rosenberg, A. (1977). Action of intrinsic sialidase of rat brain synaptic membranes on membrane sialolipid and sialoprotein components *in situ*. *J. Biol. Chem.* **252**, 2412–2418.

Yu, J., Fischman, D. A., and Steck, T. L. (1973). Selective solublization of proteins and phospholipids from red blood cell membranes by non-ionic detergents. *J. Supramol. Struct.* **1**, 233–248.

Zanetta, J. P., Morgan, I. G., and Gombos, G. (1975). Synaptosomal plasma membrane glycoproteins: Fractionation by affinity chromatography on concanavalin A. *Brain Res.* **83**, 337–348.

Zanetta, J. P., Reeber, A., Vincendon, G., and Gombos, G. (1977). Synaptosomal plasma membrane glycoproteins. II. Isolation of fucosyl-glycoproteins by affinity chromatography on the *Ulex europeus* lectin specific for L-fucose. *Brain Res.* **138**, 317–328.

Zwerner, R. K., Acton, R. T., and Seeds, N. W. (1977). The developmental appearance of thy-1 in mouse reaggregating brain cell cultures. *Dev. Biol.* **60**, 331–335.

AXONAL TRANSPORT OF MACROMOLECULES

Jan-Olof Karlsson

I. INTRODUCTION

In comparison with other cell types, neurons frequently have a relatively large surface area and enormously elongated processes, axons, whose volume can exceed that of the nerve cell body by severalfold. Consequently, axonal metabolism must constitute an important function of the total metabolism in the central nervous system. Since ribosomes have not been detected in the axons of mature mammalian neurons (Palay and Palade, 1955), it appears likely that most

Molecular Approaches to Neurobiology

of the proteins present in the axon are supplied from sources other than the axon itself. The principal source of axonal and presynaptic components is probably the nerve cell body, which provides most of the axonal macromolecules by a type of intracellular transport which is called axonal transport. Other sources of axonal or presynaptic macromolecules, such as a local (mitochondrial) axonal protein synthesis (Lasek *et al.*, 1973) or a direct transfer of macromolecules from the surrounding glial cells (Lasek *et al.*, 1974, 1977), probably make a minor contribution to the axonal content of macromolecules, at least in the mammalian nervous system.

Studies of axonal transport of macromolecules offer many experimental advantages and may give important information concerning the mechanisms behind intracellular transport in general. Such aspects of axonal transport, although important for the cell biology, will be dealt with rather superficially in this chapter. In the field of neurobiology, studies of axonal transport with radioactive tracers offer rather unique possibilities to investigate the synthesis, transport, and degradation of important neuronal constituents.

The present chapter offers a somewhat subjective and incomplete review of different aspects of axonal transport with special empahsis on rapid anterograde transport of protein in mammalian neurons. For example, intermediate phases of axonal transport and the question of mitochondrial transport will not be dealt with at all. Many reviews on axonal transport have been written during the last decade. Recent and valuable reviews have been published by Heslop (1975), Lasek and Hoffman (1976), Wilson and Stone (1979), Schwartz (1979), Rambourg and Droz (1980), and Grafstein and Forman (1980).

A. Rates of Axonal Transport

In the early studies (Weiss and Hiscoe, 1948; Droz and Leblond, 1963) one distinct phase of axonal transport was clearly detected. The rate of migration of this relatively slow phase was found to be in the order of a few millimeters per day. Since then a number of investigators working with quite different neurons and techniques have reported the existence of many different transport rates in the axon. The most rapidly transported components in the axon are usually carried in an anterograde direction with a net velocity of about 100–400 mm/day (Lasek, 1966; Dahlström and Häggendal, 1966; Grafstein, 1967; Karlsson and Sjöstrand, 1968; Ochs and Johnson, 1969). In many mammalian nerves the rapid components seem to be transported at a fairly constant rate of about 400 mm/day (Ochs, 1972a). The velocities of rapid axonal transport in many poikilothermic animals will fall in the same category if they are compensated for the effect of temperature. Rapid axonal transport in optic nerves seems to proceed at a significantly slower rate. Karlsson and Sjöstrand (1968, 1971a) found a transport rate of about 150 mm/day in rabbit retinal ganglion cells, and Hendrickson and

Cowan (1971) found a rate of about 220–240 mm/day in the same nerve. These differences in rate of axonal transport in optic nerves are probably due to the use of barbiturate anaesthesia during the isotope administration in the former studies (Gustavsson and Karlsson, unpublished).

It was early suggested that axonal transport can occur at rates which are slower than that of rapid axonal transport but far in excess of slow axonal transport (Lasek, 1968a,b). Karlsson and Sjöstrand (1971a) showed that in the rabbit optic nerve axonal transport proceeded with at least four different rates (150, 40, 6–12, and 2 mm/day, respectively). These four different phases of axonal transport also had different subcellular distributions. Willard *et al.* (1974) clearly demonstrated that these different phases of axonal transport each had a unique polypeptide composition.

Recently Willard and Hulebak (1977) and Black and Lasek (1979) have shown the existence of a fifth very slow rate of axonal transport. The concept of two single categories of axonal transport (rapid and slow) thus seems to be an over-simplification. Invesitgations using more sensitive techniques to resolve minor labeled polypeptides will perhaps reveal an even more complex situation.

B. Amounts of Axonally Transported Material

The absolute amounts of macromolecules transported via axonal transport is not known. If we assume that the axon, the axon terminals, and the neuronal cell body have the same catabolic rate, then the fraction of axonally transported material is proportional to the relative axonal volume. For example, in neurons with long axons, more than 99% of the protein-synthesizing capacity may be devoted to proteins subjected to an axonal transport. Most studies using radioisotopic labeling of axonally transported proteins have shown that the amount of radioactivity transported with the slow phases ofaxonal transport is larger than that transported with the rapid phases (McEwen and Grafstein, 1968; Karlsson and Sjöstrand, 1968, 1971a). If one assumes that rapidly and slowly transported proteins are synthesized from amino acids in the same pool and that they need approximately the same time for synthesis, these data indicate that the major portion of axonal protein is transported at a slow rate. The functional state of the mature neuron may affect the amount of axonally transported material (Lux *et al.*, 1970; Norström and Sjöstrand, 1971b).

C. Retrograde Axonal Transport

The existence of a retrograde axonal transport of different substances was clearly shown a decade ago (Kristensson and Olsson, 1971; Lubińska and Niemierko, 1971). In fact, most of the axonal particles visible with the light microscope seem to move in a retrograde direction (Smith, 1971; Forman *et al.*,

1977). Most studies on retrograde axonal transport have used administration of exogenous macromolecules to the axon or axon terminal. Following endocytosis of the applied substances there were later traced in the retrograde direction with various techniques. Neurotrophic viruses like Herpex simplex and poliovirus, and toxins such as tetanus toxin and cholera toxin (Bodian and Howe, 1941; Kristensson *et al.*, 1971; Price *et al.*, 1975; Stöckel *et al.*, 1977), as well as foreign macromolecules such as horseradish peroxidase and ferritin, have been shown to be retrogradely transported. Of particular interest is the finding that the endogenous protein nerve growth factor was transported in a retrograde direction to reach the nerve cell body (Stöckel *et al.*, 1974). Transmitter substances may also be subjected to a selective uptake in the nerve terminal and transported in a retrograde direction (Streit *et al.*, 1979).

The functional significance of retrograde axonal transport remains unknown. Experiments involving injection of exogenous substances demonstrate only that there is a mechanism for uptake and retrograde transport in the axon. Experiments using artificial nerve blocks like ligation or cold blocks may not always reflect the true *in vivo* conditions. Very recently a study using a radioactive tracer substance capable of binding to already formed proteins has demonstrated that there is a retrograde axonal transport of endogenous protein (Fink and Gainer, 1980).

Uptake and the subsequent retrograde axonal transport of macromolecules seems to be a very selective process (Schwab *et al.*, 1979).

Retrograde axonal transport occurs at rates close to or less than rapid anterograde transport. It has been widely held that retrograde axonal transport takes place in the smooth andoplasmic reticulum in the axon. Recent results presented by LaVail *et al.* (1980) indicate that the axonal organelles transporting horseradish peroxidase are short and discrete structures different from the smooth endoplasmic reticulum.

D. Mechanisms of Axonal Transport

The mechanisms behind axonal transport are largely unknown. Investigators of axonal transport have presented elaborate models for the mechanisms of axonal transport. Many of these models are based on the interaction between different cytoskeletal elements in the axon, or point to the smooth endoplasmic reticulum in the axon as the crucial factor (Schmitt, 1968; Ochs, 1971; Droz *et al.*, 1973, 1975; Gross, 1975; Lasek and Hoffman, 1976). However, the direct experimental evidence in favor of any of these models is relatively weak.

Rapid anterograde axonal transport may, directly or indirectly, depend upon microtubular function. It has been shown that rapid axonal transport is very sensitive to drugs such as colchicine which interacts with microtubular protein (Kreutzberg, 1969; Dahlström, 1968; Karlsson and Sjöstrand, 1969). However, the results from studies involving administration of drugs must be interpreted with caution since secondary effects are likely to occur. The drug may also have unknown side effects. The literature on mitosis inhibitors and axonal transport has recently been summarized (Hanson and Edström, 1978).

Rapid axonal transport is dependent upon temperature (Gross and Beidler, 1973, 1975), and needs Ca^{2+} (Ochs *et al.*, 1977). In a series of reports it has clearly been shown that rapid axonal transport is independent of axonal continuity with the cell body, but highly dependent upon local oxidative phosphorylation in the axon (Ochs, 1972b).

The mechanisms behind slow axonal transport have not been subjected to as much attention as that of rapid transport. Weiss and Hiscoe (1948) and Weiss (1972) suggested that the axoplasm was propelled inside the axon via peristaltic waves of contraction, which originated in or close to the axonal membrane. Slow axonal transport appears to be independent of protein synthesis in the cell body (Peterson *et al.*, 1967), but dependent on contact between cell body and axon (McLean *et al.*, 1976). Lasek (1980) has recently suggested that in slow axonal transport microtubules and neurofilaments move as associated structures and continuously advance from the cell body to the nerve terminal. However, Komiya and Kurokawa (1980) have recently shown that colchicine can block tubulin transport with little effect on the transport of the neurofilament triplet of protein.

II. SLOW AXONAL TRANSPORT

A. Components in Slow Axonal Transport

Early studies on the subcellular distribution of axonally transported material revealed that an appreciable portion of slowly transported proteins was found in the soluble fraction (McEwen and Grafstein, 1968; Bray and Austin, 1969; Kidwai and Ochs, 1969; Sjöstrand and Karlsson, 1969). This finding distinguished slow axonal transport from the rapid transport phase in which only a small portion of the transported protein could be found in the soluble fraction. Thus the slow phase of axonal transport seemed to consist of soluble "cytoplasmic" proteins and membrane-bound proteins. In the rabbit retinal ganglion cell, approximately one-half of the leucine-labeled slowly transported proteins was found among the soluble proteins, and one-half in the membrane fractions (Karlsson and Sjöstrand, 1971a).

It should be borne in mind that many proteins in the cell exist in a polymeric form in equilibrium with its monomer and that subcellular fractionation of a tissue may easily change this equilibrium. Tubulin is a typical example of a soluble protein that can rapidly polymerize to form microtubules. Radioisotopic labeling of slowly transported proteins have clearly shown that tubulin is transported at a slow rate in the axon (James and Austin, 1970; Karlsson and Sjöstrand, 1971b; Feit *et al.*, 1971; Hoffman and Lasek, 1975). Neurofilaments are very prominent cytoskeletal elements in most axons and consist of three closely associated polypeptides (Hoffman and Lasek, 1975; Liem *et al.*, 1978). The neurofilament triplet has been shown to be a constituent of slow axonal transport (Hoffman and Lasek, 1975; Willard and Hulebak, 1977). Microfilaments with a diameter of 5 nm consisting of actin are observed in all axons. The actin polypeptide (MW 43,000) has also been shown to be transported at a slow rate in the axon (Black and Lasek, 1979; Willard *et al.*, 1979), but a somewhat more rapid rate than microtubular protein and the neurofilament triplet.

Among the soluble proteins suggested to be transported with the slow phase of axonal transport is tryptophan 5-hydroxylase (Meek and Neff, 1972), glutamic acid decarboxylase (Hariri and Hammerschlag, 1977), $Mg^{2+} + Ca^{2+}$-activated ATPase (Khan and Ochs, 1974), lactic dehydrogenase and monoamine oxidase (Khan and Ochs, 1975), polypeptides resembling myosin (Willard, 1977), the neuron-specific enolase (Erickson and Moore, 1980), and calmodulin (Erickson *et al.*, 1980).

B. Function of Slowly Transported Components

From Section II,A it is obvious that all the important cytoskeletal elements in the axon, i.e., microtubules, neurofilaments (intermediate filaments), and actin filaments (microfilaments) are transported at a slow rate toward the axon terminal. In addition, most cytoplasmic components in the axon are probably also transported with the slow phases of axonal transport. The functions of these different components in the axon must be very divergent and not necessarily different from that in non-neuronal cells. In this context it is interesting to note that the rate of slow axonal transport is often comparable with the rate of axon regeneration (Frizell and Sjöstrand, 1974). A major fraction of the slowly transported proteins is not degraded in the axon en route but reaches the axon terminals (Karlsson and Sjöstrand, 1971a). In the nerve terminals there is a relatively slow turnover of microtubular protein (Karlsson and Sjöstrand, 1971b), actin (Black and Lasek, 1979), and neurofilaments (Hoffman and Lasek, 1975). In addition to serving as cytoskeletal elements and as enzymes for different metabolic processes in the axon, the components of the slow phase of axonal transport may also have other important functions in the nerve terminal (*vide infra*).

C. Degradation of Slowly Transported Components

The half-life for proteins of the slow axonal transport in the nerve terminals has been found to be in the order of 1–4 weeks (Karlsson and Sjöstrand, 1971a;

Black and Lasek, 1979). As the volume of the axon and axon terminal is relatively constant in a nongrowing neuron, all of the transported material remaining in the axon must be locally degraded and/or secreted.

It is possible that Ca ions in the nerve terminal are of importance in this degradation process. Calcium may promote disassembly of microtubules and activate proteases which specifically act upon microtubule-associated proteins (Sandoval and Weber, 1978). A Ca^{2+}-activated protease in the axoplasm which selectively degrades neurofilament proteins has recently been described (Pant and Gainer, 1980). Lasek and Hoffman (1976) have made the very interesting suggestion that an activation of presynaptic proteases and a selective degradation of cytoskeletal elements are dependent on the interaction of the axon terminal with the postsynaptic cell. Axonal growth and regeneration would thus be a consequence of a lack of contact with postsynaptic cells.

With the exception of those components which may be carried backward via the retrograde axonal transport, the proteins which reach the nerve terminal will be locally degraded or secreted. The quantitative significance of slow axonal transport leads to the conclusion that a considerable portion of neuronal macromolecules are degraded outside the nerve cell body. The large fraction of small molecular weight compounds, which are formed during degradation of axonally transported proteins, cannot be used locally as protein precursors, since protein synthesis machinery is largely missing in the axon terminal. In a recent study of the superior colliculus it was shown (Sandberg *et al.*, 1980a) that slowly transported proteins from the retinal ganglion cells were degraded to acid-soluble components in the nerve terminal. A perfused tissue slice responded to depolarization with a Ca^{2+}-dependent release of acid-soluble material. It is possible that a part of the released material may have physiological functions and serve as transmitter or neuromodulating substances. The regulation of proteolytic activity in the nerve terminal may also be of crucial importance in this context.

III. RAPID AXONAL TRANSPORT

A. Rapid Transport of Low Molecular Weight Compounds

1. Amino Acids

Rapid axonal transport of free amino acids is not well documented. This is an important issue in view of the increasing interest in amino acids and small peptides as putative transmitters in the nervous system. For a correct evaluation of the extent of proteolytic activity in the nerve terminal (see the proceeding sections) it is also necessary to know if free amino acids are supplied via axonal transport.

In the cat sciatic nerve it has been reported that free leucine, as well as small polypeptides, are rapidly transported in the axon (Kidwai and Ochs, 1969; Ochs *et al.*, 1970). Similarly, Csanyi *et al.* (1973) reported that several amino acids were transported rapidly in the optic nerve of the carp. Roberts *et al.* (1973) found no evidence for a rapid axonal transport of any labeled material following injection of [^{14}C]glucose into the dorsal root ganglion of the rat. On the other hand an accumulation of several amino acids was observed following ligation of the dorsal root *in vivo*. In goldfish optic nerve, Elam and Agranoff (1971) found no evidence for a rapid transport of leucine and other amino acids with the possible exception of proline. In the optic nerves of newly hatched chicks, Bondy (1971) found evidence for a rapid transport of γ-aminobutyric acid (GABA), leucine, and lysine but not of proline or glutamic acid. In the optic nerve of the rabbit there is probably no anterograde axonal transport of free amino acids or soluble leucine-labeled molecule of MW below 20,000 (Karlsson and Sjöstrand, 1972; Karlsson, 1977; Sandberg *et al.*, 1980b). Evidence for rapid axonal transport of amino acids have also been reported by Cancalon and Beidler (1975) and Gross and Kreutzberg (1978).

These conflicting observations may be due to several factors.

1. Type of neuron and amino acid investigated.
2. Insufficient control of systemic or extra-axonal labeling.
3. Acid solubility of some labeled macromolecules.
4. Proteolysis of rapidly transported proteins in the nerve terminals *in vivo*.
5. Postmortal proteolysis during handling of tissue and homogenates.

In my view many of the reports claiming a rapid anterograde transport of free amino acids have failed to provide adequate controls for systemic labeling or of proteolysis *in vivo*. One possibility is that free amino acids may be axonally transported, bound to carrier molecules in a protected form such as inside dense core vesicles. Evidence for this has been found in an *Aplysia* neuron (Price *et al.*, 1979).

2. Peptides

Evidence supporting a relatively rapid transport in the axon of substance P (Holton, 1959; Hökfelt *et al.*, 1975; Gamse *et al.*, 1979; Gilbert *et al.*, 1980; Brimijoin *et al.*, 1980), carnosine (β-alanyl-histidine) (Margolis and Grillo, 1977), vasoactive intestinal peptide (Giachetti and Said, 1979), and uncharacterized peptides in *Aplysia* neurons (Berry and Geinisman, 1979) in the axon have been obtained. It has, however, not completely been ruled out that the peptides may be formed proximal to the ligature or in the nerve by proteolysis of a larger precursor. The intra-axonal processing of a large precursor to smaller components is well known in the hypothalamo-hypophyseal system (Gainer *et al.*, 1977). In addition, it is also possible that in this case the peptides (or its precursors) are transported protected in various types of vesicles.

3. Other Substances

The polyamines spermine and spermidine and their precursors putrescine and ornithine are of importance in studies of cellular growth and development. There are conflicting views concerning the possible axonal transport of these compounds. Fischer and Schmatolla (1972) claimed that putrescine was rapidly transported in the optic nerve of embryonal zebra fishes. Siegel and McClure (1975b), however, found no evidence for a transport in the optic nerve of the adult rat. Ingoglia et al. (1977) showed that putrescine, spermine, and spermidine were transported in regenerating goldfish optic nerves and that no putrescine was transported rapidly in nonregenerating nerves. A recent careful study by Harik et al. (1979) on adult rat sciatic nerve showed no signs of a rapid transport of ornithine, putrescine, spermine, or spermidine.

Taurine is a sulfonic amino acid present in large amounts in the nervous system and in the retina. It is not used for protein synthesis or a source of metabolic energy (Hayes, 1976), but it has been suggested that it may function as a modulator, stabilizing axonal membrane potentials (Gruener and Bryant, 1975). There is now rather good evidence for a rapid axonal transport of this compound (Ingoglia et al., 1976; Sturman, 1979; Politis and Ingoglia, 1979a).

Rapid anterograde transport of transmitter substances such as noradrenaline and acetylcholine, probably enclosed in synaptic vesicles, have been recognized for many years (Livett et al., 1968; Laduron and Belpaire, 1968; Koike et al., 1972; Schafer, 1973; Dahlström et al., 1974). There are also reports on rapid transport of calcium ions (Hammerschlag et al., 1975; Knull and Wells, 1975; Neale and Barker, 1977; Iqbal and Ochs, 1978). It seems likely that Ca^{2+} is transported bound to some macromolecules with an affinity for divalent cations.

Following administration of labeled nucleosides to nerve cell bodies it has been possible to show a relatively rapid axonal transport of radioactivity bound to nucleosides, nucleotides, and also to some extent RNA (Austin et al., 1966; Autilio-Gambetti et al., 1973; Schubert and Kreutzberg, 1974; Hunt and Künzle, 1976; Lindquist and Ingoglia, 1979; Wise et al., 1978; Politis and Ingoglia, 1979b; Gunning et al., 1979). From these studies it appears that most of the radioactivity is transported to the nerve terminals in the form of nucleosides or nucleotides. At least some part of the labeled material may be transported as fast as the most rapidly transported proteins. At longer time intervals following isotope injection there is some labeling of RNA, especially 4 S RNA in the nerve terminal region. This has been taken as evidence for a preferential axonal transport of 4 S RNA (transfer RNA) (Politis and Ingoglia, 1979b; Gunning et al., 1979). However, RNA synthesis in glial or postsynaptic cells from axonally supplied RNA precursors has not been completely ruled out. The compartment in which the nucleosides and nucleotides are transported in the axon remains unknown.

B. Rapid Transport of Lipids

Rapid anterograde axonal transport of phospholipids (Miani, 1963; Abe *et al.*, 1973; Grafstein *et al.*, 1975; Currie *et al.*, 1978; Haley *et al.*, 1979; Toews *et al.*, 1979), cholesterol (McGregor *et al.*, 1973; Rostas *et al.*, 1975, 1979), and glycolipids including gangliosides (Forman and Ledeen, 1972; Rösner, 1975; Landa *et al.*, 1979) is well established. Choline and ethanolamine phosphoglycerides seem to be the major transported phospholipids (Toews *et al.*, 1979). In general, experiments involving administration of lipid precursors have revealed that a part of the labeled lipids is transported together with the most rapidly transported proteins. The major part of the labeled lipids appears somewhat later in the axon and axon terminal, probably as a consequence of a prolonged period of release from the cell body into the axon. Recent experiments on single neurons in *Aplysia* and in frog sciatic nerves have indicated that the synthesis of lipids and protein in the cell body and the initiation of rapid axonal transport are two closely associated and correlated processes (Sherbany *et al.*, 1978; Longo and Hammerschlag, 1980). It then appears likely that the lipid and protein components of rapid axonal transport are synthesized and assembled into membranes in the cell body before transport into the axon. These findings support the autoradiographic studies by Droz (Droz *et al.*, 1973; Rambourg and Droz, 1980), and further endorse the view that the axonal smooth endoplasmic reticulum is the primary subcellular component of rapid axonal transport.

C. Rapid Transport of Glycosaminoglycans

The presence of glycosaminoglycans in neurons has been demonstrated with histochemical techniques (Abood and Abul-Haj, 1956; Castejón, 1970) as well as with cell separation methods (Margolis and Margolis, 1974). Clear evidence for the presence of chondroitin sulfate, hyaluronidate, and heparan sulfate in the neuronal cell body and in the nerve have been presented (Margolis and Margolis, 1974; Margolis *et al.*, 1979; Branford White, 1979). Very little is, however, known about the synthesis, transport, turnover, and function of these compounds in the neuron. Elam *et al.* (1970) first demonstrated a rapid axonal transport of sulfated mucopolysaccharides in retinal ganglion cells of the goldfish following an intraocular injection of $^{35}SO_4$. They found evidence for a rapid transport of heparan sulfate and chondroitin sulfate. Using a similar approach, Karlsson and Linde (1977) found a rapid but no slow axonal transport of $^{35}SO_4$ in retinal ganglion cells of the rabbit. A considerable part of the transported radioactivity had solubility characteristics of glycosaminoglycans, primarily chondroitin sulfate. Similar findings have also been obtained from the developing rat optic system (Goodrum *et al.*, 1979).

D. Rapid Transport of Proteins

Autoradiographic studies have convincingly showed that the profiles of the smooth endoplasmic reticulum in the axon are heavily labeled by the rapid phase of axonal transport (Schonbach *et al.*, 1971; Hendrickson, 1972; Droz *et al.*, 1973, 1975; Rambourg and Droz, 1980). Labeling of the axolemma is probably a secondary phenomenon (Markov *et al.*, 1976). Subcellular fractionation experiments have shown that the proteins of rapid axonal transport are predominantly membrane bound in a large number of nerves in a variety of species (McEwen and Grafstein, 1968; Bray and Austin, 1969; Sjöstrand and Karlsson, 1969; Elam and Agranoff, 1971; Cuénod and Schonbach, 1971). The relatively small fraction of soluble proteins found in the rapid phase of axonal transport (Karlsson, 1976) may easily be derived from membrane fractions during homogenization of the tissue. The different membrane proteins that are constituents of the rapid phase of axonal transport have, due to methodological problems, been very difficult to purify and characterize. In general, the approaches to studies of rapidly transported proteins have fallen into two categories. The first is to block axonal transport in a nerve, and then, by histochemical or biochemical techniques, to measure an accumulation of different enzymes proximal to the nerve block. The second approach is to label rapidly transported proteins with radioactive isotopes, and then extract the nerve with effective solvents in order to analyze the labeled components via electrophoresis under denaturing conditions. Experiments designed to purify relatively undenatured rapidly transported axonal proteins are few.

1. Accumulation of Enzymes in Blocked Nerves

Dopamine-β-hydroxylase, an important constituent of the transmitter-containing vesicles in adrenergic nerves, has clearly been shown to be transported rapidly in the axon (Laduron and Belpaire, 1968; Jarrott and Geffen, 1972; Brimijoin *et al.*, 1973; Wooten, 1973). There is also evidence that a part of the largely soluble enzyme tyrosine hydroxylase is transported at the same rate (Jarrot and Geffen, 1972; Brimijoin and Wiermaa, 1977). In the cholinergic neuron there is a rapid transport of acetylcholinesterase (Lubińska and Niemierko, 1971; Partlow *et al.*, 1972; Fonnum *et al.*, 1973; Di Giamberardino and Couraud, 1978) and probably also a fraction of choline acetyltransferase (Fonnum *et al.*, 1973).

2. Electrophoretic Characterization of Rapidly Transported Proteins

The first attempts to make an electrophoretic characterization of rapidly transported proteins were made by Karlsson and Sjöstrand (1971d,e) and Marko *et al.* (1971). They found that rapidly transported proteins differed from those of the slow phase, and that they had a relatively low-electrophoretic mobility depending on large size and/or low charge. The electrophoretic system used did not allow for further discrimination. Subsequent studies using polyacrylamide gel electrophoresis in buffers containing sodium dodecyl sulfate have given a more complete picture with estimation of an apparent molecular weight for the separated polypeptides. Two dimensional gel electrophoresis of rapidly transported polypeptides have now revealed that a very large number (probably more than 150) of different polypeptides may be transported rapidly in the axon to reach the nerve terminal (Stone *et al.*, 1978; Wagner *et al.*, 1979). The many published reports on rapidly transported polypeptides are difficult to compare, partly due to differences in the species and nerves examined, and partly to the type of radioactive precursor used. The main source of variation between the results obtained by different investigators may be due to the particular electrophoresis system used. Many reports have indicated that a major portion of rapidly transported polypeptides contain covalently bound carbohydrates (Edström and Mattsson, 1973; Barker *et al.*, 1975). It is well known that glycoproteins migrate at an anomalously slow rate in sodium dodecyl sulfate polyacrylamide gel electrophoresis (Segrest *et al.*, 1971; Bretscher, 1971). The resulting overestimation of the molecular weight for a carbohydrate-containing polypeptide will be more marked when a low concentration of acrylamide is used in the electrophoretic system.

However, it is now possible to point out some similarities between the numerous rapidly transported polypeptides from various sources. Careful studies of electrophoretically separated labeled polypeptides have shown that, except for a few minor exceptions (Black and Lasek, 1978; Kelly *et al.*, 1980), there are no marked differences between the polypeptides transported in different axonal branches from the same neuron or between the polypeptides transported in a variety of different neurons (Siegel and McClure, 1975a; Barker *et al.*, 1976; 1977; Bisby, 1977; White and White, 1977; Wagner *et al.*, 1979; Padilla *et al.*, 1979).

A common feature in these studies was that a single electrophoretic system was used to analyze the molecular weight of labeled axonal polypeptides from different neurons. Marked similarities were thus found in the labeling pattern. However, the estimated values for the molecular weights of the transported polypeptides did not correspond exactly to that obtained by other investigators using different electrophoretic system. The major labeled polypeptides in one study may not have even the approximate molecular weight of the major transported polypeptides from another study. Due to the insensitivity of one-dimensional electrophoresis, a group of minor labeled high molecular weight polypeptides may be registered as one single component if a gel with a high-acrylamide concentration is used. Similar artifacts may occur for low molecular weight components if a gel with low-acrylamide content is used. Another potential source of variation is the use of different labeled amino acids to label the transported proteins. For technical reasons [^{35}S]methionine has been used in increasing frequency in studies of axonally transported proteins. It is likely that the content of methionine in a number of different proteins is subject to more variation than, for example, the content of leucine. Keeping these sources of artifacts in mind, it seems nevertheless possible to arrange the molecular weights of axonally transported polypeptides in certain basic groups. Investigations using leucine, proline or lysine (Karlsson and Sjöstrand, 1971e; Edström and Mattsson, 1973; Krygier-Brévart et al., 1974; Cancalon et al., 1976; Bisby, 1977; White and White, 1977; Weiss et al., 1978, 1979; Black and Lasek, 1978), methionine (Willard et al., 1974; Barker et al., 1975, 1976, 1977; Theiler and McClure, 1977; Bisby, 1978; Lorenz and Willard, 1978; Estridge and Bunge, 1978; Stone et al., 1978; Wagner et al., 1979; Padilla et al., 1979), and fucose (Karlsson and Sjöstrand, 1971c; Edström and Mattsson, 1973; Ambron et al., 1974a,b; Goodrum et al., 1979; Karlsson, 1980) have revealed that the major labeled components could be found in at least three different groups: group A polypeptides with an apparent molecular weight of 95–125 K (K = 1000 daltons); group B polypeptides with an apparent molecular weight of 60–70 K; and group C polypeptides with a molecular weight of 18–30 K. No polypeptide with a molecular weight below 15 K seems to be a constituent of the rapid phase of axonal transport in non-neurosecretory neurons. In neurosecretory neurons it is well documented that neurophysins (MW 12,000) are subjected to a rapid axonal transport (Norström and Sjöstrand, 1971a; Gainer et al., 1977). Numerous single polypeptides have, of course, been found between these categories, especially with a higher molecular weight than that of group A.

These similarities may be somewhat surprising in view of the many different specific functions executed and different transmitters used by different neuronal systems. It seems obvious that most neurons in the nervous system use a common basic set of polypeptides for fundamental functions in the axon and in the nerve terminal. The specific features of various synaptic connections and functions may be dependent upon relatively minor post-translational events such as phosphorylation, varying degrees of glycosylation, or proteolytic modification of synaptic macromolecules. For example, it has been reported that axonal polypeptides may be glycosylated locally in the axon (Ambron and Treistman, 1977).

3. Isolation of Undenatured Axonally Transported Proteins

From Section III,D,2 it is clear that investigators concerning the electrophoretic characteristics of denatured rapidly axonally transported proteins have been numerous. Experiments designed to purify relatively undenatured rapidly transported axonal proteins are few. Ochs and co-workers (Kidwai and Ochs, 1969; Ochs et al., 1979) and Karlsson (1976) subjected the relatively small proportion of soluble (operational definition) rapidly transported proteins to gel filtration and ion-exchange chromatography, but it was not possible to isolate any defined component. Iqbal and Ochs (1978) claimed that there was a rapid axonal transport of a calcium binding protein in the sciatic nerve of the cat. This calcium binding protein had a molecular weight of about 15,000, i.e., similar to other calcium binding proteins such as calmodulin. Calmodulin has, however, been shown to be transported slowly in the axon (Erickson *et al.*, 1980). It is not clear from the report by Iqbal and Ochs (1978) if the authors had isolated the calcium binding protein in a pure form.

Methodological difficulties have delayed progress in the purification of rapidly transported proteins. The difficulties arise from the fact that most, if not all, rapidly transported proteins are associated with whole or integral parts of intra-axonal membranes. This severely restricts the number of available separation methods. The small fraction of "soluble" proteins in the rapid phase of axonal transport may be derived from membraneous fractions during the preparative steps. Relatively simple manipulations with ionic strength, pH, and the presence of a chelating agent in the extraction solutions can solubilize a certain fraction of rapidly transported proteins (Karlsson, 1976). In the latter study it was also found that lithium diiodosalicylate solubilized most of the rapidly transported membrane porteins. This agent has been used by Marchesi and Andrews (1971) to solubilize the major glycoprotein from the erythrocyte membrane. It is possible to extract highly labeled rapidly transported proteins from membrane fractions with various types of detergents (Karlsson and Sjöstrand, 1971d; Siegel and McClure, 1975a).

It was demonstrated that rapidly transported material could be completely solubilized by sequential extraction with isotonic buffer, hypotonic buffer, nonionic detergents, deoxycholate, and dodecyl sulfate (Karlsson, 1978). Anionic and nonionic detergents were much more effective than a cationic detergent in solubilizing rapidly transported membrane-bound proteins (Karlsson, 1978). It is generally considered that treatment with a nonionic detergent will preserve the conformation of a solubilized protein (Helenius and Simons, 1975). In principle then, it is possible to subject a sample of rapidly transported proteins, solubilized with nonionic detergents, to a variety of separation methods.

Gel filtration in detergent-containing buffers of rapidly transported proteins has shown that the major labeled component(s) is eluted as a complex with a Stokes radius of about 7 nm (Karlsson, 1979). It was not possible to obtain a valid measure for the molecular weight of the labeled components due to the fact that the amount of bound detergent was unknown. The finding that the major portion of solubilized rapidly transported proteins could bind to a hydrophobic matrix (Karlsson, 1980) supports previous autoradiographic and subcellular frac-tionation data and gives biochemical evidence for a membrane localization of rapidly transported axonal proteins. The major portion of rapidly transported proteins seems to carry a negative net charge (Karlsson and Sjöstrand, 1971d; Karlsson, 1976).

It has recently been demonstrated that rapidly transported glycoproteins may bind to lectins such as concanavalin A, wheat germ agglutinin (Karlsson, 1979), and lentil lectin (Karlsson, 1980) (Table I). These findings indicate that rapidly transported glycoproteins may contain exposed mannose and N-acetyl-D-glucosamine residues. Besides providing some information concerning the structure of these proteins, these findings show that lectins may be used as valuable tools in a purification procedure for rapidly transported axonal proteins. A variety of lectins is available with different specificities toward various carbohydrate groups. It will thus be possible to perform experiments with sequential affinity chromatography on different lectin columns in order to obtain a pure glycoprotein. Affinity chromatography of rapidly transported glycoproteins on lectins specific for galactose of N-acetyl-D-galactosamine showed no binding to the lectins (Karlsson, 1979). Very small amounts of fucose-labeled rapidly transported glycoproteins did bind to some fucose-specific lectins from *Ulex eur.* and *Lotus tetragonolobus* (Gustavsson and Karlsson, in preparation; Table I). Thus rapidly transported proteins may lack these sugars in an exposed position. Biochemical investigations have clearly shown that isolated synaptic membranes from whole brain and proteins solubilized from such preparations can bind to lectins such as concanavalin A, wheat germ agglutinin, and *Lotus* lectin (Zanetta *et al.*, 1975; Gurd, 1977, 1979; Kelly and Cotman, 1977).

It has been reported (Churchill *et al.*, 1976) that synaptic membranes were

Table I. Binding of Rapidly Axonally Transported Glycoproteins to Different Substances[a]

Substance	Percent bound	Reference
Lentil lectin	61	a
Concanavalin A	50	b
Wheat germ lectin	50	b
Peanut agglutinin	8	b
Soybean lectin	4	b
Ricin	11	c
Lectin from *Phytolacca am.*	0	c
Lectin from *Ulex eur.*	6	c
Lectin from *Lotus tetragonolobus*	9	c
Asialofetuin	2	b
Ovomucoid	8	b
N-acetyl-D-galactosamine	1	b
D-galactosamine	4	b
L-fucose	1	b

[a] Labeled rapidly transported fucose-containing glycoproteins were subjected to affinity chromatography on different substances coupled to Sepharose. The amount of bound radioactivity in percent of the applied amount was determined. Data from Karlsson, 1979b, 1980a; and Gustavsson and Karlsson, in preparation, 1981c.

able to bind to simple carbohydrates. In order to test the possibility that rapidly transported glycoproteins themselves had properties of lectins, and were able to interact with, and bind to, different carbohydrates, labeled axonally transported glycoproteins were subjected to affinity chromatography on columns containing different glycoproteins and sugars. The results showed that the major labeled components did not show any lectin behavior when tested against asialofetuin (rich in terminal D-galactose), ovomucoid (rich in terminal D-mannose), D-galactosamine, N-acetylgalactosamine, and L-fucose (Karlsson, 1979) (Table I).

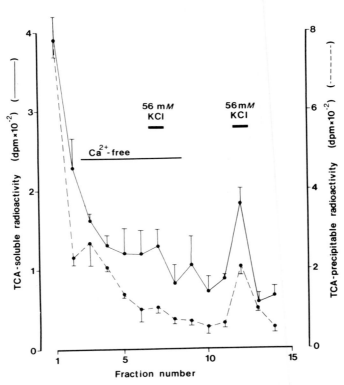

Fig. 1. Efflux of TCA-soluble (—) and TCA-precipitable (---) radioactivity from slices of superior colliculus labeled by an intraocular injection of [³H]glycine 18 h earlier. Each fraction was collected during 20 min. After perfusion of the tissue with Ca^{2+}-free media supplemented with 10 mM $MgCl_2$ for 80 min, the tissue was exposed to 56 mM KCl for 20 min (fraction 7). After subsequent perfusion with normal medium for 60 min, a 56 mM KCl pulse was introduced for 20 min (fraction 12). (From Sandberg *et al.,* 1980b.)

E. Degradation of Rapidly Transported Components

The half-life for proteins of rapid axonal transport in the nerve terminals is considerably shorter than that of the slowly transported proteins. At least some components appear to have half-lives in the order of a few hours (Karlsson and Sjöstrand, 1971a; Willard et al., 1974; Goodrum et al., 1979). As is the case for slow axonal transport, those components which reach the nerve terminal and are not subjected to a retrograde axonal transport must be locally degraded or secreted. In cholinergic or adrenergic neurons it is well known that the content of the transmitter-containing vesicles is released via exocytosis during nerve activity. Similarly, neurosecretory neurons release their axonally transported hormones and other substances upon functional stimulation (Norström and Sjöstrand, 1971b; Berry, 1979). Other noncholinergic and nonadrenergic neurons with no known neurosecretory function may also secrete different components from the nerve terminal upon stimulation. Rapidly transported proteins have been found to be released from frog sciatic nerve in vitro (Hines and Garwood, 1977). It has recently been shown that rapidly transported proteins from retinal ganglion cells of the rabbit in an in vitro system may be released upon chemical depolarization (Sandberg et al., 1980b) (Fig. 1). However, it is evident that the major part of the rapidly transported proteins must be degraded in the nerve terminals. This degradation to low molecular weight components was found to be highly Ca^{2+}-dependent and inhibited by sulfhydryl blockers. Gel filtration of the soluble components formed from rapidly transported proteins in the presence of calcium has shown the formation of peptides and amino acids (Sandberg et al., 1980b). Some of these small molecular weight components were released from the nerve terminals upon chemical depolarization (Fig. 1). It is thus possible that nerve activity with an increase in intraterminal Ca^{2+} activity (Blaustein et al., 1978) stimulates proteolytic activities in the nerve terminal. The newly formed low molecular weight components may then be released from the terminal and subserve important physiological functions.

IV. CONCLUSIONS

The concept of two single categories of anterograde axonal transport (rapid and slow) seems to be an oversimplification. Recent investigations have shown the presence of several intermediate phases of axonal transport as well as at least two distinct phases of slow axonal transport. Future investigations will perhaps reveal an even more complex situation. Rapid retrograde axonal transport of endogenous substances probably occurs in all neurons. Uptake and retrograde transport of macromolecules and small molecular weight substances seem to be a very selective and specific process. The mechanisms behind both anterograde and retrograde axonal transport are unknown.

The components of slow axonal transport include many soluble proteins as well as most of the proteins of the cytoskeletal elements in the axon (microtubules, neurofilaments, and microfilaments). The quantity of slowly transported components is relatively large. Degradation of slowly transported proteins in the nerve terminal may supply the terminal with small molecular weight compounds of potentially physiological importance.

Small molecular weight substances may be subjected to rapid axonal transport in a protected form, such as in vesicles of various types. Proteins and lipids are probably synthesized and assembled into membranes in the nerve cell body before transport into the axon. In the axon, the smooth endoplasmic reticulum is the primary subcellular component of rapid axonal transport.

The components of the rapid phase of axonal transport in noncholinergic or nonadrenergic neurons is not known. Electrophoretic separations of denatured rapidly transported polypeptides have shown that similar polypeptides are axonally transported in a wide variety of different neurons. The major labeled polypeptides may be grouped in at least three main groups (group A, MW 95–125 K; B, 60–70 K; and C, 18–30 K). Isolation of a pure undenatured rapidly transported protein has not yet been accomplished. Solubilization of rapidly transported proteins with nonionic detergents and a subsequent separation of different glycoproteins via affinity chromatography seems to be a promising technique.

Studies of radioactively labeled rapidly transported axonal proteins can make it possible to distinguish synthesis, transport, and degradation of important neuronal components in a defined neuronal population. Such studies may complement other studies dealing with biochemical characterization of axonal or synaptic components derived from whole brain.

ACKNOWLEDGMENTS

I thank Gull Crönstedt for excellent secretarial help, and the Swedish Medical Research Council (grant No. 05932), Torsten and Ragnar Södersberg Stiftelser, and Magn. Bergvalls Stiftelse for financial support.

REFERENCES

Abe, T., Haga, T., and Kurokawa, M. (1973). Rapid transport of phosphatidylcholine occuring simultaneously with protein transport in the frog sciatic nerve *Biochem. J.* **136,** 731–740.

Abood, L. G., and Abul-Haj, S. K. (1956). Histochemistry and characterization of hyaluronic acid in axons of peripheral nerve. *J. Neurochem.* **1,** 119–125.

Ambron, R. T., and Triestman, S. N. (1977). Glycoproteins are modified in the axon of R2, the giant neuron of *Aplysia Californica,* after intra-axonal injection of [³H]N-acetylgalactosamine. *Brain Res.* **121,** 287–309.

Ambron, R. T., Goldman, J. E., Barnes Thompson, E., and Schwartz, J. H. (1974a). Synthesis of glycoproteins in a single identified neuron of *Aplysia Californica*. *J. Cell Biol.* **61**, 649–664.

Ambron, R. T., Goldman, J. E., and Schwartz, J. E. (1974b). Axonal transport of newly synthesized glycoproteins in a single identified neuron of *Aplysia California*. *J. Cell Biol.* **6**, 665–675.

Austin, L., Bray, J. J., and Young, R. J. (1966). Transport of proteins and ribonucleic acid along nerve axons. *J. Neurochem.* **13**, 1267–1269.

Autilio-Gambetti, L., Gambetti, P., and Shafer, B. (1973). RNA and axonal flow. Biochemical and autoradiographic study in the rabbit optic system. *Brain Res.* **53**, 387–398.

Barker, J. L., Hoffman, P. N., Gainer, H., and Lasek, R. J. (1975). Rapid transport of proteins in the sonic motor system of the toadfish. *Brain Res.* **97**, 291–301.

Barker, J. L., Neale, J. H., and Gainer, H, (1976). Rapidly transported proteins in sensory, motor and sympathetic nerves of the isolated frog nervous system. *Brain Res.* **105**, 497–515.

Barker, J. L., Neale, J. H., and Bonner, W. M. (1977). Slab gel analysis of rapidly transported proteins in the isolated frog nervous system. *Brain Res.* **124**, 191–196.

Berry, R. W. (1979). Secretion of axonally transported neural peptides from the nervous system of *Aplysia*. *J. Neurobiol.* **10**, 499–508.

Berry, R. W., and Geinisman, Y. (1979). Interganglionic axonal transport of neural peptides within the nervous system of *Aplysia*. *J. Neurobiol.* **10**, 489–498.

Bisby, M. A. (1977). Similar polypeptide composition of fast-transported proteins in rat motor and sensory axons. *J. Neurobiol.* **8**, 303–314.

Bisby, M. A. (1978). Fast axonal transport of labeled protein in sensory axons during regeneration. *Exp. Neurol.* **61**, 281–300.

Black, M. M., and Lasek, R. J. (1978). A difference between the proteins conveyed in the fast component of axonal transport in guinea pig hypoglossal and vagus motor neurons. *J. Neurobiol.* **9**, 433–443.

Black, M. M., and Lasek, R. J. (1979). Axonal transport of actin: slow component b is the principal source of actin for the axon. *Brain Res.* **171**, 401–413.

Blaustein, M. P., Ratzlaff, R. W., and Kendrick, N. K. (1978). The regulation of intracellular calcium in presynaptic nerve terminals. *Ann. N.Y. Acad. Sci.* **307**, 195–211.

Bodian, D., and Howe, H. A. (1941). The rate of progression of poliomyelitis virus in nerves. *Bull. Johns Hopkins Hosp.* **69**, 79–85.

Bondy, S. C. (1971). Axonal transport of macromolecules I. Protein migration in the central nervous system. *Exp. Brain Res.* **13**, 127–134.

Branford White, C. J. (1979). Identification of glycosaminoglycans in nerve terminals. *J. Neurol. Sci.* **41**, 261–169.

Bray, J. J., and Austin, L. (1969). Axoplasmic transport of ^{14}C proteins at two rates in chicken sciatic nerve. *Brain Res.* **12**, 230–233.

Bretscher, M. S. (1971). Major human erythrocytes glycoprotein spans the cell membrane. *Nature (London) New Biol.* **231**, 229–232.

Brimijoin, S., and Wiermaa, M. J. (1977). Rapid axonal transport of tyrosine hydroxylase in rabbit sciatic nerves. *Brain Res.* **120**, 77–96.

Brimijoin, S., Capek, P., Dyck, P. J. (1973). Axonal transport of dipomine-β-hydroxylase by human sural nerves *in vitro*. *Science* **180**, 1295–1297.

Brimijoin, S., Lundberg, J. M., Brodin, E., Hökfelt, T., and Nilsson, G. (1980). Axonal transport of substance P in the vagus and sciatic nerves of the guinea pig. *Brain Res.* **191**, 443–457.

Cancalon, P., and Beidler, L. M. (1975). Distribution along the axon and into various subcellular fractions of molecules labeled with [^{3}H]leucine and rapidly transported in the garfish olfactory nerve. *Brain Res.* **89**, 225–244.

Cancalon, P., Elam, J. S., and Beidler, L. M. (1976). SDS gel electrophoresis of rapidly transported proteins in garfish olfactory nerve. *J. Neurochem.* **27**, 687–693.

Castejón, H. V. (1970). Histochemical demonstration of acid glycosaminoglycans in the nerve cell cytoplasm of mouse central nervous system. *Acta Histochem. Bd.* **35**, 161–172.

Churchill, L., Cotman, C., Banker, G., Kelly, P., and Shannon, L. (1976). Carbohydrate composition of central nervous system synapses. Analysis of isolated synaptic junctional complexes and postsynaptic densities. *Biochim. Biophys. Acta* **448**, 57–72.

Csanyi, V., Gervai, J., and Lajtha, A. (1973). Axoplasmic transport of free amino acids. *Brain Res.* **56**, 271–284.

Cuénod, M., and Schonbach, J. (1971). Synaptic proteins and axonal flow in the pigeon visual pathway. *J. Neurochem.* **18**, 809–816.

Currie, J. R., Grafstein, B., Whitnall, M. H., and Alpert, R. (1978). Axonal transport of Lipid in goldfish optic axons. *Neurochem. Res.* **3**, 479–492.

Dahlström, A. (1968). Effect of colchicine on transport of amine storage granules in sympathetic nerves of rat. *Eur. J. Pharmacol.* **5**, 111–113.

Dahlström, A., and Häggendal, J. (1966). Studies on the transport and life-span of amine storage granules in a peripheral adrenergic neuron system. *Acta Physiol. Scand.* **67**, 278–288.

Dahlström, A. B., Evans, C. A. N., Häggendal, J., Heiwall, P.-O., and Saunders, N. R. (1974). Rapid transport of acetylcholine in rat schiatic nerve proximal and distal to a lesion. *J. Neural Transm.* **35**, 1–11.

Di Giamberardino, L., and Couraud, J. Y. (1978). Rapid accumulation of high molecular weight acetylcholinesterase in transected sciatic nerve. *Nature (London)* **271**, 170–172.

Droz, B., and Leblond, C. P. (1963). Axonal migration of proteins in the central nervous system and peripheral nerves as shown by radioautography. *J. Comp. Neurol.* **121**, 325–337.

Droz, B., Koenig, H. L., and Di Giamberardino, L. (1973). Axonal migration of protein and glycoprotein to nerve endings. I. Radioautographic analysis of the renewal of protein in nerve endings of chicken ciliary ganglion after intracerebral injection of [³H] lysine. *Brain Res.* **60**, 93–127.

Droz, B., Rambourg, A., and Koenig, H. L. (1975). The smooth endoplasmic reticulum: structure and role in the renewal of axonal membrane and synaptic vesicles by fast axonal transport. *Brain Res.* **93**, 1–13.

Edström, A., and Mattsson, H. (1973). Electrophoretic characterization of leucine-glucosamine- and fucose-labelled proteins rapidly transported in frog sciatic nerve. *J. Neurochem.* **21**, 1499–1507.

Elam, J. S., and Agranoff, B. W. (1971). Rapid transport of protein in the optic system of the goldfish. *J. Neurochem.* **18**, 375–387.

Elam, J. S., Goldberg, J. M., Radin, N. S., and Agranoff, B. W. (1970). Rapid axonal transport of sulfated mucopolysaccharide proteins. *Science* **170**, 458–460.

Erickson, P. F., and Moore, B. W. (1980). Investigation of the axonal transport of three acidic, soluble proteins (14-3-2, 14-3-3, and S-100) in the rabbit visual system. *J. Neurochem.* **35**(1), 232–241.

Erickson, P. F., Seamon, K. B., Moore, B. W., Lasher, R. S., and Minier, L. N. (1980). Axonal transport of the Ca^{2+}-dependent protein modulator of 3′:5′-cyclic-AMP phosphodiesterase in the rabbit visual system. *J. Neurochem.* **35**(1), 242–248.

Estridge, M., and Bunge, R. (1978). Compositional analysis of growing axons from rat sympathetic neurons. *J. Cell Biol.* **79**, 138–155.

Feit, H., Dutton, G. R., Barondes, S. H., and Shelanski, M. L. (1971). Microtubule protein. Identification and transport to nerve endings. *J. Cell Biol.* **51**, 138–147.

Fink, D. J., and Gainer, H. (1980). Axonal transport of proteins. A new view using *in vivo* covalent labeling. *J. Cell Biol.* **85**, 175–186.

Fischer, H. A., and Schmatolla, E. (1972). Axonal transport of tritium-labeled putrescine in the embryonic visual system of Zebrafish. *Science* **176**, 1327–1329.

Fonnum, F., Frizell, M., and Sjöstrand, J. (1973). Transport, turnover and distribution of choline acetyltransferase and acetylcholin esterase in the vagus and hypoglossal nerves of the rabbit. *J. Neurochem.* **21**, 1109–1120.

Forman, D. S., and Ledeen, R. W. (1972). Axonal transport of gangliosides in the goldfish optic nerve. *Science* **177**, 630–632.

Forman, D. S., Padjen, A. L., and Siggins, G. R. (1977). Axonal transport of organelles visualized by light microscopy: Cinemicrographic and computer analysis. *Brain Res.* **136**, 197–213.

Frizell, M., and Sjöstrand, J. (1974). The axonal transport of slowly migrating [³H] leucine labelled proteins and the regeneration rate in regenerating hypoglossal and vagus nerves of the rabbit. *Brain Res.* **81**, 267–283.

Gainer, H., Sarne, Y., and Brownstein, M. J. (1977). Biosynthesis and axonal transport of rat neurohypophysial proteins and peptides. *J. Cell Biol.* **73**, 366–381.

Gamse, R., Lembeck, F., and Cuello, A. C. (1979). Substance P in the vagus nerve. *Naunyn-Schmiedeberg's Arch. Pharmacol.* **306**, 37–44.

Giachetti, A., and Said, S. I. (1979). Axonal transport of vasoactive intestinal peptide in sciatic nerve. *Nature (London)* **281**, 574–575.

Gilbert, R. F. T., Emson, P. C., Fahrenkrug, J., Lee, C. M., Penman, E., and Wass, J. (1980). Axonal transport of neuropeptides in the cervical vagus nerve of the rat. *J. Neurochem.* **34**(1), 108–113.

Goodrum, J. F., Toews, A. D., and Morell, P. (1979). Axonal transport and metabolism of [³H]fucose- and [³⁵S]-sulfate-labeled macromolecules in the rat visual system. *Brain Res.* **176**, 255–272.

Grafstein, B. (1967). Transport of protein by goldfish optic nerve fibres. *Science* **157**, 196–198.

Grafstein, B., and Forman, D. S. (1980). Intracellular Transport in Neurons. *Physiol. Rev.* **60**, 1167–1283.

Grafstein, B., Miller, J. A., Ledeen, R. W., Haley, J., and Specht, S. C. (1975). Axonal transport of phospholipid in goldfish optic system. *Exp. Neurol.* **46**, 261–281.

Gross, G. V. (1975). The microstream concept of axoplasmic and dendritic transport. *Adv. Neurol.* **12**, 283–296.

Gross, G. W., and Beidler, L. M. (1973). Fast axonal transport in the C-fibers of the garfish olfactory nerve. *J. Neurobiol.* **4**, 413–428.

Gross, G. W., and Beidler, L. M. (1975). A quantitative analysis of isotope concentration profiles and rapid transport velocities in the C-fibers of the garfish olfactory nerve. *J. Neurobiol.* **6**, 213–232.

Gross, G. W., and Kreutzberg, G. W. (1978). Rapid axoplasmic transport in the olfactory nerve of the pike: I. Basic transport parameters for proteins and amino acids. *Brain Res.* **139**, 65–76.

Gruener, R., and Bryant, H. J. (1975). Excitability modulation by taurine: action on axon membrane permeabilities. *J. Pharmacol. Exp. Ther.* **194**, 514–521.

Gunning, P. W., Por, S. B., Langford, C. J., Scheffer, J., Austin, L., and Jeffrey, P. L. (1979). The direct measurement of the axoplasmic transport of individual RNA species: Transfer but not ribosomal RNA is transported. *J. Neurochem.* **32**, 1737–1743.

Gurd, J. W. (1977). Synaptic plasma membrane glycoproteins: Molecular identification of lectin receptors. *Biochemistry* **16**, 369–374.

Gurd, J. W. (1979). Molecular and biosynthetic heterogeneity of fucosyl glycoproteins associated with rat brain synaptic junctions. *Biochim. Biophys. Acta* **555**, 221–229.

Haley, J. E., Tirri, L. J., and Ledeen, R. W. (1979). Axonal transport of lipids in the rabbit optic system. *J. Neurochem.* **32**, 727–734.

Hammerschlag, R., Dravid, A. R., and Chin, A. Y. (1975). Mechanisms of axonal transport: A proposed role for calcium ions. *Science* **188**, 273–275.

Hanson, M., and Edström, A. (1978). Mitosis inhibitors and axonal transport. *Int. Rev. Cytol.* *Suppl.* **7**, 373–402.

Harik, S. I., Wehle, S. U., and Schwerin, F. (1979). Axonal transport of polyamines: A reappraisal. *Exp. Neurol.* **63**, 311–321.

Hariri, M., and Hammerschlag, R. (1977). Axonal transport of glutamic acid decarboxylase in crayfish peripheral nerve: Dependence on contact between soma and axon. *Neurosci. Lett.* **7**, 319–325.

Hayes, K. C. (1976). A review on the biological function of taurine. *Nutr. Rev.* **34**, 161–165.

Helenius, A., and Simons, K. (1975). Solubilization of membranes by detergents. *Biochim. Biophys. Acta* **415**, 29–79.

Hendrickson, A. E. (1972). Electron microscopic distribution of axoplasmic transport. *J. Comp. Neurol.* **144**, 381–398.

Hendrickson, A. E., and Cowan, W. M. (1971). Changes in the rate of axoplasmic transport during postnatal development of the rabbits optic nerve and tract. *Exp. Neurol.* **30**, 403–422.

Heslop, J. P. (1975). Axonal flow and fast transport in nerves. *Adv. Comp. Physiol. and Biochem.* **6**, 75–163.

Hines, J. F., and Garwood, M. M. (1977). Release of protein from axons during rapid axonal transport: An *in vitro* preparation. *Brain Res.* **125**, 141–148.

Hoffman, P. N., and Lasek, R. J. (1975). The slow component of axonal transport. Identification of major structural polypeptides of the axon and their generality among mammalian neurons. *J. Cell Biol.* **66**, 351–366.

Hökfelt, T., Kellerth, J.-O., Nilsson, G., and Pernow, B. (1975). Experimental immunohistochemical studies on the localization and distribution of substance P in cat primary sensory neurons. *Brain Res.* **100**, 235–252.

Holton, P. (1959). Further observations on substance P in degenerating nerve. *J. Physiol.* (*London*) **149**, 35–36.

Hunt, S. P., and Künzle, H. (1976). Bidirectional movement of label and transneuronal transport phenomena after injection of [³H] adenosine into the central nervous system. *Brain Res.* **112**, 127–132.

Ingoglia, N. A., Sturman, J. A., Lindquist, T. D., and Gaull, E. (1976). Axonal migration of taurine in the garfish visual system. *Brain Res.* **115**, 535–539.

Ingoglia, N. A., Sturman, J. A., and Eisner, R. A. (1977). Axonal transport of putrescine, spermidine and spermine in normal and regenerating goldfish optic nerves. *Brain Res.* **130**, 433–445.

Iqbal, Z., and Ochs, S. (1978). Fast axoplasmic transport of a calcium-binding protein in mammalian nerve. *J. Neurochem.* **31**, 409–418.

James, K. A. C., and Austin, L. (1970). The binding in *vitro* of colchicine to axoplasmic proteins from chicken sciatic nerve. *Biochem. J.* **117**, 773–777.

Jarrott, B., and Geffen, L. B. (1972). Rapid axoplasmic transport of tyrosine hydroxylase in relation to other cytoplasmic constituents. *Proc. Natl. Acad. Sci. U.S.A.* **69**, 3440–3442.

Karlsson, J.-O. (1976). Proteins of axonal transport: investigation of solubility characteristics and behaviour in gel filtration. *J. Neurochem.* **27**, 1135–1143.

Karlsson, J.-O. (1977). Is there an axonal transport of amino acids? *J. Neurochem.* **29**, 615–617.

Karlsson, J.-O. (1978). Proteins of axonal transport: gel filtration in sodium dodecyl sulphate of components with different solubility characteristics. *J. Neurochem.* **31**, 257–260.

Karlsson, J.-O. (1979). Proteins of axonal transport: Interaction of rapidly transported proteins with lectins. *J. Neurochem.* **32**, 491–494.

Karlsson, J.-O. (1980). Proteins of rapid axonal transport: Polypeptides interacting with the lectin from *Lens Culinaris*. *J. Neurochem.* **34**(5), 1184–1190.

Karlsson, J.-O., and Linde, A. (1977). Axonal transport of [^{35}S] sulphate in retinal ganglion cells of the rabbit. *J. Neurochem.* **28**, 293–297.

Karlsson, J.-O., and Sjöstrand, J. (1968). Transport of labelled proteins in the optic nerve and tract of the rabbit. *Brain Res.* **11**, 431–439.

Karlsson, J.-O., and Sjöstrand, J. (1969). The effect of colchicine on the axonal transport of protein in the optic nerve and tract of the rabbit. *Brain Res.* **13**, 617–619.

Karlsson, J.-O., and Sjöstrand, J. (1971a). Synthesis, migration and turnover of protein in retinal ganglion cells. *J. Neurochem.* **18**, 749–767.

Karlsson, J.-O., and Sjöstrand, J. (1971b). Transport of microtubular protein in axons of retinal ganglion cells. *J. Neurochem.* **18**, 975–982.

Karlsson, J.-O., and Sjöstrand, J. (1971c). Rapid intracellular transport of fucose-containing glycoproteins in retinal ganglion cells. *J. Neurochem.* **18**, 2209–2216.

Karlsson, J.-O., and Sjöstrand, J. (1971d). Characterization of the fast and slow components of axonal transport in retinal ganglion cells. *J. Neurobiol.* **2**, 135–143.

Karlsson, J.-O., and Sjöstrand, J. (1971e). Electrophoretic characterization of rapidly transported proteins in axons of retinal ganglion cells. *FEBS Lett.* **16**, 329–332.

Karlsson, J.-O., and Sjöstrand, J. (1972). Axonal transport of proteins in retinal ganglion cells. Amino acid incorporation into rapidly transported proteins and distribution of radioactivity to the lateral geniculate body and the superior colliculus. *Brain Res.* **37**, 279–285.

Kelly, P. T., and Cotman, C. W. (1977). Identification of glycoproteins and proteins at synapses in the central nervous system. *J. Biol. Chem.* **252**, 786–793.

Kelly, A. S., Wagner, J. A., and Kelly, R. B. (1980). Properties of individual nerve terminal proteins identified by two-dimensional gel electrophoresis. *Brain Res.* **185**, 192–197.

Khan, M. A., and Ochs, S. (1974). Magnesium or calcium activated ATPase in mammalian nerve. *Brain Res.* **81**, 413–426.

Khan, M. A., and Ochs, S. (1975). Slow axoplasmic transport of mitochondria (MAO) and lactic dehydrogenase in mammalian nerve fibers. *Brain Res.* **96**, 267–277.

Kidwai, A. M., and Ochs, S. (1969). Components of fast and slow phases of axoplasmic flow. *J. Neurochem.* **16**, 1105–1112.

Knull, H. R., and Wells, W. W. (1975). Axonal transport of cations in the chick optic system. *Brain Res.* **100**, 121–124.

Koike, H., Eisenstade, M., and Schwartz, J. H. (1972). Axonal transport of newly synthesized acetylcholine in an identified neuron of *Aplysia*. *Brain Res.* **37**, 152–159.

Komiya, Y., and Kurokawa, M. (1980). Preferential blockade of the tubulin transport by colchicine. *Brain Res.* **190**, 505–516.

Kreutzberg, G. W. (1969). Neuronal dynamics and axonal flow. IV. Blockage of intra-axonal enzyme transport by colchicine. *Proc. Natl. Acad. Sci. U.S.A.* **62**, 722–728.

Kristensson, K., and Olsson, Y. (1971). Retrograde axonal transport of protein. *Brain Res.* **29**, 363–365.

Kristensson, K., Lycke, E., and Sjöstrand, J. (1971). Spread of Herpes simplex virus in peripheral nerves. *Acta Neuropathol.* **17**, 44–53.

Krygier-Brévart, V., Weiss, D. G., Mehl, E., Schubert, P., and Kreutzberg, W. (1974). Maintenance of synaptic membranes by the fast axonal flow. *Brain Res.* **77**, 97–110.

Laduron, P., and Belpaire, F. (1968). Transport of noradrenaline and dopamine-β-hydroxylase in sympathetic nerves. *Life Sci.* **7**, 1–7.

Landa, C. A., Maccioni, H. J. F., and Caputto, R. (1979). The site of synthesis of gangliosides in the chick optic system. *J. Neurochem.* **33**, 825–838.

Lasek, R. J. (1966). Axoplasmic streaming in the cat dorsal root ganglion cell and the rat ventral motoneuron. *Anat. Rec.* **154**, 373–374.

Lasek, R. J. (1968a). Axoplasmic transport in cat dorsal root ganglion cells: As studied by [³H]-l-leucine. *Brain Res.* **7**, 360–377.

Lasek, R. J. (1968b). Axoplasmic transport of labeled proteins in rat ventral motoneurons. *Exp. Neurol.* **21**, 41–51.

Lasek, R. J. (1980). Axonal transport: A dynamic view of neuronal structures. *Trends Neurosci. (Pers. Ed.)* **3**, 87–91.

Lasek, R. J., and Hoffman, P. N. (1976). The neuronal cytoskeleton, axonal transport and axonal growth. *In* "Cell Motility" (R. Goldman, T. Pollard, and J. Rosenbaum, eds.) pp. 1021–1049. Cold Spring Harbor Lab., Cold Spring Harbor, New York.

Lasek, R. J., Dabrowski, C., and Nordlander, R. (1973). Analysis of axoplasmic RNA from invertebrate giant axons. *Nature (London) New Biol.* **244**, 162–165.

Lasek, R. J., Gainer, H., and Pryzbylski, R. J. (1974). Transfer of newly synthesized proteins from Schwann cells to the squid giant axon. *Proc. Natl. Acad. Sci. U.S.A.* **71**, 1188–1192.

Lasek, R. J., Gainer, H., and Barker, J. L. (1977). Cell-to-cell transfer of glial proteins to the squid giant axon. *J. Cell Biol.* **74**, 501–523.

LaVail, J. H., Rapisardi, S., and Sugino, I. K. (1980). Evidence against the smooth endoplasmic reticulum as a continuous channel for the retrograde axonal transport of horseradish peroxidase. *Brain Res.* **191**, 3–20.

Liem, R. K., Yen, S.-H., Salomon, G. D., and Shelanski, M. L. (1978). Intermediate filaments in nervous tissue. *J. Cell Biol.* **79**, 637–645.

Lindquist, T. D., and Ingoglia, N. A. (1979). Evidence that 4S RNA is axonally transported in normal and regenerating rat sciatic nerves. *Brain Res.* **166**, 95–112.

Livett, B. G., Geffen, L. B., Austin, L. (1968). Axoplasmic transport of ¹⁴C-noradenaline and protein in splenic nerves. *Nature (London)* **217**, 278–279.

Longo, F. M., and Hammerschlag, R. (1980). Relation of somal lipid synthesis to the fast axonal transport of protein and lipid. *Brain Res.* **193**, 471–485.

Lorenz, T., and Willard, M. (1978). Subcellular fractionation of intraaxonally transported polypeptides in the rabbit visual system. *Proc. Natl. Acad. Sci. U.S.A.* **75**, 505–509.

Lubińska, L., and Niemierko, S. (1971). Velocity and intensity of bidirectional migration of acetylcholinesterase in transected nerves. *Brain Res.* **27**, 329–342.

Lux, H. D., Schubert, P., Kreutzberg, G. W., and Globus, A. (1970). Excitation and axonal flow: Autoradiographic study on motoneurons intracellularly injected with a ³H-amino acid. *Exp. Brain. Res.* **10**, 197–204.

McEwen, B. S., and Grafstein, B. (1968). Fast and slow components in axonal transport of protein. *J. Cell Biol.* **38**, 494–508.

McGregor, Jeffrey, P. L., Klingman, J. D., and Austin, L. (1973). Axoplasmic flow of cholesterol in chicken sciatic nerve. *Brain Res.* **63**, 466–469.

McLean, W. G., Frizell, M., and Sjöstrand, J. (1976). Slow axonal transport of labelled proteins in sensory fibres of rabbit vagus nerve. *J. Neurochem.* **26**, 1213–1216.

Marchesi, V. T., and Andrews, E. P. (1971). Glycoproteins: Isolation from cell membranes with lithium diiodosalicylate. *Science* **174**, 1247–1248.

Margolis, F. L., and Grillo, M. (1977). Axoplasmic transport of carnosine (β-alanyl-L-histidine) in the mouse olfactory pathway. *Neurochem. Res.* **2**, 507–519.

Margolis, R. U., and Margolis, R. K. (1974). Distribution and metabolism of mucopolysaccharides and glycoproteins in neuronal perikarya, astrocytes, and oligodendroglia. *Biochemistry* **13**, 2849–2852.

Margolis, R. K., Thomas, M. D., Crockett, C. P., and Margolis, R. U. (1979). Presence of chondroitin sulfate in the neuronal cytoplasm. *Proc. Natl. Acad. Sci. U.S.A.* **76**, 1711–1715.

Marko, P., Susz, J.-P., and Cuénod, M. (1971). Synaptosomal proteins and axoplasmic flos: Fractionation by SDS polyacrylamide gel electrophoresis. *FEBS Lett.* **17**, 261–264.

Markov, D., Rambourg, A., and Droz, B. (1976). Smooth endoplasmic reticulum and fast axonal transport of glycoproteins, an electron microscope radioautographic study of thick sections after heavy metals impregnation. *J. Microsc. Biol. Cell* **125**, 57–60.

Meek, J. L., and Neff, N. H. (1972). Tryptophan 5-hydroxylase: Approximation of half-life and rate of axonal transport. *J. Neurochem.* **19**, 1519–1525.

Miani, N. (1963). Analysis of the somato-axonal movement of phospholipids in the vagus and hypoglossal nerves. *J. Neurochem.* **10**, 859–874.

Neale, J. H., and Barker, L. (1977). Bidirectional axonal transport of $^{45}Ca^{2+}$: Studies in isolated frog sensory motor and sympathetic neurons, *Aplysia* cerebral ganglion and the goldfish visual system. *Brain Res.* **129**, 45–59.

Norström, A., and Sjöstrand, J. (1971a). Axonal transport of proteins in the hypothalamo-neurohypophysial system of the rat. *J. Neurochem.* **18**, 29–39.

Norström, A., and Sjöstrand, J. (1971b). Effect of haemorrhage on the rapid axonal transport of neurohypophysial proteins of the rat. *J. Neurochem.* **18**, 2017–2026.

Ochs, S. (1971). Characteristics and a model for fast axoplasmic transport in nerve. *J. Neurobiol.* **2**, 331–345.

Ochs, S. (1972a). Rate of fast axoplasmic transport in mammalian nerve fibers. *J. Physiol.* **227**, 627–645.

Ochs, S. (1972b). Fast transport of materials in mammalian nerve fibers. *Science* **176**, 252–260.

Ochs, S., and Johnson, J. (1969). Fast and slow phases of axoplasmic flow in ventral root nerve fibers. *J. Neurochem.* **16**, 845–853.

Ochs, S., Sabri, M. I., and Ranish, N. (1970). Somal site of synthesis of fast transported materials in mammalian nerve fibers. *J. Neurobiol.* **1**, 329–344.

Ochs, S., Worth, R. M., and Chan, S. Y. (1977). Calcium requirement for axoplasmic transport in mammalian nerve. *Nature (London)* **270**, 748–750.

Padilla, S. S., Roger, L. J., Toews, A. D., Goodrum, J. F., and Morell, P. (1979). Comparison of proteins transported in different tracts of the central nervous system. *Brain Res.* **176**, 407–411.

Palay, S. L., and Palade, G. E. (1955). The fine structure of neurons. *J. Biophys. Biochem. Cytol.* **10**, 69–88.

Pant, H. C., and Gainer, H. (1980). Properties of a calcium-activated protease in squid axoplasm which selectively degrades neurofilament proteins. *J. Neurobiol.* **11**, 1–12.

Partlow, L. M., Ross, C. D., Motwani, R., and McDougal, Jr., D. B. (1972). Transport of axonal enzymes in surviving segments of frog sciatic nerve. *J. Gen. Physiol.* **60**, 388–405.

Peterson, R. P., Hurwitz, R. M., and Lindsay, R. (1967). Migration of axonal protein: absence of a protein concentration gradient and effect of inhibition of protein synthesis. *Exp. Brain. Res.* **4**, 138–145.

Politis, M. J., and Ingoglia, N. A. (1979a). Axonal transport of taurine along neonatal and young adult rat optic axons. *Brain Res.* **166**, 221–231.

Politis, M. J., and Ingoglia, N. A. (1979b). Axonal transport of nucleosides, nucleotides and 4S RNA in the neonatal rat visual system. *Brain Res.* **169**, 343–356.

Price, C. H., McAdoo, D. J., Farr, W., and Okuda, R. (1979). Bidirectional axonal transport of free glycine in identified neurons R3-R14 of *Aplysia*. *J. Neurobiol.* **10**, 551–571.

Price, D. L., Griffin, J., Young, A., Peck, K., Stocks, A. (1975). Tetanus toxin: Direct evidence for retrograde intraaxonal transport. *Science* **188**, 945–947.

Rambourg, A., and Droz, B. (1980). Smooth endoplasmic reticulum and axonal transport. *J. Neurochem.* **35**(1), 16–25.

Roberts, P. J., Keen, P., and Mitchell, J. F. (1973). The distribution and axonal transport of free amino acids and related compounds in the dorsal sensory neuron of the rat, as determined by the dansyl reaction. *J. Neurochem.* **21**, 199, 209.

Rostas, J. A., McGregor, A., Jeffrey, P. L., and Austin, L. (1975). Transport of cholesterol in the chick optic system. *J. Neurochem.* **24**, 295–302.

Rostas, J. A., Austin, L., and Jeffrey, P. L. (1979). Selective labelling of two phases of axonal transport of cholesterol in the chick optic system. *J. Neurochem.* **32**, 1461–1466.

Rösner, H. (1975). Incorporation of silalic acid into gangliosides and glycoproteins of the optic pathway following an intraocular injection of (N-^{3}H) acetylmannosamine in the chicken. *Brain Res.* **97**, 107–116.

Sandberg, M., Hamberger, A., Karlsson, J.-O., and Tirillini, B. (1980a). Potassium stimulated release of axonally transported radioactivity from slices of rabbit superior colliculus. *Brain Res.* **188**, 175–183.

Sandberg, M., Hamberger, A., Jacobson, I., and Karlsson, J.-O. (1980b). The role of calcium ions in the formation and release of small molecular weight substances from optic nerve terminals. *Neurochem. Res.* **5**, 1185–1198.

Sandoval, I. V., and Weber, K. (1978). Calcium-induced inactivation of microtubule formation in brain extracts. Presence of a calcium-dependent protease acting on polymerization-stimulating microtubule-associated proteins. *Eur. J. Biochem.* **92**, 463–470.

Schafer, R. (1973). Acetylcholine: Fast axoplasmic transport in insect chemoreceptor fibers. *Science* **180**, 315–316.

Schmitt, F. O. (1968). Fibrous proteins—neuronal organelles. *Proc. Natl. Acad. Sci. U.S.A.* **60**, 1092–1101.

Schonbach, J., Schonbach, Ch., and Cuénod, M. (1971). Rapid phase of axoplasmic flow and synaptic proteins: An electron microscopical autoradiographic study. *J. Comp. Neurol.* **141**, 485–498.

Schubert, P., and Kreutzberg, G. W. (1974). Axonal transport of adenosine and uridine derivatives and transfer to postsynaptic neurons. *Brain Res.* **76**, 526–530.

Schwab, M. E., Suda, K., and Thoenen, H. (1979). Selective retrograde transsynaptic transfer of a protein, Tetanus Toxin, subsequent to its retrograde axonal transport. *J. Cell Biol.* **82**, 798–8.0.

Schwartz, J. H. (1979). Axonal transport: Components, mechanisms, and specificity. *Annu. Rev. Neurosci.* **2**, 467–504.

Segrest, J. P., Jackson, R. L., Andrews, E. P., and Marchesi, V. T. (1971). Human erythrocyte membrane glycoprotein: A re-evaluation of the molecular weight as determined by SDS polyacrylamide gel electrophores.s *Biochem. Biophys. Res. Commun.* **44**, 390–394.

Sherbany, A. A., Ambron, R. T., and Schwartz, J. H. (1978). Membrane glycolipids: Regional synthesis and axonal transport in a single identified neuron of *Aplysia californica*. *Science* **203**, 78–81.

Siegel, L. G., and McClure, W. O. (1975a). Fractionation of protein carried by axoplasmic transport II. Comparison in the rat of proteins carried to the optic relay nuclei. *Neurobiology* **5**, 167–177.

Siegel, L. G., and McClure, W. O. (1975b). Putrescine: Effect on axoplasmic transport. *Brain Res.* **93**, 543–547.

Sjöstrand, J., and Karlsson, J.-O. (1969). Axoplasmic transport in the optic nerve and tract of the rabbit: A biochemical and radioautographic study. *J. Neurochem.* **16**, 833–844.

Smith, R. S. (1971). Centripetal movement of particles in myelinated neurons. *Cytobios* **3**, 259–262.

Stöckel, K., Paravicini, U., and Thoenen, H. (1974). Specificity of the retrograde axonal transport of nerve growth factor. *Brain Res.* **76**, 413–421.

Stöckel, K., Schwab, M., and Thoenen, H. (1977). Role of gangliosides in the uptake and regrograde axonal transport of cholera and tetanus toxin as compared to nerve growth factor and wheat germ agglutinin. *Brain Res.* **132**, 273–285.

Stone, G. C., Wilson, D. L., and Hall, M. E. (1978). Two-dimensional gel electrophoresis of proteins in rapid axoplasmic transport. *Brain Res.* **144,** 287–302.

Streit, P., Knecht, E., and Cuénod, M. (1979). Transmitter-specific retrograde labeling in the striato-nigral and raphe-nigral pathways. *Science* **205,** 306–308.

Sturman, J. A. (1979). Taurine in the developing rabbit visual system: changes in concentration and axonal transport including a comparison with axonally transported proteins. *J. Neurobiol.* **10,** 221–237.

Theiler, R. F., and McClure, W. O. (1977). A comparison of axonally transported protein in the rat sciatic nerve by *in vitro* and *in vivo* techniques. *J. Neurochem.* **28,** 321–330.

Toews, A. D., Goodrum, J. F., and Morell, P. (1979). Axonal transport of phospholipids in rat visual system. *J. Neurochem.* **32,** 1165–1173.

Wagner, J. A., Kelly, A. S., and Kelly, R. B. (1979). Nerve terminal proteins of the rabbit visual relay nuclei identified by axonal transport and two-dimensional gel electrophoresis. *Brain Res.* **168,** 97–117.

Weiss, P. A. (1972). Neuronal dynamics and axonal flow: Axonal peristalsis. *Proc. Natl. Acad. Sci. U.S.A.* **69,** 1309–1312.

Weiss, P., and Hiscoe, H. B. (1948). Experiments on the mechanism of nerve growth. *J. Exp. Zool.* **107,** 315–396.

Weiss, D. G., Krygier-Brévart, V., Gross, G. W., and Kreutzberg, G. W. (1978). Rapid axoplasmic transport in the olfactory nerve of the pike: II. Analysis of transported proteins by SDS gel electrophoresis. *Brain Res.* **139,** 77–87.

Weiss, D. G., Krygier-Brévart, V., Mehl, E., and Kreutzberg, G. W. (1979). Subcellular distribution of proteins delivered to the synapse by rapid axoplasmic transport. *Biol. Cell.* **34,** 59–64.

White, F. P., and White, S. R. (1977). Characterization of proteins transported at different rates by axoplasmic flow in the dorsal root afferents of rats. *J. Neurobiol.* **8,** 315–324.

Willard, M. (1977). The identification of two intra-axonally transported polypeptides resembling myesin in some respects in the rabbit visual system. *J. Cell Biol.* **75,** 1–11.

Willard, M. B., and Hulebak, K. L. (1977). The intra-axonal transport of polypeptide H: Evidence for a fifth (very slow) group of transported proteins in the retinal ganglion cells of the rabbit. *Brain Res.* **136,** 289–306.

Willard, M., Cowan, W. M., and Vagelos, P. R. (1974). The polypeptide composition of intra-axonally transported proteins: Evidence for four transport velocities. *Proc. Natl. Acad. Sci. U.S.A.* **71,** 2183–2187.

Willard, M., Wiseman, M., Levine, J., and Skene, P. (1979). Axonal transport of actin in rabbit reginal ganglion cells. *J. Cell Biol.* **81,** 581–591.

Wilson, D. L., and Stone, G. C. (1979). Axoplasmic transport of proteins. *Annu. Rev. Biophys. Bioeng.* **8,** 27–45.

Wise, S. P., Jones, E. G., and Berman, N. (1978). Direction and specificity of the axonal and transcellular transport of nucleosides. *Brain Res.* **139,** 197–217.

Wooten, G. F. (1973). Subcellular distribution and rapid axonal transport of dopamine-βhydroxylase. *Brain Res.* **55,** 491–494.

Zanetta, J. P., Morgan, I. G., and Gombos, G. (1975). Synaptosomal plasma membrane glycoproteins: Fractionation by affinity chromatography on concanavalin A. *Brain Res.* **83,** 337–348.

6

MECHANISTIC STUDIES ON THE CELLULAR EFFECTS OF NERVE GROWTH FACTOR

David E. Burstein and Lloyd A. Greene

Identification of the chemical signals that regulate cell growth and maturation is an activity attracting great current interest in developmental neurobiology. The mechanisms of cellular responses to such "growth factors" are under intense investigation. We discuss in this chapter studies conducted in our laboratory and in others on the mechanism of action of nerve growth factor (NGF), one of the first such signals to be identified and chemically characterized. In particular, we shall stress current attempts to understand NGF mechanism of action at the molecular level.

I. NGF AND ITS TARGET CELLS

NGF activity was first detected in sarcoma tumors which evoked outgrowth from dorsal root ganglia of chick embryos into which they were implanted (for

Molecular Approaches to Neurobiology

reviews see Levi-Montalcini, 1966; Levi-Montalcini and Angeletti, 1968). Activity was subsequently found in snake venom, in the adult male mouse salivary gland (this being an extremely potent source of the factor), in tissue culture medium conditioned by various types of cells such as glia, fibroblasts, skeletal and cardiac muscle (for review see Bradshaw, 1978), and recently, in guinea pig prostate (Harper *et al.*, 1979). Mouse salivary NGF has been purified and is presently the most used and best-characterized form of the factor (for reviews, see Bradshaw, 1978; Greene and Shooter, 1980).

Classic experiments done by Levi-Montalcini and co-workers established a critical role for NGF in the development and survival of sympathetic and dorsal root ganglionic (DRG) neurons of vertebrates (for reviews see Levi-Montalcini, 1966; Levi-Montalcini and Angeletti, 1968). Animals and embryos given NGF develop enlargement of ganglia, which is due in large part to neuronal hypertrophy as well as massively increased neurite outgrowth. Newborn animals given antibody to NGF undergo degeneration of the sympathetic nervous system. Exposure of embryos to anti-NGF antibodies also results in degeneration of DRG neurons (Gorin and Johnson, 1979). Another important action of NGF is to regulate the levels of several neurotransmitters (norepinephrine and substance P) in its target cells (Thoenen *et al.*, 1971; Kessler and Black, 1980).

The effects of NGF can also be demonstrated *in vitro*. Explanted ganglia cultured in the presence of NGF undergo rapid regeneration of neurites, and survive many weeks in culture. In contrast, neurons of ganglia cultured without NGF do not regenerate neurites and do not survive. The successful *in vitro* cultivation of explanted DRG and sympathetic neurons provided the first accessible experimental system for examining the mechanism of action of NGF on its target cells. In early studies that compared the capacities of ganglia cultured with and without NGF to incorporate labeled leucine and uridine, it was hypothesized that NGF worked by selectively regulating the transcription of specific genes, which in turn led to the synthesis of specific proteins (Levi-Montalcini and Angeletti, 1968). However, Partlow and Larrabee (1971) showed that sympathetic neurons cultured with NGF and actinomycin D still underwent complete regeneration of neurites, despite over 90% blockade of RNA synthesis. In contrast, inhibitors of protein synthesis were found to block neurite outgrowth. These workers concluded that the mechanism of NGF action was post-transcriptional.

The above and other experiments on the NGF mechanism which used explanted ganglionic neurons as target cells have two significant difficulties in interpretation. First, the neurons used as "control" cells—those not receiving NGF—were dying cells. Second, the cells studied were presumably exposed to NGF *in vivo* before explantation into culture; hence the role of long-term pre-exposure to NGF in responses such as cell survival and neurite outgrowth could not be assessed.

The discovery (Tischler and Greene, 1975) of a new target of NGF, the pheochromocytoma cell, has been the basis for new insights into the molecular changes induced by the factor. In 1976, a single cell clonal line (designated PC12) was established from a transplantable rat pheochromocytoma (Greene and Tischler, 1976). In the absence of NGF, PC12 cells display the features of their non-neoplastic counterparts, adrenal chromaffin cells (Greene and Tischler, 1976; Tischler and Greene, 1978). For example, PC12 cells contain chromaffin-like granules and synthesize, store, and release catecholamines (Greene and Rein, 1977a,b). The response of PC12 cells to NGF converts them into cells that are strikingly similar to sympathetic neurons. Within 4–7 days of NGF exposure, they cease cell division, increase threefold in volume, aggregate into ganglion-like clusters, and slowly extend long neurites, both as individual fibers and in bundles (Greene and Tischler, 1976; Tischler and Greene, 1978). In addition, they acquire electrical excitability (Dichter *et al.*, 1977). Neuronal differentiation of PC12 cells, unlike that of DRG or sympathetic neurons, is reversible; upon removal of NGF from cultures that have undergone neuronal differentiation, the neurites degenerate, and in 2–3 days, the cells resume division (Greene and Tischler, 1976).

PC12 cells have properties that provide certain unique experimental advantages for studying NGF mechanism of action.

1. Unlike ganglionic neurons, PC12 cells do not require NGF for survival. Hence control cells, those not cultured in the presence of NGF, are not dying cells.

2. Unlike DRG and sympathetic neurons that are exposed to NGF *in vivo* prior to explantation, PC12 cells can be used for examining the initial effects of NGF on previously unexposed target cells.

3. Large quantities of PC12 cells can easily be obtained for biochemical studies.

4. Mutant cells can be isolated, allowing for genetic approaches to the mechanism of action of NGF (Section VII).

5. The cells dedifferentiate as well as differentiate. The changes accompanying dedifferentiation yield insights into the process of neuronal differentiation (Section III).

II. ROLE OF TRANSCRIPTION IN NGF MECHANISM OF ACTION ON NEURITE OUTGROWTH

A major question regarding the molecular mechanism of NGF action is the role of transcription. That is, does NGF work at the level of the genome? As noted above, resolution of this issue with cultured ganglia has been difficult. Because of their properties, PC12 cells have been extremely useful in this regard. As discussed below, differences in behavior of PC12 cells with or without pretreatment with NGF have been particularly revealing.

PC12 cells exposed to NGF for 1 week or longer can be mechanically divested of neurites by trituration and replated. Such cells recultured in the absence of NGF survive, but do not regenerate neurites (Greene, 1977; Burstein and Greene, 1978). If recultured in the presence of NGF, on the other hand, such cells rapidly regenerate long neurites at a rate of 200–300 μm/day in the first 24 h (Burstein and Greene, 1978; Greene *et al.*, 1980). If such NGF-pretreated cells are passaged into medium with NGF and with actinomycin D or other RNA synthesis inhibitors at doses that completely block transcription, neurite regeneration occurs with nearly the same efficiency as with cells passaged into the presence of NGF alone. Hence, PC12 cells pre-exposed to NGF for 1 week or longer behave analogously to explanted DRG or sympathetic neurons. That is, regeneration requires the presence of NGF, occurs at a rapid rate, and does not require *de novo* RNA synthesis. In addition, as with the neurons, neurite regeneration by PC12 cells is completely blocked by protein synthesis inhibitors at translation-blockading doses (D. E. Burstein and L. A. Greene, unpublished data).

In contrast to regeneration by NGF pretreated cells, initiation of neurite outgrowth by PC12 cells without pre-exposure to NGF is slow. After a lag period of about 18 h, neurite elongation by such cells occurs at a linear rate of about 20–40 μm/day (Burstein and Greene, 1978; Greene *et al.*, 1980). Significantly, in contrast to regeneration from pre-exposed cells, NGF-induced neurite outgrowth from previously unexposed cells is completely blocked by drugs that inhibit RNA synthesis (Burstein and Greene, 1978). One of these drugs—camptothecin—is a rapidly reversible RNA synthesis inhibitor. Upon removal of camptothecin but not of NGF, neurite outgrowth commences at the usual rate of 20–40 μm/day, but without the 18-h lag period. Hence, inhibition of neurite outgrowth is not due to inhibitor-induced loss of cell viability. Interestingly, with camptothecin, neurite initiation is completely blocked at doses that block as little as 20% of total RNA synthesis. The mechanistic implications of this are discussed below.

III. PRIMING MODEL OF NEURITE OUTGROWTH

From the above observations, we have formulated the "priming" model of the mechanism of NGF-induced neurite outgrowth (Burstein and Greene, 1978; Greene *et al.*, 1980). The components of this model are as follows.

1. Cells not previously exposed to NGF (which we call unprimed), when grown in the presence of NGF, undergo an essential transcriptional change,

which results in the accumulation of a pool of transcription-dependent material required for neurite outgrowth.

2. This pool accumulates slowly, hence initiation of neurite outgrowth is slow.

3. Cells that have undergone neuronal differentiation (primed cells) possess a maximal pool of transcription-dependent material. This allows for rapid regrowth of long neurites when the cells are mechanically divested of neurites and passaged with NGF. Because primed cells have such a pool of transcription-dependent material, they can regrow neurites when *de novo* synthesis of RNA is completely blocked.

4. Upon removal of NGF, the pool gradually declines and the cells hence gradually lose the capacity for neurite regeneration and again become sensitive to inhibitors of RNA transcription. This aspect of the model is consistent with the following observation. If primed cells are passaged into medium lacking NGF (under which condition regeneration does not occur) for increasing lengths of time, the cells display progressively less capacity for transcription–inhibitor-resistant neurite regrowth upon re-exposure to NGF. By 16 h of NGF deprivation, one-half of this capacity for inhibitor-resistant regrowth is lost (Burstein and Greene, 1978).

5. NGF-induced transcriptional changes alone are not sufficient for neurite outgrowth. Even though primed cells possess a pool of transcription-dependent material sufficient in size to sustain neurite regeneration, such regeneration will only occur if NGF is present. Hence, NGF must play an additional nontranscription-mediated role in neurite outgrowth. This nontranscription-mediated role can be mimicked to some extent by dBcAMP and by phosphodiesterase inhibitors (Burstein and Greene, in preparation), whereas the transcriptional changes cannot be induced by any known compound other than NGF (Section V).

6. While generation and regeneration of neurites requires both the transcriptional and nontranscriptional pathway of NGF action, priming, on the other hand, appears to require only the transcriptional pathway of the NGF mechanism.

We have defined a primed cell as one that has the capacity for lag-free rapid transcription–inhibitor-resistant growth. Priming is the process by which the cell acquires this capacity. The following observations show that priming and neurite outgrowth are distinct and separable events.

1. Cells grown in suspension with NGF do not initiate neurites, but acquire the capacity for immediate rapid transcription–inhibitor-resistant regeneration when passaged onto dishes.

2. Cells grown in nontoxic doses of the reversible microtubule inhibitory drugs colchicine or nocodazole do not initiate processes, but do acquire the

capacity for rapid transcription–inhibitor-resistant outgrowth immediately upon removal of these compounds (Greene *et al.*, in preparation).

3. Cells can lose priming, but maintain their neurites. If cells that have undergone neuronal differentiation are treated with low doses of camptothecin for 24 h, they lose their capacity to rapidly regrow neurites, presumably because they have blocked new synthesis of the priming-specific substance(s), and have depleted their pool of this material. However, such cells maintain their neurites and remain viable for as long as 5 days (Burstein and Greene, 1978).

4. Compounds that mimic NGF ability to cause regeneration of neurites by primed cells (dBcAMP and phosphodiesterase inhibitors; Burstein and Greene, in preparation) cannot cause the cells to acquire the capacity for rapid transcription–inhibitor resistant regrowth.

By our definition, DRG and sympathetic neurons, which can undergo rapid transcription–inhibitor-resistant neurite outgrowth when placed into culture, are primed cells. We propose that pre-exposure of these neurons *in vivo* to NGF results in the analogous transcriptional changes and pool accumulation that are responsible for rapid regeneration by primed PC12 cells.

What then could be the molecular basis of priming? One possibility is that NGF induces the selective synthesis of specific mRNAs, which in turn leads to synthesis of particular proteins. Although PC12 cells undergo a dramatic change in phenotype in response to NGF, very few compositional changes take place. Of the 1000 or so most abundant PC12 peptides and glycopeptides detected by one- and two-dimensional gel electrophoresis, no qualitative changes occur and only a few changes in relative abundance take place as the cells become primed (McGuire *et al.*, 1978; McGuire and Greene, 1980). Among the species that change in relative abundance are a 230,000-MW surface glycoprotein (McGuire *et al.*, 1978), a 25,000- to 30,000-MW glycoprotein (McGuire *et al.*, 1978), and an 80,000-MW protein designated p80 (McGuire and Greene, 1980). All three are present in unprimed cells, but increase in relative abundance significantly in response to NGF. However, even after NGF treatment, these peptides still constitute only a very minor proportion of total cell protein.

Several findings indicate that the above changes in synthesis could be related to priming. These increases can occur in the absence of neurite outgrowth such as when the cells are primed by NGF in suspension. Increases in synthesis are first detectable by 48 h of NGF exposure. This is also the earliest time at which the capacity for rapid regeneration appears. Increases in the synthesis of two of these three proteins are blocked by low doses of camptothecin. (The sensitivity of the 25,000- to 30,000-MW-glycoprotein to low doses of camptothecin has not yet been examined.) In striking contrast, the relative abundances of the other peptides that are unchanged between the unprimed and primed state are unaffected by low doses of camptothecin (McGuire and Greene, 1980).

The above data suggest that NGF specifically regulates the synthesis of at least several PC12 proteins via a transcriptional mechanism. However, the possible causal role of such proteins in priming, while suggestive, is presently far from being conclusive. Further experiments (such as of the genetic type discussed below) will be required to test for such a functional relationship to priming. If the transcriptional component of priming does indeed require regulation of the synthesis of specific proteins, several molecular mechanisms might be envisioned. One mechanistic possibility is that NGF regulates the synthesis of the mRNAs required for each of the proteins required for priming and that synthesis of each of these messages is highly sensitive to camptothecin and actinomycin D. Another possibility is that NGF alters the synthesis of either a regulatory RNA or mRNA for a regulatory protein that is in turn responsible for increasing the pools of the priming-specific proteins; such RNA would be inhibitable by low doses of camptothecin and actinomycin D.

Aside from mRNA another possible target of NGF is ribosomal RNA. Priming is blocked by actinomycin D at doses as low as 0.1 μg/ml (Burstein and Greene, 1978). In other cells, 0.4 μg/ml of actinomycin D blocks transcription of rRNA but does not demonstrably affect mRNA transcription (Perry and Kelley, 1970). Perhaps consistent with the possibility that NGF-induced priming requires *de novo* rRNA synthesis is reported evidence that internalized NGF can be localized to PC12 nucleoli, the site of rRNA synthesis (Marchisio *et al.*, 1980). One point to consider, however, in this possible interpretation is that camptothecin completely blocks priming and neurite initiation at doses that block only 20% of total RNA synthesis. Separation of mRNA from rRNA with oligo-dT sepharose shows that this dose of camptothecin inhibits only about 20% of rRNA and 20% of mRNA synthesis (Burstein and Greene, 1978). This finding suggests that if rRNA is the only transcriptional target RNA of NGF, then only 20% inhibition is sufficient to block priming. This would seem to be an unlikely possibility.

A final, although remote, possibility is that the RNA species affected by NGF plays a structural role. Such structural RNA could conceivably play a role in reorganizing the cell cytoskeleton. For example, there is evidence that RNA may play an important role in the microtubule organizing capacities of centrioles in other cell types (Heidemann *et al.*, 1977).

Whatever the case for the type of RNA species involved, it is particularly intriguing that priming is blocked by low doses of actinomycin D and camptothecin which leave a major portion of cellular RNA synthesis intact. It is conceivable that a certain threshold of transcription must be surpassed for priming to occur, and that reduction of this threshold by low doses of actinomycin D or camptothecin could block priming. Evidence that priming requires such a threshold, however, is lacking. An alternative is that transcription of the RNA(s) affected by NGF is selectively sensitive to the actions of the drugs employed. If this is so, then drugs such as camptothecin could be powerful tools for recognizing the RNA involved and for probing the mechanism by which NGF alters its synthesis.

IV. TRANSCRIPTIONAL REGULATION OF NGF RESPONSES OTHER THAN NEURITE OUTGROWTH

As mentioned above, several proteins have been thus far identified whose syntheses in PC12 cells appear to be transcriptionally regulated by NGF (Table I). Little is yet known about the function of these three proteins. The 230,000-MW component, termed the NGF-inducible large external, or NILE, glycoprotein is present on the cell surfaces of both NGF-treated and NGF-untreated PC12 cells, but is present in considerably higher abundance after NGF treatment (McGuire *et al.*, 1978). NILE glycoprotein is also released from the cells in a soluble form (Richter-Landsberg *et al.*, in preparation). Although NILE glycoprotein is similar in apparent molecular weight to fibronectin, the two species are immunologically noncrossreacting (McGuire *et al.*, 1978). Experiments with antisera against PC12 cells and cultured rat sympathetic neurons have shown that NILE crossreacting material is present in sympathetic neurons and in brain, but not in a variety of other non-neural tissues (Lee *et al.*, in press). NILE glycoprotein has been purified and antiserum against it has been obtained (Salton *et al.*, in preparation). It is hoped that such antiserum will be useful for assessing the functional significance of this molecule. Less information is available about the 25,000- to 30,000-MW glycoprotein. Recent experiments suggest that in NGF-treated PC12 cells, this glycoprotein may be enriched in the cytoskeleton-associated fraction remaining after extraction with a membrane-solubilizing detergent (C. Richter-Landsberg *et al.*, unpublished data).

Another class of proteins regulated by NGF are those associated with neurotransmitter metabolism. Tyrosine hydroxylase (TOH) activity can be increased in sympathetic ganglia *in vivo* and *in vitro* in response to NGF (Thoenen *et al.*, 1971; Lewis *et al.*, 1978). The role of transcription in this effect, however, is presently controversial (for review, see Greene and Shooter, 1980). TOH activity is present in unprimed PC12 cells, but is not increased in response to NGF and hence the PC12 line cannot be used to evaluate the role of transcription in regulation of this enzyme. However, a cell line has been isolated from the same rat pheochromocytoma from which PC12 cells were derived, that responds to NGF by induction of TOH activity. This line, however, does not cease cell division or undergo neuronal differentiation in response to NGF. Interestingly, EGF also induces a large increase in TOH activity in this cell line (Goodman *et al.*, 1979). The role of transcription in these effects on TOH have yet to be reported.

The cholinergic enzymes choline acetyltransferase (CAT) and acetylcholines-terase (AChE) are also present in PC12 cultures. Unlike TOH, these enzyme activities are both induced to higher levels after exposure of PC12 cells to NGF. The transcriptional requirement of CAT induction is presently unknown. Induction of AChE appears to require transcription (Greene *et al.*, 1981); low doses of camptothecin and actinomycin D both inhibit this effect of NGF. Recent experiments indicate that it is possible to induce increases in AChE activity without inducing priming. Using PC12 variants that are not primed by NGF under normal growth conditions it was found that AChE was inducible by NGF although cell division proceeded and priming did not occur (Section VII). It appears, thus, that NGF can induce nuclear-mediated functions that are unrelated to priming, but that are also sensitive to doses of camptothecin which permit as much as 80% of total RNA synthesis. These findings suggest that the transcriptional mechanism by which NGF regulates AChE activity may be similar to that by which NGF regulates priming and synthesis of NILE glycoprotein and p80. Moreover, there appear to be multiple pathways by which NGF leads to such effects on transcription.

An additional enzyme regulated by NGF is ornithine decarboxylase (ODC). *In vivo* and *in vitro* induction of ODC by NGF has been demonstrated in sympathetic ganglia (MacDonnell *et al.*, 1977) and in PC12 cultures (Greene and McGuire, 1978). In PC12 cultures, ODC activity peaks about 4–6 h after addition of NGF to a level 20- to 40-fold higher than in untreated cultures. It was demonstrated that priming can occur in the absence of ODC induction. Thus, inhibitors of ODC activity or of its induction do not prevent PC12 cells from becoming primed. Furthermore, PC12 variants with high basal ODC activity can, under growth-arresting culture conditions, become primed (Section VII). ODC induction can be blocked in PC12 cultures by fully inhibiting RNA synthesis and hence appears to be transcriptional in nature. However, low doses of camptothecin, which inhibit priming, do not block induction of ODC (L. A. Greene and J. C. McGuire, unpublished data). Hence the transcriptional pathway by which NGF induces ODC appears to be distinct from the transcriptional pathway involved in priming. Lastly, it is of interest to note that ODC induction in PC12 cultures is not specific to NGF; replacing spent medium with fresh serum-containing medium (Greene and McGuire, 1978) or exposure to EGF (Huff and Guroff, 1979) all induce ODC in PC12 cultures although they do not cause priming.

V. NGF RECEPTORS AND TRANSCRIPTION

Sympathetic neurons, DRG neurons, and PC12 cells have specific high affinity external receptors for NGF (for review see Bradshaw, 1978; Greene and

Shooter, 1980). How then might interaction of NGF with such receptors lead to the specific transcriptional changes described above? One possibility is that binding of NGF to its cell surface receptors triggers a separate "second messenger." The present status of data regarding second messengers for NGF has been recently reviewed in detail (Bradshaw, 1978; Greene and Shooter, 1980). Briefly, Ca^{2+} and cyclic AMP (Schubert *et al.*, 1978) are among agents that have been proposed to play second messenger roles for NGF. However, experiments with PC12 cells reported to support this action of Ca^{2+} have not been reproducible by other laboratories (Landreth *et al.*, 1980). Moreover, in direct contrast to reports claiming to support a cyclic AMP mechanism, we have found that dBcAMP and phosphodiesterase inhibitors do not cause priming or neurite initiation and do not induce NILE glycoprotein, p80, or AChE in PC12 cultures (Burstein and Greene, in preparation). Moreover, these reagents can cause process regeneration by NGF-primed PC12 cells, but do so by a mechanism that is different from that by which NGF causes neurite regeneration (Burstein and Greene, in preparation). Recently, evidence has been presented that NGF causes, with short latency, an extrusion of Na^+, which is proposed to set into motion a cascading set of events that ultimately ends in regulation of gene transcription (Varon and Skaper, 1980). This model awaits testing in the PC12 system in which non-NGF-treated cells are not losing viability. In general, then, a second messenger mechanism for NGF has yet to be definitively identified.

Investigations from several laboratories have documented that NGF can be internalized by its target cells (for reviews see Bradshaw, 1978; Greene and Shooter, 1980). This process is believed to be similar to that postulated for other polypeptides (Goldstein *et al.*, 1979). That is (1) NGF binds to its surface receptor; (2) the NGF–receptor complexes cluster in "coated pits" in the cell surface; (3) the coated pits bud off to form coated vesicles, which in turn fuse with secondary lysosomes. The function of such internalization is unclear. One possibility is that it serves to remove occupied NGF receptors from the cell surface, thereby "down-regulating" them. Such down-regulation of receptors clearly occurs within the first several hours of NGF exposure (Calissano and Shelanski, 1980). Another tantalizing possibility is that internalized NGF might serve as its own second messenger. Of relevance to this, evidence has been presented that internalized NGF, either via lysosomes or another route, can bind to specific receptors in the nucleus of DRG neurons (Andres *et al.*, 1977) and nuclear envelope of PC12 cells (Yankner and Shooter, 1979). It is thus possible that NGF could directly trigger its transcriptional effects in the nucleus or on the nuclear membrane. However, definitive evidence that internalization is an essential step in NGF mechanism of action is currently lacking. No doubt many future efforts will be needed to sort out the cellular locus from which and mechanism by which NGF directs changes in RNA transcription.

VI. NONTRANSCRIPTIONAL, ANABOLIC, AND RAPID EFFECTS OF NGF

As discussed above in the priming model, NGF-induced neurite outgrowth appears to have a nontranscriptional as well as transcriptional requirement (Table I). The nature of this nontranscriptional action of NGF is at present also far from clear. Possible mechanisms could include modulation of ion flux (such as Na^+ extrusion; Varon and Skaper, 1980) and post-translational modification of proteins (by means such as glycosylation, phosphorylation, or methylation).

In addition to effects on neurite outgrowth, another class of major actions of NGF is anabolic in nature. For example, NGF induces increased uptake of amino acids by PC12 cells within 15 min of addition of NGF (McGuire and Greene, 1979). A similar effect has been demonstrated in cultured DRG and sympathetic neurons (Horii and Varon, 1977). Protein content per cell also increases three- to

Table I. Transcription-Dependent and -Independent Effects of NGF on PC12 Cells

	Blocked by high doses of actinomycin D and camptothecin (1–10 μg/ml)	Blocked by low doses of camptothecin (0.2–0.5 μg/ml)
Neurite initiation	+	+
NGF-mediated regeneration	– [a]	– [a]
dBcAMP-mediated regeneration	+	–
Increases in cell size and protein content	NT [b]	–
Stabilization of neuronal phenotype	–	–
NGF-induced survival in serum-free medium	NT	–
Increased amino acid transport	–	–
Rapid cell-surface changes	–	–
NILE glycoprotein induction	NT	+
p80 protein induction	+	+
30,000-MW glycoprotein induction	NT	NT
Acetylcholinesterase induction	NT	+
Choline acetyltransferase induction	NT	NT
Ornithine decarboxylase induction	+	–

[a] Transcription-independent but requires priming which is transcription dependent.
[b] NT, not tested.

fourfold in PC12 cultures in response to NGF (Greene and Tischler, 1976). Again, similar effects have been demonstrated *in vivo* and *in vitro* for DRG and sympathetic neurons (Levi-Montalcini and Angeletti, 1968). These two anabolic effects do not appear to require the transcription-requiring pathway by which NGF causes priming. Thus, increased uptake of amino acids takes place in a time course too short to be a result of transcriptional regulation and is unimpeded by inhibition of transcription (McGuire and Greene, 1979). Furthermore, unprimed PC12 cells cultured in the presence of NGF and low doses of camptothecin that block priming but are not toxic demonstrate a dramatic NGF-dependent increase in size and protein content (Greene *et al.*, 1980). Aside from demonstrating that priming and anabolic effects have separable mechanisms the latter experiment also clearly underscores the point that priming is not simply due to nonspecific effects of NGF on the protein content of its target cells. Finally, it is worth noting here that whereas priming and anabolic effects of NGF may have separable mechanisms, the two actions are not necessarily unrelated. Teleologically, anabolic effects would be very useful during development for neurons that are in the process of extending long neurites.

Another nontranscriptional effect of NGF is the apparent stabilization of cellular components of the neuronally differentiated PC12 cell. Cells cultured with NGF and high or low doses of camptothecin for more than 24 h maintain their neuronal differentiation although they have lost priming and, presumably, their pool of transcription-dependent priming-associated material. The neuronal phenotype is present for at least 5 days in these culture conditions. In contrast, when NGF is removed from neuronally differentiated PC12 cultures, most cells lose their neuronal phenotype within 48 h. Hence NGF appears to stabilize the cell components of the neuronally differentiated cells by a nontranscriptional mechanism, despite depletion of the transcription-dependent pool, which is required for regeneration and priming (Burstein and Greene, 1978).

VII. GENETIC APPROACHES TO NGF MOLECULAR MECHANISM

One of the major advantages of studying a replicating clonal cell line that responds to NGF is that it is possible to apply genetic techniques. For example, mutants that are defective in specific steps of NGF mechanism of action can be selected and analyzed.

Genetic approaches used in our laboratory have thus far begun to yield certain mechanistic insights. An advantage of the PC12 model is the relative ease with which at least one class of variant cells can be selected. Since the normal PC12 line ceases cell division in response to NGF, variants that continue to divide in NGF-containing medium may be easily identified and propagated. Using this technique we have selected a number of stable variant cell lines that arose spontaneously in cultures of PC12 cells (Burstein and Greene, in preparation). All such variants tested maintain noradrenergic functions and have growth rates in complete medium (15% serum) that are unaffected by NGF. Under such conditions, the variants do not undergo neuronal differentiation. However, if such cells are cultured in medium lacking serum, they survive if NGF is present and will die in its absence. Normal PC12 cells also die in serum-free medium, but will survive under such conditions if NGF is included, and undergo neuronal differentiation (Greene, 1978). When the variant cultures are maintained for one or more weeks in serum-free medium containing NGF, they too very slowly undergo neurite outgrowth, with the relative abundance of neurites varying between different lines. Thus NGF elicits neuronal differentiation of the variants in serum-free medium, but not in complete growth medium. One reason for this phenomenon could relate to cell proliferation. Although NGF allows variants to survive in the absence of serum, this culture condition will not sustain cell division. The arrest of cell division hence may be required for variant cells to undergo NGF-induced neuronal differentiation. This concept is corroborated by the observations that variant cells cultured in complete growth medium containing the division-arresting compounds cytosine arabinoside or hydroxyurea, begin NGF-induced neuronal differentiation in a time period (4–7 days) equivalent to that of normal PC12 cells. The extent of neurite outgrowth under such conditions varies from one variant line to the next in a manner consistent with the observations made for variant cultures maintained with NGF in serum-free medium. Furthermore, the neuronally differentiated variants produced by the above means can undergo rapid NGF-dependent transcription–inhibitor-resistant neurite outgrowth. To summarize then, NGF does not induce cessation of cell division and neuronal differentiation when the variant lines are cultured in growth-promoting medium. However, when proliferation of the variants is exogenously arrested by appropriate drugs or by culturing in serum-free medium, such cells are able to undergo NGF-promoted neuronal differentiation and priming. This suggests the possibility that in normal PC12 cells, arrest of cell division may be a necessary precondition for NGF-induced priming and neuronal differentiation to take place and that in such cells, NGF itself can shut off cell division. The variants, on the other hand, have lost the necessary cell machinery to cease cell division in response to NGF, but clearly retain the cell machinery to undergo neuronal differentiation in response to NGF once proliferation has been arrested by other means.

Another interesting facet of our variants regards the reversibility of their NGF response. If the variant cells are caused to undergo neuronal differentiation, they will stay locked into an NGF-requiring differentiated nondividing state, even when growth conditions permitting variant cell proliferation are restored. Thus, when hydroxyurea is removed from variant cultures that have differentiated in response to NGF, they will *not* resume cell division, but will retain their neuronal phenotype. Similarly, variants that have differentiated in serum-free medium containing NGF will retain their neuronal phenotype and not begin dividing when restored to medium containing NGF and 15% serum. However, in either case, if NGF is removed, leaving the neuronally differentiated variant cells in medium with 15% serum, the latter lose their neuronal phenotype and resume cell division. Several interesting implications emerge from these observations. First, the variant studies support the notion that cell division and cell differentiation may be mutually opposing processes. The dividing state may thus prevent NGF from inducing differentiation in the variants. Conversely, the NGF-dependent differentiated state appears to prevent the resumption of cell proliferation when cells are restored to growth conditions under which they are otherwise capable of proliferating.

An additional point of interest that has arisen from work on the variant lines (Greene *et al.*, 1981) is that NGF can cause a transcription-dependent increase in AChE activity in the variants under growth conditions (complete growth medium) in which proliferation is not affected and in which priming and neurite outgrowth do not occur. Thus, at least part of the pathway by which NGF

Table II. **Effects of NGF on Proliferation and Differentiation of Parent and Variant Lines of PC12**[a]

	Complete medium (15% serum)	Low serum (0.5–2%)	No serum
PC12			
− NGF	Maximal proliferation	Slow growth or stationary	Cell death
+ NGF	Division arrested Differentiation	Division arrested Differentiation	Cells survive Differentiation
− NGF + ara C	Division arrested	NT[b]	NT
+ NGF + ara C	Division arrested Differentiation	NT	NT
Variants			
− NGF	Maximal proliferation	Slow growth or stationary	Cell death
+ NGF	Maximal proliferation	Near-maximal proliferation	Cells slowly differentiate
− NGF + ara C	Division arrested	NT	NT
+ NGF + ara C	Division arrested Differentiation	NT	NT

[a] Cells were cultured with or without 50 ng/ml NGF and/or 10–30 μM cytosine arabinoside (ara C).

[b] NT, not tested.

causes transcriptional regulation of enzyme activity appears to be separate from the pathway by which it regulates cell division and priming.

The mechanistic bases underlying the difference in behavior between normal PC12 cells and its variant sublines are not presently known. One apparent biochemical difference that has been found is in induction of ODC activity. In preliminary studies most sublines tested have basal ODC activities that are considerably higher than that of normal PC12 cells. In contrast to the 20- to 40-fold induction of ODC by NGF in the parent cell line, in variant cultures ODC is enhanced two- to sixfold by NGF, and in some cases, not at all (Burstein and Greene, in preparation). It is not clear, however, if the lack of ODC inducibility is related to the unresponsiveness of variant cells to NGF in growth-supporting medium.

The difference in behavior between the variant line and the parent line does not appear to be a consequence of the number of cell surface NGF receptors. All the variant lines tested thus far show numbers of cell surface receptors equivalent to the number of receptors on normally responsive PC12 cells (D. E. Burstein and L. A. Greene, unpublished data). A possibility not yet tested is that the processing of NGF–cell surface receptor complexes or the triggering of essential cellular events subsequent to binding of NGF to its receptor may be defective in the variant cells.

VIII. NGF AND CELL PROLIFERATION

The mechanism by which NGF mediates the cessation of cell division in normal PC12 cultures is poorly understood. It has recently been shown that addition of NGF to PC12 cultures causes down-regulation of receptors for epidermal growth factor (EGF) (Huff and Guroff, 1979), a well-described peptide (Carpenter and Cohen, 1979) that can stimulate proliferation of PC12 cells (Burstein and Greene, in preparation). On this basis it was proposed that down-regulation by NGF of receptors for mitogens such as EGF might eliminate responsiveness to the usual stimuli for PC12 cell division.

Recent work in our laboratory suggests that the growth-regulating effects of NGF may be more complex than suggested above. We have demonstrated with several of our variant PC12 lines described above that NGF has intrinsic division-stimulating activity (Burstein and Greene, in preparation) (Table II). When PC12 cells or their variants are grown in medium containing 0.5–2% serum without NGF, the cells survive but divide at a very slow rate, with doubling times of about 2.5–4 days. Addition of NGF to PC12 cultures under these conditions causes, predictably, cessation of cell division and neuronal differentiation. In striking contrast, addition of NGF to variant cells cultured under these conditions does not cause neuronal differentiation and results in a near-doubling of the rate of proliferation. Variant cells growing in complete growth medium containing 15% serum are already growing at their maximal rate and addition of NGF causes no further stimulation under these conditions. It appears, therefore, that NGF is not simply a survival factor, but a proliferation stimulating factor. Such mitogenic activity is observable in the variants because they have apparently lost the mechanism responsible for cessation of cell division in response to NGF.

The above studies on PC12 variant lines suggest the intriguing possibility that NGF may have intrinsic proliferation-inducing activity on all of its target cells and that the parent line of PC12 cells and perhaps neuroblasts have, as part of their cellular machinery, a means for overriding the intrinsic mitogenic properties of NGF. These observations also suggest the intriguing possibility that hyperplasia of ganglionic neurons and adrenal medulla chromaffin cells (Aloe and Levi-Montalcini, 1979) and persistence of ectopic catecholamine-containg cells in embryos given large doses of NGF (Kessler *et al.*, 1979), may be a consequence of NGF-induced proliferation as well as NGF-induced prevention of cell death.

IX. CONCLUDING REMARKS

Whereas some progress has been made in understanding the mechanism of action of NGF at the molecular level, a number of critical questions remain to be answered. For instance, at what locus (or loci) does NGF act? What mechanism links receptor binding of NGF to subsequent actions of the factor? At what cellular level and by what molecular mechanism does NGF alter transcription? How do the transcriptional products whose syntheses are altered by NGF lead to complex morphologic changes such as neurite outgrowth?

The PC12 line promises to be a valuable system for approaching the above and other questions regarding NGF mechanism of action. However, although the responses of PC12 cells and their variants to NGF closely resemble the *in vivo* responses of NGF normal embryonic targets, the ultimate verification of our observations will lie in direct examination of the early neural crest tissue from which the adrenal medullary cells and DRG and sympathetic neurons arise. One promising development is the cultivation of neural crest and immature embryonic ganglia *in vitro* (Cohen, 1977; Coughlin *et al.*, 1977). These or similar experimental systems might be useful for extending studies on neoplastic PC12 cells to the true embryonic targets of NGF.

Lastly, it is of interest to point out that research directions on the mechanism of action of NGF have thus far drawn heavily from parallel studies being done on the mechanisms of action of other peptide hormones and growth factors. Conversely, we believe that current and future studies on NGF, particularly those regarding binding and internalization of NGF, the use of chemical probes such as low doses of camptothecin, and exploitation of genetic variants such as NGF-unresponsive PC12 mutants will have applications that will contribute to the general understanding of the responses of cells to a variety of chemical signals that influence growth and differentiation.

ACKNOWLEDGMENTS

A portion of the research described above was supported by grants from the United States Public Health Service (NS 16036) and the March of Dimes Birth Defects Foundation. We thank Yvel Calderon for her assistance in preparation of this manuscript.

REFERENCES

Aloe, L., and Levi-Montalcini, R. (1979). Nerve growth factor-induced transformation of immature chromaffin cells *in vivo* into sympathetic neurons: Effect of antiserum to nerve growth factor. *Proc. Natl. Acad. Sci. U.S.A.* **76**, 1246–1250.

Andres, R. Y., Jeng, I., and Bradshaw, R. A. (1977). Nerve growth factor receptors: Identification of distinct classes in plasma membranes and nuclei of embryonic dorsal root neurons. *Proc. Natl. Acad. Sci. U.S.A.* **74**, 2785–2789.

Bradshaw, R. A. (1978). Nerve growth factor. *Annu. Rev. Biochem.* **47**, 191–216.

Burstein, D. E., and Greene, L. A. (1978). Evidence for both RNA-synthesis-dependent and independent pathways in stimulation of neurite outgrowth by nerve growth factor. *Proc. Natl. Acad. Sci. U.S.A.* **75**, 6059–6063.

Calissano, P., and Shelanski, M. L. (1980). Interaction of nerve growth factor with pheochromocytoma cells. Evidence for tight binding and sequestration. *Neuroscience* **5**, 1033–1040.

Carpenter, G., and Cohen, S. (1979). Epidermal growth factor. *Annu. Rev. Biochem.* **48**, 193–216.

Cohen, A. M. (1977). Independent expression of the adrenergic phenotype by neural crest cells *in vitro. Proc. Natl. Acad. Sci. U.S.A.* **74**, 2899–2903.

Coughlin, M. D., Boyer, D. M., and Black, I. B. (1977). Embryological development of a mouse sympathetic ganglion *in vivo* and *in vitro. Proc. Natl. Acad. Sci. U.S.A.* **74**, 3438–3442.

Dichter, M. A., Tischler, A. S., and Greene, L. A. (1977). Nerve growth factor-induced increase in electrical excitability and acetylcholine sensitivity of a rat pheochromocytoma cell line. *Nature (London)* **268**, 501–504.

Goldstein, J. L., Andersen, R. G. W., and Brown, M. S. (1979). Coated pits, coated vesicles and receptor-mediated endocytosis. *Nature (London)* **279**, 679–685.

Goodman, R., Chandler, C., and Herchman, H. R. 1979. Pheochromocytoma cell lines as models of neuronal differtiation. In ''Hormones and Cells in Culture'' (R. Roth and G. Sato, eds.), Vol. 6, pp. 653–669. Cold Spring Harbor Lab., Cold Spring Harbor, New York.

Gorin, P. D., and Johnson, E. M. (1979). Experimental autoimmune model of nerve growth factor deprivation: Effects on developing peripheral sympathetic and sensory neurons. *Proc. Natl. Acad. Sci. U.S.A.* **76**, 5382–5386.

Greene, L. A. (1977). A quantitative bioassay for nerve growth factor (NGF) activity employing a clonal pheochromocytoma cell line. *Brain Res.* **133**, 350–353.

Greene, L. A. (1978). Nerve growth factor prevents the death and stimulates neuronal differentiation of clonal PC12 pheochromocytoma cells in serum-free medium. *J. Cell Biol.* **78**, 747–755.

Greene, L. A., and McGuire, J. C. (1978). Induction of ornithine decarboxylase by nerve growth factor dissociated from effects on survival and neurite outgrowth. *Nature (London)* **276**, 191–194.

Greene, L. A., and Rein, G. (1977a). Release, storage and uptake of catecholamines by a clonal cell line of nerve growth factor (NGF) responsive pheochromocytoma cells. *Brain Res.* **129,** 247–263.

Greene, L. A., and Rein, G. (1977b). Release of (^{3}H)-norepinephrine from a clonal line of pheochromocytoma (PC12) by nicotinic cholinergic stimulation. *Brain Res.* **138,** 521–528.

Greene, L. A. and Rukenstein, A. (1981). Regulation of acetylcholinesterase activity by nerve growth factor. Role of transcription and dissociation from effects on proliferation and neurite outgrowth. *J. Biol. Chem.* **256,** 6363–6367.

Greene, L. A., and Shooter, E. M. (1980). The nerve growth factor. *Annu. Rev. Neurosci.* **3,** 353–402.

Greene, L. A., and Tischler, A. S. (1976). Establishment of a noradrenergic clonal line of rat adrenal pheochromocytoma cells which respond to nerve growth factor. *Proc. Natl. Acad. Sci. U.S.A.* **73,** 2424–2428.

Greene, L. A., Burstein, D. E., and Black, M. M. (1980). The priming model for the mechanism of action of nerve growth factor: Evidence derived from clonal PC12 pheochromocytoma cells. *In* "Tissue Culture in Neurobiology" (E. Giacobini and A. Vernadakis, eds.), pp. 313–319. Raven, New York.

Harper, G. P., Barde, Y. A., Burnstock, G. A., Carstavis, J. R., Dennison, M. E., Suda, K., and Vernon, C. A. (1979). Guinea pig prostate is a rich source of nerve growth factor. *Nature (London)* **279,** 160–162.

Heidemann, S. R., Sander, G., and Kirschner, M. W. (1977). Evidence for a functional role of RNA in centrioles. *Cell* **10,** 337–350.

Horii, Z.-I., and Varon, S. (1977). Nerve growth factor action on membrane permeation to exogenous substrates in dorsal root ganglion dissociates from the chick embryo. *Brain Res.* **124,** 121–133.

Huff, K. R., and Guroff, G. (1979). Nerve growth factor induced reduction in epidermal growth factor responsiveness and epidermal growth factor receptors in PC12 cells aspect of cell differentiation. *Biochem. Biophys. Res. Commun.* **89,** 175–180.

Kessler, J. A., and Black, I. B. (1980). Nerve growth factor stimulates the development of substance P in sensory ganglia. *Proc. Natl. Acad. Sci. U.S.A.* **77,** 649–652.

Kessler, J. A., Cochard, P., and Black, I. B. (1979). Nerve growth factor alters the rate of embryonic neuroblasts. *Nature (London)* **280,** 141–142.

Landreth, G., Cohen, P., and Shooter, E. M. (1980). Ca^{2+} transmembrane fluxes and nerve growth factor action on a clonal cell line of rat pheochromocytoma. *Nature (London)* **283,** 202–204.

Lee, V. M., Greene, L. A., and Shelanski, M. L. (1981). Identification of neural and adrenal medullary surface membrane glycoproteins recognized by antisera to cultured sympathetic neurons and PC12 pheochromocytoma cells. *Neuroscience,* in press.

Levi-Montalcini, R. (1966). The nerve growth factor, its mode of action on sensory and sympathetic nerve cells. *Harvey Lect.* **60,** 217–259.

Levi-Montalcini, R., and Angeletti, P. U. (1968). Nerve growth factor. *Physiol. Rev.* **48,** 534–569.

Lewis, M. E. Lakshmanan, J., Nagaiah, K., MacDonnell, P. C., and Guroff, G. (1978). Nerve growth factor increases activity of ornithine decarboxylase in rat brain. *Proc. Natl. Acad. Sci. U.S.A.* **75,** 1021–1023.

MacDonnell, P. C., Nagaiah, K., Lakshmanan, J., and Guroff, G. (1977). Nerve growth factor increase activity of ornithine decarboxylase in superior cervical ganglia of young rats. *Proc. Natl. Acad. Sci. U.S.A.* **74,** 4681–4684.

McGuire, J. C., and Greene, L. A. (1979). Rapid stimulation by nerve growth factor of amino acid uptake by clonal PC12 pheochromocytoma cells. *J. Biol. Chem.* **254,** 3362–3367.

McGuire, J. C., and Greene, L. A. (1980). Nerve growth factor stimulation of specific protein synthesis by rat PC12 pheochromocytoma cells. *Neuroscience* **5,** 179–189.

McGuire, J. C., Greene, L. A., and Furano, A. V. (1978). NGF stimulates incorporation of glucose or glucosamine into an external glycoprotein in cultured rat PC12 pheochromocytoma cells. *Cell* **15**, 357–365.

Marchisio, P. C., Naldini, L., and Calissano, P. (1980). Intracellular distribution of nerve growth factor in rat pheochromocytoma PC12 cells: Evidence for a perinuclear and intranuclear location. *Proc. Natl. Acad. Sci. U.S.A.* **77**, 1656–1660.

Partlow, L. M., and Larrabee, M. G. (1971. Effects of a nerve growth factor, embryonic age and metabolic inhibitors on synthesis of ribonucleic acid and protein in embryonic sensory ganglia. *J. Neurochem.* **18**, 2101–2118.

Perry, R. B., and Kelley, D. E. (1970). Inhibition of RNA synthesis by actinomycin D: Characteristic dose responses of different RNA species. *J. Cell. Physiol.* **76**, 127–140.

Schubert, D. S., LaCorbiere, M., Whitlock, C., and Stallcup, W. (1978). Alterations in the surface properties of cells responsive to nerve growth factor. *Nature (London)* **273**, 718–723.

Thoenen, H., Angeletti, P. U., Levi-Montalcini, R., and Kettler, R. (1971). Selective induction by nerve growth factor of tyrosine hydroxylase and dopamine hydroxylase in rat superior cervical ganglion. *Proc. Natl. Acad. Sci. U.S.A.* **68**, 1598–1602.

Tischler, A. S., and Greene, L. A. (1975). Nerve growth factor-induced process formation by cultured rat pheochromocytoma cell. *Nature (London)* **258**, 341–342.

Tischler, A. S. and Greene, L. A. (1978). Morphologic and cytochemical properties of a clonal line of rat adrenal pheochromocytoma cells which respond to nerve growth factor. *Lab. Invest.* **39**, 77–89.

Varon, S., and Skaper, S. D. (1980). Short-latency effects of nerve growth factor: An ionic view. *In* "Tissue Culture in Neurobiology" (E. Giacobini and A. Vernadakis, eds.), pp. 333–348, Raven, New York.

Yankner, B. A., and Shooter, E. M. (1979). Nerve growth factor in the nucleus: Interaction with receptors on the nuclear membrane. *Proc. Natl. Acad. Sci. U.S.A.* **76**, 1269–1273.

7

CELL INTERACTIONS IN EMBRYONIC NEURAL RETINA: ROLE IN HORMONAL INDUCTION OF GLUTAMINE SYNTHETASE

Paul Linser and A. A. Moscona

I. INTRODUCTION

It is a generally accepted concept that cells in the developing embryo interact with one another and communicate by means of long-range and short-range signals, and that such cell interactions play a crucial role in differentiation and morphogenesis. Cell–cell interactions represent only one in a spectrum of mechanisms controlling embryonic differentiation; however, their role and nature have been recently attracting increasing interest because of growing awareness of their critical importance. The need for precision in cell communication is undoubtedly greatest in the morphogenesis and differentiation of the nervous system, where even localized perturbations in cell relationships can have potentially profound consequences on the functionality of the whole system.

Several kinds of mechanisms for cell–cell communication in development

Molecular Approaches to Neurobiology

have been postulated (Moscona, 1974; Saxen *et al.*, 1976; Lash and Vasan, 1977). Some of these mechanisms may depend on metabolic cell cooperation, i.e., one-way transfer or two-way exchange between cells of precursor substances or regulatory metabolites. Others may involve the action of more specific "inducers," "signals," etc., and these can be categorized, for convenience of discussion, into three classes.

One class includes cases in which specific activator molecules released from effector cells reach receptive cells (target cells), either by way of circulatory conduits (blood and lymphatic vessels) or through the intercellular space. This kind of humoral communication can take place across relatively long distances, as in the cases of hormonal actions and inductions; or it can occur across short distances, as in the function of neurotransmitters.

Another class of cell–cell communication is that mediated by so-called gap or low-resistance junctions; these intercellular connections enable direct transfer between cells of molecules or ions by way of physical channels (Sheridan, 1974; Lo and Gilula, 1979).

Still another means of cell communication may depend on direct surface contacts between cells or between cells and elements in the intercellular matrix. This involves positive or negative interactions between surface molecules on adjacent cells or between molecules on the cell surface and in the extracellular matrix. The role of such surface-mediated molecular interactions in the mechanism of embryonic cell–cell recognition and in "positional guidance" of migrating cells during morphogenesis has been discussed in detail elsewhere (Sidman, 1974; Lash and Vasan, 1977; Monroy and Moscona, 1979). It is generally believed that modifications in the characteristics of cell surface molecules caused by changes in cell contacts may generate transmembrane signals which are relayed into the cells and can alter aspects of biosynthesis and cell behavior (Moscona, 1974, 1980).

Although it has long been held that cell interactions play an important role in the organization and biochemical differentiation of cells in the embryonic nervous system (before the formation of synapses), the complexity of the central nervous system has made it very difficult to study this problem in detail. Therefore, investigators have turned to various relatively simpler "models," which are more amenable to experimental analysis. Such a model system must meet two essential conditions. It should have developmentally regulated cell type specific biochemical markers, and it should lend itself to experimental modification of cell contacts and cell interactions. In our work we have concentrated on the neural retina of the chick embryo, since it meets these conditions and offers still other advantages. This chapter describes some of our results and views, without attempting to present a comprehensive review of the general subject of cell interactions in development, or to discuss all the aspects of the retina system.

II. NEURAL RETINA OF CHICK EMBRYO

The neural retina arises in the embryo as an extension of the forebrain, and its development follows a program of growth and differentiation that has been investigated in considerable detail (Meller and Glees, 1965; Moscona, 1972; Kahn, 1974; Moscona *et al.,* 1980b). The mature neural retina contains a relatively small number of neuronal cell types and only one class of glia cells (Müller cells). It has an almost diagrammatic histological architecture in that it consists of several layers (strata) of cell bodies and cell processes. As a "model system," the chicken retina has the added advantage of being nonvascularized; therefore, its biochemical analysis is not complicated by the presence of plasma, blood cells, and vascular elements. Unlike the eyes of most mammalian embryos, the eyes of the chick embryo are, relatively, very large and the retina is quite easy to isolate. Furthermore, retinas isolated from early chick embryos can be maintained *in vitro,* in organ culture and, under appropriate conditions, continue to develop and to differentiate in a manner similar to retina *in vivo.* Retina tissue can be readily dissociated into suspensions of single live cells; these can be maintained in monolayer cultures or can be reaggregated. Reaggregates of cells derived from retinas of young embryos reconstruct retinotypic tissue architecture, and such cell aggregates are an effective experimental–analytical system for a variety of studies (Moscona, 1974, 1980; Linser and Moscona, 1981).

Cell type specific biochemical markers have been identified in the neural retina and some of these are developmentally regulated or can be experimentally manipulated. Of particular interest to the subject matter of this chapter is the enzyme glutamine synthetase.

III. GLUTAMINE SYNTHETASE (GS) IN AVIAN NEURAL RETINA

Glutamine synthetase (E.C. 6.3.1.2), referred to for brevity as GS, catalyzes the amidation of glutamate to glutamine. Glutamine is of major importance in a variety of biosynthetic pathways including the synthesis of proteins, mucopolysaccharides, and nucleic acids (Meister, 1974). GS is present throughout the nervous system of higher vertebrates and is a key enzyme in the "small glutamate compartment" (Van den Berg, 1970) that functions in salvaging neuronally released neurotransmitter metabolites (such as glutamate). Because of the importance of GS in these metabolic pathways and cycles, its role in neural tissues and its control mechanisms during neurodifferentiation are of growing interest.

The level of GS is uniquely high in the adult neural retina and particularly in the retina of adult birds (Moscona and Piddington, 1966; Moscona and Degenstein, 1981). On the other hand, in the retina of early embryos the level of GS is very low. This difference led us originally to investigate the mechanisms that control the development of this enzyme in the neural retina.

An essential step in this study was the purification of GS from chicken retina, and its characterization (Sarkar *et al.*, 1972). The active enzyme is an octamer with a subunit MW of 42,000 ± 2000. The purified GS was found to be antigenic in rabbits and to yield a highly specific antiserum (γ-globulin) that has served as an effective probe in various aspects of our work.

IV. INDUCTION OF GS

During embryonic development of the chick retina, the specific activity of GS begins to rise sharply on about the sixteenth day of incubation, increases 100-fold in 5–6 days, and then plateaus at this high level (Moscona, 1972; Moscona *et al.*, 1980b). The sharp increase in GS level begins 4 days after cell multiplication and growth in the retina have virtually ceased; and it is induced by adrenal corticosteroids, which become elevated in the circulation shortly beforehand. However, the retina is competent for GS induction long before it normally occurs in the embryo. It has been demonstrated that this competence develops progressively during the fifth to eighth day of embryogenesis, and that GS can be induced already in the retinas of 8-day-old embryos by administering a 11-β-OH-corticosteroid, such as cortisol, to the embryo or by adding cortisol to organ cultures of isolated retina tissue in a synthetic medium (Moscona and Piddington, 1966; Piddington and Moscona, 1967; Moscona and Moscona, 1979). Studies on organ cultures of retinas from 10-day-old embryos demonstrated that cortisol enters into cells and promptly elicits accumulation of mRNA for GS; this results within 1–2 h, in a rapid and progressive increase in the rate of enzyme synthesis and in the enzyme level (Moscona *et al.*, 1972; Sarkar and Moscona, 1973; Moscona and Wiens, 1975; Soh and Sarkar, 1978). This induction is not prevented by inhibition of DNA synthesis, but is blocked by agents that stop RNA and protein synthesis. It is important to note that only in the retina can GS be induced so precociously and to this extent; and that only cortisol and closely related 11β-hydroxycorticosteroids can induce GS in the retina (Moscona and Piddington, 1967). Therefore, this induction represents a highly specific "long-range cell interaction," between the adrenal cortex and the retina; it results in stimulation of differential gene expression by the adrenal hormone and in GS synthesis. It is noteworthy that a very similar situation has been described also for GS in the retina of the quail (Moscona and Degenstein, 1981).

Immunohistochemical studies with antiserum specific for retina GS have shown that the accumulation of GS during its precocious induction takes place exclusively in the Müller cells (Linser and Moscona, 1979; Moscona *et al.*,

1980a); also in the mature retina GS is localized only in these glial cells (Linser and Moscona, 1979) (Fig. 1).

Localization of GS in Müller cells has been corroborated by finding that these cells can be selectively damaged or destroyed in chick embryo retina by the gliatoxic agent α-aminoadipic acid (AAA). Treatment of retina with AAA at a susceptable age damages or destroys most of the glia cells and results in almost complete loss of inducibility for GS (Moscona *et al.*, 1980b; Linser and Moscona, 1981). The cytotoxic specificity of AAA for Müller cells was further confirmed by finding that AAA greatly reduced also the level of carbonic anhydrase (CAH). The retina contains a high level of CAH (Clark, 1951) and this

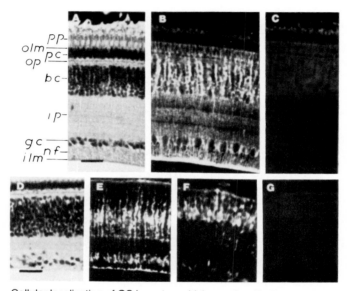

Fig. 1. Cellular localization of GS in mature chicken retina (A–C) and in cortisol-induced embryonic chicken retina (D–G): immunostaining with GS antiserum and indirect immuno-fluorescence with FITC-GAR. (A) Section of mature retina (6-week-old chicken) stained with hematoxylin and eosin. pp, Photoreceptor processes; olm, outer limiting membrane; pc, photoreceptor cell layer; op, outer plexiform layer; bc, bipolar cell layer; ip, inner plexiform layer; gc, ganglion cell layer; nf, nerve fiber layer; film, inner limiting membrane. (B) Section similar to that in A; immunostained with GS antiserum and FITC-GAR. Light areas represent immunofluorescence in Müller cells. (C) Control section treated with nonimmune rabbit serum and FITC-GAR. (D) Section of 13-day-old embryo retina stained with hematoxylin and eosin. (E) 13-day-old embryo retina induced *in vivo* with cortisol section treated with GS antiserum and FITC-GAR. Fluorescence is localized to Müller fibers. (F) Section of 13-day-old embryo retina induced in organ culture with cortisol; immunofluorescence shows localization of induced GS in Müller cells. (G) Section of control, noninduced 13-day-old embryo retina treated as in E and F; no immunofluorescence. All magnifications × 350. Bars represent 25 μM. (From Linser and Moscona, 1979.)

enzyme is also localized predominantly in Müller cells (Musser and Rosen, 1973); however, unlike GS, CAH is not subject to regulation by cortisol and thus provides an independent biochemical marker for Müller cells. In contrast to its effects on glial cells, AAA had no detectable effects on the level of enzymes located in retina neurons, e.g., cholineacetyltransferase (CAT) and γ-aminobutyric acid transaminase (GABA-T) (Sarthy and Lam, 1978; Linser and Moscona, 1981) (Fig. 2).

Therefore, in the chick embryo retina, competence for GS induction by specific corticosteroid hormones and accumulation of GS represent developmentally regulated features of glia elements, i.e., Müller cells.

With respect to competence for GS induction, differentiation of Müller cells can be divided into three phases (Moscona and Moscona, 1979): (1) the early phase at which the future Müller cells are still in a prospective precursor stage (pre-competence phase); (2) competence acquisition phase, during which these cells become programmed with responsiveness to GS induction; (3) maturation

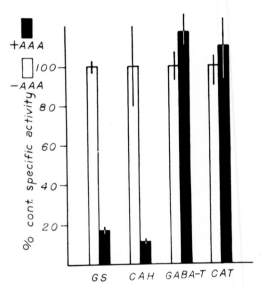

Fig. 2. Effects of pretreatment with α-aminoadipic acid (AAA) of retina cell aggregates on levels of the enzymes GS, CAH, GABA-T, and CAT. Cell aggregates were prepared from 6-day-old embryo retinal cells. After 4 days, AAA (200 μg/ml) was added to one-half the cultures for 48 h. All the cultures were then transferred to cortisol-containing culture medium for 48 h and then assayed for the enzymes. The specific activity for each of the controls (open bars) represents 100%, and the experimental values (black bars) are expressed accordingly. Each bar represents the mean of two or three determinations. (From Linser and Moscona, 1981.)

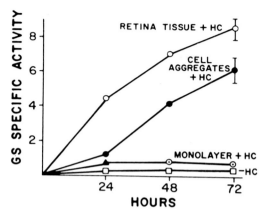

Fig. 3. Induction of GS by cortisol (hydrocortisone; + HC) in aggregates of retina cells, compared with absence of induction in cell monolayers. Also shown, GS induction in cultures of retina tissue. (From Moscona, 1974.)

phase, during which responsiveness to GS induction progressively increases to a maximal level.

V. CELL INTERACTIONS AND GS INDUCTION

Experiments begun some years ago (Morris and Moscona, 1970, 1971) suggested that, in addition to the effect of the specific hormonal inducer, still another mechanism may be involved in the control of GS induction in the embryonic neural retina. In these experiments, we examined if GS was inducible not only in intact retina tissue, but also in separated individual retina cells; in other words, whether GS could be induced in dissociated retina cells plated as a monolayer culture. Essentially, these experiments asked if specific cell–cell contacts such as exist in the intact tissue were required for GS induction.

Retina tissue from 10-day-old embryos was dissociated by mild trypsinization into a cell suspension, and the cells were plated in culture dishes under conditions which favored their remaining dispersed and in random contacts. Cortisol was added at plating time and GS activity was determined after 24 h and daily thereafter. There was no measurable induction of GS in the dispersed cells at any time (Fig. 3). The absence of induction was not due to secondary reasons: there was no significant cell loss; overall protein and RNA synthesis were not markedly reduced; the steroid hormone was taken up by the cells; there was no detectable leakage of the enzyme into the culture medium.

Since in the intact retina GS is induced and localized in Müller cells, special attention was paid to their fate in monolayer cultures. Previous work (Kaplowitz and Moscona, 1976) demonstrated that dissociated Müller cells assume in monolayer cultures the shape of large flattened epithelioid cells (LER cells). Immunostaining with GS antiserum of cortisol-treated monodispersed retina cells failed to reveal accumulation of GS in these, or in any type of cells present in these monolayer cultures. In contrast, immunostaining of such cultures with antiserum to carbonic anhydrase (which, as stated above, also is a marker for Müller cells, but is not regulated by cortisol) showed that this enzyme was present in the Müller cell-derived LER cells, at least for several days.

When cells in these monolayer cultures were plated at higher densities, some became associated in small clusters that consisted of closely juxtaposed neurons and LER cells. In such clusters (in cortisol-containing cultures) GS could often be detected immunohistochemically in the LER cells (Fig. 4) (Linser and Moscona, 1979; Moscona *et al.*, 1980a). After 2–3 days, such clusters tended to disperse and then the cells no longer contained detectable GS.

These results raised the possibility that specific cell–cell associations, possibly juxtapositions between glia and neurons, were prerequisite for responsiveness of Müller cells to GS induction by cortisol; absence or low frequency of such associations in monolayer cultures would explain their failure to show induction. An alternative possibility was that dissociation of retina tissue into a cell suspension by treatment with trypsin, rapidly and permanently abolished the competence of most Müller cells for GS induction. These alternatives were examined as follows.

The possible adverse effect of tissue trypsinization (to dissociate the cells) was tested by using a calcium chelator, EGTA, instead of trypsin, to disrupt retina tissue into cells. Monolayer cultures prepared from such cells also were not inducible for GS. Therefore, trypsin in itself was not the responsible factor for the noninducibility of the dispersed cells.

In a more direct test, trypsin-dissociated cells (from retinas of 8- or 10-day-old embryos) were reaggregated and the aggregates were examined for GS inducibility. As previously described (Moscona, 1974, 1980), when a suspension of dissociated embryonic retina cells is gently swirled in a flask on a rotary shaker, the cells aggregate into multicellular clusters. Within these clusters, the cells progressively sort out, become organized and reestablish histological relationships; this results, in 1 to 2 days, in reconstruction of retinotypic tissue architecture.

Treatment of such cell aggregates with cortisol resulted in GS induction; inducibility increased coordinately with the progressive histological organization of the cells and resumption of their characteristic spatial relationships (Morris and Moscona, 1971). When histological sections of such aggregates were immunostained for GS, the enzyme was found to be localized in Müller cells (Linser and Moscona, 1979) (Fig. 5).

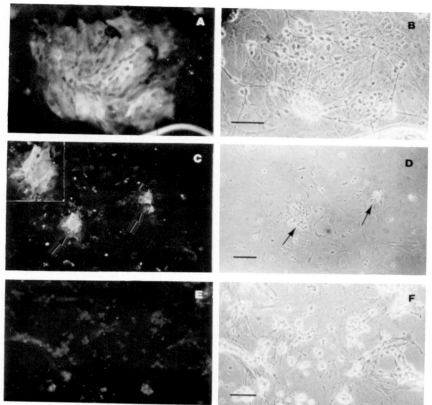

Fig. 4. Immunofluorescent analysis of GS localization in monolayer cultures of neural retina cells derived from retinas of 10-day-old chick embryos (A, B, E, F) and 18-day-old embryos (C, D). Retinas were dissociated with trypsin into cell suspensions, and the cells were plated at 1.3×10^6 cells/cm^2 in tissue culture dishes, in medium 199 with 10% fetal calf serum, 1% Pen-Strep mixture, and cortisol (0.33 μg/ml). Medium was changed daily; after 7 days the cultures were fixed and examined for the presence of GS by indirect immuno fluorescence, following treatment either with GS antiserum (A, C), or with nonimmune rabbit serum (E), as described (Linser and Moscona, 1979). (A, B) Culture of retina cells (from 10-day-old embryo) stained with GS antiserum; A, note fluorescence indicative of the presence of GS in LER cells (Müller glia-derived cells) only in region immediately beneath network of neuronal cells and cell processes, seen in the adjacent (B) phase micrograph of the same field. Peripheral LER cells (i.e., outside the area covered by the neuronal network) do not stain ($\times$ 245; bar represents 50 μM). (C, D) Culture of retina cells (from 18-day-old embryos) stained with GS antiserum; (C) note presence of GS only in LER cells (arrows) and only when these cells are juxtaposed (overlaid) with neuronal cells, as seen in the accompanying phase micrograph of the same field (D). Also note the abundance of LER cells which are not in contact with neurons and do not stain for GS ($\times$ 85; bar represents 100 μM). Inset in C is a higher magnification of the GS-positive plaque of LER cells seen on the left side in C and D. (E, F) Parallel control culture to that in A and B, treated with nonimmune rabbit serum (E); F, phase micrograph of same field; note light background autofluorescence in neuronal cells, but absence of staining in LER cells ($\times$ 200; bar represents 50 μM).

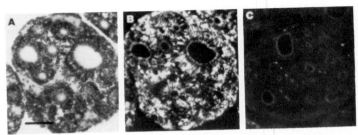

Fig. 5. Cellular localization of cortisol-induced GS in retina cell aggregates (see text). (A) Section of an aggregate stained with hematoxylin and eosin showing reconstruction of re- tinotypic tissue architecture characterized by the presence of retinal rosettes. (B) Section similar to A, immunostained for GS; immunofluorescence shows localization of induced GS in cells identified as Müller fibers. (C) Section of noninduced aggregate immunostained for GS. All magnification × 225. Bar represents 25 μM. (From Linser and Moscona, 1979.)

Therefore, the absence of GS induction in dissociated and dispersed retina cells is not due to a rapid, permanent loss of competence in Müller cells; as demonstrated by the above results, when the dispersed cells were reaggregated and they reconstructed retinotypic tissue architecture, Müller cells "recovered" inducibility for GS.

These results are consistent with the working hypothesis that specific cell–cell contacts, most likely between Müller cells and neurons are an essential condition for Müller cell responsiveness to the induction of GS by the steroidal inducer. One speculative possibility is that such specific cell–cell contacts or interactions reflect a need for "metabolic cell cooperation" in GS induction; they might enable transfer of substances that function as regulators (promoters or derepressors) at some control level in the mechanism of GS induction. Possible involvement of such postulated regulators in the mechanism of cortisol-mediated GS induction has been suggested by various experimental results and this concept was included in a heuristic model for this mechanism (Moscona, 1972). Although GS is localized in Müller cells, it is theoretically conceivable that the hormone acts also on neurons and elicit in them a "promoting" substance whose transfer into Müller cells enables GS induction; such neuron–glia interaction could require specific cell–cell contacts.

Another hypothetical possibility is that histotypic contacts between Müller cells and neurons are essential for maintaining the normal characteristics of the cell surface in Müller cells. Conceivably, "signals" relayed from the cell surface into the cell result in "permissive" conditions for GS induction; cell dissociation and cell separation modify the cell surface and thereby abrogate or change these "signals" and this, in turn, alters intracellular conditions necessary for GS induction by cortisol.

VI. CORTISOL RECEPTORS

Among the intracellular candidate mechanisms for regulation by cell contacts are the cytoplasmic receptors for cortisol. As in other target tissues for steroid hormones, also in the retina the effect of cortisol involves first, binding of the hormone to specific receptors in the cytoplasm and formation of cortisol-receptor complexes. The presence of cytoplasmic cortisol receptors in chick embryo neural retina cells has been demonstrated (Koehler and Moscona, 1975). The receptor–cortisol complexes are rapidly translocated into the nuclei where they associate with chromatin (Sarkar and Moscona, 1975, 1977); it is assumed that this elicits changes in gene expression resulting in accumulation of mRNA for GS and in GS induction (Moscona *et al.*, 1980). Hence, a change in the state of cortisol receptors triggered by changes in contact relationships of the cells could render the cells non-inducible. Therefore, we examined if the level of cortisol receptors in retina cells (specific cortisol-binding activity in the cytosol) was affected by dissociating the cells and maintaining them in a separated state, i.e., in monolayer cultures (Saad *et al.*, 1981).

Retina tissue (from 10-day-old chick embryos) was dissociated by trypsinization into single cells and these were plated as monolayer cultures. It was found (Fig. 6) that very soon after the dissociation, the level of detectable cortisol receptors in the cells began to decline; within 24 h the level dropped to below 30% of that present in intact retina tissue controls, and remained at this low level during further time in culture (Saad *et al.*, 1981).

The possibility that loss of receptors was due to their selective leakage from

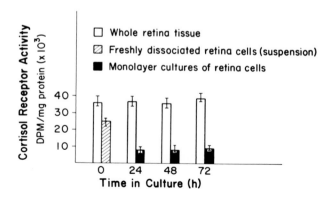

Fig. 6. Levels of cortisol receptors in whole retina tissue (from 10-day-old chick embryos), and in freshly dispersed retina cells in monolayer cultures. Receptor activity was measured at 0 h and after 24–72 h in culture. (From Saad *et al.*, 1981.)

cells into the culture medium could not be substantiated. The assumption that the receptors were degraded by residual trypsin was excluded by showing that, also in cells dissociated without trypsin, by treatment with a calcium chelating agent, EGTA, there was a similarly rapid loss of receptors. Inactivation of receptors by proteases that might be released from cells during preparation of cytosols (for receptor assay) also could not be confirmed. Therefore, the most likely possibility is that the decline in receptor level is triggered by separation of the cells from their mutual contacts within the tissue.

This possibility is further supported by the following results. When the separated cells were reaggregated and they reconstructed histotypic associations, the decline in receptor level was reversed (Fig. 7); within 48–72 h receptor level in such cell aggregates increased to 75% of values present in intact retina controls (Saad *et al.,* 1981).

The above results strongly suggest that histotypic cell–cell associations are involved in the maintenance (or activity) of cortisol receptors in embryonic retina cells. However, it cannot be stated with certainty that the loss of GS inducibility in dissociated-separated cells is due directly, or exclusively, to the decline in the level of cortisol receptors. Although there are definite correlations between loss of receptors and of GS inducibility, on one hand, and between recovery of receptors and of GS inducibility on the other, a direct cause–effect relationship has yet to be established. The possibility remains that cell separation, in addition to triggering loss of cortisol receptors, causes still other changes in the cells' regulatory mechanisms required for GS induction. It should also be pointed out that receptor loss in the separated cells is not complete; failure of the remaining

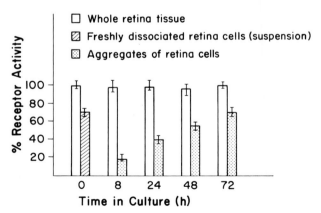

Fig. 7.   Level of cortisol receptors in whole retina tissue (from 10-day-old chick embryos) and in aggregates of retina cells after various culture times. Receptor activity (DPM/mg protein) is expressed as percent of the level present in freshly dissected whole retina tissue. (From Saad *et al.,* 1981.)

receptors to mediate GS induction cannot be presently explained. These residual receptors could be localized in neurons; if they are contained in Müller cells, they may represent a subclass of cortisol receptors that can not elicit GS induction. Finally, disruption of cell–cell contacts may have rendered the cells unable, for still other reasons, to respond to the hormonal inducer, as discussed earlier in this chapter. These questions remain to be explored.

VII. COMMENT

The results described and discussed above support the conclusion that, in the embryonic retina, specific contact-dependent cell interactions are prerequisite for responsiveness of glia cells to GS induction by the steroid hormone. This is consistent with the broader view that neurons and glia are functionally interdependent not only in their mature state, but also during their embryonic differentiation (for other examples consistent with this view see Sidman, 1974; Bignami et al., 1980). This concept raises certain general considerations. It focuses attention to the mechanisms that mediate specific cell–cell contacts and associations during retina morphogenesis; and it suggests that impairment of these mechanisms (due to genetic or environmental causes) could adversely affect glia–neuron interactions and thereby reduce retina responsiveness to induction of differentiation-related gene products. In the case of GS, the result would be a subnormal level of the enzyme; in turn, this would lower the effectiveness of metabolic processes requiring GS, such as the "recycling" of neuronally released neurotransmitter substances. It can be readily envisaged that, even a localized lesion of this kind, resulting from a failure in contact-dependent cell interactions during embryonic development, could seriously impair the normal function of the retina.

It seems unlikely that regulation by cell–cell contacts of GS inducibility in the retina represents a unique example of such a mechanism. It would be surprising if similar mechanisms did not play a role also in the control of other aspects of cell differentiation in the retina and in the brain. Accordingly, the results discussed here might encourage increased interest in the role of cell interactions in the biochemical differentiation of the nervous system, especially of those aspects that are regulated by steroid hormones. Such studies could deepen not only our understanding of normal neurodifferentiation, but might provide new insights into the nature and causality of various congenital defects in the nervous system.

ACKNOWLEDGMENTS

The original research described here is part of a research program supported by grant HD01253 from the National Institute of Child Health and Human Development, and by Grant 1-733 from the March of Dimes–Birth Defects Foundation.

REFERENCES

Bignami, A., Kozak, L. P., and Dahl, D. (1980). Molecular markers for the analysis of neural differentiation in culture. *In* "Tissue Culture in Neurobiology" (E. Giacobini, A. Vernadakis, and A. Shahar, eds.), pp. 63–73. Raven, New York.

Clark, A. M. (1951). Carbonic anhydrase activity during embryonic development. *J. Exp. Biol.* **28,** 332–343.

Kahn, A. J. (1974). An autoradiographic analysis at the time of appearance of neurons in the developing chick neural retina. *Dev. Biol.* **38,** 30–40.

Kaplowitz, P. B., and Moscona, A. A. (1976). Stimulation of DNA synthesis by ouabain and concanavalin A in cultures of embryonic neural retina cells. *Cell Differ.* **5,** 109–119.

Koehler, D. E., and Moscona, A. A. (1975). Corticosteroid receptors in the neural retina and other tissues of the chick embryo. *Arch. Biochem. Biophys.* **170,** 102–113.

Lash, J. W., and Vasan, N. S. (1977). Tissue interactions and extracellular matrix components. *In* "Cell and Tissue Interactions" (J. W. Lash and M. M. Burger, eds.), pp. 101–113. Raven, New York.

Linser, P., and Moscona, A. A. (1979). Induction of glutamine synthetase in embryonic neural retina: Localization in Müller fibers and dependence on cell interactions. *Proc. Natl. Acad. Sci. U.S.A.* **76,** 6476–6480.

Linser, P. J., and Moscona, A. A. (1981). Induction of glutamine synthetase in embryonic neural retina: Its suppression by the gliatoxic agent α-aminoadipic acid. *Dev. Brain Res.* **1,** 103–119.

Lo, C. W., and Gilula, N. B. (1979). Gap junctional communication in the preimplantation mouse embryo. *Cell* **18,** 399–409.

Meister, A. (1974). Glutamine synthetase of mammals. *In* "The Enzymes," (P. D. Boyer, ed.), Vol. 10, 3rd ed. pp. 699–754. Academic Press, New York.

Meller, K., and Glees, P. (1965). The differentiation of neuroglia-Müller-cells in the retina of chick. *Z. Zellforsch. Mikrosk. Anat.* **66,** 321–332.

Monroy, A., and Moscona, A. A. (1979). *In* "Introductory Concepts in Developmental Biology." Chapters 7 and 8, pp. 128–205. Univ. of Chicago Press, Chicago, Illinois.

Morris, J. E., and Moscona, A. A. (1970). Induction of glutamine synthetase in embryonic retina: Its dependence on cell interactions. *Science* **167,** 1736–1738.

Morris, J. E., and Moscona, A. A. (1971). The induction of glutamine synthetase in aggregates of embryonic neural retina cells: Correlations with differentiation and multicellular organization. *Dev. Biol.* **25,** 420–444.

Moscona, A. A. (1972). Induction of glutamine synthetase in embryonic neural retina: A model for the regulation of specific gene expression in embryonic cells. *In* "Biochemistry of Cell Differentiation" (A. Monroy and R. Tsanev, eds.), Vol. 24, pp. 1–23. Academic Press, New York.

Moscona, A. A. (1974). Surface specification of embryonic cells: Lectin receptors, cell recognition, and specific cell ligands. *In* "The Cell Surface in Development" (A. A. Moscona, ed.), pp. 67–99. Wiley, New York.

Moscona, A. A. (1980). Embryonic cell recognition: Cellular and molecular aspects. *In* "Membranes, Receptors, and the Immune Response" (E. P. Cohen and H. Köhler, eds.), Vol. 42, pp. 171–188. Alan R. Liss, New York.

Moscona, A. A., and Degenstein, L. (1981). Normal development and precocious induction of glutamine synthetase in the neural retina of the quail embryo. *Dev. Neurosci.,* **4,** 211–219.

Moscona, M., and Moscona, A. A. (1979). The development of inducibility for glutamine synthetase in embryonic neural retina: Inhibition of BrdU. *Differentiation* **13,** 165–172.

Moscona, A. A., and Piddington, R. (1966). Stimulation by hydrocortisone of premature changes in the developmental pattern of glutamine synthetase in embryonic retina. *Biochim. Biophys. Acta* **121**, 409–411.

Moscona, A. A., and Piddington, R. (1967). Enzyme induction by corticosteroids in embryonic cells: Steroid structure and inductive effect. *Science* **158**, 496–497.

Moscona, A. A., and Wiens, A. W. (1975). Proflavine as a differential probe of gene expression: Inhibition of glutamine synthetase induction in embryonic retina. *Dev. Biol.* **44**, 33–45.

Moscona, M., Frenkel, N., and Moscona, A. A. (1972). Regulatory mechanisms in the induction of glutamine synthetase in the embryonic retina: Immunochemical studies. *Dev. Biol.* **28**, 229–241.

Moscona, A. A., Mayerson, P., Linser, P., and Moscona, M. (1980a). Induction of glutamine synthetase in the neural retina of the chick embryo: Localization of the enzyme in Müller fibers and effects of BrdU and cell separation. *In* "Tissue Culture in Neurobiology" (E. Giacobini, A. Vernadakis, and A. A. Shahar, eds.), pp. 111–127. Raven, New York.

Moscona, A. A., Linser, P., Mayerson, P., and Moscona, M. (1980b). Regulatory aspects of the induction of glutamine synthetase in embryonic neural retina. *In* "Glutamine: Metabolism, Enzymology, and Regulation" (J. Mora and R. Palacios, eds.), pp. 299–313. Academic Press, New York.

Musser, G. L., and Rosen, S. (1973). Localization of carbonic anhydrase activity in the vertebrate retina. *Exp. Eye Res.* **15**, 105–109.

Piddington, R., and Moscona, A. A. (1967). Precocious induction of retinal, glutamine synthetase by hydrocortisone in the embryo and in culture: Age-dependent differences in tissue response. *Biochim. Biophys. Acta* **141**, 429–432.

Saad, A. D., Soh, B. M., and Moscona, A. A. (1981). Modulation of cortisol receptors in embryonic retina cells by changes in cell-cell contacts: Correlations with induction of glutamine synthetase. *Biochem. Biophys. Res. Commun.* **98**, 701–708.

Sarkar, P. K., and Moscona, A. A. (1973). Glutamine synthetase induction in embryonic neural retina: Immunochemical identification of polysomes involved in enzyme synthesis. *Proc. Natl. Acad. Sci. U.S.A.* **70**, 1667–1671.

Sarkar, P. K., and Moscona, A. A. (1975). Nuclear binding of hydrocortisone-receptors in the embryonic chick retina and its relationship to glutamine synthetase induction. *Am. Zool.* **15**, 241–247.

Sarkar, P. K., and Moscona, A. A. (1977). Glutamine synthetase induction in embryonic neural retina: Interactions of receptor–hydrocortisone complexes with cell nuclei. *Differentiation* **7**, 75–82.

Sarkar, P. K., Fischman, D. A., Goldwasser, E., and Moscona, A. A. (1972). Isolation and characterization of glutamine synthetase from chicken neural retina. *J. Biol. Chem.* **247**, 7743–7749.

Sarthy, P. V., and Lam, D. K. (1978). Biochemical studies of isolated glial (Müller) cells from the turtle retina. *J. Cell Biol.* **78**, 675–684.

Saxen, L., Lehtonen, E., Karkinen-Jääskeläinen, M., Nordling, S., and Wartiovaara, J. (1976). Are morphogenetic tissue interactions mediated by transmissible signal substances or through cell contacts? *Nature (London)* **259**, 662–663.

Sheridan, J. D. (1974). Low resistance junctions: Some functional considerations. *In* "The Cell Surface in Development" (A. A. Moscona, ed.), pp. 187–206. Wiley, New York.

Sidman, R. L. (1974). Contact interaction among developing mammalian brain cells. *In* "The Cell Surface in Development" (A. A. Moscona, ed.), pp. 221–253. Wiley, New York.

Soh, B. M., and Sarkar, P. K. (1978). Control of glutamine synthetase mRNA by hydrocortisone in the embryonic chick retina. *Dev. Biol.* **64**, 316–328.

Van den Berg, C. J. (1970). Glutamate and glutamine. *In* "Handbook of Neurochemistry" (A. Lajtha, ed.), Vol. III, pp. 355–379. Plenum, New York.

8

SEXUAL DIFFERENTIATION OF THE BRAIN: GONADAL HORMONE ACTION AND CURRENT CONCEPTS OF NEURONAL DIFFERENTIATION

Bruce S. McEwen

I. INTRODUCTION

The differentiation of cells within the various organs of a multicellular organism is influenced by chemical signals from nearby cells and from other parts of the body, notably the endocrine glands. For example, gonadal hormones function as chemical signals to trigger and coordinate the orderly acquisition of sex-specific characteristics such as the reproductive tract, mammary glands, and vocal organs (Burns, 1961). The brain is also subject to hormonal influences during development which determine special features of its structure and function

Molecular Approaches to Neurobiology

that are related to the reproductive process. This chapter will assess the present state of knowledge regarding the cellular and molecular events that underlie gonadal hormone action on the sexual differentiation of the brain of a number of representative vertebrate species. In this chapter, we shall refer to the actions of gonadal steroids on brain sexual differentiation as ''organizational'' influences (Phoenix *et al.,* 1959) in order to emphasize their permanence and their involvement with fundamental aspects of neuronal structure, connectivity, and function. The usually reversible hormonal influences on the adult brain will be referred to as ''activational'' effects. Recent reviews of various aspects of brain sexual differentiation include articles by Gorski (1979), Arnold (1980), Baum (1979), and McEwen (1980), as well as a conference report (Goy and McEwen, 1980).

II. GENERAL PLAN OF SEXUAL DIFFERENTIATION FOR MAMMALS

In mammals, the process of masculine sexual differentiation is imposed by testicular secretions on developing tissues that would otherwise acquire the feminine phenotype (Jost, 1970). Sexually dimorphic behavioral and neuroendocrine features follow this plan, for the most part. However, several variations and exceptions must be noted (Table I). The predominant plan of sexual differentiation of behavioral characteristics is one in which testicular secretions both organize the capability for displaying the behavior during development and later activate it in adult life (type Ia). An example is the capacity to display male copulatory behavior (Table II). In some but not all mammalian species testicular secretions act during development also to suppress feminine characteristics such as cyclic gonadotropin discharge and feminine sexual receptivity so that the males of these species lack the capability to display these characteristics even when castrated and primed with female sex hormones (Table II). The display of these feminine characteristics is normally activated by estrogens (and in some

Table I.   Classification of Sexual Differences: Hormonal Dependence[a]

Type	Organizational phase	Activational phase
Ia	Gonadally mediated ↑	Gonadally mediated ↑
Ib	Gonadally mediated ↓	Gonadally mediated ↑
II	None	Gonadally mediated ↑
III	Gonadally mediated ↑	Gonadally independent
IV	Gonadally independent ↑	Gonadally mediated ↑

[a] Arrow ↑ or ↓ indicates direction of effect to increase or decrease response.

Table II. Effects of Perinatal Administration of Androgen to Females on the Development of Sexually Dimorphic Traits[a]

Species	Critical[b] period	Dimorphic characteristic[c]			Reference
			Type of sexual behavior		
		Ovulation	Female	Male	
Rat	Post	↓↓↓↓	↓↓↓	↑	See text
Mouse	Post	↓↓↓↓	↓↓↓	↑↑	Barraclough and Leathem, 1954; Edwards and Burge, 1971
Hamster	Post	↓	↓	↑↑↑↑	Carter et al., 1972; Paup et al., 1972, 1974; Whitsett and Vandenbergh, 1975
Ferret	Post	?	0	↑↑↑	Baum, 1976
Dog	Pre + post	?	↓↓	↑↑	Beach and Kuehn, 1970; Beach et al., 1972
Guinea pig	Pre	↓↓	↓↓↓	↑↑	Phoenix et al., 1959; Goy et al., 1964; Brown-Grant and Sherwood, 1971
Sheep	Pre	↓↓	↓↓	↑↑	Short, 1974; Clarke et al., 1976
Rhesus	Pre	Onset delayed	?	↑↑	Goy, 1970a,b

[a] Reprinted from Goy and McEwen, 1980, by permission.
[b] Pre, prenatal; post, postnatal.
[c] Arrows indicate direction and relative ease of obtaining effect: ↓, defeminization; ↑, masculinization; 0, no effect.

species by the synergistic action of progesterone as well; Baum, 1979), and so this aspect of sexual differentiation is classified as type Ib because hormones are involved both in organization and in activation (Tables I and III). Another mode of sexual differentiation is one in which a hormonal influence occurs for organization, but is lacking for activation (type III, Table I). Examples of this mode are fewer in number than for type I: e.g., the juvenile play and mounting behavior of rhesus monkeys (Goy, R. W. in Goy and McEwen, 1980) and the urination pattern of the dog (Beach, 1974). The third category in Table I deals with behaviors which are not subject to organizational influences of hormones, but which are activated by gonadal hormones in adulthood (type II). Examples of this category are not plentiful (e.g., the yawning behavior of rhesus monkeys) and do not represent features of sexual differentiation because they can be manifested in

Table III.   Relationships between Organizational and Activational Effects of Hormones

Category of behavior	Organizational phase hormones and effect		Activational phase hormones and effect
Masculine	A[c]	↑[a]	A ↑
	T→E[d]	↑[a]	T→E ↑
	A + T→E	↑[a]	A + T→E ↑
Feminine	T→E	↓[b]	E + P ↑

[a] Masculinization.
[b] Defeminization.
[c] A, androgen pathway involving androgen receptors.
[d] T→E, aromatization pathway involving estrogen receptors.

either sex only during the time the influence of the activating hormone is present. A final, provisional category of sexual differentiation (type IV, Table I) is one in which permanent sex differences develop which are subject to activation and yet which are not apparently organized by gonadal hormones acting during development. There are no unequivocal examples of this category, but one report for the guinea pig is highly suggestive (Goldfoot and van der Werff ten Bosch, 1975). In this study, female guinea pigs exposed to testosterone propionate (TP) or dihydrotestosterone propionate (DHTP) became masculinized and displayed masculine sexual behavior in adulthood when treated with TP. Yet no androgenized female responded to DHTP in adulthood, although castrated males of this strain did show activation of male copulatory behavior by DHTP as well as by TP. These results, in the words of the authors, "raise the possibility that mechanisms determining sensitivity to specific steroids may not be mediated exclusively by steroids during critical periods of embryological differentiation" (Goldfoot and van der Werff ten Bosch, 1975). Regarding the possible role of genetic factors, it may be noted that several studies on reciprocal crosses between inbred strains of mice have revealed a contribution of the Y chromosome to the level of aggression shown by males (Selmanoff *et al.*, 1975; Stewart *et al.*, 1978).

Although it is tempting to speculate that genetic factors per se may play a role in neural sexual differentiation, the major function of the genetic sex is to determine the sex of the gonads (Mittwoch, 1970). Other contributions of genetic factors are summarized in Fig. 1. Neural sensitivity to gonadal hormones is one aspect, because gonadal steroid receptors and the steroid-metabolizing enzymes are gene products. For androgens, the various forms of testicular feminization represent mutations in the steroid-metabolizing enzymes (5α-reductase mutation) or the androgen receptors (complete androgen insensitivity) (J. D. Wilson in Goy and McEwen, 1980). Some consequences of these mutations will be discussed below.

The genome is also the repository of the multiple genetic factors (Fig. 1) that

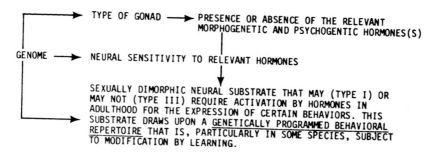

TYPE OF GONAD ⟶ PRESENCE OR ABSENCE OF THE RELEVANT
MORPHOGENETIC AND PSYCHOGENTIC HORMONES(S)

GENOME ⟶ NEURAL SENSITIVITY TO RELEVANT HORMONES

SEXUALLY DIMORPHIC NEURAL SUBSTRATE THAT MAY (TYPE I) OR
MAY NOT (TYPE III) REQUIRE ACTIVATION BY HORMONES IN
ADULTHOOD FOR THE EXPRESSION OF CERTAIN BEHAVIORS. THIS
SUBSTRATE DRAWS UPON A GENETICALLY PROGRAMMED BEHAVIORAL
REPERTOIRE THAT IS, PARTICULARLY IN SOME SPECIES, SUBJECT
TO MODIFICATION BY LEARNING.

Fig. 1. Role of genetic factors in hormone–brain behavior interactions. (Reproduced from Goy and McEwen, 1980, by permission of MIT Press.)

underlie species and strain differences in the types of behavior patterns and their hormonal sensitivity (e.g., the categories summarized in Table I and the patterns of sexual differentiation shown in Table II). One example of strain differences that are accentuated as a result of sexual differentiation deals with aggressive behavior in mice, i.e., inherited strain differences in aggression are latent in the genetic female and are brought forth when the female pups of these strains are treated with testosterone (Vale *et al.*, 1972).

Another aspect of the diversity among mammals with respect to hormone sensitivity concerns the metabolic pathways by which testosterone (T) produces its organizational and activational effects. These two pathways (Fig. 2) involve, on the one hand, 5α-reduction and interaction of the product, 5α-dihydrotestosterone (DHT), with androgen receptors and, on the other, aromati-

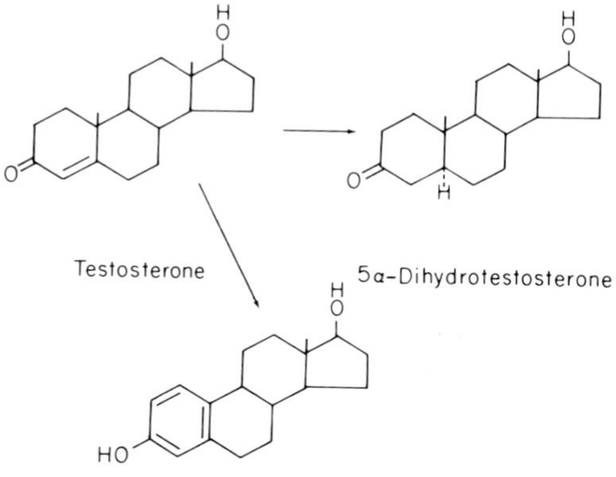

Testosterone

5α-Dihydrotestosterone

17β-Estradiol

Fig. 2. Aromatization and 5α-reduction of testosterone.

zation and interaction of the product, estradiol (E_2), with estrogen receptors. Masculinization of hamsters involves the aromatization pathway, since the administration of estrogens like diethylstilbestrol to newborn females masculinizes their behavior (Paup *et al.*, 1972). In contrast, masculinization of guinea pigs and rhesus monkeys involves the androgen pathway, since exposure *in utero* to nonaromatizable androgens like DHT masculinizes their behavior (Goldfoot and Van der Werff ten Bosch, 1975; Goy, 1978). In rats, a combination of aromatization and androgen pathways may be involved in masculinization (see below). A generalization which appears to connect organization with activation is that the same hormones which organize are also the ones involved in activation (R. W. Goy, personal communication) (Table III). This appears to hold for the hamster (Noble and Alsum, 1975), guinea pig (Goldfoot and Van der Werff ten Bosch, 1975), rhesus monkey (Goy, 1978; Phoenix, 1974), and rat (see below). The inference from this information is that the basis of the diversity among species is the variation in the numbers of androgen- or estrogen-sensitive neurons which have been laid down within the neural circuits destined to subserve masculine sexual behavior.

It should be noted that defeminization, where it occurs (Table II), also appears to follow the generalization summarized above and in Table III, in that estrogens (resulting from aromatization) are responsible for organization and are also involved in activation of feminine behavior and the LH surge. It appears likely that those species in which defeminization occurs are ones in which progesterone synergizes with estradiol to activate feminine sexual behavior and the LH surge (see Baum, 1979).

III. SEXUAL DIFFERENTIATION IN RATS

A. Ontogenesis of Gonadal Function and of Neural Steroid Sensitivity

The most detailed view of brain sexual differentiation may be obtained from an examination of the information pertaining to one species, the albino rat, in which both masculinization and defeminization occur during the perinatal period of development.

At the end of the second week of fetal life in rats, a genetically controlled signal related to the presence of the Y chromosome imposes testicular organogenesis on a presumptive gonad which would otherwise become an ovary (Jost, 1970). Leydig cells of the testes show two distinct phases of activity, the first commencing in the rat around the seventeenth day of gestation (birth = day 21) and ending in the second postnatal week. The second phase begins in the third postnatal week and continues into adulthood (Niemi and Ikonen, 1963;

Lording and De Kretser, 1972). The ability of the testes to produce testosterone is detected as early as the fourteenth to fifteenth day of gestation, and this ability precedes the appearance of typical Leydig cells (see J. Weisz in Goy and Mc-Ewen, 1980, pp. 90–91). That the brain or pituitary is controlling this early phase of testicular secretion is indicated by decreases in Leydig cell volume following decapitation of fetuses (Eguchi *et al.*, 1975). Gonadotropins are detected in the male pituitary on day 17 of gestation, but are not detectable in female pituitary until birth (Chowdhury and Steinberger, 1976). Testosterone levels in blood and testes are elevated at birth in the rat and decline, as does Leydig cell activity, during the first and second postnatal weeks (Resko *et al.*, 1968).

According to Jost (1970), the process of masculine differentiation of body sex is imposed by testicular secretions on developing tissues which would otherwise acquire the feminine phenotype. This process has two components: regression of the müllerian ducts and masculinization of the Wolffian ducts, the urogenital sinus, and the urogenital tubercle and swelling. All of this occurs in the rat from approximately the seventeenth postnatal day until the first or second day after birth. Müllerian duct regression is regulated by an unidentified testicular hormone, possibly a peptide (Josso *et al.*, 1975). The transformation of the Wolffian duct into epididymis, vas deferens, and seminal vesicles, the induction of prostate development from the urogenital sinus, and the differentiation of the urogenital tubercle and swelling into the male genitalia are all caused by testicular androgens. In the rat, as in most mammals, it appears to be T itself that stimulates Wolffian duct development, and a testosterone metabolite, DHT, that stimulates development of the urogenital sinus and external genitalia (Schultz and Wilson, 1974; Goldstein and Wilson, 1975). Evidence for this generalization is that DHT formation precedes virilization in the urogenital sinus and external genitalia and follows virilization in the Wolffian ducts. It should be noted, however, that the receptor system mediating virilization of the Wolffian ducts does respond to high doses of DHT in the rat (Schultz and Wilson, 1974). The differential role of T and DHT leads to an important difference between two forms of androgen insensitivity in man (Goldstein and Wilson, 1975). In the disorder of familial incomplete male pseudohermaphroditism, type II, deficient DHT formation results in a selective failure of masculinization of urogenital sinus and external genitalia. In the syndrome of complete androgen insensitivity (i.e., testicular feminization) there is a complete failure in the masculine development of Wolffian duct, urogenital sinus, and external genitalia, which has been traced to an absence of androgen receptors in all target tissues.

Neural sensitivity to gonadal steroids in the rat becomes evident toward the end of gestation. Testosterone 5α reductase is established in brain and pituitary tissue at birth (Denef *et al.*, 1974; Martini, 1976); and aromatizing enzymes are present in hypothalamus before birth (Reddy *et al.*, 1974), possibly as early as day 15 of fetal life (MacLusky, Naftolin, and Phillip, unpublished). Androgen

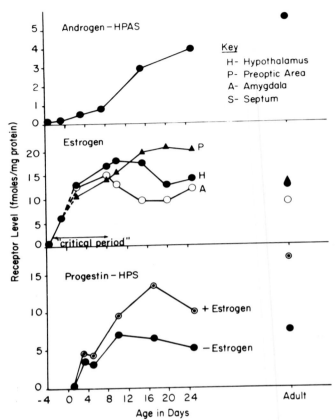

Fig. 3. Developmental time-course of gonadal steroid receptors in female rat brains. Receptor levels are expressed as fmoles/mg cytosol protein versus age in days post- or prenatal. Brain structures are denoted by letters (see key); note that they were studied individually (estradiol) or as tissue pools (e.g., HPAS, HPS). Androgen receptors were measured with [³H]testosterone and were isolated by DNA-cellulose chromatography (data from Lieberburg *et al.,* 1980). Estrogen receptors were measured with ³H-R2858 (moxestrol) by the LH-20 method (data from MacLusky *et al.,* 1979). Progestin receptors were measured with ³H-R5020 by the LH-20 method (MacLusky and McEwen, 1980). Arrows denote critical period for defeminization.

receptors are detectable a few days before birth, although they increase in concentration only toward the end of the first postnatal week (Lieberburg *et al.,* 1980) (Fig. 3). Neural estrogen receptors are also detectable a few days before birth (Vito and Fox, 1979; MacLusky *et al.,* 1979) and their concentration increases sharply between 1 or 2 days before birth and 6 or 7 days after birth (MacLusky *et al.,* 1979) (Fig. 3). Neural progestin receptors are not detected until a few days after birth and their postnatal increase is similar to that of the neural estrogen receptors (MacLusky and McEwen, 1980) (Fig. 3).

A method designed to estimate occupancy of neural estrogen receptors (Roy and McEwen, 1977) has revealed that by day 21 of gestation there is occupation of estrogen receptors that is substantial in male brain and barely detectable in fetal female brains (MacLusky et al., 1979) (Fig. 4). There is, however, no marked sex difference in estrogen receptor levels (MacLusky et al., 1979). During the first postnatal week of life the sex difference in occupancy is maintained and it parallels the substantially higher levels of testosterone in male serum (Lieberburg et al., 1979). Recent estimates of the capacity of the limbic (i.e., hypothalamic, preoptic area, amygdala) estrogen receptors to accumulate aromatized testosterone as estradiol reveal that it is approximately one-half of the capacity of the receptors to accumulate estradiol (Krey et al., 1980) (Fig. 5). This has led to the hypothesis that the number of estrophilic neurons in the limbic brain capable of aromatizing testosterone is approximately one-half of the total number of estrophilic neurons. For those cells which aromatize testosterone, the level of estrogen receptor occupation in normal male pups during the period of sexual differentiation is on the order of 50% of their capacity, a figure that is in

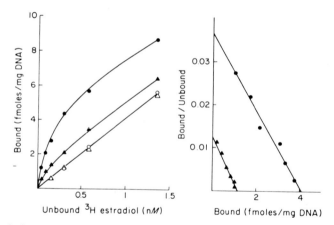

Fig. 4. Left, isotherms for the binding of [³H]estradiol in 0.4 M KCl extracts of purified cell nuclei from fetal rat brain tissue. Cell nuclei were prepared from limbic brain tissues of male (●) or female (▲) fetuses taken on day 21 of gestation. Estrogen–receptor complexes were extracted from the nuclei as described in the text, and equilibrated with a range of [³H]estradiol concentrations for 4 h at 25°C, in the presence (open symbols) or absence (closed symbols) of 0.3 μM unlabeled moxestrol. Bound radioactivity was measured in each incubate using Sephadex LH-20 gel filtration. Right, limited capacity binding, calculated from the data presented in the left-hand panel as the difference between the results in the presence and absence of unlabeled moxestrol, and represented by the method of Scatchard (1949). (Reprinted from MacLusky et al., 1979, by permission.)

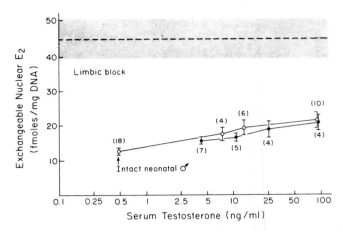

Fig. 5. Semilogarithmic plot of serum testosterone levels versus exchangeable cell nuclear estradiol in the limbic block (containing hypothalamus, preoptic area, and amygdala) of neonatal male and female rats of 3–4 days of age. Rats were grouped according to treatment and mean serum testosterone levels were: females (●) 50 μg TP, 4.5 ± 0.6 ng/ml (mean ± SE); 100 μg TP, 10.7 ± 0.9; 200 μg TP, 25.3 ± 6.5; 500 μg TP, 93.4 ± 2.2 (n = 4); males (○) 50 μg TP, 7.5 ± 1.1; 100 μg TP, 13.2 ± 1.6; 500 μg TP, 91.9 ± 22.6. Error bars representing SEM are indicated for each data point. The data for males were subjected to a one-way analysis of variance (F = 2.969; df 3/32; $p < 0.05$). Estrogen receptor capacity is presented as a dashed line at the top. Shaded area abutting the line represents the SEM (number of determinations = 6). (Reprinted from Krey *et al.*, 1980, by permission.)

agreement with the operating range of other steroid receptors under physiological conditions (Krey *et al.*, 1980).

B. Pharmacological and Genetic Evidence Pertaining to Defeminization

The extensive discussion of aromatization and estrogen receptor occupation is based on the fact that this pathway mediates the defeminization of the rat brain by testosterone. Two kinds of evidence have led to this conclusion. The first is pharmacological. Genetic evidence will be summarized below. It was first established that estrogens and aromatizable androgens are capable of defeminizing the developing rat brain, whereas nonaromatizable androgens like DHT are not. (For review of literature and discussion, see Plapinger and McEwen, 1978; Goy and McEwen, 1980.) Subsequently, it was possible to show that a competitive inhibitor of aromatization (ADT; 1,4,6-androstatriene-3,17-dione) blocks defeminization in males and in T-treated females (Vreeburg *et al.*, 1977; McEwen *et al.*, 1977; Clemens and Gladue, 1978; Davis *et al.*, 1978). Similar results were obtained with another inhibitor (ADT, androst-4-ene-3,6,17-trione) (Booth,

1978). Antiestrogens such as MER25 (McDonald and Doughty, 1972) and CI628 (McEwen *et al.*, 1977) attenuate the defeminization of females by T. Antiestrogens (Doughty *et al.*, 1975; McEwen *et al.*, 1979), but not ATD (McEwen *et al.*, 1979), block the defeminization of the female rat by estrogens.

The overall scheme demonstrated by these results is summarized in Fig. 6, which attributes a primary role to the estrogen receptors and to aromatization in the defeminization of the rat brain. The major role of α-fetoprotein, a fetal and neonatal estrogen binding protein (fEBP) present in serum (Raynaud *et al.*, 1971) and also in cerebrospinal fluid (Plapinger *et al.*, 1973), is to afford protection against estrogens present in the fetal, neonatal, and maternal circulations, including that which may be reaching the pups through the mothers milk (Fig. 6). Recently, fEBP has been found in brain cells by immunocytochemistry (Benno and Williams, 1978; Toran-Allerand, 1980b). The distribution of fEBP in the neonatal mouse brain (Toran-Allerand, 1980b) is remarkable, in that it is absent in certain brain regions, like the arcuate nucleus, which has intracellular estrogen receptors and which is believed to participate in sexual differentiation (see below). The significance of this distribution is obscure, and it is not yet clear if some fEBP is made by brain cells or is taken up from the cerebrospinal fluid. However, it is worth noting that fEBP inhibits multiplication of estrogen-sensitive tumor cells in culture even in the presence of an estrogen, R2858, which does not bind to fEBP (Soto and Sonnenschein, 1980). This observation, if it could be extended to brain cells, would have important implications for the mechanism by which estrogens promote brain sexual differentiation (see below).

Supporting evidence for this scheme shown in Fig. 6 has been obtained from studies on the complete androgen insensitivity (Tfm) mutation in rats, in which androgen receptors are lacking but aromatization and estrogen receptors function

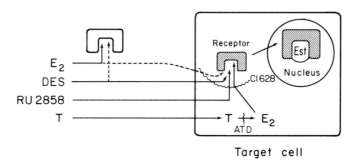

Target cell

Fig. 6. Events involved in the interaction of testosterone (T) and estrogens with the rat brain. fEBP, α-fetoprotein; R2858, moxestrol; DES, diethylstilbestrol; E_2, estradiol; ATD, aromatase inhibitor; CI628, antiestrogen; Est, receptor-bound estrogens in nucleus. (Reprinted from McEwen *et al.*, 1975, by permission.)

normally (Naess *et al.,* 1976; L. C. Krey, unpublished). It has been demonstrated that genetic males bearing the *Tfm* gene are normally defeminized by their testes (Shapiro *et al.,* 1975; Beach and Buehler, 1977; Olsen, 1979a; Shapiro *et al.,* 1980). Moreover, neonatal castration of such Tfm males prevents defeminization (Olsen, 1979a).

C. Pharmacological and Genetic Evidence Pertaining to Masculinization in the Rat

The blockade of defeminization in neonatal rats with inhibitors such as ATD does not prevent masculinization from taking place (Vreeburg *et al.,* 1977; Davis *et al.,* 1978; Thomas *et al.,* 1980). One reason for this may be that the process of masculinization begins somewhat earlier than defeminization (for references, see McEwen, 1980). Another reason is that the process of masculinization may involve a combination of aromatization and androgen receptor pathways. Briefly, the evidence for this statement is as follows. Treatment of male rats castrated on the day of birth with the combination of an estrogen and DHT has been reported to result in better masculinization than that produced by either estrogen or DHT alone (Booth, 1977a; Hart, 1979; van der Schoot, 1980). [It should be noted that another study found that estrogen treatment of neonatally castrated males was as effective in masculinizing as treatment with testosterone or with estrogen plus DHT (Södersten and Hansen, 1978), indicating perhaps that strains of rats differ in their dependence on DHT.]

Pharmacological studies with the aromatase inhibitor ADT (Booth, 1978) and with the antiestrogen MER25 (Booth, 1977b; Södersten and Hansen, 1978) indicate that these substances attenuate testosterone-induced masculinization. [Experiments involving postnatal treatment of male rats with ATD, another aromatase inhibitor that is effective in blocking defeminization (see above), have not shown marked effects on the process of masculinization (Vreeburg *et al.,* 1977; Davis *et al.,* 1978; Thomas *et al.,* 1980). It is, however, possible to invoke strain differences in the perinatal timing of masculinization (see above) and thus avoid an outright contradiction, which would be unwarranted given the weight of evidence.]

Genetic information obtained on the Tfm rats strongly supports a partial role for the aromatization pathway in masculinization of the rat brain. In both of two studies, Tfm rats castrated as adults exhibited signs of perinatal masculinization by showing male copulatory behavior in response to estrogen as well as testosterone replacement (Olsen, 1979b; Shapiro *et al.,* 1980).

The final study to be mentioned in this section is one that ties together the dual involvement of aromatization and androgen pathways in terms of neuroanatomical loci (Hart, 1979). In this study, neonatally castrated male rats treated with estrogen plus DHT exhibited marked masculinization as adults and more so than rats treated with estrogen or DHT alone. After testing masculine sexual behavior, spinal transection was carried out for examination of penile reflexes. Rats treated with DHT alone exhibited normal penile reflexes, whereas those treated with estrogens alone did not, leading to the conclusion that the masculinization of the spinal cord is an androgen-mediated process (Hart, 1979). It will be seen below that a group of androgen-concentrating cells in the cells in the spinal cord project to the muscles of the penis and that these cells show marked sexual dimorphism (Breedlove and Arnold, 1980). These cells are absent in Tfm rats, according to Breedlove and Arnold (1980).

It can be inferred from the evidence cited above that neural mechanisms in the brain undergo masculinization under the influence of testosterone via the aromatization pathway and that there may be some contribution of the androgen pathway, but judging from the occurrence of some complete masculine sexual behavior in rats treated neonatally with estrogen alone (in which spinal mechanisms are not masculinized), the normal spinal penile reflexes are not strictly required (Hart, 1979).

IV. MORPHOLOGICAL SEX DIFFERENCES IN THE BRAIN

Experiments involving testosterone implantation in neonatal rat brain present a strong argument in favor of a neural target for sexual differentiation located in the hypophysiotropic area (Nadler, 1968, 1972, 1973; Hayashi and Groski, 1974; Christensen and Gorski, 1978). Further evidence comes from morphological studies indicating that brain cells change as a result of their neonatal hormonal environment. For example, Dörner and Staudt (1969a,b) have reported a diminution of cell nuclear volume in neurons of the preoptic area and ventromedial nucleus in male (as compared with female) rats during the last 2 days of fetal life. Other studies indicate that neonatal castration and gonadal hormone treatment of males produce effects on brain cell nuclear and nucleolar size in adult life. Pfaff (1966) found smaller cell nuclear and nucleolar areas in neonatally castrated male rats (as compared with controls) in a wide variety of brain areas: ventromedial hypothalamus, dentate gyrus, neocortex, and reticular formation. There was also an increase in nuclear area in the habenular nucleus of neonatal castrates compared with controls (Pfaff, 1966). In another study of cell size parameters, Gregory (1975) reported a sex difference in the size of pyramidal cells and cell nuclei of the somatosensory cortex of the adult rat under conditions in which hippocampal pyramidal cell nuclear volume was not sexually dimorphic. A third study, on adult squirrel monkeys, revealed a sex difference in cell nuclear size of the medial amygdala, but did not uncover any differences in the medial preoptic area, arcuate nucleus, suprachiasmatic nucleus, or cerebral cortex (Bubenik and Brown, 1973). In none of these studies is there any indication if adult hormone secretion contributes to the morphological differences. This is not an unreasonable possibility in view of one demonstration that testosterone promotes growth of brain vocal control areas in adult songbirds (Nottebohm, 1980). This question is considered in studies by Dörner and Staudt (1968, 1969a,b) who treated all rats in adulthood with androgen in order to equalize hormonal states. In spite of this treatment, neonatal castration increased nuclear volume in preoptic–anterior hypothalamic neurons and in neurons of the ventromedial nucleus, and testosterone replacement 2 days after neonatal castration attenuated the increases.

More detailed information as to the morphological consequences of sexual differentiation comes from Raisman and Field (1973), who describe a sexual dimorphism in the relative number of synapses from nonamygdaloid projections to shafts and spines of preoptic area dendrites. In these studies, castration of male rats at 12 h (but not 7 days) after birth partially inhibited the male patterns from appearing, whereas treatment of females with TP on day 4 (but not on day 16) shifted the pattern of preoptic morphology toward that of the male (Raisman and Field, 1973). In a related light microscopic study of the hamster dorsomedial preoptic area, Greenough *et al.* (1977) identified sex differences in the distribution of the dendritic processes. Males tended to have more centralized concentrations of processes, whereas females had an irregular dendritic density distribution surrounding the point of highest density in males. Such differences might account for differences in the ratio of synaptic types reported by Raisman and Field (1973). More recent work has revealed a sex difference in the numbers of axosomatic and axodendritic spine synapses in the hypothalamic arcuate nucleus of the rat (Matsumoto and Arai, 1980). The male arcuate nucleus contains more of the former and fewer of the latter than the female, but both sexes contain the same density of axodendritic shaft synapses. The sex differences are reversed by castrating males on day 1 and by giving females testosterone propionate on day 5 (Matsumoto and Arai, 1980). In view of the proximity of the arcuate nucleus to the median eminence, it is noteworthy that adult female rats have a larger neurovascular contact surface in the median eminence than males (Rethelyi, 1979).

Several reports indicated striking morphological sex differences at the levels of entire cell groups, or nuclei, within brains of songbirds and rats. In canaries and zebra finches, three areas associated with vocal control are larger in males than in females, and this is positively correlated with proficiency in singing (Nottebohm and Arnold, 1976). The female vocal control areas of the zebra finch are enlarged as a result of treatment with estradiol at hatching (Gurney and Konishi, 1980). In two of these brain areas of the zebra finch, males contain more neurons that concentrate [^{3}H]testosterone than females, whereas in the third brain area there is a sex difference in the size distribution of labeled cells (Arnold, 1980). In rats, a part of the medial preoptic area just anterior to the suprachiasmatic nucleus is larger in males than in females (Gorski *et al.*, 1978, 1980). Adult gonadal secretion plays little role in the size differences, whereas neonatal testosterone increases the size of the nucleus in females (Gorski *et al.*, 1978).

With respect to the sexually dimorphic morphological characteristics of the rat preoptic area, it is relevant to note the electrophysiological studies of Dyer *et al.* (1976), who found that males have more functional amygdala projections to the preoptic area than females. Neonatally castrated males resemble females in this respect, and neonatally androgenized females are intermediate between males and females. These results differ from those of Raisman and Field (1973), who found their morphological sex differences in synapses of nonamygdaloid origin. Yet the two sets of data are not incompatible.

As suggested by the results of Pfaff (1966) and Gregory (1975) and Bubenik and Brown (1973) cited above, the hypophysiotrophic area may not be the only neural target of neonatal gonadal steroid action. One of the most important areas in the spinal cord, which in the rat contains a sexually dimorphic nucleus, the spinal nucleus of the bulbocavernosus (Breedlove and Arnold, 1980). This nucleus, which contains motoneurons of the levator ani and bulbocavernosus, is virtually absent in females and in androgen-insensitive genetic males, and it contains cells that concentrate [^{3}H]androgens but not [^{3}H]estrogens (Breedlove and Arnold, 1980).

V. SEXUAL DIFFERENTIATION IN RELATION TO CURRENT CONCEPTS OF NEURONAL DEVELOPMENT AND DIFFERENTIATION

The sex differences in brain structure described above provide a basis for understanding the permanence of mammalian brain sexual differentiation. Insofar as is presently known, these structural changes are stable in relation to the endocrine environment of the adult. Therefore, it is important to consider how sex differences in brain structure may arise during the critical period when sexual differentiation occurs. We shall consider what is known about the ontogeny of these sex differences and also make some speculations in relation to current concepts of neuronal development and differentiation.

The life history of a neuron may be divided into four phases (Lund, 1978) as follows.

1. The period up to the final cell division.
2. The phase of growth, migration, and differentiation leading to the formation of connection with other neurons.
3. The period during which "good" connections survive and are strengthened and "bad" connections are eliminated. The elimination of bad connections is frequently accompanied by the death of those cells which have been unsuccessful in forming good connections.

4. In the final phase, which overlaps with the previous one, the neuron assumes its mature function.

With respect to neurons of the preoptic area and hypothalamus of the rat, phase 1 involving the final cell division appears to be completed by fetal day 17 prior to the onset of the critical period of hormonal sensitivity for sexual differentiation (Ifft, 1972; Altman and Beyer, 1979). Therefore, the hormonally influenced events leading to sexual differentiation of these brain regions probably do not include cell division, unless the hormonal stimulus reinitiates this event. Since evidence is lacking on this important point, we shall focus on the role of the events that constitute phases 2 and 3 of neuronal development.

Regarding phase 2, neurons of the medial preoptic area (Reier *et al.*, 1977) and arcuate nucleus (Walsh and Brawer, 1979) have an immature appearance at birth with sparse cytoplasmic organelles characteristic of immature neurons, a paucity of synapses, and evidence of growth cones. The postnatal critical period in these two brain regions is characterized by a rapid increase of synapses and dendritic spines (Arai and Matsumoto, 1978; Reier *et al.*, 1977; Lawrence and Raisman, 1980). In the arcuate nucleus, estrogen treatment during this period has been shown to increase the density of synapses of certain types (Arai and Matsumoto, 1978). This observation may be interpreted either as a hormonal influence on the process of synapse formation or as a hormonal influence on the growth of neurites which would secondarily lead to increased synapse formation through increased cell–cell interactions.

Taking these possibilities in reverse order, there is evidence that estrogens and testosterone *in vitro* both increase the outgrowth of neurites from the cut edges of preoptic area and hypothalamic blocks maintained in organ culture from newborn mouse brains (Toran-Allerand, 1976, 1978, 1980a). This outgrowth is not directed at a neural target. Moreover, it has been traced in several instances and shown to emanate from groups of neurons which contain estrogen receptors as judged by autoradiography with [³H]estrogens (Toran-Allerand *et al.*, 1980). Therefore, it would appear that neurite outgrowth is a primary effect of estrogens (including estrogens derived from aromatization) in cells which have acquired estrogen receptors, and that it is independent of the presence of a neural target. Nevertheless, regarding the first possibility described above, the hormones that influence neurite outgrowth may also trigger a program within these cells which would increase or decrease interactions with other cells that lead to synapse formation. Such a mechanism might explain the selectivity of synaptic types that are increased by hormone (e.g., Matsumoto and Arai, 1980). There is no evidence pointing directly to this type of hormone effect.

However, an examination of other work on neuronal differentiation provides further insights into other cellular events where hormonal signals may play a role. As neurons grow and form connections, they acquire their characteristic shapes. Judging by studies on Purkinje cells of the cerebellum of genetic mutants and of animals treated to destroy granule cells early in development (summarized by Lund, 1978, Chapter 11), the shape of the dendritic field is distorted by the lack of normal input, but the basic features of the neurons are present, including dendritic spines which apparently develop in the absence of synaptic contacts. These observations lead one to wonder if the decrease in dendritic spine synapses observed in the preoptic area and arcuate nucleus of males and neonatally androgenized females (Raisman and Field, 1973; Matsumoto and Arai, 1980) may result at least in part from a hormonal signal which reduces the autonomous production of dendritic spines and thereby leads to a reduction in the number of spine synapses which can form. On the other hand, from the same body of information one may speculate that the changes in the orientation of the dendritic fields of preoptic neurons produced by testicular androgens during early development of the hamster (Greenough *et al.*, 1977) reflects the interactions between brain cells and might be due to the influence of ingrowing afferents.

With respect to phase 3 of neuronal development (see above) the period of synapse formation for cells making good synapses is accompanied by the marked elaboration of cytoplasmic organelles and by the enlargement of the neuron soma as the full growth of the synaptic and dendritic fields is attained (e.g., Lund, 1978; Reier *et al.*, 1977). For cells unsuccessful in making enough good contacts, cell atrophy and death occur, and there appears to be considerable cell loss in many regions of the nervous system at the time of synapse formation, since overproduction of neurons with later cell loss is apparently a characteristic of neural development (Cowan, 1973; Landmesser and Pilar, 1978; Lund, 1978). With regard to brain regions undergoing sexual differentiation, there is as yet no complete analysis of cell loss during the course of development. Yet there is some suggestive information that lends itself to the cell loss interpretation. The virtual absence of cell groupings in brain regions of female songbirds and in the spinal cord of female rats, which are found in males of these species, is strongly suggestive of a loss of cells lacking either a hormonal factor to maintain them (by analogy with wolffian duct regression in the female reproductive tract) or lacking a normal target (by analogy with many studies of cell loss following removal of neural targets; see Lund, 1978). Particularly suggestive in relation to the second possibility are the studies on the rat spinal cord, since females lack the muscles which are innervated in the male by the neurons of the nucleus of the bulbocavernous (Breedlove and Arnold, 1980). The first possibility, involving hormonal stabilization of cell types, is only suggested by analogy with examples such as the wolffian ducts which regress in the absence of testicular androgens. A more complex variant on the theme of humoral stabilization of cells is suggested by the studies which indicate that the stabilization of developing adrenergic neurons may be due in part to the retrograde axoplasmic transport of NGF from the target neurons with which synaptic contacts are forming (Hendry, 1976).

[Hence a good synapse is one in which a stabilizing substance is obtained from the postsynaptic cell or its environs—see Lund, 1978.] It is possible to imagine ways in which gonadal hormones might determine if a good synapse is formed, e.g., they could do so by stimulating or retarding the formation of NGF-like stabilizing factors.

Another variation on the phenomenon of cell loss and hormonal regulation is the notion of programmed cell death. There is no direct evidence bearing on this mechanism in relation to sexual differentiation. Yet it remains a possibility by analogy with the programmed destruction of the müllerian ducts of the developing reproductive tract by a testicular product other than testosterone (Josso *et al.*, 1975) and by analogy with the programmed loss of neural and other tissues brought about by thyroid hormones during metamorphosis of the tadpole into the frog (Weber, 1969).

Another aspect of phase 3 of neuronal development involves cell enlargement after formation of ''good'' synapses and as the dendritic and axonal fields increase further in size. The work of Gorski and co-workers (Gorski *et al.*, 1980; Jacobson *et al.*, 1980) indicates that the sexually dimorphic nucleus (SDN) of rat preoptic area contains larger neurons in males than in females. (The same is true for cells with androgen receptors in at least one of the sexually dimorphic brain areas of the zebra finch; Arnold, 1980). The ontogeny of the rat SDN is characterized by a marked increase in the volume of the male SDN with a relatively smaller increase in the volume of the female SDN. This developmental profile has not yet been analyzed in terms of the relative contributions of cell number and cell size, and this distinction is important because the adult male SDN also appears to have more cells than the adult female SDN (Gorski *et al.*, 1980). Thus, we do not yet know if cell body enlargement is the only factor in the increased volume of the SDN and whether either cell loss in the female or a late mitotic phase in the male may make some contribution. If increased cell size is a major feature in the larger male SDN, it would appear likely, based upon the discussion above, that larger cell size may be a consequence of the faster growth of those neurons stimulated by testicular androgens which enables them to achieve a larger dendritic or axonal field, presumably with larger numbers of synaptic connections. It should be recalled in this regard that Dyer *et al.* (1976) found by electrophysiological recording that male rats have more functional projections from the amygdala to the preoptic area than females.

VI. CONCLUSIONS

Sexual differentiation of the brain appears to be a widespread event among mammals, judging from behavioral and neuroendocrine studies summarized above and in Table II. In one mammalian species, the rat, we are beginning to get a good idea of how the brains of adult males and females differ from each other; and undoubtedly our knowledge of sex differences in the brains of other mammalian species will increase now that we know what to look for. One fact which has emerged from the analysis of how hormone sensitivity develops in the rat brain is that receptors for estrogens and androgens appear late in fetal life. Their appearance, together with the capacity for aromatization and 5α-reduction of testosterone, signals the onset of the critical period for sexual differentiation.

A major question for future research in this area concerns the neuronal mechanisms responsible for producing sex differences. This question has two aspects: first, what are the principal features of neuronal differentiation that take place during the critical period when gonadal steroids act to trigger sexual differentiation? Second, which of these processes are influenced by these hormones? Much of the previous section of this chapter dealt with these two questions and that analysis revealed the extent of our ignorance, whereas at the same time providing speculations on possible sites of hormonal influence.

Perhaps the most difficult task will be to distinguish between primary and secondary or tertiary influences of gonadal hormones (e.g., discussion above of cell body size in the SDN of the rat). As we have seen, one primary effect of E_2 and T is to stimulate neurite outgrowth from hormone-sensitive neurons (Toran-Allerand, 1980a; Toran-Allerand *et al.*, 1980). Does this process represent a sufficient condition for producing sex differences in cell number, cell size, and numbers of synaptic endings? Is hormone-stimulated growth universal among all hormone-sensitive neurons participating in sexual differentiation? We do not yet know the answers to these questions. Among other, possibly primary, effects of gonadal steroids (noted above) are the reinitiation of cell division; the alteration of autonomous programs of gene expression and differentiation such as elaboration of dendritic spines; the induction of factors that increase cell–cell interactions and synapse formation; the induction of factors which stabilize afferent neurons that form connections with hormone-sensitive cells; and the triggering of cell death. Each of these possibilities deserves thorough investigation.

ACKNOWLEDGMENTS

Research from the author's laboratory referred to in this chapter was supported by grants from the United States Public Health Service (NS07080) and Rockefeller Foundation (Institutional Grant RF70095 for research in reproductive biology). The author thanks Mrs. Oksana Wengerchuk for editorial assistance.

REFERENCES

Altman, J., and Beyer, S. (1979). Development of the diencephalon of the rat. *J. Comp. Neurol.* **182**, 945–1016.

Arai, Y., and Matsumoto, A. (1978). Synapse formation of the hypothalamic arcuate nucleus during postnatal development in the female rat and its modification by neonatal estrogen treatment. *Psychoneuroendocrinology (Oxford)* **3**, 31–45.

Arnold, A. P. (1980). Quantitative analysis of sex differences in hormone accumulation in the Zebra Finch brain: Methodological and theoretical issues. *J. Comp. Neurobiol.* **189**, 421–436.

Arnol , A. P. (1980). Sexual differences in the brain. *Am. Sci.* **68**, 165–173.

Barraclough, C. A., and Leathem, J. H. (1954). Infertility induced in mide by a single injection of testosterone propionate. *Proc. Soc. Exp. Biol. Med.* **85**, 673–674.

Baum, M. J. (1976). Effects of testosterone propionate administered perinatally on sexual behavior of female ferrets. *J. Comp. Physiol. Psychol.* **99**, 399–410.

Baum, M. J. (1979). Differentiation of coital behavior in mammals: A comparative analysis. *Neurosci. Biobehav. Rev.* **3**, 265–284.

Beach, F. A. (1974). Effects of gonadal hormones in urinary behavior in dogs. *Physiol. Behav.* **12**, 1005–1013.

Beach, F. A., and Buehler, M. G. (1977). Male rats with inherited insensitivity to androgen show reduced sexual behavior. *Endocrinology (Baltimore)* **100**, 197–200.

Beach, F. A., and Kuehn, R. E. (1970). Coital behavior in dogs. X. Effects of androgenic stimulation during development on feminine mating reponses in females and males. *Horm. Behav.* **1**, 347–367.

Beach, F. A., Kuehn, R. E., Sprague, R. H., and Anisko, J. J. (1972). Coital behavior in dogs. XI. Effects of androgenic stimulation during development on masculine mating responses in females. *Horm. Behav.* **3**, 143–168.

Benno, R. H., and Williams, T. H. (1978). Evidence for intracellular localization of alpha-fetoprotein in the developing rat brain. *Brain Res.* **142**, 182–186.

Booth, J. E. (19771). Sexual behaviour of neonatally castrated rats injected during infancy with oestrogen and dihydrotestosterone. *J. Endocrinol.* **72**, 135–141.

Booth, J. E. (1977b). Sexual behaviour of male rats injected with the anti-oestrogen MER-25 during infancy. *Physiol. Behav.* **19**, 35–39.

Booth, J. E. (1978). Effects of the aromatization inhibitor androst-4-ene-3,6,17-trione on sexual differentiation induced by testosterone in the neonatally castrated rat. *J. Endocrinol.* **79**, 69–76.

Breedlove, S. M., and Arnold, A. P. (1980). Hormone accumulation in a sexually dimorphic motor nucleus in the rat spinal cord. *Science* **210**, 564–566.

Brown-Grant, K., and Sherwood, M. R. (1971). The "early androgen syndrome" in the guinea pig. *J. Endrocrinol.* **49**, 277–291.

Bubenik, G. A., and Brown, G. M. (1973). Morphologic sex differences in primate brain areas involved in regulation of reproductive activity. *Experientia* **29**, 619–621.

Burns, R. K. (1961). Role of hormones in the differentiation of sex. *In* "Sex and Internal Secretions" (W. C. Young, ed.), pp. 76–158. Williams & Wilkins, Baltimore, Maryland.

Carter, C. S., Clemens, L. G., and Hoekema, D. J. (1972). Neonatal androgen and adult sexual behavior in the golden hamster. *Physiol. Behav.* **9**, 89–95.

Chowdhury, M., and Steinberger, E. (1976). Pituitary and plasma levels of gonadotrophins in foetal and newborn male and female rats. *J. Endocrinol.* **69**, 381–384.

Christensen, L. W., and Gorski, R. A. (1978). Independent masculinization of neuroendocrine systems by intracerebral implants of testosterone or estradiol in the neonatal female rat. *Brain Res.* **146**, 325–340.

Clarke, I. J., Scaramuzzi, R. J., and Short, R. V. (1976). Sexual differentiation of the brain: endocrine and behavioral responses of androgenized ewes to oestrogen. *J. Endocrinol.* **71**, 175–176.

Clemens, L. G., and Gladue, B. A. (1978). Feminine sexual behavior in rats enhanced by prenatal inhibition of androgen aromatization. *Horm. Behav.* **11**, 190–201.

Cowan, M. (1973). Neuronal death as a regulative mechanism in the control of cell number in the nervous system. *In* "Development and Aging in the Nervous System" (M. Rochstein, ed.), pp. 19–41. Academic Press, New York.

Davis, P. G., Chaptal, C. V., and McEwen, B. S. (1978). Independence of the differentiation of masculine and feminine sexual behavior in rats. *Horm. Behav.* **12**, 12–19.

Denef, C., Magnus, C., and McEwen, B. S. (1974). Sex-dependent changes in pituitary 5α-dihydrotestosterone and 3α-androstanediol formation during postnatal development and puberty in the rat. *Endocrinology (Baltimore)* **94**, 1265–1274.

Dörner, G., and Staudt, J. (1968). Structural changes in the preoptic anterior hypothalamic area of the male rat, following neonatal castration and androgen substitution. *Neuroendocrinology* **3**, 136–140.

Dörner, G., and Staudt, J. (1969a). Structural changes in the hypothalamic ventromedial nucleus of the male rat, following neonatal castration and androgen treatment. *Neuroendocrinology* **4**, 278–281.

Dörner, G., and Staudt, J. (1969b). Perinatal structural sex differentiation of the hypothalamus in rats. *Neuroendocrinology* **5**, 103–106.

Doughty, C., Booth, J. E., McDonald, P. G., and Parrott, R. F. (1975). Inhibition, by the anti-oestrogen MER-25, of defeminization induced by the synthetic oestrogen RU2858, *J. Endocrinol.* **67**, 459–460.

Dyer, R. G., MacLeod, N. K., and Ellendorf, F. (1976). Electrophysiological evidence for sexual dimorphism and synaptic convergence in the preoptic and anterior hypothalamic areas of the rat. *Proc. R. Soc. London Ser. B.* **193**, 421–440.

Edwards, D. A., and Burge, K. G. (1971). Early androgen treatment and male and female sexual behavior in mice. *Horm. Behav.* **2**, 49–58.

Eguchi, Y., Sakamoto, Y., Arishima, K., Morikawa, Y., and Hashimoto, Y. (1975). Hypothalamic control of the pituitary-testicular relation in fetal rats: measurement of collective volume of leydig cells. *Endocrinology (Baltimore)* **96**, 504–507.

Goldfoot, D. A., and van der Werff ten Bosch, J. J. (1975). Mounting behavior of female guinea pigs after prenatal and adult administration of the propionate of testosterone, dihydrotestosterone, and androstanedione. *Horm. Behav.* **6**, 139–148.

Goldstein, J. L., and Wilson, J. D. (1975). Genetic and hormonal control of male sexual differentiation. *J. Cell Physiol.* **85**, 365–378.

Gorski, R. A. (1979). Long-term hormonal modulation of neuronal structure and function. *In* "The Neurosciences: 4th Study Program" (F. O. Schmitt, ed.), pp. 969–982. MIT Press, Cambridge, Massachusetts.

Gorski, R. A., Gordon, J. H., Shryne, J. E., and Southam, A. M. (1978). Evidence for a morphological sex difference within the medial preoptic area of the rat brain. *Brain Res.* **148**, 333–346.

Gorski, R. A., Harlan, R. E., Jacobson, C. D., Shryne, J. E., and Southam, A. M. (1980). Evidence for the existence of a sexually dimorphic nucleus in the preoptic area of the rat. *J. Comp. Neurol.*, **193**, 529–539.

Goy, R. W. (1970a). Experimental control of psychosexuality. *Philos. Trans. R. Soc. London Ser. B.* **259**, 149–162.

Goy, R. W. (1970b). Early hormonal influences on the development of sexual and sex-related behavior. *In* "The Neurosciences: Second Study Program" (F. O. Schmitt, ed.), pp. 196–206. Rockefeller Univ. Press, New York.

Goy, R. W. (1978). Development of play and mounting behavior in female rhesus monkeys virilized prenatally with esters of testosterone and dihydrotestosterone. *In* "Recent Advances in Primatology, Vol. 1: Behaviour." (D. J. Chivers and J. Herbert, eds.), pp. 449–462. Academic Press, New York.

Goy, R. W., and McEwen, B. S. (1980). "Sexual Differentiation of the Brain." MIT Press, Cambridge, Massachusetts.

Goy, R. W., Bridson, W. E., and Young, W. C. (1964). Period of maximal susceptibility of the prenatal female guinea pig to masculinizing actions of testosterone propionate. *J. Comp. Physiol. Psychol.* **57,** 166–174.

Greenough, W. T., Carter, C. S., Steerman, C., and DeVoogd, T. (1977). Sex differences in dendritic patterns in hamster preoptic area. *Brain Res.* **126,** 63–72.

Gregory, E. (1975). Comparison of postnatal CNS development between male and female rats. *Brain Res.* **99,** 152–156.

Gurney, M. E., and Konishi, M. (1980). Hormone-induced sexual differentiation of brain and behavior in zebra finches. *Science* **208,** 1380–1382.

Hart, B. L. (1979). Sexual behavior and penile reflexes of neonatally castrated male rats treated in infancy with estrogen and dihydrotestosterone. *Horm. Behav.* **13,** 256–268.

Hayashi, S., and Gorski, R. A. (1974). Critical exposure time for androgenization by intracranial crystals of testosterone propionate in neonatal female rats. *Endocrinology (Baltimore)* **94,** 1161–1167.

Hendry, I. A. (1976). Control in the development of the vertebrate sympathetic nervous system. *In* "Reviews of Neuroscience" (S. Ehrenpreis and I. Kopin, eds.), Vol. 2, pp. 149–194. Raven, New York.

Ifft, J. D. (1972). An autoradiographic study of the time of final division of neurons in rat hypothalamic nuclei. *J. Comp. Neurol.* **144,** 193–204.

Jacobson, C. D., Shryne, J. E., Shapiro, F., and Gorski, R. A. (1980). Ontogeny of the sexually dimorphic nucleus of the preoptic area. *J. Comp. Neurol.* **193,** 541–548.

Josso, N., Forest, M. G., and Picard, J. Y. (1975). Mullerian-inhibiting activity of calf fetal testis: Relationship of testosterone and protein synthesis.

Jost, A. (1970). Hormonal factors in the sex differentiation of the mammalian foetus. *Philos. Trans. R. Soc. London.* **259,** 119–130.

Krey, L. C., Kamel, F., and McEwen, B. S. (1980). Parameters of neuroendocrine aromatization and estrogen receptor occupation in the male rat. *Brain Res.* **193,** 277–283.

Landmesser, L., and Pilar, G. (1978). Interactions between neurons and their targets during *in vivo* synaptogenesis. *Fed. Proc., Fed. Am. Soc. Exp. Biol.* **37,** 2016–2022.

Lawrence, J. M., and Raisman, G. (1980). Ontogeny of synapses in a sexually dimorphic part of the preoptic area in the rat. *Brain Res.* **183,** 466–471.

Lieberburg, I., Krey, L. C., and McEwen, B. S. (1979). Sex differences in serum testosterone and in exchangeable brain cell nuclear estradiol during the neonatal period in rats. *Brain Res.* **178,** 207–212.

Lieberburg, I., MacLusky, N. J., and McEwen, B. S. (1980). Androgen receptors in the perinatal rat brain. *Brain Res.* **196,** 125–138.

Lording, D. W., and De Kretser, D. M. (1972). Comparative ultrastructural and histochemical studies of the interstitial cells of the rat testis during fetal and postnatal development. *J. Reprod. Fertil.* **29,** 261–269.

Lund, R. D. (1978). "Development and Plasticity of the Brain." Oxford Univ. Press, London and New York.

McEwen, B. S. (1980). Gonadal influences on the developing brain. *In* "Handbook of Biological Psychiatry, Part III, Brain Mechanisms and Abnormal Behavior—Genetics and Neuroendocrinology" (H. M. van Praag, M. H. Lader, O. J. Rafaelsen, and E. J. Sachar, eds.), pp. 241–278. Dekker, New York.

McEwen, B. S., Lieberburg, I., Chaptal, C., and Krey, L. C. (1977). Aromatization: important for sexual differentiation of the neonatal rat brain. *Horm. Behav.* **9,** 249–263.

McEwen, B. S., Lieberburg, I., Chaptal, C., Davis, P. G., Krey, L. C., MacLusky, N. J., and Roy, E. J. (1979). Attentuating the defeminization of the neonatal rat brain: Mechanisms of action of cyproterone acetate, 1,4,6-androstatriene-3,17-dione and a synthetic progestin, R5020. *Horm. Behav.* **13**, 269–281.

MacLusky, N. J., Lieberburg, I., and McEwen, B. S. (1979). The development of estrogen receptor systems in the rat brain: Perinatal development. *Brain Res.* **178**, 129–142.

MacLusky, N. J., and McEwen, B. S. (1980). Progestin receptors in the developing rat brain and pituitary. *Brain Res.* **189**, 262–268.

Martini, L. (1976). Androgen reduction by neuroendocrine tissues: Physiological significance. *In* "Subcellular Mechanisms in Reproductive Neuroendocrinology" (F. Naftolin, K. J. Ryair, and J. Davies, eds.), pp. 327–355. Elsevier, Amsterdam.

Matsumoto, A., and Arai, Y. (1980). Sexual dimorphism in 'wiring pattern' in the hypothalamic arcuate nucleus and its modification by neonatal hormonal environment. *Brain Res.* **190**, 238–242.

McDonald, P. G., and Doughty, C. (1972). Inhibition of androgen-sterilization in the female rat by administration of an antioestrogen. *J. Endocrinol.* **55**, 455–456.

McEwen, B. S. (1980). Gonadal influences on the developing brain. *In* "Handbook of Biological Psychiatry, Part III, Brain Mechanisms and Abnormal Behavior—Genetics and Neuroendocrinology" (H. M. van Praag, M. H. Lader, O. J. Rafaelsen, and E. J. Sachar, eds.), pp. 241–278. Dekker, New York.

Mittwoch, U. (1970). How does the Y chromosome affect gonadal differentiation? *Philos. Trans. R. Soc. London Ser. B.* **259**, 113–117.

Nadler, R. D. (1968). Masculinization of female rats by intracranial implantation of androgen in infancy. *J. Comp. Physiol. Psychol.* **66**, 157–167.

Nadler, R. D. (1972). Intrahypothalamic exploration of androgen-dependent brain loci in neonatal female rats. *Trans. N.Y. Acad. Sci. Ser. II*, **34**, 572–581.

Nadler, R. D. (1973). Further evidence on the intrahypothalamic locus for androgenization of female rats. *Neuroendocrinology* **12**, 110–119.

Naess, O., Haug, E., Attramadal, A., Aakvaag, A., Hansson, V., and French, F. (1976). Androgen receptors in the anterior pituitary and central nervous system of the androgen "insensitive" (Tfm) rat: Correlation between receptor binding and effects of androgens on gonadotropin secretion. *Endocrinology (Baltimore)* **99**, 1295–1303.

Niemi, M., and Ikonen, M. (1963). Histochemistry of the Leydig cells in the postnatal prepubertal testis of the rat. *Endocrinology (Baltimore)* **72**, 443–448.

Noble, R. G., and Alsum, P. B. (1975). Hormone dependent sex dimorphisms in the golden hamster (*Mesocricetus auratus*). *Physiol. Behav.* **14**, 567–574.

Nottebohm, F. (1980). Testosterone triggers growth of brain vocal control nuclei in adult female canaries. *Brain Res.* **189**, 429–436.

Nottebohm, F., and Arnold, A. P. (1976). Sexual dimorphism in vocal control areas of the songbird brain. *Science* **194**, 211–212.

Olsen, K. L. (1979a). Androgen-insensitive rats are defeminized by their testes. *Nature (London)* **279**, 238–239.

Olsen, K. L. (1979b). Induction of male mating behavior in androgen-insensitive (Tfm) and normal (King-Holtzman) male rats: Effect of testosterone propionate, estradiol benzoate, and dihydrotestosterone. *Horm. Behav.* **13**, 66–84.

Paup, D. C., Coniglio, L. P., and Clemens, L. G. (1972). Masculinization of the female golden hamster by neonatal treatment with androgen or estrogen. *Horm. Behav.* **3**, 123–132.

Paup, D. C., Coniglio, L. P., and Clemens, L. G. (1974). Hormonal determinants in the development of masculine and feminine behavior in the female hamster. *Behav. Biol.* **10**, 353–363.

Pfaff, D. W. (1966). Morphological changes in the brains of adult male rats after neonatal castration. *J. Endocrinol.* **36**, 415–416.

Phoenix, C. H. (1974). Effects of dihydrotestosterone on sexual behavior of castrated male rhesus monkeys. *Physiol. Behav.* **12**, 1045–1055.

Phoenix, C. H., Goy, R. W., Gerall, A. A., and Young, W. C. (1959). Organizing action of prenatally administered testosterone propionate on the tissues mediating mating behavior in the female guinea pig. *Endocrinology (Baltimore)* **65**, 369–382.

Plapinger, L., and McEwen, B. S. (1978). Gonadal steroid-brain interactions in sexual differentiation. *In* "Biological Determinants of Sexual Behavior" (J. Hutchinson, ed.), pp. 193–218. Wiley, New York.

Plapinger, L., McEwen, B. S., and Clemens, L. E. (1973). Ontogeny of estradiolbinding sites in rat brain II. Characteristics of a neonatal binding macromolecule. *Endocrinology (Baltimore)* **93**, 1129–1139.

Raisman, G., and Field, P. M. (1973). Sexual dimorphism in the neuropil of the preoptic area of the rat and its dependence on neonatal androgen. *Brain Res.* **54**, 1–29.

Raynaud, J. P., Mercier-Bodard, C., and Baulieu, E. E. (1971). Rat estradiol binding plasma protein (EBP). *Steroids* **18**, 767–788.

Reddy, V. V. R., Naftolin, F., and Ryan, K. J. (1974). Conversion of androstenedione to estrone by neural tissues from fetal and neonatal rats. *Endocrinology (Baltimore)* **94**, 117–121.

Reier, P. J., Cullen, M. J., Froelich, J. S. and Rothchild, I. (1977). The ultrastructure of the developing medial preoptic nucleus in the postnatal rat. *Brain Res.* **122**, 415–436.

Resko, J. A., Feder, H. H., and Goy, R. W. (1968). Androgen concentrations in plasma and testis of developing rats. *J. Endocrinol.* **40**, 485–491.

Rethelyi, M. (1979). Regional and sexual differences in the size of the neurovascular contact surface of the rat median eminence and pituitary stalk. *Neuroendocrinology* **28**, 82–91.

Roy, E. J., and McEwen, B. S. (1977). An exchange assay for estrogen receptors in cell nuclei of the adult rat brain. *Steroids* **30**, 657–669.

Scatchard, G. (1949). The attractions of proteins for small ions and molecules. *Ann. N.Y. Acad. Sci.* **51**, 660–672.

Schultz, F. M., and Wilson, J. D. (1974). Virilization of the wolffian duct in the rat fetus by various androgens. *Endocrinology (Baltimore)* **94**, 979–986.

Selmanoff, M. K., Jumonville, J. E., Maxson, S. C., and Ginsburg, B. E. (1975). Evidence for a Y chromosome contribution to an aggressive phenotype in inbred mice. *Nature (London)* **253**, 529–530.

Shapiro, B. H., Goldman, A. S., and Gustafsson, J. A. (1975). Masculine-like hypothalamic-pituitary axis in the androgen-insensitive genetically male rat pseudohermaphrodite. *Endocrinology (Baltimore)* **97**, 487–492.

Shapiro, B. H., Levine, D. C., and Adler, N. T. (1980). The testicular feminized rat: a naturally occurring model of androgen independent brain masculinization. *Science* **209**, 418–420.

Short, R. V. (1974). Sexual differentiation of the brain of the sheep. *In* "Endocrinologie Sexuelle de la Periode Perinatale" (M. G. Forest and J. Bertrand, eds.), Vol. 32, pp. 121–142. Colloque de *INSERM*.

Södersten, P., and Hansen, S. (1978). Effects of castration and testosterone dihydrotestosterone or oestradiol replacement treatment in neonatal rats on mounting behavior in the adult. *J, Endocrinol.* **76**, 251–260.

Soto, A. M., and Sonnenschein, C. (1980). Control of growth of estrogen-sensitive cells: role for α-fetoprotein. *Proc. Natl. Acad. Sci. U.S.A.* **77**, 2084–2087.

Stewart, A. D., Manning, A., and Batty, J. (1978). Effects of the Y-chromosome in mice: a study of testis weight plasma testosterone and behaviour. *Heredity,* **40**, 326–327.

Thomas, D. A., McIntosh, T. K., and Barfield, R. J. (1980). Influence of androgen in the neonatal period on ejaculatory and postejaculatory behavior in the rat. *Horm. Behav.* **14,** 153–162.

Toran-Allerand, C. D. (1976). Sex steroids and the development of the newborn mouse hypothalamus and preoptic area *in vitro:* Implications for sexual differentiation. *Brain Res.* **106,** 407–412.

Toran-Allerand, C. D. (1978). Gonadal hormones and brain development; cellular aspects of sexual differentiation. *Am. Zool.* **18,** 553–565.

Torand-Allerand, C. D. (1980a). Sex steroids and the development of the newborn mouse hypothalamus and preoptic area *in vitro.* II. Morphological correlates and hormonal specificity. *Brain Res.* **189,** 413–427.

Toran-Allerand, C. D. (1980b). Coexistence of α-fetoprotein, albumin and transferrin immunoreactivity in neurones of the developing mouse brain. *Nature (London)* **286,** 733–735.

Toran-Allerand, C. D., Gerlach, J. L., and McEwen, B. S. (1980). Autoradiographic localization of ^{3}H estradiol related to steroid responsiveness in cultures of the newborn mouse hypothalamus and preoptic area. *Brain Res.* **184,** 517–522.

Vale, J. R., Ray, D., and Vale, C. A. (1972). The interaction of genotype and exogenous neonatal androgen: Agonistic behavior in female mice. *Behav. Biol.* **7,** 321–334.

van der Schoot, P. (1980). Effects of dihydrotestosterone and oestradiol on sexual differentiation in male rats. *J. Endocrinol.* **84,** 397–407.

Vito, C. C., and Fox, T. O. (1979). Embryonic rodent brain contains estrogen receptors. *Science* **204,** 517–519.

Vreeburg, J. T. M., van der Vaart, P. D. M., and van der Schoot, P. (1977). Prevention of central defeminization but masculinization in male rats by inhibition neonatally of oestrogen biosynthesis. *J. Endocrinol.* **74,** 375–382.

Walsh, R. J., and Brawer, J. R. (1979). Cytology of the arcuate nucleus in newborn male and female rats. *J. Anat.* **128,** 121–133.

Weber, R. (1969). Tissue involution and lysosomal enzymes during asruran metamorphosis. *In* "Lysosomes in Biology and Pathology" (J. T. Dingle and H. B. Fell, eds.), pp. 437–461. North-Holland Publ., Amsterdam.

Whitsett, J. M., and Vandenbergh, J. G. (1975). Influence of testosterone propionate administered neonatally on puberty and bisexual behavior in female hamsters. *J. Comp. Physiol. Psychol.* **88,** 248–255.

ANALYSIS OF PROTEIN SYNTHESIS IN THE MAMMALIAN BRAIN USING LSD AND HYPERTHERMIA AS EXPERIMENTAL PROBES

Ian R. Brown, John J. Heikkila, and James W. Cosgrove

Molecular Approaches to Neurobiology

I. INTRODUCTION

At present the analysis of the molecular biology of the mammalian brain is a
relatively new field. Considerable knowledge has accumulated on the brain at the
ultrastructural and physiological level, however, comparatively little is known
concerning underlying molecular processes such as regulation of transcription
and translation. Advances in the understanding and treatment of brain disorders
requires that considerable progress be made in our knowledge of the molecular
basis of brain function. Applying molecular techniques to the analysis of brain
function is an interesting prospect, but given the overwhelming complexity of the
brain it has been difficult to know how to approach the subject. One avenue has
been to determine how biochemical processes in the brain are affected by the
introduction of physiologically relevant treatments. As will be illustrated in this
chapter, protein synthesis in the mammalian brain has been shown to be sensitive
to perturbation by a wide range of physical and chemical treatments.

The approach we have taken is to experimentally manipulate biochemical
processes in the rabbit brain by intravenous (iv) injection of the potent psychotropic drug lysergic acid diethylamide (LSD) or by the elevation of body temperature, i.e., hyperthermia. These treatments rapidly induce specific changes in
protein synthesis in the brain, i.e., activation of a translational control
mechanism which causes a global inhibition of translation in the brain and
induction of synthesis of proteins similar in molecular weight to two of the major
"heat shock" proteins reported in tissue culture cell lines. Our investigation of
possible brain-specific mechanisms for the regulation of protein synthesis has
involved the use of a spectrum of techniques, i.e., kinetics of polysome disaggregation, *in vivo* labeling of specific brain subcellular fractions, fractionation
of brain cell-free translation systems, and translation of brain polysomes in a
heterologous system. The use of several techniques to analyze protein synthesis
permits the confirmation of results by independent methods and extends the
range of approaches that may be used to probe translational mechanisms.

II. PERTURBATION OF PROTEIN SYNTHESIS IN THE BRAIN

A. Physiological Treatments

Protein synthesis in the brain appears to be sensitive to a wide range of
physiological treatments as evidenced by a reduction in the incorporation of

labeled amino acids *in vivo* and a disaggregation of brain polysomes to monosomes. Acute treatments such as spreading cortical depression (Krivanek, 1970; Bennett and Edelman, 1969), epileptic seizures (Wasterlain, 1977; Fando *et al.*, 1979), intracranial hypertension (Wasterlain, 1974), hyperthermia (Murdock *et al.*, 1978; Millan *et al.*, 1979; Heikkila and Brown, 1979a,b, 1981), hypoxia (Morimoto *et al.*, 1978), fever (Heikkila and Brown, 1979b), ischemia (Kleihues and Hossman, 1971; Kleihues *et al.*, 1975; Cooper *et al.*, 1977), and electroshock (Cotman *et al.*, 1971; MacInnes *et al.*, 1970; Dunn *et al.*, 1971; Dunn and Bergert, 1976; Metafora *et al.*, 1977) have been reported to disrupt brain protein synthesis.

In many of these systems examination of the inhibitory mechanism has revealed that RNase activation is not the cause of the disaggregation of brain polysomes. Inhibition of initiation has been implicated as a possible mechanism for the decrease in brain protein synthesis, which is induced by hyperthermia (Murdock *et al.*, 1978) epileptic seizures (Fando *et al.*, 1979), intracranial hypertension (Wasterlain, 1974), and hypoxia (Morimoto *et al.*, 1978). Experiments carried out by Cooper *et al.* (1977) utilizing poly(I), a specific inhibitor of initiation, have suggested that the decrease in brain protein synthesis in rats after ischemia treatment is due mainly to inhibited initiation. An analysis of the effect of electroshock on adult rabbits (Metafora *et al.*, 1977) revealed that this treatment resulted in brain polysome disaggregation with no loss of polysomal mRNA or ribosomes, no alterations in the activity of elongation factors or ribonuclease, and no change in the ability of ribosomes to elongate, terminate, and release polypeptide chains. Although no direct examination of initiation was carried out, an effect of electroshock on initiation of translation was inferred.

B. Amino Acids

The intraperitoneal injection of large doses of amino acids such as phenylalanine, methionine, isoleucine, leucine, and valine induces a transient disaggregation of brain polysomes and inhibition of incorporation of labeled amino acid either *in vivo* (Aoki and Siegal, 1970; Agrawal *et al.*, 1970; Siegal *et al.*, 1971; MacInnes and Schlesinger, 1971; Copenhaver *et al.*, 1973; Roberts and Morelos, 1976; Hughes and Johnson, 1978) or *in vitro* (Aoki and Siegal, 1970). One explanation for the effect of excess phenylalanine on brain protein synthesis is that this treatment causes labilization of brain lysosomes resulting in the release of lytic enzymes such as ribonucleases (Roberts and Morelos, 1976). Taub and Johnson (1975), however, found that the increased pool of 80 S monosomes, which is produced following phenylalanine, was dissociable into subunits in the presence of high concentrations of salt. The production of monomeric ribosomes by RNase would result in monosomes with mRNA fragments and peptidyl-tRNA still attached, which would stabilize the complex in high salt. Inhibition of brain protein synthesis by phenylalanine does not, therefore, appear to be due to RNase activation. Hughes and Johnson (1978) suggested that inhibition is primarily due to inhibited reinitiation with a lesser effect on elongation. An examination of *in vivo* levels of aminoacyl-tRNA in brain after phenylalanine demonstrated a decrease in the aminoacylation of the initiator methionyl-tRNA species (Hughes and Johnson, 1978). Injection of a mixture of neutral amino acids following phenylalanine injection was found to reverse the brain polysome shift, increase the acylation levels of methionyl-tRNA (in particular the initiator methionyl-tRNA species), and stimulate elongation (Hughes and Johnson, 1978).

C. 5-HTP and L-Dopa

The intraperitoneal injection of large doses of either L-dihydroxyphenylalanine (L-Dopa) or 5-hydroxytryptophan (5-HTP) to rats has been shown to induce a transient disaggregation of brain polysomes (Weiss *et al.*, 1971, 1972, 1973, 1975; Roel *et al.*, 1974) and inhibit amino acid incorporation *in vivo* (Roel *et al.*, 1974). The disaggregation of brain polysomes by excess L-Dopa or 5-HTP is dependent upon the conversion of these compounds to the neurotransmitters dopamine (DA) or serotonin (5-HT) since the introduction of inhibitors that prevent the formation of these amines inhibits the response (Weiss *et al.*, 1972, 1973). Furthermore, blocking the breakdown of the two neurotransmitters with monoamine oxidase accentuates the brain polysome shift. The disaggregation of brain polysomes by L-Dopa or 5-HTP appears to be mediated via neurotransmitter receptors since DA and 5-HT receptor blocking agents could prevent disaggregation (Weiss *et al.*, 1975).

D. Drugs

A number of drugs such as morphine, ethanol, and *d*-amphetamine have been reported to inhibit brain protein synthesis. Acute morphine treatment markedly inhibits the incorporation of labeled amino acid into brain protein (Clouet and Ratner, 1967; Harris *et al.*, 1974; Loh and Hitzemann, 1974; Hitzemann and Loh, 1976). The effect of chronic morphine treatment on brain protein synthesis appears to be unresolved, i.e., the development of morphine tolerance has been reported to increase (Lang *et al.*, 1975; Craves *et al.*, 1978), decrease (Clouet and Ratner, 1967; Loh and Hitzemann, 1974), or have no effect (Harris *et al.*, 1974) on brain protein synthesis. Chronic morphine treatment has been reported to differentially alter the turnover of specific proteins of the synaptic junctional and nonjunctional membranes and synaptic vesicles of rat brain (Clouet and Ratner, 1979). The molecular mechanism underlying the effect of either acute or chronic morphine treatment on brain protein synthesis is not known.

Chronic ethanol ingestion has been shown to inhibit brain protein synthesis *in vitro* and decrease the aminoacylation of tRNA (Tewari and Nobel, 1971; Tewari *et al.*, 1978). The latter effect of chronic ethanol ingestion is presumably due to an effect on aminoacyl-tRNA synthetase activity (Fleming *et al.*, 1975). Lindholm and Khawaja (1979) have reported a differential effect of chronic ethanol ingestion on the protein synthesis capacity of free and membrane-bound polysomes *in vitro*. Prolonged ethanol consumption had a stimulatory effect on the protein synthetic activity of free polysomes while significantly inhibiting the activity of membrane-bound brain polysomes. It was suggested in this study that ethanol disrupts the endoplasmic reticulum membrane, thus inhibiting the elongation step in protein synthesis. Ethanol withdrawal in physically dependent rats leads to a transient inhibition of protein synthesis in a brain cell-free system (Tewari *et al.*, 1977). This group has suggested that ethanol-induced changes at the ribosomal level may result in defective association of mRNA causing an inhibition of brain protein synthesis.

The administration of large doses of *d*-amphetamine induces brain polysome disaggregation (Moskowitz *et al.*, 1975; Widelitz *et al.*, 1975) and inhibits amino acid incorporation both *in vivo* (Roel *et al.*, 1978) and *in vitro* (Widelitz *et al.*, 1975, 1976). *d*-Amphetamine-induced rat brain polysome disaggregation can be inhibited by pretreatment with DA neurotransmitter receptor blockers (Moskowitz *et al.*, 1975; Widelitz *et al.*, 1977). Although the action of *d*-amphetamine on protein synthesis may involve the release of dopamine, differences between the responses to *d*-amphetamine and L-Dopa have been reported. For example, a large dose of *d*-amphetamine (15 mg/kg) has no effect on young rats, whereas it causes brain polysome disaggregation in older animals (Moskowitz *et al.*, 1975). Furthermore, *d*-amphetamine has been shown to disaggregate liver polysomes, whereas L-Dopa does not (Baliga *et al.*, 1976).

Since *d*-amphetamine is taken up intracellularly, a number of studies have examined the effect of amphetamine on protein synthesis *in vitro* (Widelitz *et al.*, 1976; Nowak, and Munro, 1977; Baliga *et al.*, 1976). The direct addition of *d*-amphetamine to a noninitiating brain cell-free system does not appear to affect elongation (Baliga *et al.*, 1976; Nowak, and Munro, 1977). *d*-Amphetamine added directly to an initiating poly(U)-dependent system derived from wheat germ results in an inhibition of initiation (Widelitz *et al.*, 1976; Baliga *et al.*, 1976). A similar effect on initiation has been found in a reticulocyte system using globin mRNA (Nowak and Munro, 1977). In this latter report inhibited initiation was suggested as being due to a decreased aminoacylation of tRNA. It should be noted that the concentration of amphetamine used to inhibit *in vitro* protein synthesis $(2-4 \text{ m}M)$ is many fold greater than that required to produce the effects *in vivo*.

E. Developmental Changes

A system that has been very useful for the analysis of translational mechanisms in the brain is the well-documented phenomenon of a decrease in protein synthesis during maturation of the brain (Johnson, 1976). Decreases in brain protein synthesis with development have been observed at several levels of organization including *in vivo* studies (Dainat *et al.*, 1970; Szijan *et al.*, 1971; Dunlop *et al.*, 1975; Fando *et al.*, 1980), brain slices (Cain *et al.*, 1972; Dunlop *et al.*, 1977; Fando *et al.*, 1980), brain cell suspensions (Gilbert and Johnson, 1972a,b), and cell-free systems (Lerner and Johnson, 1970; MacInnes and Schlesinger, 1971; Andrews and Tata, 1971; Gilbert and Johnson, 1974; Yamagani and Mori, 1970; Johnson and Belytschko, 1969; Fando *et al.*, 1980).

The decrease in brain protein synthesis during maturation does not appear to be due to major structural alterations in brain ribosomes since no differences in physical chemical properties could be measured when purified ribosomes isolated from newborn or adult brain tissue were compared (Lerner and Johnson, 1970; Johnson, 1976). The maturation-dependent decline in brain protein synthesis may involve an alteration in the soluble components of protein synthesis. For example, a decline in aminoacyl-tRNA synthetase activity has been reported with development (Fellous *et al.*, 1973; Harris and Maas, 1974; Wender and Zgorzalwicz, 1976). In contrast to the above findings, Fando *et al.* (1980), using a brain cell-free protein synthesis system, found that supplementation with tRNA or amino acids had no effect on the age-dependent decline in translational activity *in vitro*.

It has been suggested that one site of translational regulation in the brain may be at the level of ribosome–mRNA–cytosol interactions (Johnson, 1976; Fando *et al.*, 1980). For example, Zomzely *et al.* (1971) suggested that a decrease in the stability of mRNA–ribosomal complexes noted with brain development may account for the decrease in protein synthesis with aging. Lim and White (1974) suggested that an alteration in the ability of certain brain ribosomal proteins to bind mRNA may be involved. Schmidt and Sokoloff (1973), on the other hand, suggested that chemical modification of ribosomal proteins such as phosphorylation by a cAMP-dependent protein kinase may act as a translational regulatory mechanism in brain. While brain ribosomal proteins have been shown to be phosphorylated (Ashby and Roberts, 1975), the effect of these phosphorylation events on brain protein synthesis has not been demonstrated. This latter possibility, however, is tenable since protein synthesis inhibition in reticulocytes due to hemin deficiency has been suggested to involve the phosphorylation of a subunit of the initiation factor eIF-2 (Das *et al.*, 1979).

III. EFFECT OF LSD ON THE TRANSLATIONAL APPARATUS OF THE BRAIN

A. Background

A possible approach to the study of macromolecular mechanisms in the mammalian brain is through the manipulation of biochemical processes by the use of specific drugs. Information on the effect of the potent psychotropic drug LSD on macromolecular events in the mammalian brain is limited as research has concentrated on psychological and physiological aspects of this drug (Sankar, 1975). Following intravenous administration, LSD binds to neurotransmitter receptors in the brain inducing profound perceptual and behavioral alterations (Kalant and Khanna, 1975). Through interaction with neurotransmitter receptors, LSD has been reported to influence adenylate cyclase activity, which in turn can affect cytoplasmic levels of cAMP (Von Hungen *et al.*, 1974, 1975; Brockaert *et al.*, 1976). This provides a possible mechanism for the transmission of signals to the intracellular environment of brain cells, which might affect macromolecular events. At a physiological level the precise mode of action of LSD is not clear at present, however, a number of investigations examining neurotransmitter synthesis and catabolism (Freedman *et al.*, 1970, Lin *et al.*, 1969; Persson, 1977); nerve firing (Aghajanian *et al.*, 1969; Wang and Aghajanian, 1977; Christoph *et al.*, 1977); adenylate cyclase activity (Von Hungen *et al.*, 1975; Brockaert *et al.*, 1976); and receptor binding studies (Bennett and Aghajanian, 1974; Bennett and Snyder, 1976; Lovell and Freedman, 1976; Burt *et al.*, 1976) have demonstrated the ability of LSD to interact with both DA and 5-HT receptor sites.

In our laboratory we have shown that LSD is a useful tool with which to probe macromolecular synthesis in the brain (Brown, 1977). We have sought to determine how intravenous injection of LSD into rabbits affects the biochemistry of brain cells when doses are employed that are in a range comparable to those causing distortion of sensory perception in man (Rossi, 1971; Kalant and Khanna, 1975). Our initial studies examined the effect of LSD on the modification of brain chromosomal proteins and on brain RNA synthesis. Most of our recent studies have concentrated on the effects of this drug on protein synthesis in the brain.

The intravenous administration of LSD at a dose of either 10 μg/kg or 100 μg/kg was found to increase the acetylation of histones in rabbit cerebral hemispheres and midbrain 30 min after drug administration (Brown and Liew, 1975). These results suggested that LSD may affect gene activity in brain since a modification of chromosomal proteins may be prerequisite to changes in transcription (Allfrey, 1971). This hypothesis is supported by the observation that 2.5 h after an *in vivo* injection of LSD, a stimulation of α-amanatin-sensitive

RNA synthesis in isolated brain nuclei was noted (Brown, 1975). Since the direct addition of LSD to control nuclei did not have any effect, it was suggested that the nuclear effect might require initial binding of LSD to plasma membrane or cytoplasmic receptors.

LSD has also been reported to affect protein synthesis in the brain. In an early study by Krawczynski (1961), the intracranial injection of 250 μg/kg of LSD to rats induced a 30% decrease in the incorporation of intracranially administered [^{35}S]methionine. This study should be viewed with caution since the intracranial injection procedure alone has been shown to inhibit protein synthesis (Dunn, 1975; Dunn and Bergert, 1976). In an autoradiographic analysis, Von Gatzke (1977) found that LSD given intramuscularly to rats markedly depressed the incorporation of [^{3}H]leucine into neuronal and glial cells of the parietal cortex after 2 h.

B. Disaggregation of Brain Polysomes following Intravenous Injection of LSD

Our initial series of investigations into the effect of LSD on protein synthesis in the rabbit brain involved an examination of brain polysomes following drug administration. The intravenous injection of LSD (10–100 μg/kg) induced a marked disaggregation of polysomes to monosomes in all brain regions examined (Holbrook and Brown, 1976; Heikkila and Brown, 1979a; Heikkila et al., 1979). The effect appeared to be organ-specific as no polysome shift was seen in spleen or kidney (Holbrook and Brown, 1976). The kinetics of the effect of LSD on brain polysomes involved a rapid shift to monosomes between 15 and 30 min followed by a gradual return to normal polysome levels by 4 h (Holbrook and Brown, 1976). Doses over the range of 10–100 μg/kg resulted in an increasing degree of brain polysome disaggregation 30–60 min after drug injection (Holbrook and Brown, 1976).

These initial studies on LSD-induced brain polysome disaggregation utilized polysome preparations derived from postmitochondrial supernatant which consisted of both free and membrane-bound polysomes. Since free and membrane-bound polysomes have been reported to behave differently under conditions that inhibit protein synthesis (Sarma et al., 1969; Hemminki, 1972; Raghupathy et al., 1971; Taub and Johnson, 1975), it was of interest to compare the effect of LSD on these two classes of polysomes. Intravenous administration of LSD (100 μg/kg) induced a disaggregation of both free and membrane-bound brain polysomes (Fig. 1). The degree of brain polysome shift was significantly greater in the free polysome population compared to the membrane-bound polysome fraction (Heikkila and Brown, 1981). In other studies the administration of excess phenylalanine (Taub and Johnson, 1975) or hypothermia (Raghupathy et al., 1971) also induced a greater degree of disaggregation of free rat brain poly-

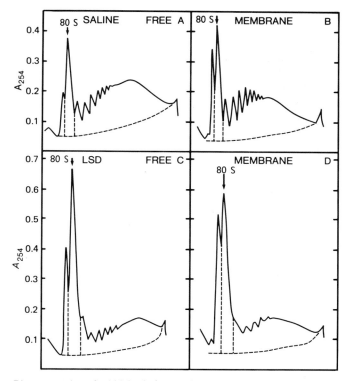

Fig. 1. Disaggregation of rabbit brain free and membrane-bound polysomes after administration of LSD *in vivo*. Rabbits were injected intravenously with LSD at 100 μg/kg, whereas control animals received saline. After 1 h the rabbits were sacrificed and free and membrane-bound polysomes from the cerebral hemispheres were isolated and analyzed on 15–45% sucrose density gradients following centrifugation for 75 min in a Beckman SW-41 rotor at 40,000 rev/min at 4°C. The vertical dashed line indicates the separation between monosomes and polysomes while the horizontal dashed line is the A_{254} in a blank gradient. (From Heikkila and Brown, 1981, by permission from Raven Press, New York.)

somes. Attachment of ribosomes to membranes may provide a stabilizing influence with respect to brain polysome disaggregation.

C. Mechanism of LSD-Induced Brain Polysome Disaggregation

1. Receptor Mediation

LSD appears to exert an effect on brain metabolism via interaction with serotonin and dopamine neurotransmitter receptors (Bennett and Snyder, 1976; Von Hungen *et al.*, 1975). Since LSD induces disaggregation of brain polysomes, agents which block these receptors might be expected to alter the poly-some response to LSD. We have found that disaggregation of brain polysomes is effectively inhibited by the prior injection of the DA receptor blockers haloperi-dal and chlorpromazine; the antiserotonin agent pizotyline; and the adrenergic blockers propranolol and phentolamine (Holbrook and Brown, 1977b). These results imply that a range of neurotransmitter receptors are involved in the mechanism by which LSD influences brain protein synthesis. Induction of a brain polysome shift in rats by L-Dopa or 5-HTP also appears to be mediated via neurotransmitter receptors since dopamine and serotonin receptor blocking agents could prevent polysome disaggregation induced by L-Dopa and 5-HTP, respectively (Weis *et al.*, 1975).

2. Molecular Mechanism

Theoretically the mechanism by which polysomes can disaggregate to monosomes could involve (1) the degradation of mRNA by RNase activation; (2) premature termination of translation; or (3) decreased reinitiation of protein synthesis. RNase hydrolysis of polysomal mRNA was discounted on the basis of the following tests. The elevated pool of monosomes generated by LSD was dissociable to ribosomal subunits after high-salt treatment (Holbrook and Brown, 1976; Heikkila *et al.*, 1978) indicating that the monosomes were not stabilized against dissociating conditions by associated nascent peptides (Blobel and Saba-tini, 1971; Christman, 1973). The possibility of RNase activation was also dis-counted by the results of homogenizing together control and drug-treated cerebral hemispheres. The polysome percentage in the combined fraction was precisely the mean of separate control and drug polysomes (Holbrook and Brown, 1976). Ribonuclease activity measured in brain postmitochondrial supernatants from control and LSD-treated animals showed no significant differences in the levels of either alkaline or acid ribonuclease activity (Mahony and Brown, 1979). Finally, mRNA relocalized to the monosome complex following LSD can be translated in a mRNA-dependent cell-free system (Mahony and Brown, 1979). These results indicate that the elevated pool of monosomes which is produced after LSD is not due to induced RNase activity either *in vivo* or during the course of polysome purification.

An increase in monosomes could arise during the LSD response either from premature termination and ribosome fall-off or from normal run-off with inhib-ited reinitiation. Monosomes produced by premature fall-off or by RNase are associated with peptidyl tRNA, whereas run-off monosomes accumulated by normal termination and decreased reinitiation are relatively free of this class of RNA. The large-monosome peak produced after LSD administration was col-lected from sucrose gradients and RNA purified from this fraction was subjected to polyacrylamide gel electrophoresis (Holbrook and Brown, 1977a). Characteri-zation of the tRNA content revealed that these monosomes were the result of normal termination and not premature fall-off or RNase. A further experiment ruling out premature termination was the examination of the size range of newly released proteins present in the postribosomal supernatant of cerebral hemispheres from control and LSD-treated animals. No shift to short peptides was observed as would be expected if premature termination occurred (Holbrook and Brown, 1977a).

Relocalization of mRNA from polysomes to monosomes has been reported to accompany polysome disaggregation due to inhibited initiation of translation by NaF or hemin deficiency in rabbit reticulocytes or lysates (Terada *et al.*, 1972; Jagus and Safer, 1979). After LSD administration the disaggregation of brain polysomes is also characterized by a relocalization of mRNA from polysomes to monosomes as determined by [³H]polyuridylate hybridization (Mahony and Brown, 1979; Heikkila and Brown, 1979a,b) and translation of mRNA associated with the monosome complex (Mahony and Brown, 1979). Metafora *et al.* (1977) noted that mRNA was conserved following the electroshock-induced disaggregation of rabbit brain polysomes and inhibition of protein synthesis.

Based on the observation that LSD induces the accumulation of a brain monosome complex consisting of intact mRNA and both 40 S and 60 S ribosomal subunits it is tenable that protein synthesis is blocked at a step in initiation (Mahony and Brown, 1979; Heikkila and Brown, 1979a). Since the polysome profile is a reflection of the relative rates of initiation, elongation, and termination (Villa-Trevino *et al.*, 1964; Noll, 1969; Vassart *et al.*, 1971; Kisilevsky, 1972; Bergmann and Lodish, 1979) a decrease in the rate of initiation relative to elongation and termination would result in a shift of polysomes to monosomes. A number of other studies in which brain polysome disaggregation and inhibition of protein synthesis occurs, i.e., electroshock (Metafora *et al.*, 1977), *d*-amphetamine (Baliga *et al.*, 1976), epileptic seizures (Wasterlain, 1977; Fando *et al.*, 1979), excess phenylalanine (Taub and Johnson, 1975), and postischemia (Cooper *et al.*, 1977), have also suggested that the lesion in brain protein synthesis is at a step in initiation.

D. Effect of Environment, Stress, and Developmental Age

The LSD-induced brain polysome shift can be altered by holding cage environment, pre-LSD sedation, and post-LSD handling (Holbrook and Brown, 1977a). These results suggested that certain elements of environment and physiological arousal were involved in the macromolecular effect of the drug on the brain protein synthesis apparatus. A synergistic interaction of LSD and stress was evident since physiological stress applied in combination with low doses of LSD significantly increased the shift of brain polysomes to monosomes relative to that observed with LSD alone (Heikkila *et al.*, 1978). Three forms of stress, i.e., restraint, food deprivation, and epinephrine injection, accentuated the

LSD-induced disaggregation of brain polysomes. Polysome disaggregation was dependent on the administration of LSD, since none of the stressing procedures alone, could induce a brain polysome shift. Furthermore this phenomenon required the psychoactive form of LSD, since BOL, the nonhallucinogenic derivative of LSD which binds to the same neural receptors as LSD (Lovell and Freedman, 1976; Burt et al., 1976), could not induce a polysome shift with or without stress. Each of the above-mentioned stressful treatments in combination with LSD did in fact produce a measurable stresslike response as evidenced by an elevation in plasma corticosteroids. In agreement with our results using LSD, Blackshear et al. (1979) have found that disaggregation of brain polysomes in mice by d-amphetamine can be accentuated if the animals are placed in a stressful environment (i.e., crowding).

The administration of LSD at a dose as low as 1 μg/kg was sufficient to induce a massive brain polysome shift when combined with the restraining procedure (Heikkila et al., 1978). This dose is well within levels utilized by man (Kalant and Khanna, 1975, Rossi, 1971). In contrast, studies in other laboratories have shown that very high doses of agents such as phenylalanine, L-Dopa, 5-HTP, and d-amphetamine are required to induce disaggregation of brain polysomes (Aoki and Siegal, 1970; Moskowitz et al., 1975; Taub and Johnson, 1975; Weiss et al., 1975).

If LSD is administered to a pregnant female rabbit a disaggregation of polysomes was observed in the maternal brain and in fetal kidney, liver, and brain (Fig. 2). This result is in contrast to the ineffectiveness of LSD on brain polysomes in 3-week-old rabbits (Holbrook and Brown, 1977a). The LSD-induced fetal organ polysome shift is not due to RNase and does not appear to be a brain-specific phenomenon as found in the adult rabbit. Pretreatment of the pregnant female rabbit with neurotransmitter receptor blocking agents effectively inhibited the LSD-induced disaggregation of polysomes in the maternal brain and all three fetal organs (Heikkila et al., 1979). These agents may inhibit polysome disaggregation of the fetal organs directly, but very likely these receptor antagonists act indirectly via neurotransmitter receptors in the maternal central nervous system. LSD administered to pregnant mice has been shown to pass into the fetal circulation, but the amount of drug present in the fetus is very low compared to the amount in maternal circulation (Idanpaan-Heikkila and Schoolar, 1969). Also the number of LSD receptor sites and synaptic connections in fetal brain is likely to be very low since studies done with 1-day-old rats show only 7–8% of the number of high-affinity ^{3}H-d-LSD or ^{3}H-5-HT binding sites relative to the adult (Bennett and Snyder, 1976). A number of groups have found very little stereospecific LSD binding in tissues other than brain (Bennett and Aghajanian, 1974; Fillion et al., 1976; Lovell and Freedman, 1976) Since LSD disaggregates polysomes in fetal kidney and liver as well as in fetal brain, it is possible that the effect on fetal organs is a result of drug-induced changes of

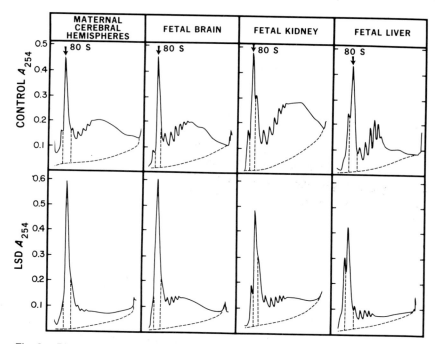

Fig. 2. Disaggregation of polysomes from maternal brain and fetal organs after administration of LSD *in vivo*. Pregnant female rabbits were injected intravenously with LSD at 50 μg/kg (control animals received saline). After 1 h rabbits were sacrificed and total polysomes from maternal brain and fetal organs were analyzed on 15–45% surcrose gradients as described in Fig. 1. (From Heikkila *et al.*, 1979, by permission from Raven Press, New York.)

factors in maternal circulation or physiology. Polysome disaggregation in fetal brain has been reported following increased phenylalanine levels in the maternal plasma (Copenhaver *et al.*, 1973). This study suggested that free amino acid pools in fetal brain may regulate the degree of polysome aggregation. LSD, however, did not induce any major alterations in the levels of free amino acids in either fetal brain, maternal brain, or plasma (Heikkila *et al.*, 1979).

IV. EFFECT OF HYPERTHERMIA ON THE TRANSLATIONAL APPARATUS OF THE BRAIN

A. LSD-Induced Hyperthermia

While our initial results elucidated the parameters and possible mechanism of the LSD-induced brain polysome shift, evidence in the literature pointed toward a possible involvement of LSD-induced hyperthermia. Horita and Hill (1972) have reported that the administration of LSD to rabbits induces an elevation in body temperature (41°–42°C). This was of interest since it has been shown in reticulocytes (Townes and Fuhr, 1977; Sprechman and Ingram, 1972), ascites cells (Austin and Kay, 1975), L cells (Schochetman and Perry, 1972), and HeLa cells (McCormick and Penman, 1969) that elevated temperature (42°C) induces polysome disaggregation and inhibition of protein synthesis. A number of experiments were, therefore, designed to examine if LSD-induced hyperthermia is involved in the disaggregation of brain polysomes.

If LSD-induced hyperthermia was involved in the phenomenon of brain polysome disaggregation, one would expect that the extent of brain polysome shift would vary with the degree of hyperthermia. Rectal temperature measurements were, therefore, taken in conjunction with an examination of brain polysome profiles. Other studies have shown that rectal temperature in rabbits is similar to temperature in the brain and that the temperature at these two sites increases in parallel during hyperthermia (Kluger *et al.*, 1973; Baker, 1979). Measurement of rectal temperature is thus an appropriate indicator of brain temperature.

In all experimental situations, the degree of LSD-induced brain polysome shift was correlated with the extent of LSD-induced hyperthermia (Heikkila and Brown, 1979a). Treatments that accentuated the extent of LSD-induced brain polysome shift such as increasing LSD dose or stressing the animal after drug administration (i.e., restraint) also accentuated the degree of LSD-induced hyperthermia (Table IA). Conversely, pretreatment with neurotransmitter receptor blockers (i.e., chlorpromazine and pizotyline) prevented LSD-induced hyperthermia and inhibited LSD-induced brain polysome disaggregation (Table IB). The attentuation of LSD-induced hyperthermia by DA and 5-HT receptor blockers is in agreement with a number of other studies (Roszell and Horita, 1975; Horita and Hill, 1972; Consroe *et al.*, 1977; Hashimoto *et al.*, 1977). It is also of interest that pretreatment of rabbits with BOL (a psychotropically inactive brominated derivative of LSD) and sodium pentobarbitol, which have been shown to block the LSD-induced brain polysome shift (Holbrook and Brown, 1977b), also attenuates LSD-induced hyperthermia (Horita and Hill, 1972; Horita and Dille, 1954). The hallucinogenic form of LSD was required for both hyperthermia and polysome disaggregation since BOL, the nonpsychoactive derivative of LSD, did not have any effect on body temperature or brain polysomes.

An involvement of LSD-induced hyperthermia in the brain polysome shift is shown in experiments in which the ability of the rabbit to dissipate body heat was altered (Heikkila and Brown, 1979a). For example, placement of the animal at an ambient temperature of 33°C following LSD accentuated both brain polysome disaggregation and hyperthermia, whereas an ambient temperature of 4°C after drug injection blocked both effects (Table IC). The time course of LSD-induced hyperthermia and brain polysome shift was very similar (Heikkila and Brown,

Table I.   **Effect of Various Parameters on LSD-Induced Hyperthermia and Disaggregation of Brain Polysomes**[a]

Treatment	Percent monosomes/total ribosomes	Rectal temperature (°C)
A. Saline (9)	23.6 ± 3.0	39.7 ± 0.2
LSD 10 μg/kg (9)	32.3 ± 6.5*	41.1 ± 0.1*
LSD 25 μg/kg (12)	37.1 ± 3.1*	41.2 ± 0.2*
LSD 10 μg/kg + restraint (4)	47.0 ± 5.3*	41.5 ± 0.6*
LSD 25 μg/kg + restraint (4)	51.4 ± 2.8*	42.3 ± 0.2*
B. LSD 50 μg/kg (9)	41.9 ± 4.5*	41.4 ± 0.3*
Chlorpromazine 2 mg/kg + LSD 50 μg/kg (4)	25.0 ± 2.6	40.4 ± 0.3
Pizotyline 500 μg/kg + LSD 50 μg/kg (4)	22.0 ± 1.0	39.7 ± 0.5
C. Elevated ambient temperature (33°C) + LSD 10 μg/kg (5)	48.6 ± 2.9*	42.0 ± 0.7*
Decreased ambient temperature (4°C) + LSD 50 μg/kg (9)	26.3 ± 2.5	40.3 ± 0.4

[a] Rabbits were given LSD intravenously at the dose indicated (controls received saline) and sacrificed after 1 h. Total cerebral hemisphere polysomes were isolated and the percent monosomes/total ribosomes determined by analysis on sucrose gradients (Fig. 1). Rectal temperatures were recorded 1 h after LSD administration. Chlorpromazine and pizotyline were injected intravenously 25–30 min prior to LSD. The level of statistical significance ($p < 0.01$) relative to the saline value is indicated with an asterisk. The number of separate experiments is shown in parentheses. (From Heikkila and Brown, 1979a.)

1979a). The onset of hyperthermia, however, precedes any detectable brain polysome disaggregation, an expected temporal arrangement if hyperthermia is involved in the brain polysome shift. Unlike rats and mice, the rabbit is extremely sensitive to the hyperthermic effects of LSD (Horita and Hill, 1972).

Apomorphine, a dopamine agonist, has been shown to induce severe hyperthermia in rabbits (Roszell and Horita, 1975). The intravenous administration of apomorphine to rabbits also induced marked hyperthermia, which was associated with an extensive disaggregation of brain polysomes (Heikkila and Brown, 1979a). These findings reinforce the contention that hyperthermia may affect the protein synthesis apparatus of brain.

Since these results show a strong correlation between LSD-induced hyperthermia and brain polysome disaggregation in adult rabbits, it is possible that LSD-induced hyperthermia may cause a temperature-sensitive lesion in protein synthesis. This is a reasonable possibility since brain polysome disaggregation is a global phenomenon (Heikkila *et al.*, 1979), which parallels brain vascularization rather than reflecting sites of ^{3}H-LSD binding (Lovell and Freedman, 1976; Bennett and Aghajanian, 1974). Both neuronal and glial cells are involved in the disaggregation of polysomes since regions containing various amounts of white and gray matter show similar degrees of polysome shift (Heikkila *et al.*, 1979) and the synthesis of S-100, a glial specific protein, is also depressed (Mahony and Brown, 1979).

LSD-induced hyperthermia was also associated with LSD-induced brain poly-some disaggregation in specific stages of pre- and postnatal development (Heikkila *et al.*, 1979; Heikkila and Brown, 1979a). The extent of LSD-induced brain polysome shift and hyperthermia increased with postnatal maturation from 3-week-old rabbits to adults. In the 3-week-old rabbit the administration of LSD induced behavioral effects, but it did not affect brain polysomes or rectal temperature. It is possible that the lack of insulation (i.e., body hair or fat) in these young rabbits allowed them to dissipate heat more efficiently and thus prevent severe hyperthermia. It is also tenable that the synaptic pathways leading to LSD-induced hyperthermia in the 3-week-old rabbit are not fully developed. This ineffectiveness of LSD on brain polysomes in the 3-week-old rabbit was in contrast to the effect of LSD on maternal brain and fetal organs. The administration of LSD to a pregnant female rabbit induced hyperthermia and a disaggregation of polysomes in maternal brain and fetal kidney, liver, and brain (Heikkila *et al.*, 1979). Since others have found that rapidly proliferating cells (i.e., embryonic or cancer cells) are more susceptible to the effects of hyperthermia (Edwards, 1979; Edwards *et al.*, 1976; Dickson and Shah, 1972; Cavaliere *et al.*, 1967), it is possible that the protein synthesis apparatus of fetal organs such as kidney is more susceptible to elevated body temperatures than polysomes in adult organs.

Pretreatment of the pregnant female rabbit with the receptor blocking agents haloperidol and pizotyline effectively inhibited both LSD-induced hyperthermia and the LSD-induced disaggregation of polysomes in the maternal brain and all three fetal organs (Heikkila *et al.*, 1979). Roszell and Horita (1975) have also shown that pretreatment of a rabbit with haloperidol decreases the extent of LSD-induced hyperthermia. It is, therefore, possible that haloperidol and pizotyline block the LSD-induced maternal brain and fetal organ polysome shift by interferring with LSD-induced hyperthermia.

Hyperthermia according to Bligh and Johnson (1973) is a term used to describe any unspecified or unidentified elevation in body temperature above the normal resting range of body temperature designated for any given species of animal. In hyperthermic animals, heat production exceeds heat loss so that heat is stored within the body and core temperature rises (Stitt, 1979). This can be due to either an increase in metabolic heat production or a reduction in heat loss mechanisms. Hyperthermia has been divided into four categories by Stitt (1979), i.e., (1) fever induced by bacterial pyrogens; (2) hyperthermias due to inadequate means of heat dissipation; (3) exercise hyperthermia; and (4) hyperthermias resulting from pharmacological or pathological impairment of thermoregulatory mechanisms.

The hyperthermic action of LSD falls into this last category. LSD has been shown to induce dose-dependent hyperthermia in rabbits (Horita and Dille, 1954) with a minimal response detected at an intravenous dose of 0.25 μg/kg (Horita and Hill, 1972). The intravenous administration of 50–100 μg/kg of LSD to rabbits induces transient hyperthermia reaching a peak (i.e., rectal temperature: 41.5°C) by 1–1.5 h and returning to normal levels by 6 h (Horita and Hill, 1972). This type of hyperthermic response varies among different mammals. The rabbit has been found to be very sensitive to LSD while higher doses (250–1000 μg/kg) must be given to rats and mice in order to obtain a statistically significant increase in rectal temperature (Hashimoto *et al.*, 1977; Rewerski *et al.*, 1972). Severe hyperthermia has also been observed with man after LSD overdoses (Friedman and Hirsch, 1971; Klock *et al.*, 1975).

LSD-induced hyperthermia results from an inhibitory action on evaporative heat loss mechanisms coupled with an excitatory action on heat production (Bligh, 1979). The blockage of evaporative heat loss by LSD in rabbits is apparently due to an interaction with central nervous system receptors involved in vasoconstriction. In the rabbit this is primarily a reduction of blood movement through the ears, which is the major factor in the regulation of arterial blood temperature since this species does not possess sweat glands. The exact neuronal pathway by which LSD induces hyperthermia is not fully resolved, but it has been suggested that LSD by acting as a 5-HT receptor blocker in the central nervous system exerts an inhibitory effect on the pathway involved in evaporative heat loss while releasing an inhibitory action on heat production (Bligh, 1979). Since both pathways may have catecholaminergic intermediaries (Bligh, 1979), this may explain the attenuation of LSD-induced hyperthermia by DA and NA receptor antagonists. Since LSD is known to have both DA agonist and antagonist properties (Cresse *et al.*, 1976), this hallucinogen may also act independently on DA receptors involved in the heat production and evaporative heat loss pathways.

B. Hyperthermia Induced by Fever or Elevation of Ambient Temperature

Since the previously mentioned studies suggested an involvement of LSD-induced hyperthermia in the brain polysome shift, it was of interest to directly elevate body temperature to levels found after LSD (41°–42°C) and determine the effect on brain polysomes. Elevation of the body temperature by either placing the animals at a high ambient temperature (37°–38°C) or induction of fever by bacterial pyrogen injection induced the disaggregation of polysomes in brain but not in kidney (Heikkila and Brown, 1979a,b). Both free and membrane-bound brain polysomes are disaggregated following elevation of body temperature by placement of the animals at a high-ambient temperature (37°–38°C) (Heikkila and Brown, 1981). In agreement with these results Murdock *et al.* (1978) have shown that infant rats, whose body temperature was directly elevated by high-ambient temperatures demonstrate brain polysome disaggregation and inhibited brain protein synthesis (Murdock *et al.*, 1978; Millan *et al.*, 1979). Moskowitz *et al.* (1977) have found that hyperthermia plays a permissive role in the disaggregation of rat brain polysomes following large doses of L-Dopa or *d*-amphetamine and that direct elevation of the body temperature to 42°C also induced a brain polysome shift.

The disaggregation of brain polysomes by elevated ambient temperature treatment or bacterial pyrogen injection is not due to RNase activation since the elevated pool of monosomes was dissociable to 40 S and 60 S ribosomal subunits after high-salt treatment (Heikkila and Brown, 1979a,b, 1981). The induction of a brain polysome shift by either direct heat or bacterial pyrogen results in a conservation of mRNA and a relocalization of mRNA from polysomes to the monosome region as determined by [^{3}H]polyuridylate hybridization (Heikkila and Brown, 1979a,b). The fact that a monosome complex consisting of mRNA and both 40 S and 60 S ribosomal subunits accumulate following elevation of ambient temperature or bacterial pyrogen injection suggests that hyperthermia induced by these means may block protein synthesis at the initiation stage as suggested for LSD.

Hyperthermia is a clinical complication associated with physiological disturbances such as bacterial infections, hypophyseal gland tumors, disturbances of central monoamine metabolism, malignant hyperthermia, and hypermetabolic disorders (Britt, 1979; Stitt, 1979). At a clinical level it is apparent that brain is sensitive to hyperthermia. Fever in susceptible children can induce convulsions, hallucinations, and other neurological disorders (Aecardi and Chevrie, 1976). Severe hyperthermia may lead to cerebral dysfunction and damage, but little is known of the consequences of elevated body temperature on macromolecular events in the brain. Our results suggest that the protein synthesis apparatus in the mammalian brain may be sensitive to physiologically relevant increases in body temperature. Hyperthermia produced by three different procedures (i.e., LSD, elevation of ambient temperature, and fever induced by a bacterial pyrogen) result in a brain polysome shift in which similarities were observed in the mechanism of disaggregation.

V. EFFECTS OF LSD AND HYPERTHERMIA ON BRAIN PROTEIN SYNTHESIS *IN VIVO*

A. Decreased Incorporation of Labeled Amino Acids

Intravenous administration of LSD to young adult rabbits induces a decrease in the incorporation of labeled amino acids *in vivo* into a brain postmitochondrial supernatant fraction during a pulse interval which was chosen to coincide with the period of maximal polysome disaggregation (Holbrook and Brown, 1976). Analysis of more discrete brain subcellular fractions demonstrated a parallel reduction in specific activities of microsomal fractions and synaptic plasma membranes (Freedman *et al.*, 1981). A psychotropically active form of the drug was required to elicit this effect since no decrease was observed with BOL, an inactive brominated derivative that binds to the same neural receptors (Lovell and Freedman, 1976). This inhibitory effect of LSD on the incorporation of amino acids *in vivo* appears to be independent of LSD-induced hyperthermia since the effect is still observed when hyperthermia was prevented by maintaining the rabbits at 4°C, a procedure that does not affect the LSD-induced behavioral changes.

B. Selective Effects on Translation

Electrophoretic analysis on one-dimensional gels of brain proteins labeled *in vivo* following LSD, demonstrated that the drug induced selective effects on translation. The relative labeling of a 75 K protein (K = 1000 daltons) was selectively increased in the postmitochondrial supernatant and microsomal fractions during the period of general decrease in incorporation of labeled amino acid (Freedman *et al.*, 1981). These selective labeling effects were not observed when LSD-induced hyperthermia was prevented by maintaining rabbits at 4°C. Recently we have further resolved the 75 K protein into 74 K and 75 K components by two-dimensional gel electrophoresis and fluorography. As shown in Fig. 3, labeling of the 74 K component was observed following LSD-induced hyperthermia.

The above results suggested that the selective labeling effects induced by LSD were related to hyperthermic effects of the drug. Induction of hyperthermia by means other than LSD, i.e., elevation of ambient temperature to 37°C, also increased labeling of the 75 K brain protein and in addition an increase in labeling of a 95 K protein was noted (Freedman *et al.*, 1981). These two brain proteins whose synthesis is stimulated by hyperthermia are similar in molecular weight to two of the major ''heat-shock'' proteins whose synthesis is activated by elevation of temperature in *Drosophila,* chicken, and mammalian tissue culture cells while synthesis of other proteins is reduced (Tissieres *et al.*, 1974; Kelley and Schlesinger, 1978). It has been suggested that ''heat-shock'' proteins may be important for cell survival under stressful conditions such as temperature elevation and energy deprivation (Kelley and Schlesinger, 1978). It is of interest that in the intact animal, temperature increases in a physiologically relevant range, i.e., attained during fever reactions, result in selective increases in the relative synthesis of brain proteins of molecular weights 75 K and 95 K.

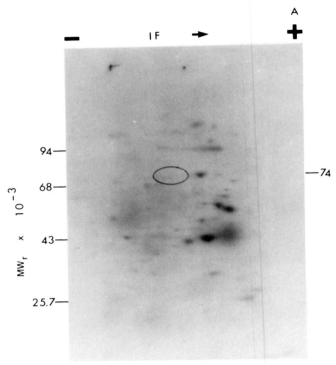

Fig. 3. Two-dimensional gel-electrophoresis of [^{35}S]methionine-labeled proteins from rabbit brain following intravenous injection of LSD. Twenty minutes after the administration of LSD (100 μg/kg), 0.5 mCi of [^{35}S]methionine was injected into the lateral ventricles of the brain through stereotaxically implanted cannulae. Following a 1-h pulse period, animals were sacrificed and a postmitochondrial supernatant fraction was prepared from the cerebral hemispheres as described by Cosgrove and Brown (1981). Equal amounts of acid-precipitable radioactivity (50,000 cpm) from control and LSD-treated animals were analyzed by two-dimensional gel electrophoresis and fluorography as described by Heikkila *et al.* (1981). The first dimension (horizontal axis) consisted of isoelectric focusing in a pH gradient (5.5–8.0), followed by 7–17% gradient polyacrylamide/sodium dodecyl sulfate electrophoresis in the second dimension. The position of the 74 K group of proteins is encircled. (A) Saline control; (B) LSD treatment.

C. Synthesis of Synaptic Proteins and Glycoproteins

Since LSD interacts with synaptic receptors and affects patterns of translation and nerve firing in the brain, we undertook a project to analyze the effect of the drug on the synthesis of synaptic proteins and glycoproteins (Freedman *et al.*, 1981). The effect of the intravenous administration of LSD on the synthesis of

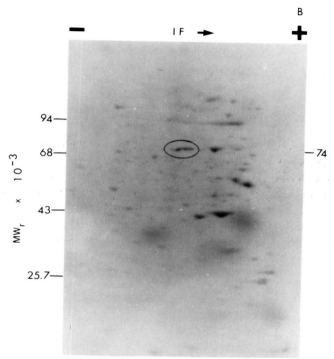

Fig. 3. (*Continued*)

microsomal and synaptic glycoproteins was analyzed by fractionation of [^{35}S]methionine-labeled proteins on lectin affinity columns and by labeling experiments with [^{3}H]fucose, a specific precursor for glycoproteins.

In contrast to the overall decrease in labeling of synaptic proteins, the synthesis of synaptic glycoproteins was not altered by LSD (Freedman *et al.*, 1981). The synthesis of microsomal glycoproteins was, however, reduced. Inhibition of brain protein synthesis by as much as 50% did not appear to alter the synthesis, transport, and incorporation of glycoproteins into components of the synapse. The differential effects of LSD on the synthesis of synaptic glycoproteins may suggest a selective control mechanism for the synthesis of these proteins.

Electrophoretic analysis on one-dimensional gels of synaptic proteins labeled *in vivo* following LSD injection, demonstrated relative increases in labeling of synaptic proteins of molecular weight 75 K and 95 K at a time when overall incorporation of amino acid into synaptic protein was reduced by 50% (Freedman *et al.*, 1981). These specific labeling effects were not observed when LSD-induced hyperthermia was blocked. Increased labeling of the 75 K and 95 K

synaptic proteins were also observed when hyperthermia was induced by elevation of ambient temperature. It appears that hyperthermia results in relative
increases in the synthesis of two synaptic proteins, which are similar in molecular
weight to major "heat-shock" proteins previously reported in tissue culture
systems (Kelley and Schlesinger, 1978).

VI. EFFECT OF LSD AND HYPERTHERMIA ON SUBSEQUENT CELL-FREE PROTEIN SYNTHESIS IN THE BRAIN

A. Characterization of an Initiating Cell-Free System

Various disruptive treatments have been shown to result in decreased incorporation of amino acids using brain cell-free protein synthesis systems (Aoki and
Siegel, 1970; MacInnes and Schlesinger, 1971; Siegel *et al.*, 1971; Wasterlain,
1974; Widelitz *et al.*, 1976; Cooper *et al.*, 1977; Metafora *et al.*, 1977; Schotman *et al.*, 1980; Cosgrove *et al.*, 1981). It has often been suggested that
translational lesions in brain are the result of a decreased rate of initiation of
protein synthesis (Holbrook and Brown, 1976; Cooper *et al.*, 1977; Metafora *et
al.*, 1977). In order to establish which translation steps are affected during
alterations in brain protein synthesis, a cell-free system derived from brain tissue
is required which is capable of initiating protein synthesis. Such a system would
allow the analysis of individual steps in initiation and would provide a means of
assaying either changes in initiation and elongation factor activity or the formation of inhibitory substances.

To date *in vitro* translation systems derived from brain tissue have generally
not demonstrated a capacity to initiate protein synthesis (Goodwin *et al.*, 1969;
Zomzely-Neurath, 1972; Zomzely-Neurath and Roberts, 1974; Nowak and
Munro, 1977; Fando and Wasterlain, 1980). Recently, we have characterized an
initiating cell-free protein synthesis system derived from rabbit brain (Cosgrove
and Brown, 1981). The following criteria were used to verify the occurrence of
initiation in this brain cell-free translation system: (1) sensitivity of translation to
initiation inhibitors such as aurintricarboxylic acid (ATA); (2) binding of labeled
initiator tRNA to 40 S and 80 S initiation complexes; (3) incorporation of labeled
initiation methionine into high molecular weight proteins; and (4) association of
exogenous labeled mRNA with monosomes and polysomes.

B. Inhibitory Effects of LSD and Hyperthermia

The initiating cell-free translation system derived from brain was used to
demonstrate that intravenous injection of LSD into rabbits maintained at room
temperature (23°C) results in a subsequent transient decrease in capacity to

incorporate amino acids *in vitro* following an initial stimulatory phase (Table II). Previous results have suggested that under these injection conditions, LSD results in a rapid rise in body temperature (hyperthermia), which is correlated with the LSD-induced polysome disaggregation (Heikkila and Brown, 1979a). Injection of LSD to rabbits maintained at 4°C prevents the LSD-induced hyperthermia and LSD-induced polysome disaggregation, however, behavioral effects of the drug are still apparent (Heikkila and Brown, 1979a). Lack of polysome disaggregation does not reflect that protein synthesis capacity is unaffected since there is still a transient inhibition of *in vitro* protein synthesis following an initial stimulation after LSD at 4°C (Table II). A decrease in overall protein synthesis capacity is, therefore, observed in either the presence or absence of LSD-induced hyperthermia. Similar inhibitory results have been obtained using *in vivo* analysis of brain protein synthesis (Freedman *et al.*, 1981).

Brain postmitochondrial supernatant can be separated into cell sap and microsome fractions in order to localize the lesion in protein synthesis following LSD or hyperthermia (Cosgrove *et al.*, 1981). Mixing experiments indicate that under conditions of hyperthermia induced by LSD or elevated ambient temperature, there is an inhibitory substance formed in the cell sap fraction. When LSD-induced hyperthermia is prevented by placing animals at 4°C, no inhibitor is detected in the cell sap even though inhibition is observed in the unfractionated

Table II. Effect of Intravenous Administration of LSD on Subsequent Protein Synthesis in an Initiating Brain Cell-Free System[a]

Time after LSD	Cell-free protein synthesis capacity (%)[b]	
	Hyperthermia present (23°C)	Hyperthermia absent (4°C)
7 min	129 ± 16 (4)	116 ± 21 (3)
15 min	87 ± 22 (5)	114 ± 22 (2)
30 min	47 ± 10 (7)	122 ± 18 (4)
1 h	—	53 ± 6 (3)
2 h	48 ± 6 (4)	69 ± 9 (4)
4 h	66 ± 12 (6)	129 ± 21 (3)
6 h	100 ± 11 (3)	—

[a] Brain postmitochondrial supernatant (PMS) was isolated at timed intervals following the intravenous injection of LSD (100 µg/kg) to rabbits which were either maintained at a room temperature of 23°C (LSD-induced hyperthermia present) or placed in a cold room at 4°C (LSD-induced hyperthermia absent). Aliquots (50 µl) of the PMS were analyzed for cell-free protein synthesis capacity relative to saline-injected control animals using [^{35}S]methionine in an inititating translation assay as described by Cosgrove and Brown (1981). The number of experimental trials is shown in parentheses. The 100% value was approximately 35,000 cpm/aliquot. (From Cosgrove *et al.*, 1981, by permission of Raven Press, New York.)

[b] cpm/mg protein LSD PMS/cpm/mg protein saline PMS × 100% ± SD.

system (Table II). These results and the lack of a polysome shift following LSD at 4°C suggest that the inhibition of translation which is observed with LSD at 4°C is the result of a different mechanism from that seen under hyperthermic conditions.

C. Selective Labeling Effects

In addition to the overall inhibition of translation in the cell-free system, we have noted differential effects on the labeling of specific brain proteins. Following LSD-induced hyperthermia greater than average relative decreases were observed on one-dimensional gels in the *in vitro* labeling of proteins of molecular weight 17 K, 29 K, 43 K, and 55 K, whereas proteins of 75 K and 95 K were observed to increase in relative labeling (Cosgrove *et al.,* 1981). These differential labeling effects were also noted in the translation products of the fractionated cell-free protein synthesis system (i.e., LSD cell sap plus control microsomes). These selective labeling effects were not observed when LSD-induced hyperthermia was prevented by placing the animals at 4°C (Cosgrove *et al.,* 1981). We have established using two-dimensional gel electrophoresis that the increased labeling of the 75 K protein is due to the appearance of a 74 K protein, which is not resolved from the more abundant 75 K brain protein on one-dimension gels (Heikkila *et al.,* 1981). The appearance of this 74 K protein is dependent on the induction of hyperthermia by LSD or elevated ambient temperature (Cosgrove *et al.,* 1980).

D. Translation in a Heterologous Cell-Free System

In order to examine whether LSD or elevated ambient temperature treatment induced any specific effects on the relative abundance of brain mRNAs associated with either free or membrane-bound polysomes, intact polysomes were translated in a mRNA-dependent rabbit reticulocyte cell-free system (Heikkila *et al.,* 1981). Programming the reticulocyte lysate assay with either free or membrane-bound brain polysomes and resolution of labeled translation products on two-dimensional gels revealed the labeling of a 74 K protein following hyperthermia induced by either LSD or elevated ambient temperature. This protein was similar in electrophoretic mobility to the 74 K protein which was induced *in vivo* (Fig. 3). The 74 K protein was observed only in the cell-free translation products of free and membrane-bound polysomes isolated from hyperthermic animals (i.e., animals administered LSD at an ambient temperature of 23°C or subjected to direct heat treatment). Brain polysomes isolated from animals given an intravenous injection of LSD and maintained at 4°C, which prevents the hyperthermic response, did not demonstrate labeling of the 74 K protein.

Synthesis of the LSD or direct heat-induced 74 K protein may be the result of newly synthesized mRNA or due to an influx of mRNA from a preexisting pool of cytoplasmic ribonucleoprotein particles. In an early study we noted a stimulation of α-amanatin-sensitive RNA synthesis in brain nuclei isolated 2.5 h after an *in vivo* injection of LSD (Brown, 1975). If the hyperthermia-induced 74 K protein is in fact the result of a newly synthesized mRNA, then this phenomenon is quite similar to the heat-shock phenomenon found in tissue culture (Tissieres *et al.*, 1974; Kelley and Schlesinger, 1978). It has been suggested that these proteins are important for cell survival under stressful conditions such as temperature elevation (Kelley and Schlesinger, 1978).

VII. CONCLUDING REMARKS

An experimental approach to the analysis of regulation of macromolecular synthesis in the brain has been to investigate the effects of the introduction of physiologically relevant treatments. Protein synthesis in the mammalian brain has been found to be sensitive to perturbation by a wide range of physical and chemical treatments (see Section II). Inhibition of initiation has frequently been implicated as a possible mechanism for observed decreases in brain protein synthesis.

The psychotropic drug LSD has been a useful tool to probe translational mechanisms in the rabbit brain. Using several *in vivo* and *in vitro* experimental techniques, we have observed that this drug rapidly activates a translational mechanism which causes a global inhibition of protein synthesis in the brain and a selective increase in synthesis of certain brain proteins. Since LSD induces a rapid rise in body temperature attempts were made to differentiate between translational effects attributable to hyperthermia and those attributable to more psychotropic effects of the drug.

LSD administered under conditions where hyperthermia is induced is associated with an organ-specific disaggregation of brain polysomes (Holbrook and Brown, 1976; Heikkila and Brown, 1981; Heikkila *et al.*, 1978), a conservation of mRNA and relocalization of mRNA from polysomes to monosomes (Mahony and Brown, 1979; Heikkila and Brown, 1979a,b, 1981), a decreased incorporation of labeled amino acid both *in vivo* (Freedman *et al.*, 1981) and in an homologous brain cell-free translation system (Cosgrove *et al.*, 1981), and selective increases in the labeling of specific brain proteins both *in vivo* and *in vitro* (Cosgrove *et al.*, 1981; Freedman *et al.*, 1981; Heikkila *et al.*, 1981). These events appear to be associated with hyperthermic effects of LSD since they can also be induced by hyperthermia, which is induced by other means, i.e., elevation of ambient temperature or fever produced by injection of a bacterial pyrogen (Cosgrove *et al.*, 1981; Freedman *et al.*, 1981; Heikkila and Brown, 1979a,b; Heikkila *et al.*, 1981). These results suggest that translation in the mammalian brain is sensitive to elevations in body temperature and that a lesion in reinitiation of protein synthesis is induced.

Mechanistic differences are apparent between hyperthermic and more psychotropic effects of LSD on brain protein synthesis. Injection of LSD to rabbits at 4°C prevents hyperthermia but the behavioral effects of the drug are still apparent. The effect of LSD in the absence of hyperthermia is characterized by a decreased incorporation of labeled amino acid both *in vivo* (Freedman *et al.*, 1981) and in an homologous brain cell-free translation system (Cosgrove *et al.*, 1981) however, disaggregation of brain polysomes is not observed (Heikkila and Brown, 1979a) nor are the selective increases in the labeling of specific brain proteins. Since the polysome profile is a reflection of the relative rates of initiation, elongation, and termination (Noll, 1969; Vassart *et al.*, 1971; Kisilevsky, 1972, Bergmann and Lodish, 1979), an overall decrease in the rates of all three steps in translation would account for the observation of a decreased translational capacity without a brain polysome shift.

The brain proteins which are selectively increased in relative labeling following hyperthermia induced by LSD or by other means, are similar in molecular weight to two of the major heat-shock proteins which are induced in several tissue culture cell lines following elevation of ambient temperature at a time when the synthesis of other proteins is reduced (Tissieres *et al.*, 1974; Kelley and Schlesinger, 1978). These heat-shock proteins may be important for cell survival under stressful conditions such as temperature elevation and energy deprivation (Kelley and Schlesinger, 1978). Brain functions are particularly sensitive to such stresses (Ritter and Ritter, 1977; Millan *et al.*, 1979). In addition to providing information on the mechanism of the effect of the psychotropic drug LSD on brain protein synthesis our studies demonstrate the sensitivity of the translational apparatus in the mammalian brain to physiologically relevant increases in body temperature. Future research will concern the localization and function of the brain proteins that are induced by hyperthermia. In addition the initiating brain cell-free translation system will be utilized to determine which precise steps in protein synthesis are affected by LSD and hyperthermia.

ACKNOWLEDGMENTS

Studies mentioned in this chapter were supported by grants to Ian R. Brown from the Medical Research Council of Canada.

REFERENCES

Aecardi, J. and Chevrie, J. T. (1976). Febrile convulsions: Neurological sequelae and mental retardation. *In* ''Brain Dysfunction in Infantile Febrile Convulsions'' (M. A. Brazier and F. Coceani, eds.). pp. 247–257. Raven, New York.

Aghajanian, G. K., Foote, W. E. and Sheard M. H. (1969). LSD: Sensitive neuronal units in the midbrain raphe. *Science* **161,** 706–708.

Agrawal, H. C., Bone, A. H., and Davison A. N. (1970). Effect of phenylalanine on protein synthesis in the developing rat brain. *Biochem. J.* **117**, 325–331.

Allfrey, V. G. (1971). *In* "Histones and Nucleohistones" (D. P. Phillip, ed.), p. 241. Plenum, New York.

Andrews, T. M., and Tata, J. R. (1971). Protein synthesis by membrane-bound and free ribosomes of the developing rat cerebral cortex. *Biochem. J.* **124**, 883–889.

Aoki, K., and Siegal, F. (1970). Hyperphenylalanemia: Disaggregation of brain polysomes in young rats. *Science* **168**, 129–130.

Ashby, C. D., and Roberts, S. (1975). Phosphorylation of ribosomal proteins in rat cerebral cortex *in vitro*. *J. Biol. Chem.* **250**, 2546–2555.

Austin, S. A., and Kay, J. E. (1975). Defective initiation on natural messenger RNA by cell-free systems from Krebs ascites cells incubated at elevated temperatures. *Biochim. Biophys. Acta* **395**, 468–477.

Baker, M. A. (1979). A brain cooling system in mammals. *Sci. Amer.* **240**, 130–139.

Baliga, B. S., Zahringer, J., Trachtenberg, M., Moskowitz, M. A., and Munro, H. N. (1976). Mechanisms of D-amphetamine inhibition of protein synthesis. *Biochim. Biophys. Acta* **442**, 239–250.

Bennett, G. S., and Edelman G. M. (1969). Amino acid incorporation into rat brain during spreading cortical depression. *Science* **163**, 393–395.

Bennett, J. L., and Aghajanian, G. K. (1974). d-LSD binding to brain homogenates: Possible relationship to serotonin receptors. *Life Sci.* **15**, 1935–1944.

Bennett, J. L., and Snyder, S. H. (1976). Serotonin and LSD binding in rat membranes: Relationship to post-synaptic serotonin receptors. *Mol. Pharmacol.* **12**, 373–389.

Bergmann, J., and Lodish, H. F. (1979). A kinetic model of protein synthesis. *J. Biol. Chem.* **254**, 11227–11237.

Blackshear, M. A. U., Wade, L. H., and Proctor, C. D. (1979). Effect of crowding on amphetamine-induced disaggregation of brain polyribosomes. *Arch. Int. Pharmacodyr. Ther.* **241**, 180–189.

Bligh, J. (1979). The central neurology of mammalian thermoregulation. *Neuroscience* **4**, 1212–1236.

Bligh, J., and Johnson, K. G. (1973). Glossary terms for thermal physiology. *J. Appl. Physiol.* **35**, 941–961.

Blobel, G., and Sabatini, D. (1971). Dissociation of mammalian polyribosomes into subunits by puromycin. *Proc. Natl. Acad. Sci. U.S.A.* **68**, 30–39.

Britt, B. A. (1979). Etiology and pathophysiology of malignant hyperthermia. *Fed. Proc., Fed. Am. Soc. Exp. Biol.* **38**, 44–48.

Brockaert, J., Premont, J., Glowinski, J., Thierry, A. M., and Tassin J. P. (1976). Topographical distribution of dopaminergic receptors in the rat striatum. II. Distribution and characteristics of dopamine adenylate cyclase interaction of D-LSD with dopaminergic receptors. *Brain Res.* **107**, 303–315.

Brown, I. R. (1975). RNA synthesis in isolated brain nuclei after administration of D-lysergic acid diethylamide (LSD) *in vivo*. *Proc. Natl. Acad. Sci. U.S.A.* **72**, 837–839.

Brown, I. R. (1977). Analysis of gene activity in the mammalian brain. In "Mechanisms, Regulation and Special Functions of Protein Synthesis in the Brain" (S. Roberts, A. Lajtha, and W. H. Gispen, eds.), pp. 29–46. Elsevier/North-Holland, New York.

Brown, I. R. and Liew, C. C. (1975). Lysergic acid diethylamide: effect on histone acetylation in rabbit brain. *Science* **188**, 1122–1123.

Burt, D. R., Creese, I., and Synder, S. H. (1976). Binding interactions of lysergic acid diethylamide and related agents with dopamine receptors in the brain. *Mol. Pharmacol.* **12**, 631–638.

Cain, D. F., Ball, E. D., and Dekaban, A. S. (1972). Brain proteins: Qualitative and quantitative changes, synthesis and degradation during fetal development of the rabbit. *J. Neurochem.* **19**, 2031–2042.

Cavaliere, R., Riocatte, E., Giovanella, B., Heidelberger, C., Johnson, R., Margottini, M., Mondovi, B., Moricca, G., and Rossi-Fanelli, A. (1967). Selective heat sensitivity of cancer cells. *Cancer* **20**, 1351–1381.

Christman, J. K. (1973). Effect of elevated potassium level and amino acid deprivation on polysome distribution and rate of protein synthesis in L cells. *Biochim. Biophys. Acta* **294**, 138–152.

Christoph, G. R., Kuhn, D. M., and Jacobs, B. L. (1977). Electrophysiological evidence for a dopaminergic action of LSD. Depression of unit activity in the substantia nigra of the rat. *Life Sci.* **21**, 1585–1596.

Clouet, D. H., and Ratner, M. (1967). The effect of the administration of morphine on the incorporation of [^{14}C]-leucine into proteins of the rat brain *in vivo*. *Brain Res.* **4**, 33–43.

Clouet, D. H., and Ratner, M. (1979). The effect of morphine tolerance on the incorporation of ^{3}H-leucine into proteins of rat synaptic membranes. *J. Neurosci. Res.* **4**, 93–103.

Consroe, P., Jones, B., and Martin, P. (1977). Lysergic acid diethylamide antagonism by chlorpromazine, haloperidol diazepam and pentobarbitol in the rabbit. *Toxicol. Appl. Pharmacol.* **42**, 45–54.

Cooper, H. K, Zalewska, T., Kawakami, S., Hossman, K. A., and Kleihues, P. (1977). The effect of ischemia and recirculation on protein synthesis in the rat brain. *J. Neurochem.* **28**, 929–934.

Copenhaver, J. H., Vacanti, J. P., and Carver, M. (1973). Experimental hyperphenylalanemia: Disaggregation of fetal brain ribosomes. *J. Neurochem.* **21**, 273–280.

Cosgrove, J. W., and Brown, I. R. (1981). Characterization of an initiating cell-free protein synthesis system derived from rabbit brain. *J. Neurochem.* **36**, 1026–1036.

Cosgrove, J. W., Heikkila, J. J., and Brown, I. R. (1980). Cell-free synthesis of a heat shock protein by polysomes isolated from mammalian brain following hyperthermia induced by LSD. *J. Cell. Biol.* **87**, 284a.

Cosgrove, J. W., Clark, B. D., and Brown, I. R. (1981). Effect of intravenous administration of D-lysergic acid diethylamide on subsequent protein synthesis in a cell-free system derived from brain. *J. Neurochem.* **36**, 1037–1045.

Cotman, C. W., Banker, G., Zornetzer, S. F., and McGaugh, J. L. (1971). Electroshock effects on brain protein synthesis: relation to brain seizures and retrograde amnesia. *Science* **173**, 454–456.

Craves, F. B., Loh, H. H., and Meyerhoff, J. L. (1978). The effect of morphine tolerance and dependence on cell-free protein synthesis. *J. Neurochem.* **31**, 1309–1316.

Creese, I., Burt, D. R., and Snyder, S. H. (1976). The dopamine receptor: differential binding of D-LSD and related agents to agonist and antagonist states. *Life Sci.* **17**, 1715–1720.

Dainat, J., Rebiere, A., and Legrand, J. (1970). The effect of thyroid deficiency on the incorporation of L-[^{3}H]-leucine into proteins of the cerebellum in the young rat. *J. Neurochem.* **17**, 581–586.

Das, A., Ralston, R. O., Grace, M., Ray, R., Ghosh-Dastidar, P., Das, H. K., Yaghmai, B., Palmieri, S., and Gupta, N. K. (1979). Protein synthesis in rabbit reticulocytes: Mechanism of protein synthesis inhibition by heme-regulated inhibitor. *Proc. Natl. Acad. Sci. U.S.A.* **76**, 5076–5079.

Dickson, J. A., and Shah, D. M. (1972). The effects of hyperthermia (40°C) on the biochemistry and growth of a malignant cell line. *Eur. J. Cancer* **8**, 561–571.

Dunlop, D. S., Van Elden, W., and Lajtha, A. (1975). A method for measuring brain protein synthesis rates in young and adult animals. *J. Neurochem.* **24**, 337–344.

Dunlop, D. S., Van Elden, W., and Lajtha, A. (1977). Developmental effects on protein synthesis rates in regions of the CNS *in vivo* and *in vitro*. *J. Neurochem.* **29**, 939–945.

Dunn, A. J. (1975). Intracerebral injections inhibit amino acid incorporation. *Brain Res.* **99**, 405–408.

Dunn, A. J., and Bergert, B. (1976). Effects of electroconvulsive shock and cycloheximide on the incorporation of amino acids into proteins of mouse brain subcellular fractions. *J. Neurochem.* **26,** 369–375.

Dunn, A. J., Giuditta, A., and Pagliuca, N. (1971). The effect of electroconvulsive shock on protein synthesis in mouse brain. *J. Neurochem.* **18,** 2093–2099.

Edwards, M. J. (1979). Is hyperthermia a human teratogen? *Am. Heart J.* **98,** 277–280.

Edwards, M. J., Warner, P., and Mulley, R. C. (1976). Growth and development of the brain in normal and heat-retarded guinea pigs. *Neuropath. Appl. Neurobiol.* **2,** 439–450.

Fando, J. L., and Wasterlain, C. G. (1980). A simple, reproducible cell-free system for measuring brain protein synthesis. *Neurochem. Res.* **5,** 197–207.

Fando, J. L., Conn, M., and Wasterlain, C. G. (1979). Brain protein synthesis during neonatal seizures: An experimental study. *Exp. Neurol.* **63,** 220–228.

Fando, J. L., Salinas, M., and Wasterlain, C. G. (1980). Age-dependent changes in brain protein synthesis in the rat. *Neurochem. Res.* **5,** 373–383.

Fellous, A., Franem, J., Nunez, J., and Sokoloff, L. (1973). Protein synthesis by highly aggregated and purified polysomes from young and adult rat brain. *J. Neurochem.* **21,** 211–222.

Fillion, G., Fillion, M., Spirakis, C., Bakers, J., and Jacob, J. (1976). 5-Hydroxytryptamine binding to synaptic membranes from rat brain. *Life Sci.* **18,** 65–74.

Fleming, E. W., Tewari, S., and Noble, E. P. (1975). Effects of chronic ethanol ingestion on brain aminoacyl-tRNA synthetases and tRNA. *J. Neurochem.* **24,** 553–560.

Freedman, D. X., Giarmann, N., and Lovell, R. (1970). Psychotomimetic drugs and brain 5-hydroxy-indole metabolism. *Biochem. Pharmacol.* **19,** 1181–1188.

Freedman, M. S., Clark, B. D., Cruz, T. F., Gurd, J. W., and Brown, I. R. (1981). Selective effects of LSD and hyperthermia on the synthesis of synaptic proteins and glycoproteins. *Brain Res.* **207,** 129–145.

Freedman, S. A., and Hirsch, S. E. (1971). Extreme hyperthermia after LSD injection. *JAMA, J. Am. Med. Assoc.* **217,** 1549–1550.

Gilbert, B. E., and Johnson, T. C. (1972a). Protein turnover during maturation of mouse brain tissue. *J. Cell Biol.* **53,** 143–147.

Gilbert, B. E., and Johnson, T. C. (1972b). The use of aminoacyl-tRNA to measure polypeptide synthesis by ribosomes isolated from neonatal and adult mouse brain tissue. *Biochem. Biophys. Res. Commun.* **46,** 2034–2037.

Gilbert, B. E., and Johnson, T. C. (1974). Effect of maturation on *in vitro* protein synthesis by mouse brain tissue. *J. Neurochem.* **23,** 811–818.

Goodwin, F., Shafritz, D., and Weissbach, H. (1969). In vitro polypeptide synthesis in brain. *Arch. Biochem. Biophys.* **130,** 183–190.

Harris, C. L., and Maas, J. W. (1974). Transfer RNA and the regulation of protein synthesis in rat brain cerebral cortex during neural development. *J. Neurochem.* **22,** 741–749.

Harris, R. A., Dunn, A., and Harris, L. S. (1974). Effect of acute and chronic morphine administration on the incorporation of ^{3}H-lysine into mouse brain and liver proteins. *Res. Commun. Chem. Pathol. Pharmacol.* **9,** 299–306.

Hashimoto, H., Hayashi, M., Nakahara, Y., Niwaguchi, T., and Ishii, H. (1977). Hyperthermic effects of D-lysergic acid diethylamide (LSD) and its derivatives in rabbits and rats. *Arch. Int. Pharmacodyn. Ther.* **228,** 314–321.

Heikkila, J. J., and Brown, I. R. (1979a). Disaggregation of brain polysomes after LSD *in vivo.* Involvement of LSD-induced hyperthermia. *Neurochem. Res.* **4,** 763–776.

Heikkila, J. J., and Brown, I. R. (1979b). Hyperthermia and disaggregation of brain polysomes induced by bacterial pyrogen. *Life Sci.* **25,** 347–352.

Heikkila, J. J., and Brown, I. R. (1981). Comparison of the effect of intravenous administration of

LSD on free and membrane-bound polysomes in the rabbit brain. *J. Neurochem.* **36,** 1219–1228.

Heikkila, J. J., Holbrook, L. A., and Brown, I. R. (1978). Stress-accentuation of the LSD-induced disaggregation of brain polysomes. *Life Sci.* **22,** 757–766.

Heikkila, J. J., Holbrook, L. A., and Brown, I. R., (1979). Disaggregation of polysomes in fetal organs and maternal brain after administration of D-lysergic acid diethylamide *in vivo. J. Neurochem.* **32,** 1793–1799.

Heikkila, J. J., Cosgrove, J. W., and Brown, I. R. (1981). Cell-free translation of free and membrane-bound polysomes and polyadenylated mRNA from rabbit brain following administration of LSD *in vivo. J. Neurochem.* **36,** 1229–1238.

Hemminki, K. (1972). Differential responses of free and bound polysomes to inhibitors and neuroactive substances *in vitro. J. Neurochem.* **19,** 2699–2702.

Hitzemann, R. J., and Loh, H. H. (1976). Influence of morphine on protein synthesis in discrete subcellular fractions of the rat brain. *Res. Commun. Chem. Pathol. Pharmacol.* **14,** 237–248.

Holbrook, L. A., and Brown, I. R. (1976). Disaggregation of brain polysomes after administration of D-lysergic acid diethylamide (LSD) *in vivo. J. Neurochem.* **27,** 77–82.

Holbrook, L. A., and Brown, I. R. (1977a). Disaggregation of brain polysomes after D-lysergic acid diethylamide administration *in vivo:* Mechanisms and effect of age and environment. *J. Neurochem.* **29,** 461–467.

Holbrook, L. A., and Brown, I. R. (1977b). Antipsychotic drugs block LSD-induced disaggregation of brain polysomes. *Life Sci.* **21,** 1037–1044.

Horita, A., and Dille, J. M. (1954). Pyretogenic effect of LSD. *Science* **120,** 1100–1101.

Horita, A., and Hill, H. F. (1972). Hallucinogens, amphetamines and temperature regulation. *In* "The Pharmacology of Thermoregulation" (E. Schonbaum and P. Lomax, eds.), pp. 416–431. Karger, Basel.

Hughes, J. V., and Johnson, T. C. (1978). Experimentally-induced and natural recovery from the effects of phenylalanine on brain protein synthesis. *Biochim. Biophys. Acta* **517,** 473–485.

Idanpaan-Heikkila, J. E., and Schoolar, J. C. (1969). LSD: Autoradiographic study on the placental transfer and tissue distribution in mice. *Science* **164,** 1295–1297.

Jagus, R., and Safer, B. (1979). Quantitation and localization of globin messenger RNA in rabbit reticulocyte lysate. *J. Biol. Chem.* **254,** 6865–6868.

Johnson, T. C. (1976). Regulation of protein synthesis during postnatal maturation of the brain. *J. Neurochem.* **27,** 17–23.

Johnson, T. C., and Belytschko, G. (1969). Alteration in microsomal protein synthesis during early development of mouse brain. *Proc. Natl. Acad. Sci. U.S.A.* **62,** 844–848.

Kalant, H., and Khanna, J. (1975). *In* "Principles of Medical Pharmacology" (P. Seeman and E. Sellers, eds.), pp. 279–284. Univ. of Toronto Press, Toronto.

Kelley, P. M., and Schlesinger, M. J. (1978). The effect of amino acid analogues and heat shock on gene expression in chicken embryo fibroblasts. *Cell* **15,** 1277–1286.

Kisilevsky, R. (1972). The regulatory parameter of protein synthesis most affected by ethionine and cycloheximide. A comparison of computer and *in vitro* studies. *Biochim. Biophys. Acta* **272,** 463–472.

Kleihues, P., and Hossman, K. A. (1971). Protein synthesis in the cat after prolonged cerebral ischemia. *Brain Res.* **35,** 409–418.

Kleihues, P., Hossman, K. A., Pegg, A., Koboyashi, K., and Zimmerman, V. (1975). Resuscitation of the monkey brain after 1 hour of complete ischemia III. indication of metabolic recovery. *Brain Res.* **100,** 61–73.

Klock, J. C., Boerner, V., and Becker, C. E. (1975). Coma, hyperthermia and bleeding associated with massive LSD overdoses: A report of eight cases. *Clin. Toxicol.* **8,** 191–203.

Kluger, M. J., Gonzalez, R., and Stolwijk, J. A. (1973). Temperature regulation in the exercising rabbit. *Am. J. Physiol.* **224**, 130–135.

Krawczynski, J. (1961). The influence of serotonin, D-lysergic acid diethylamide and 2-bromo-LSD on the incorporation of ^{35}S-methionine into brain proteins and on the level of ATP in the brain. *J. Neurochem.* **7**, 1–4.

Krivanek, J. (1970). Effects of spreading cortical depression on the incorporation of ^{14}C-leucine into proteins of rat brain. *J. Neurochem.* **17**, 531–538.

Lang, D. W., Darrah, H. K., Hedley-White, J., and Laasberg, L. H. (1975). Uptake into brain proteins of ^{35}S-methionine during morphine tolerance. *J. Pharmacol. Exp. Ther.* **192**, 521–530.

Lerner, M. P., and Johnson, T. C. (1970). Regulation of protein synthesis in developing mouse brain tissue. *J. Biol. Chem.* **245**, 1388–1393.

Lim, L., and White, J. O. (1974). On the unique dissociability of neonatal rat cerebral free polysomes and changes during development of the brain. *Biochim. Biophys. Acta* **366**, 358–363.

Lin, R. C., Ngai, S. H., and Costa, E. (1969). Lysergic acid diethylamide: Role in conversion of plasma tryptophan to brain serotonin. *Science* **166**, 237–239.

Lindholm, D. B., and Khawaja, J. A. (1979). Alterations in number and activity of cerebral free and membrane-bound ribosomes after prolonged ethanol ingestion by weanling rats. *Neuroscience* **4**, 1007–1013.

Loh, H. H., and Hitzemann, R. J. (1974). Effect of morphine on (^{14}C-choline) phosphatidylcholine and (^{3}H-leucine) protein synthesis and turnover in discrete regions of the rat brain. *Biochem. Pharmacol.* **23**, 1753–1765.

Lovell, R. A., and Freedman, D. X. (1976). Stereospecific receptor sites for d-lysergic acid diethylamide in rat brain: Effects of neurotransmitters, amine antagonists and other psychotrophic drugs. *Mol. Pharmacol.* **12**, 620–630.

MacInnes, J. W., and Schlesinger, K. (1971). Effects of excess phenylalanine on *in vitro* and *in vivo* RNA and protein synthesis and polyribosome levels in brains of mice. *Brain Res.* **29**, 101–110.

MacInnes, J. W., McConkey, E. H., and Schlesinger, K. (1970). Changes in brain polysomes following an electroconvulsive seizure. *J. Neurochem.* **17**, 457–460.

Mahony, J. B., and Brown, I. R. (1979). Fate of brain mRNA following disaggregation of brain polysomes after administration of LSD *in vivo*. *Biochim. Biophys. Acta* **565**, 161–172.

McCormick, W., and Penman, S. (1969). Regulation of protein synthesis in HeLa cells: Translation at elevated temperatures. *J. Mol. Biol.* **39**, 315–333.

Metafora, S., Persico, M., Felsani, A., Ferraiudo, R., and Giuditta, A. (1977). On the mechanism of electroshock-induced inhibition of protein synthesis in rabbit cerebral cortex. *J. Neurochem.* **28**, 1335–1346.

Millan, N., Murdock, L., Bleier, R., and Siegal, F. L. (1979). Effects of acute hyperthermia on polyribosomes, *in vivo* protein synthesis and ornithine decarboxylase activity in the neonatal rat brain. *J. Neurochem.* **32**, 311–317.

Morimoto, K., Brengman, T., and Yanagihara, T. (1978). Further evaluation of polypeptide synthesis in cerebral ischemia, anoxia and hypoxia. *J. Neurochem.* **31**, 1277–1282.

Moskowitz, M. A., Weiss, B., Lytle, L., Munro, H., and Wurtman, R. J. (1975). D-Amphetamine disaggregates brain polysomes via a dopaminergic mechanism. *Proc. Natl. Acad. Sci. U.S.A.* **72**, 834–836.

Moskowitz, M. A., Rubin, D., Liebschutz, J., Munro, H., and Wurtman, R. J. (1977). The permissive role of hyperthermia in the disaggregation of brain polysomes by L-DOPA or D-amphetamine. *J. Neurochem.* **28**, 779–782.

Murdock, L., Berlow, S., Colwell, R., and Siegal, F. L. (1978). The effects of hyperthermia on polyribosomes and amino acid levels in infant rat brain. *Neuroscience* **3**, 349–357.

Noll, H. (1969). Polysomes: Analysis of structure and function. *In* "Techniques in Protein Biosynthesis" (P. Campbell and J. Sargent, eds.), pp. 101–179. Academic Press, New York.

Nowak, Jr., T. S., and Munro, H. N. (1977). Inhibition of cell-free synthesis initiation by amphetamine: Association with reduction in tRNA aminoacylation. *Biochem. Biophys. Res. Commun.* **77**, 1280–1285.

Persson, S. (1977). The effect of LSD and 2-bromo-LSD on the striatal Dopa accumulation after decarboxylase inhibition in rats. *Eur. J. Pharmacol.* **43**, 73–83.

Raghupathy, E., Peterson, N. A., and Ko, G. K. (1971). Formation of ribosome dimers in brains of hypothermic rats. *Biochem. Biophys. Res. Commun.* **43**, 1223–1231.

Rewerski, W. J., Piechocki, T., and Rylski, M. (1972). Effects of hallucinogens on aggressiveness and thermoregulation in mice. *In* "The Pharmacology of Thermoregulation" (E. Schonbaum and P. Lomax, eds.), pp. 432–436. Karger, Basel.

Ritter, S., and Ritter, R. C. (1977). Protection against stress-induced brain norepinephrine depletion after repeated 2-deoxy-D-glucose administration. *Brain Res.* **127**, 179–184.

Roberts, S., and Morelos, B. S., (1976). Role of ribonuclease action in phenylalanine-induced disaggregation of rat cerebral polyribosomes. *J. Neurochem.* **26**, 387–400.

Roel, L. E., Schwartz, S. A., Weiss, B. F., Munro, H. N., and Wurtman, R. J. (1974). *In vivo* inhibition of rat brain protein synthesis by L-dopa. *J. Neurochem.* **23**, 233–239.

Roel, L. E., Moskowitz, M. A., Rubin, D., Markovitz, D., Lytle, L., Munro, H. N., and Wurtman, R. J. (1978). *In vivo* inhibition of rat brain protein synthesis by D-amphetamine. *J. Neurochem.* **31**, 341–345.

Rossi, G. V. (1971). LSD: a pharmacological profile. *Am. J. Pharm.* **2**, 117–123.

Roszell, D. K., and Horita, A. (1975). The effects of haloperidol and thioridazine on apomorphine- and LSD-induced hyperthermia in rabbits. *J. Psychiatr. Res.* **12**, 117–123.

Sankar, D. V. S. (1975). LSD—A Total Study. PJD Publ., Ltd., Westbury, New York.

Sarma, D., Reid, I. M., and Sidransky, S. (1969). The selective effect of Actinomycin D on free polyribosomes of mouse liver. *Biochem. Biophys. Res. Commun.* **36**, 582–588.

Schmidt, M. J., and Sokoloff, L. (1973). Activity of cyclic-AMP dependent microsomal protein kinase and phosphorylation of ribosomal protein in rat brain during postnatal development. *J. Neurochem.* **21**, 1193–1205.

Schochetman, G., and Perry, R. P. (1972). Characterization of the messenger RNA released from L cell polyribosomes as a result of temperature shock. *J. Mol. Biol.* **63**, 577–590.

Schotman, P., Heuven-Nolsen, D. V., Gispen, W. H. (1980). Protein synthesis in a cell-free system from rat brain sensitive to ACTH-like peptides. *J. Neurochem.* **34**, 1661–1670.

Siegal, F. L., Aoki, A., and Colwell, R. E. (1971). Polyribosome disaggregation and cell-free protein synthesis in preparations from cerebral cortex of hyperphenylalanemic rats. *J. Neurochem.* **18**, 537–547.

Sprechman, L. M., and Ingram, V. M. (1972). Hemoglobin synthesis at elevated temperatures. *J. Biol. Chem.* **247**, 5004–5009.

Stitt, J. T. (1979). Fever versus hyperthermia. *Fed. Proc., Fed. Am. Soc. Exp. Biol.* **38**, 39–43.

Szijan, I., Kalbermann, L. E., and Gomez, C. J. (1971). Hormonal regulation of brain development. IV Effect of neonatal thyroidectomy upon incorporation *in vivo* of L-[^{3}H]-phenylalanine into proteins of developing rat cerebral tissues and pituitary gland. *Brain Res.* **27**, 309–318.

Taub, F., and Johnson, T. C. (1975). The mechanism of polyribosome disaggregation in brain tissue by phenylalanine *Biochem. J.* **151**, 173–180.

Terada, M., Metafora, S, Banks, J., Dow, L., Bank, A., and Marks, P. A. (1972). Conservation of globin messenger RNA in rabbit reticulocyte monoribosomes after sodium fluoride treatment. *Biochem. Biophys. Res. Commun.* **47**, 766–774.

Tewari, S., and Noble, E. P. (1971). Ethanol and brain protein synthesis. *Brain Res.* **26**, 469–474.

Tewari, S., Goldstein, M. A., and Noble, E. P. (1977). Alterations in cell-free brain protein synthesis following ethanol withdrawal in physically dependent rats. *Brain Res.* **126**, 509–518.

Tewari, S., Murray, S., and Noble, E. P. (1978). Studies on the effects of chronic ethanol ingestion on the properties of rat brain ribosomes. *J. Neurosci. Res.* **3**, 375–387.

Tissieres, A., Mitchell, H. K., Tracy, U. M. (1974). Protein synthesis in salivary glands of *Drosophila melanogaster:* Relation to chromosome puffs. *J. Mol. Biol.* **84**, 389–398.

Townes, T. M., and Fuhr, J. E. (1977). Protection by cycloheximide from hyperthermic inhibition of protein synthesis in human reticulocytes. *Life Sic.* **20**, 585–592.

Vassart, G., Dumont, J. E., and Contraine, F. (1971). Translational control of protein synthesis: A simulation study. *Biochim. Biophys. Acta* **247**, 471–485.

Villa-Trevino, S., Farber, E., Staehelin, T., Wettstein, F. O., and Noll, H. (1964). Breakdown and assembly of rat liver ergosomes after administration of ethionine or puromycin. *J. Biol. Chem.* **239**, 2826–2833.

Von Gatzke, H. D. (1977). Autoradiographic studies on the possibility of influencing ^{3}H-leucine incorporation rates in different areas of rat brain by LSD. *Arzneim. Forsch.* **7**, 1399–1403.

Von Hungen, K., Roberts, S., and Hill, D. (1974). LSD as an agonist and antagonist at central dopamine receptors. *Nature (London)* **252**, 588–589.

Von Hungen, K., Roberts, S., and Hill, D. (1975). Serotonin-sensitive adenylate cyclase activity in immature rat brain. *Brain Res.* **84**, 257–267.

Wang, R. Y., and Aghajanian, G. K. (1977). Inhibition of neurons in the amygdala by dorsal raphe stimulation: Mediation through a direct serotonergic pathway. *Brain Res.* **120**, 85–102.

Wasterlain, C. G. (1974). Brain ribosomes in intracranial hypertension. *J. Neurochem.* **23**, 253–259.

Wasterlain, C. G. (1977) Effects of epileptic seizures on brain ribosomes: Mechanism and relationship to cerebral energy metabolism. *J. Neurochem.* **29**, 707–716.

Weiss, B. F., Munro, H. N., and Wurtman, R. J. (1971). L-dopa: disaggregation of brain polysomes and elevation of brain tryptophan. *Science* **173**, 838–835.

Weiss, B. F., Munro, H. N., Ordonez, L. A., and Wurtman, R. J. (1972). Dopamine: Mediator of brain polysome disaggregation after L-dopa. *Science* **177**, 613–616.

Weiss, B. F., Wurtman, R. J., and Munro, H. N. (1973). Disaggregation of brain polysomes by L-5-hydroxytryptophan: Mediation by serotonin. *Life Sci.* **13**, 411–416.

Weiss, B. F., Liebschutz, J., Wurtman, R. J., and Munro, H. N. (1975). Participation of dopamine- and serotonin-receptors in the disaggregation of brain polysomes by L-dopa and L-5-HTP. *J. Neurochem.* **24**, 1191–1195.

Wender, M., and Zgorzalwicz, B. (1976). The influence of prenatal X-irradiation on the activity of sRNA aminoacyl synthetases in the developing rabbit brain. *J. Neurochem.* **26**, 17–23.

Widelitz, M. M., Coryell, M. R., Widelitz, H., and Avadhani, N. G. (1975). Dissociation of rat brain polyribosomes *in vivo* by amphetamines. *Brain Res.* **100**, 215–220.

Widelitz, M. M., Coryell, M. R., Widelitz, H., and Avadhani, N. G. (1976). Effects of amphetamine administration *in vivo* on *in vitro* protein synthesizing system from rat brain. *J. Neurochem.* **27**, 471–475.

Widelitz, M. M., Coryell, M., and Avahani, N. G. (1977). Reversal of amphetamine-induced polysome dissociation by neuroleptic agents in rat brain. *Biochem. Biophys. Res. Commun.* **76**, 1223–1229.

Yamagani, S., and Mori, K. (1970). Changes in polysomes of the developing rat brain. *J. Neurochem.* **17**, 721–731.

Zomzely-Neurath, C. E. (1972). Cerebral protein-synthesizing systems. *Methods Mol. Biol.* **2**, 147–187.

Zomzely-Neurath, C. E., and Roberts, S. (1974). Brain Ribosomes. *In* "Research Methods in Neurochemistry" (N. Marks and R. Rodnight, eds.), Vol. 1, pp. 95–137. Plenum, New York.

Zomzely, C. E., Roberts, S., Pestka, S., and Brown D. M. (1971). Cerebral protein synthesis III. Developmental alterations in the stability of cerebral messenger ribonucleic acid-ribosome complexes. *J. Biol. Chem.* **246**, 2097–2103.

10

NEUROPEPTIDES AS PUTATIVE NEUROTRANSMITTERS: ENDORPHINS, SUBSTANCE P, CHOLECYSTOKININ, AND VASOACTIVE INTESTINAL POLYPEPTIDE

P. C. Emson and S. P. Hunt

I. INTRODUCTION

In recent years a number of biologically active peptides have been shown to be localized within neurones in the nervous system. These include peptides originally identified as hypothalamic releasing hormones or pituitary hormones [for example, thyrotropin-releasing hormone (TRH) and vasopressin], gut hormones (cholecystokinin), and peptides isolated as the result of quests for the endogenous ligands for the opiate receptor (enkephalins and endorphins). The number of these neuronally localized peptides is now between 20 and 30 (Table I) and continues to increase. The presence of many of these neuropeptides in storage granules in nerve terminals, apparently making synaptic contacts with other neurones, suggests that in addition to any hormonal role these neuropeptides may act as chemical messengers or neurotransmitters (Emson, 1979; Snyder, 1980). In support of a proposed transmitter role, electrophysiological studies indicate

Molecular Approaches to Neurobiology

Table I.   Some Neuronally Localized Peptides

1. Pituitary hormones
   Vasopressin
   Oxytocin
   Prolactin
   Adrenocorticotrophic hormone
      (corticotropin)
   Pro-opiocortin (corticotropin-
      β-lipotropin precursor)
   β-Melanotropin
   α-Melanotropin
   γ-Melanotropin

2. Hypothalamic Releasing Hormones
   Somatostatin
   Thyrotropin-releasing hormone
   Luteinizing hormone releasing hormone

3. Gastrointestinal Peptides
   Gastrin
   Cholecystokinin
   Bombesin
   Pancreatic polypeptide
   Vasoactive intestinal polypeptide

4. Endogenous Opioid Peptides
   Methionine-enkephalin
   Leucine-enkephalin
   β-Lipotropin
   α-Endorphin
   γ-Endorphin
   α-Neoendorphin
   Dynorphin.

5. Pancreatic hormones
   Glucagon
   Insulin

6. Other Peptides
   Bradykinin
   Carnosine
   Substance P
   Angiotensin
   Neurotensin

that many neuropeptides can produce a complex series of ionic and membrane changes on postsynaptic neurones (Barker, 1977, 1978).

It is perhaps simplest at this stage in our knowledge to regard all the neuropeptides listed in Table I as at least putative transmitters. This chapter will be concerned with the possible neurotransmitter roles of several neuronally localized peptides (methionine-enkephalin, substance P, vasoactive intestinal polypeptide, and cholecystokinin) in the mammalian nervous system.

II. ENDOGENOUS OPIOID PEPTIDES

The general term endorphin (endogenous morphine-like substances) refers to the group of peptides which may act as endogenous ligands for opiate receptors in the mammalian nervous system. The search for these endogenous ligands followed the biochemical characterization of stereospecific opiate-receptor binding sites in brain in the early 1970s (Goldstein et al., 1971; Simon et al., 1973; Terenius, 1973; Pert and Synder, 1974). Hughes et al. (1975) first isolated the two pentapeptides methionine- and leucine-enkephalin from pig brain and it was soon realized that the sequence of methionine enkephalin was included in the sequence of the carboxy-terminal fragment of a relatively obscure pituitary hormone β-lipotropin (Bradbury *et al.*, 1976). C fragment of β-lipotropin, or β-endorphin as it is more usually called, does not normally appear to serve as a precursor for methionine-enkephalin, as putative methionine and leucine-enkephalin precursors have been isolated from the adrenal medulla where the enkephalins coexist with catecholamines in the medullary cells (Kimura *et al.*, 1980). In the central nervous system (CNS), separate enkephalin and β-endorphin-containing neurons can be demonstrated by immunohistochemistry, and the absence of any change in brain concentration of enkephalin after hypophysectomy indicates that the enkephalins are not derived from the pituitary (Kobayashi *et al.*, 1978). The β-endorphin-containing neurons are localized in the area of the arcuate nucleus of the hypothalamus and project to a variety of midline structures including septum, periventricular thalamic nuclei, and periventricular gray (Rossier *et al.*, 1977; Watson *et al.*, 1978). Immunohistochemistry indicates that these neurones may also contain ACTH and α-MSH-like immunoreactivity. This would be expected as these peptides (and also β- and γ-MSH) are all contained within the β-lipotropin/ACTH precursor (pro-opiocortin) (Nakanishi *et al.*, 1979). Recent results indicate that these neurones also contain the N-terminal tryptophan rich fragment of pro-opiocortin (Häkanson *et al.*, 1980). It also seems likely that, as in the different lobes of the pituitary, the neurons may process and store different components of the pro-opiocortin sequence (Zakarian and Smyth, 1979). Adding to the complexity of the endorphin system, it has become clear that there are likely to be at least two and perhaps three categories of opiate receptor (μ, δ, and K) (Martin *et al.*, 1976; Lord *et al.*, 1977). One type of receptor (μ) has preferential affinity for morphine-like compounds, another (δ) has a higher affinity for the stable enkephalin analogue (D-Ala^2, D-Leu^5-enkephalin) (Lord *et al.*, 1977), and the third (K) has highest affinity for the ketocyclazocine family. Use of low concentrations of these compounds (0.5 nM) in radioligand binding experiments allows the selective labeling of these opiate receptor types (Table II) (Chang and Cuatrecasas, 1979; Chang *et al.*, 1979; Fields *et al.*, 1980). The interrelationships of these opiate receptors and their interactions with the various endogenous ligands are not yet clear. However, an interesting clue may be that the μ receptor (preferring morphine) is relatively enriched in areas of CNS such as the dorsal roots, dorsal horn of the spinal cord (Table II and III), and brain stem periventricular thalamus; areas associated with the processing of sensory information, including information relating to injury, whereas the δ receptor (preferring the stable enkephalin analogue) is relatively enriched in forebrain areas such as the hippocampus, corpus striatum, and cerebral cortex (Table II).

Studies of the distribution of Met-enkephalin by immunoassay or immunohistochemistry indicate that neurons and terminals containing this peptide are found in the dorsal horn of the spinal cord, trigeminal nuclei, brain stem, raphe nuclei, and midline thalamus, all sites at which enkephalin terminals could influence the

Table II.   Distribution of Opiate Binding Sites in Bovine Brain and Spinal Cord[a]

	Specific binding (fmoles/mg protein)		
	[³H]morphine (0.5 nM)	[³H]D-Ala², D-Leu⁵, enkephalin (0.5 nM)	Binding ratio
Frontal cortex	3.46 ± 0.33 (5)	6.85 ± 0.40 (5)	0.50
Hippocampus	3.66 ± 0.45 (5)	9.24 ± 0.47 (5)	0.40
Amygdala	9.00 ± 1.17 (5)	5.36 ± 0.57 (5)	1.68
Thalamus	7.12 ± 1.11 (5)	3.40 ± 0.17 (5)	2.10
Periaqueductal gray	8.28 ± 0.72 (5)	4.46 ± 0.22 (5)	1.85
Raphe nucleus	4.80 ± 0.47 (5)	2.70 ± 0.21 (5)	1.80
Spinal cord			
Dorsal horn (gray)	3.07 ± 0.24 (5)	1.7 ± 0.27 (5)	1.80
Ventral horn (gray)	1.92 ± 0.20 (5)	1.46 ± 0.09 (5)	1.31

[a] From M. Ninkovic and S. P. Hunt, unpublished; technique as described by Fields *et al.* (1980).

passage of sensory information (Hökfelt *et al.*, 1977; Sar *et al.*, 1978). One mechanism by which enkephalin neurones could influence the flow of sensory information (such as that relating to injury) would be by a filter mechanism involving presynaptic inhibition. Thus, an enkephalinergic axon synapsing on to a primary afferent fiber in the dorsal horn could inhibit or reduce the release of the primary afferent transmitter. Support for this idea comes from experiments using slices of the rat spinal trigeminal nucleus *in vitro* where opiates may suppress the release of substance P from the substantia gelatinosa (Jessell and Iversen, 1977), and from experiments which indicate that opiate receptors are located on primary afferent fibers (La Motte *et al.*, 1976; Hiller *et al.*, 1978; Fields *et al.*, 1980; Nagy *et al.*, 1980) (Table III) (Fig. 1). The best evidence that

Table III.   Reduction of μ and σ Binding in Dorsal Roots 1 Month following Sciatic Cut[a]

	Specific binding (fmoles/mg protein)	
	[³H]morphine (μ)	³H-D-Ala, D-Leu enkephalin (σ)
Normal control (Table II)	8.9 ± 0.45 (7)	2.7 ± 0.22 (6)
Sciatic cut	5.9 ± 1.18 (3)	1.3 ± 0.05 (3)
% Reduction	−34%	−51%

[a] From Fields *et al.*, 1980.

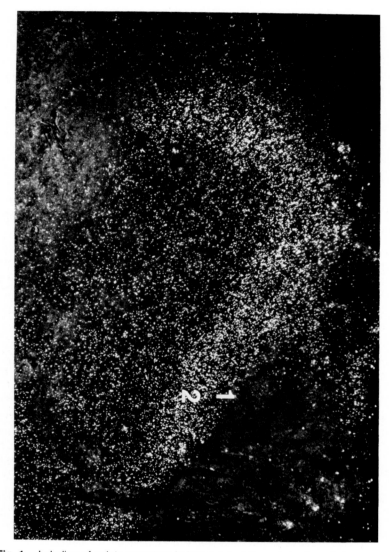

Fig. 1. Labeling of opiate receptors in layers I–II of the substantia gelatinosa of the rat with 2 n*M* [³H]etorphine (×120). Dark-field illumination (M. Ninkovic and S. P. Hunt, unpublished; technique as described by Young and Kuhar, 1979).

opiate receptors are located on the small diameter primary afferent fibers (C fibers or A δ fibers) comes from tissue culture experiments where opiate receptor binding sites can be localized autoradiographically to processes of the small dorsal root ganglionic neurons (Hiller *et al.*, 1978) and from studies of spinal cord opiate receptor content after capsaicin treatment in neonatal rats (Nagy *et al.*, 1980). Capsaicin is a powerful skin irritant, first isolated from the red pepper. When administered to neonatal rats it causes a permanent loss of up to 90% of small diameter fibers from the dorsal roots without any significant effect on the numbers of large myelinated fibers (Lawson and Nickels, 1980). Paralleling the selective loss of small-diameter fibers from the dorsal roots there is a significant reduction in opiate receptor binding in the dorsal horn (Nagy *et al.*, 1980), suggesting that the small-diameter afferent fibers may carry opiate receptors. However, although the localization of opiate receptors on primary afferents would support a presynaptic action of opiates to block pain transmission, this is unlikely to be the only site at which enkephalin influences pain transmission in the spinal cord. Thus, electron microscope studies of the localization of enkephalin terminals in the substantia gelatinosa (Fig. 2) indicate that the majority of enkephalin terminals are presynaptic to dendrites in gelatinosa and not involved in axoaxonic synapses with primary afferents (Hunt *et al.*, 1980). Thus, although the enkephalin neurons may exert presynaptic effects, they can also influence the activity of interneurons in the substantia gelatinosa.

The analgesic effects of systemic morphine appear, at least in part, to be mediated through activation of descending inhibitory serotoninergic and noradrenergic neurons located in the pons and medulla (Basbaum *et al.*, 1977; Fields and Anderson, 1978; Yaksh, 1979; Yaksh and Wilson, 1979; Dickenson, *et al.*, 1979; Ramana-Reddy and Yaksh; 1980). In view of the fact that the effects of brain stem stimulation are not inhibited by naloxone the serotonin neurons apparently synapse either on the dendrites of spinothalamic output neurons or on interneurones in layer II (Ruda and Gobel, 1980) but do not significantly activate the intrinsic substantia gelatinosa enkephalinergic neurones. Thus, removal of the descending serotoninergic projection using serotonin-directed neurotoxins (such as 5,6- or 5,7-dihydroxytryptamine) abolishes most, although not all, of the analgesic ef-

---

Fig. 2. Enkephalin in the substantia gelatinosa of the rat. (A) Enkephalin-like immunoreactivity in a nerve terminal synapsing (♦) upon a dendritic spine (d) (×43,000). (B) Morphologically similar enkephalin containing neurons in the substantia gelatinosa (×1250). Enkephalin-like immunoreactivity is demonstrated using the peroxidase–antiperoxidase (PAP) technique of Sternberger (1979). Fixed tissue slices are incubated with the primary antibody directed toward the antigen under investigation. This antibody is linked to a peroxidase–antiperoxidase antibody complex by a link antibody and produces an electron-dense reaction product when reacted with diaminobenzidine in hydrogen peroxidase. This technique can be used at both the light- and electron-microscopic level and was used for the material in Figs. 2–7.

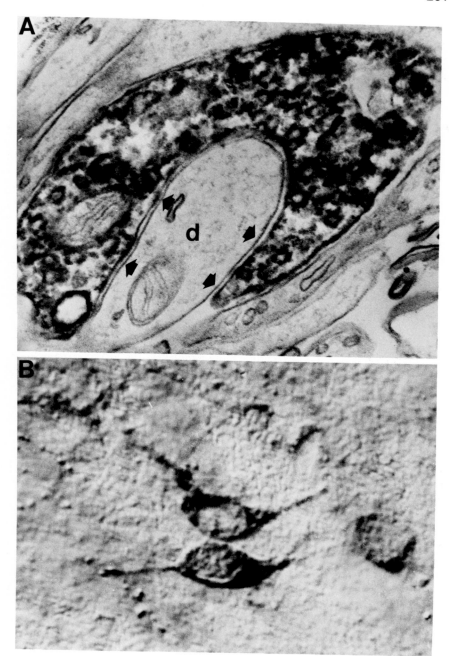

fects of systemically administered morphine or of brain stem stimulation (Fields and Anderson, 1978).

With the exception of the brain stem nuclei and the hippocampus, where enkephalins may have excitatory effects (Zieglgansberger and Fry, 1976), in most areas enkephalins and opiates exert an inhibitory effect on neuronal activity. This effect may be mediated via blockade of the sodium channel in the neuron as sodium ions strongly influence opiate receptor binding (Pert and Snyder, 1974). Another ion implicated in the effects of enkephalin is calcium. Thus in the cultured chick dorsal root ganglion opiates (and noradrenaline, serotonin, and GABA) suppress the calcium component of the action potential (Dunlap and Fischbach, 1978).

The localization of enkephalin immunoreactivity in nerve terminals synapsing directly with other neurons, and the physiological effects of applied enkephalin on mammalian neurons suggest that enkephalin may act as a neurotransmitter. Further evidence for such a role comes from experiments showing that enkephalin can be released *in vitro* from tissue slices (Henderson *et al.*, 1978; Iversen *et al.*, 1978); that enkephalin is localized in synaptic vesicles (Fig. 2); and that enkephalin is concentrated in synaptosomes (Osborne *et al.*, (1978) from which it can be released by depolarizing stimuli. The localization of enkephalins in adrenal medullary cells, from whence they can be released together with catecholamines (Viveros *et al.*, 1979), suggests that enkephalins may also have other as yet unclarified roles in physiology.

Apart from the substantia gelatinosa and periventricular areas, which may be associated with pain perception, the areas of mammalian brain richest in enkephalin content are the hypothalamus and globus pallidus where enkephalins may be involved in the regulation of both neurohypophyseal hormone release (Iversen *et al.*, 1980) and extrapyramidal motor functions. The distribution of enkephalin and ACTH/β-lipotropin-related peptides in the human brain closely parallels that in other mammals (Table IV). Detailed studies in our laboratory indicate that all the peptides listed in Table IV are relatively stable after death and that determinations of the postmortem content of neuropeptide in human brain can reflect the expected neuropathology (as in Huntington's disease, Table VI). Determination of the ratio of methionine to leucine-enkephalin immunoreactivity in human brain indicates that, as in the rat (Kobayashi *et al.*, 1978), leucine-enkephalin is relatively enriched in forebrain areas such as hippocampus, cerebral cortex, and amygdala and it has been suggested that this may reflect separate leucine-enkephalin containing neurons in the CNS (Larsson *et al.*, 1979). However, it should also be noted that one of the putative enkephalin precursors isolated from the adrenal medulla by Kimura *et al.*, (1980) contained both leucine- and methionine-enkephalin so that an alternative explanation of the histochemical and immunoassay data would be that some telencephalic neurons process the precursor containing both leucine- and methionine-enkephalin,

Table IV. Regional Distribution of Met-Enkephalin, and ACTH/LPH-Derived Peptides in Normal Human Brain[a]

	Met-enkephalin (pmoles/g)	N-ACTH (ng/g)	C-ACTH (ng/g)	N-LPH (ng/g)	C-LPH (ng/g)
Cortex Brodmann area 10	42 ± 17 (16)	4.0	10.5	2.0	10.0
Amygdala	26 ± 10 (10)	8.97	34.0	2.0	30.3
Hippocampus	56 ± 16 (3)	4.0	45.8	2.28	23.1
Ventral anterior thalamic nucleus	10 ± 9 (3)	4.0	3.6	2.8	29.1
Antero-medial thalamic nucleus	34 ± 6 (3)	16.2	40.0	11.3	47.3
Hypothalamus	141 ± 22 (18)	40.1	55.4	900	893
Superior collicus	55 ± 4 (3)	18.4	43.4	20	127
Periventricular grey	143 ± 46 (3)	10.8	41.4	193	159
Substantia nigra (pars reticulata)	641 ± 145 (15)	4.0	41.7	5.0	56.8

[a] Values for N- and C-terminal LPH-like immunoreactivity and N- and C-terminal ACTH-like immunoreactivity are preliminary and expressed as ng/g because the molecular species involved are not yet clarified. Note however, the predominance of C-terminal immunoreactivity (both C-ACTH and C-LPH) outside the hypothalamus. Preliminary data from G. M. Besser, P. J. Lowry, and L. H. Rees, using region specific radioimmunoassays directed toward ACTH or β-LPH.

whereas other neurons process additional precursors enriched in methionine-enkephalin.

The behavioral effects of intraventricular or intracerebral injections of β-endorphin in rats (Bloom *et al.*, 1976; Jacquet and Marks, 1976) aroused considerable interest. Intraventricular injections of β-endorphin in rats rendered the animals rigid and immobile, which was suggested to represent catalepsy or catatonia. Neither group reported that this type of behavioural effect was a well-established effect of classical opiate drugs, and they interpreted their data as suggesting that either overactivity (Bloom *et al.*, 1976) or underactivity (Jacquet and Marks, 1976) of the endorphin system may be involved in schizophrenia. These observations triggered a number of clinical investigations involving opiates and opiate antagonists in schizophrenia. Despite numerous clinical studies, one must conclude that at the present time there is no clear evidence to implicate endogenous opiates in either the etiopathology or treatment of mental illness (for a critical review of this literature see Mackay, 1979).

III. SUBSTANCE P

Substance P is an undecapeptide first described by von Euler and Gaddum (1931) as a vasodilatory peptidergic material in extracts of bovine brain and intestine. The isolation and purification of substance P was accomplished by Chang and Leeman (1970) and its sequence was established in 1971 (Chang *et al.*, 1971). Following the sequencing of substance P, the availability of the synthetic peptide permitted the development of sensitive radioimmunoassay and immunohistochemical methods for detection and localization of substance-P-like immunoreactivity. Such studies reveal that in mammals substance P is widely distributed in central and peripheral neurons and in endocrine cells in the gut. The neuronal localization of substance P, its calcium-dependent release, and its concentration in nerve terminals synapsing onto other neurons (Fig. 3) suggests that substance P should be considered as a neurotransmitter candidate. Indeed the evidence for a neurotransmitter role for substance P is extensive (see review by Nicoll *et al.*, 1980). Further support for a neurotransmitter role for substance P comes from the recent demonstration of substance P biosynthesis in dorsal root ganglia neurons (Harmar *et al.*, 1980). It only remains to show that the effects of applied substance P and of stimulation of the presynaptic pathway share the same pharmacology and produce the same ionic events in the postsynaptic cell to completely fulfill Dales' criteria for a neurotransmitter (Dale, 1935). However, the demonstration of pharmacological identity will require the availability of a suitable substance P antagonist, and this is not yet available. Such a substance P antagonist would be of considerable interest, as one of the important roles for substance P seems to be as a transmitter in some of the small-diameter primary afferent fibers involved in the transmission of noxious information into the CNS. A substance P antagonist might, therefore, represent a novel type of analgesic.

Lembeck (1953) first suggested that substance P might be the transmitter of primary afferent fibers terminating in the mammalian spinal cord. Subsequent immunohistochemical studies showed that substance P-containing neurons are present in the dorsal root ganglia (Hökfelt *et al.*, 1975, 1976, 1977) and that the distribution of the substance-P-containing processes of these neurons corresponded to the sites of termination of small diameter (A δ and C) fibers in the substantia gelatinosa (Barber *et al.*, 1979) (Fig. 3). Substance P is also found in the peripheral endings of primary afferents in the skin, tongue, and tooth pulp (Cuello *et al.*, 1978; Lundberg *et al.*, 1979a). Recent work suggests that substance P can be released following antidromic stimulation of peripheral nerves (Olgarth *et al.*, 1977; Bill *et al.*, 1979; Lembeck and Holzer, 1980). This peripheral release of substance P would lead to vasodilation and, indirectly, to histamine release, probably accounting for the inflammatory response seen in the skin area innervated by the relevant pain fibers (Lembeck and Holzer, 1980).

Fig. 3. Substance P in the substantia gelatinosa. (A) Substance P-like immunoreactivity in a nerve terminal synapsing on a large dendrite (d). Pigeon substantia gelatinosa (× 61,000). (B) Substance P-containing neurons in the substantia gelatinosa of the rat (layer III). Note the innervation of the cell by substance P-containing fibers (←) (× 1000).

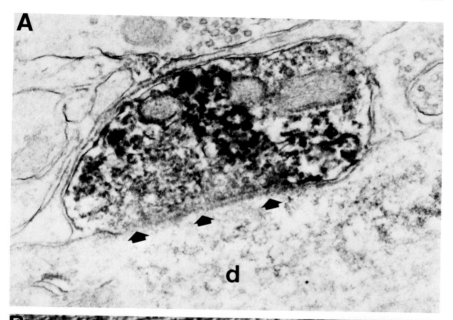

Thus rats treated neonatally with capsaicin, a drug which produces degeneration of small-diameter afferent neurons (including those containing substance P), do not show the characteristic inflammatory axon reflex response. Capsaicin was originally suggested as a selective neurotoxin for substance-P-containing primary afferent neurons. However, it is now known from histological studies of the dorsal root ganglia (Lawson and Nickels, 1980) that capsaicin damages many small-diameter sensory fibers, leading to parallel depletions not only of both substance P and somatostatin but also other peptides which may be located in small-diameter afferent fibers. Capsaicin seems to exert part of its neurotoxic effect by causing a calcium-dependent release of tissue peptide stores, resulting in a massive outflow of peptides from the central and peripheral terminals of the peptide-containing primary afferent neurones (Gamse *et al.*, 1979a). The effects of capsaicin seem to be confined to primary afferent neurons, as it does not produce any depletion of the substance P content of the gut or CNS (Gamse *et al.*, 1979b). In agreement with this observation, capsaicin did not release substance P from hypothalamic slices *in vitro* but did stimulate substance P release from slices of spinal cord (Gamse *et al.*, 1979a). In the adult rat the effects of capsaicin are slowly reversible, but in the infant rat virtually all small-diameter primary afferent fibers are irreversibly lost, leading to thermal and chemical analgesia (Nagy *et al.*, 1980). The extent of depletion of substance P from the dorsal horn after neonatal capsaicin treatment represents 45–50% of the total dorsal horn substance P, which agrees well with the reduction in substance P content produced in this region after dorsal rhizotomy (J. S. Kelly, R. F. T. Gilbert and P. C. Emson, unpublished).

Apart from the peptide present in primary afferent terminals (some 50% of dorsal horn content), the rat spinal cord also contains intrinsic substance-P-containing neurons (containing some 25% of dorsal horn substance P content) and there is also a descending substance P-containing projection, which innervates both dorsal and ventral horns. In the ventral horn the descending projection accounts for more than 90% of the total substance P content, whereas in the dorsal horn the descending projection accounts for about 25% of the total substance P content. Immunohistochemical studies using substance P or 5-hydroxytryptamine (5-HT) directed antibodies have revealed that some medullary raphe neurones stain for both substance P and 5-HT; demonstrating that substance P and 5-HT immunoreactivity can coexist in these neurons (Hökfelt *et al.*, 1978). The axons of these neurons can be damaged by the serotonin-directed neurotoxins 5,6- and 5,7-dihydroxytryptamine (Hökfelt *et al.*, 1978; Emson and Gilbert, 1980), and study of the terminal fields of these neurons suggest that the same nerve terminals may contain both substance P and 5-HT (Gilbert *et al.*, 1980; Hökfelt *et al.*, 1978; Chan-Palay *et al.*, 1978) (Fig. 4). Although the coexistence of substance P and 5-HT immunoreactivity in nerve terminals has not yet been demonstrated at the electron microscope level, the parallel depletion of both substance P and 5-HT by serotonin neurotoxins (5,7- and 5,7-dihydroxytryptamine), and by monoamine-depleting drugs such as reserpine (Gilbert et al., 1980), suggests strongly that substance P and 5-HT are likely to be stored together (Table V). This is already known to occur in carcinoid tumors (Alumets et al., 1977). To complicate this story further, additional immunohistochemical studies have shown that some medullary raphe neurons (raphe pallidus) projecting to the ventral spinal cord also contain TRH as well as substance P and 5-HT immunoreactivity (Johansson et al., 1980) although it is not yet clear if TRH, 5-HT, and substance P all coexist in the same terminals. However, immunohistochemistry has shown that substance P, TRH, and 5-HT immunoreactivities coexist in the cell bodies at least, and as with substance P and 5-HT, TRH immunoreactivity is depleted by the serotonin-directed neurotoxins and by such monoamine-depleting drugs as reserpine and tetrabenazine (Gilbert et al., 1981).

Gel chromatography and high pressure liquid chromatography (HPLC) analyses of tissue extracts indicates that the substance P in ventral spinal cord is present as the native undecapeptide. This conclusion is supported by studies using antibodies directed against either end of the substance P sequence (amino- or carboxy-terminal directed antisera; Lee et al., 1980); such antibodies detect equal amounts of substance P in the ventral horn extracts (R. F. T. Gilbert, unpublished). This would not be the case if the substance P immunoreactivity coexisting with 5-HT was merely a fragment of substance P or a cross-reacting sequence present in some material of higher molecular weight. The possible significance of three putative transmitters apparently coexisting in the same terminals, or even in the same synaptic vesicle remains to be elucidated. However, it is noteworthy that substance P and TRH applied separately to ventral cord motor neurons produce only small excitatory effects (Belcher and Ryall, 1977; Nicoll, 1978). It may be that when released together with 5-HT these weakly excitatory compounds act synergistically, resulting in a strong excitatory effect. An alternative possibility is that 5-HT is the primary neurotransmitter, with the two peptides (substance P and TRH) exerting longer-term trophic or hormonal effects.

Apart from the interesting possible coexistence of substance P with 5-HT and TRH in the descending raphe projections, substance P is not known to coexist with other monoamines. Substance P-containing neurones (not containing 5-HT) are found in a number of basal ganglia and limbic forebrain areas including the bed nucleus of the stria terminalis, the medial amygdala, the medial habenula, and the nucleus accumbens (Cuello and Kanazawa, 1978; Ljungdahl et al., 1978a,b). The axons of these substance P-containing neurons project via the major forebrain fiber pathways such as the stria medullaris, stria terminalis, and fasiculus retroflexus to interconnect these areas (Emson, 1979). The close association of many neuropeptides with limbic forebrain areas such as the bed nucleus of the stria terminalis, the habenula, and amygdala is unexplained.

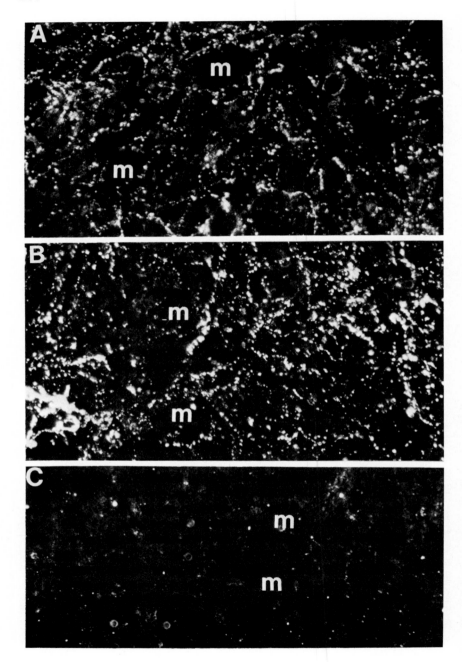

Table V. Spinal Cord Substance P and [³H]5-Hydroxytryptamine Uptake 2 Weeks after 5,7-Dihydroxytryptamine (200 μg free base icv)[a]

| | | SP content (pmoles/g wet weight) | | ³H-5-HT uptake (pmoles/5 min/mg protein) | |
		Vehicle	5,7-DHT	Vehicle	5,7-DHT
Cervical cord	Dorsal	152 ± 9 (11)	119 ± 10* (11) (−22%)		
				2.17 ± 0.19 (6)	0.20 ± 0.08*** (4) (−91%)
	Ventral	22 ± 6 (8)	6 ± 3* (9) (−73%)		
Lumbar cord	Dorsal	179 ± 13 (10)	137 ± 10** (10) (−23%)		
				2.45 ± 0.19 (5)	0.12 ± 0.03*** (4) (−95%)
	Ventral	32 ± 3 (7)	5 ±2*** (9) (−84%)		

[a] *$p<0.05$; **$p<0.02$; ***$p<0.001$ by students t-test (two-tailed) compared to vehicle injected animals. (R. Gilbert, unpublished. Substance P content determined using a C-terminal directed substance P antibody as described by Lee et al., 1980.)

Although the physiological role of most CNS substance P neurons and pathways is unknown, the presence of substance P in some retinal amacrine interneurons does suggest a role integrating retinal output. Degeneration of substance P neurons in the caudate nucleus and putamen occurs in Huntington's chorea, a human degenerative disease of the basal ganglia associated with loss of at least 50% of all neurons in the corpus striatum (Kanazawa et al., 1977; Emson et al., 1980b).

The loss of striatal substance P-containing neurons in Huntington's chorea is reflected in a substantial depletion of the substance P content of substantia nigra

Fig. 4. Substance P- and 5-hydroxytryptamine (5-HT)-like immunoreactivity in sacral ventral horn of the rat spinal cord. (A) 5-HT-like immunoreactivity (5-HT directed antibody courtesy of Dr H. Steinbusch). (B) Substance P-like immunoreactivity. (C) Disappearance of substance P-like immunoreactivity after intraventricular injection of 200 μg of 5,7-dihydroxytryptamine a 5-HT-directed neurotoxin. Dark-field illumination. m, motor neuron (×600) (R. F. T. Gilbert, unpublished).

(Table VI) the brain area to which the caudate-putamen substance-P neurons project. Treatment of rats with the neuroleptic drug haloperidol, which blocks the activity of dopamine at its receptors in the caudate nucleus and putamen, results in a reduction in the substance P content of the substantia nigra (Hong *et al.*, 1978). This probably reflects increased activity of an excitatory feedback loop, containing substance P, which innervates the dopaminergic neurons in the substantia nigra.

Studies of the actions of substance P on cellular receptors have mostly been carried out on peripheral tissues, especially on such smooth muscle preparations such as the guinea pig ileum (see Bury and Mashford, 1977; Skrabanek and Powell, 1977; Hanley and Iversen, 1980, for complete data). These studies indicate that carboxy-terminal fragments of substance P, extending down to the hexapeptide (SP_{6-11}), retain full biological activity and potency on the guinea pig ileum. However, the pentapeptide (SP_{7-11}) is some several hundredfold less potent than the undecapeptide. These observations indicate that it is primarily the carboxy-terminal hexapeptide sequence which is important for receptor recognition in the ileum. Modifications of the N-terminal sequence can be made with little effect on biological activity. However, alterations to the carboxy-terminal sequence, in particular removal of the methionine carboxy-terminal amide, substantially reduce potency on the ileum (Bury and Mashford, 1974, 1977). The preservation of biological activity in the hexapeptide (SP_{6-11}) led Bury and Mashford (1977) to suggest that this fragment of substance P might be the biologically active "transmitter," with the undecapeptide acting as a precursor. There is, however, no evidence from HPLC of rat brain extracts to suggest the existence of separate C-terminal fragments of substance P (Lee *et al.*, 1980). The use of N- and C-terminal-directed substance P antisera indicates that most of the immunoreactive substance P in rat brain, in human globus pallidus, and in plasma from patients with carcinoid tumors is present as the undecapeptide (Lee *et al.*, 1980; Emson *et al.*, 1980a) (Fig. 5). This evidence is further supported by degradation studies which indicate that the undecapeptide is stabilized to amino- and carboxypeptidase activity by the methionamide at position 11 and by the proline at position 2 (Lee *et al.*, 1980). Thus, exposure of substance P to brain membrane or cytosol enzymes results in endopeptidase attack and the resulting fragments do not retain the biologically active carboxyterminal hexapeptide sequence intact (Lee *et al.*, 1980). (Although alternative degradative enzymes may produce biologically active hexa- or heptapeptide C fragments of substance P these have not yet been detected in the rat CNS by HPLC separations.) This type of study of the mechanisms of substance P degradation together with the recent development of a radioligand binding assay for substance P using [^{3}H]substance-P (Hanley *et al.*, 1980) should greatly facilitate development of stable analogues and possibly antagonists of substance P.

Table VI. Distribution of Substance P and Met-Enkephalin in the Basal Ganglia of Normal Human Brain and in the Brains of Patients with Huntington's Disease[a]

	Substance P		Methionine-enkephalin	
	Control	Huntington's disease	Control	Huntington's disease
Caudate	138 ± 14 (13)	158 ± 28 (10)	116 ± 40 (10)	156 ± 54 (10)
Putamen	112 ± 29 (10)	77 ± 20 (10)	200 ± 71 (12)	184 ± 68 (12)
Globus pallidus (lateral segment)	197 ± 99 (10)	18 ± 9* (10)	1163 ± 216 (40)	527 ± 102 (39)
Globus pallidus (medial segment)	877 ± 253 (10)	178 ± 102* (10)	675 ± 168 (40)	317 ± 48 (40)
Substantia nigra (pars compacta)	1264 ± 239 (27)	158 ± 41* (24)	577 ± 103 (15)	229 ± 71 (15)
Substantia nigra (pars reticulata)	1535 ± 177 (30)	86 ± 25* (25)	661 ± 145 (15)	230 ± 54 (16)

[a] Each value is expressed as pmoles/g wet weight and is mean ± SEM of the number of determinations in parentheses. Substance P and methionine-enkephalin contents determined using sensitive radioimmunoassays as described in detail by Emson et al., 1980b.

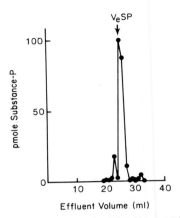

Fig. 5. High pressure liquid chromatography (HPLC) separation of SP-like immunoreactivity from human globus pallidus. Note the major peak of SP-like immunoreactivity eluting at the expected position (V_e) for authentic substance P. Separate determinations using N- and C-terminal-directed antibodies indicated that all the immunoreactivity corresponded to the undecapeptide SP.

IV. CHOLECYSTOKININ AND VASOACTIVE INTESTINAL POLYPEPTIDE

Both these peptides were originally believed to be solely gut hormones (Mutt and Jorpes, 1971). However, the discovery of substantial amounts of vasoactive intestinal polypeptide (VIP)- and cholecystokinin (CCK)-immunoreactivity in the whole brains and cerebral cortices of several mammals, together with immunohistochemical localization of both peptides in neurons and terminals, raised the possibility that both peptides may be neurotransmitters as well as hormones (Vanderhaeghen *et al.*, 1975; Dockray, 1976; Muller *et al.*, 1977; Fuxe *et al.*, 1977). Cholecystokinin in particular is widely distributed in the brains of mammals, and in terms of molar concentration it is probably the most abundant CNS neuropeptide. CCK-containing neurons are localized in greatest densities in the cerebral cortex, amygdala, hypothalamus (Muller *et al.*, 1977; Lorèn *et al.*, 1979a; Larsson and Rehfeld, 1979), and in the mesencephalon and brain stem (Vanderhaeghen *et al.*, 1980; Hökfelt *et al.*, 1980a,b) (Fig. 6). CCK is concentrated in synaptosome fractions prepared from the rat hypothalamus and cerebral

Fig. 6. (A) Cholecystokinin-/gastrin-like immunoreactivity in layers I, II, and V and the sympathetic intermedio-lateral column (←) of the rat caudal thoracic cord (×140). Dark-field illumination. (B) Cholecystokinin-like immunoreactivity in a neuron (←) of the stratum radiatum of the rat hippocampus (×1000).

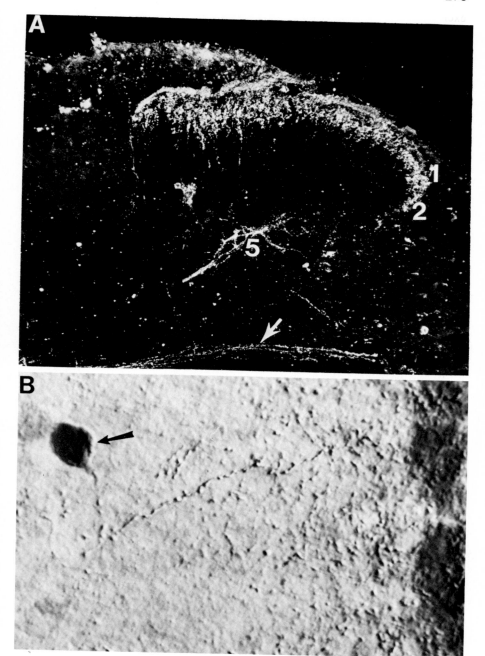

cortex, with the majority of immunoreactivity apparently corresponding to the carboxy-terminal octapeptide sequence (CCK-8) (Pinget *et al.*, 1978; Emson *et al.*, 1980). CCK-8 can be released by a calcium-dependent mechanism from rat cerebral cortex slices perfused *in vitro* (Emson *et al.*, 1980) and has potent excitatory effects when applied to hippocampal neurones *in vitro* (Dodd and Kelly, 1979). Further experiments have shown that CCK-like peptides can be synthesized rapidly, and that the biosynthesis involves processing via a high molecular weight precursor form of CCK to the probable final product, the octapeptide CCK-8 (Golterman *et al.*, 1980).

Vasoactive intestinal polypeptide, like CCK-8, is concentrated in neurons and nerve terminals in mammalian cerebral cortex (Fuxe *et al.*, 1977; Lorèn *et al.*, 1979b) from which it can be released by depolarizing stimuli in a calcium-dependent manner (Giachetti *et al.*, 1977; Emson *et al.*, 1978) (Fig. 7). The peptide which is stored and released corresponds to the native 28 amino acid VIP, and there is as yet no evidence for the formation of biologically active fragments of VIP (Fahrenkrug, 1979). VIP, like CCK, is excitatory when applied to cortical or hippocampal neurons (Phillis *et al.*, 1978; Dodd *et al.*, 1979). VIP, unlike CCK-8, stimulates cAMP formation in rat brain slices (Deschodt-Lanckman *et al.*, 1977; Quik *et al.*, 1978). Receptor binding for both VIP and CCK has been reported, using the radioiodinated peptides, with an enrichment of binding sites in forebrain areas (Taylor and Pert, 1979; Saito *et al.*, 1980).

The neuronal localization, concentration in nerve terminals, and calcium-dependent depolarization-induced release of both VIP and CCK-8 suggests that they can be considered as neurotransmitter candidates. The most interesting recent developments in the study of the physiological roles of VIP and CCK come from immunohistochemical studies which have shown that VIP and CCK may coexist with monoamines in certain neurons in the central and peripheral nervous system (Vanderhaeghen *et al.*, 1980; Hökfelt *et al.*, 1980a,b). Thus, VIP immunoreactivity is found in neurons which are also AChE positive in a number of sympathetic and parasympathetic ganglia (Lundberg *et al.*, 1979b, 1980). Although the presence of acetylcholinesterase (AChE) is not an adequate marker for a cholinergic neuron, it seems that in the sphenopalatine ganglion of the cat (in which essentially all neurons are VIP- and AChE-positive) there are parallel changes in the VIP and choline acetyltransferase (ChAT) content (the ACh synthetic enzyme) of the nasal mucosa (innervated by the sphenopalatine ganglion) after denervation (Lundberg *et al.*, 1981). This suggests that both the VIP immunoreactivity and the capacity to synthesize ACh are localized in the

---

Fig. 7. (A) Vasoactive intestinal polypeptide-like immunoreactivity in bipolar neurones of the rat parietal cortex (×300) (dendrites ←). (B) Accumulation of vasoactive intestinal polypeptide-like immunoreactivity below a knife lesion in the medial forebrain bundle (×550). (P. D. Marley, unpublished.)

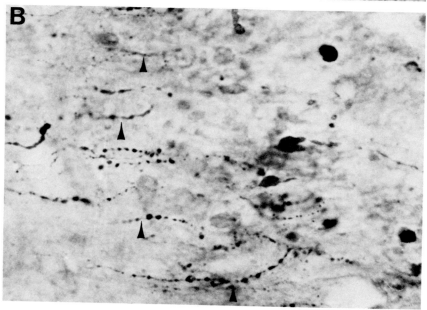

same terminals. Parallel immunohistochemical studies show that VIP- and AChE-positive terminals degenerate in parallel with the depletion of VIP and ChAT content of the innervated mucosa. Physiological studies of the cat salivary gland, which is innervated by VIP- and AChE-positive neurons, indicate that VIP and ACh may act synergistically. Infusions of low doses of VIP and ACh resulted in a marked potentiation of the atropine-resistant vasodilatation (VIP) and atropine-sensitive secretory (ACh) responses normally elicited by low doses of VIP and ACh applied separately (Lundberg *et al.*, 1980a). It is not yet known if both VIP and ACh are stored and released together, but preganglionic nerve stimulation simultaneously activates both vasodilation and secretory responses, and the ganglion blocker hexamethionium abolishes both secretory and vasodilatory responses (Lundberg *et al.*, 1980a). These results suggest that both the peptide (VIP) and the amine (ACh) are released simultaneously, as occurs with catecholamines and enkephalins from the adrenal medulla (Viveros *et al.*, 1979).

The localization of a number of CCK-positive neurons in the mesencephalon, in the regions of the dopamine-containing neurons (A10 of Dahlstrom and Fuxe, 1964) suggests that CCK immunoreactivity may coexist with dopamine in these neurons (Vanderhaeghen *et al.*, 1980; Hökfelt *et al.*, 1980a,b). Similarly, a number of neurons in the magnocellular supraoptic and paraventricular nuclei contain CCK-like immunoreactivity, suggesting a possible coexistence of CCK or gastrin with the neurohypophyseal hormones vasopressin and oxytocin. The distribution of the stained neurons indicates that such coexistence most probably occurs with oxytocin (Vanderhaeghen *et al.*, 1980). It is not yet clear if the immunoreactive material visualized in the magnocellular neurons in the hypothalamus is gastrin or CCK-like, and this will require suitable chromatographic separations (Rehfeld, 1978). However, in the rat mesencephalon and corpus striatum the majority of the immunoreactive material is likely to be present as CCK-8 (Dockray 1980, Emson *et al.*, 1980a). Lesions which interrupt the medial forebrain bundle lead to accumulation of CCK-positive material in ascending axons and 6-hydroxydopamine lesions in the ventral tegmental area deplete mesolimbic areas (such as the olfactory tubercle and nucleus accumbens) of both dopamine and CCK-like immunoreactivity (Hökfelt *et al.*, 1980a,b). It is of particular interest that it is primarily the so-called mesolimbic dopaminergic neurons that contain CCK-like immunoreactivity.

This provides a clear distinction between the nigro-striatal neurons whose loss leads to the clinical syndrome of Parkinson's disease and mesolimbic dopamineergic neurons whose hyperactivity may be involved in the etiology of schizophrenia.

V. CONCLUSION

It has not been possible to discuss in detail the distribution and physiology of all of the neuronally localized peptides listed in Table I. The peptides discussed, however, provide a representative sample of the types of problems and potentials for research in neuropeptides. The neuronal localization and synaptic concentration of many neuropeptides suggests that they have physiological roles as neurotransmitters. To satisfy completely the criteria that should be fulfilled by a putative CNS neurotransmitter will require the further development of peptide pharmacology (i.e., suitable agonists and antagonists). Apart from the development of such pharmacology it will also be of considerable interest to see how ubiquitous is the coexistence of monoamines and neuropeptides. It may be that, as predicted by Pearse's APUD theory, this is always the case (Pearse, 1979). The only certainty is that neuropeptides will continue to fascinate for the foreseeable future.

ACKNOWLEDGMENTS

We are grateful to Drs. L. L. Iversen and A. V. P. Mackay for reading and criticizing this text and to Mrs. J. Ditheridge and M. Wynn for superb secretarial assistance. We thank Drs. G. W. Bennett, G. M. Besser, A. Björklund, V. Clement-Jones, J. Fahrenkrug, R. Gilbert, T. Hökfelt, C. M. Lee, J. M. Lundberg, P. Lowry, C. A. Marsden, L. Rees, J. F. Rehfeld, M. Rossor, B. Sandberg, and H. Steinbusch for their collaboration and support. Some of the work described was supported by twinning grants from the European Training Programme in Brain and Behaviour Research.

REFERENCES

Alumets, J., Hakanson, R., Ingemansson, S., and Sundler, F. (1977). Substance P and 5-HT in granules isolated from an intestinal argentaffin carcinoid. *Histochemistry* **52,** 217–222.

Barber, R. P., Vaughn, J. E., Slemmon, J. R., Salvaterva, P. M., Roberts, E., and Leeman, S. E. (1979). The origin, distribution and synaptic relationships of substance P axons in rat spinal cord. *J. Comp. Neurol.* **184,** 331–351.

Barker, J. L. (1977). Physiological roles of peptides in the nervous system. *In* "Peptides in Neurobiology" (H. Gainer, ed.), pp. 295–343. Plenum, New York.

Barker, J. L. (1978). Evidence for diverse cellular roles of peptides in neuronal function. *In* "Peptides and Behaviour: A Critical Analysis of Research Strategies" *Neurosci. Res. Prog. Bull.* **16**(4) 535–553.

Basbaum, A.-J., Marley, N. J. E., O'Keefe, J., and Clanton, C. H. (1977). Reversal of morphine and stimulus produced analgesia by subtotal spinal cord lesions. *Pain* **3,** 43–56.

Belcher, G., and Ryall, R. W. (1977). Substance P and Renshaw cells: A new concept in inhibitory synaptic interactions. *J. Physiol. (London)* **272,** 105–109.

Besson, J. M., Chitow, D., Dickenson, A. H., and Le Bars, D. (1980). Involvement of endogenous opiates in diffuse noxious inhibitory controls. *J. Physiol. (London)* **301,** 26.

Bill, A., Stjernschantz, J., Mandahl, A., Brodin, E., Nilsson, G. (1979). Substance P: release on trigeminal nerve stimulation, effects in the eye. *Acta Physiol. Scand.* **106,** 371–373.

Bloom, F., Segal, D., Ling, N., and Guillemin, R. (1976). Endorphins: profound behavioural effects in rats suggest new aetiological factors in mental illness. *Science (Washington, D.C.)* **194,** 630–632.

Bradbury, A. F., Smyth, D. G., Snell, C. R., Birdsall, N. J. M., and Hulme, E. C. (1976). C. fragment of lipotropin has a high affinity for brain opiate receptors. *Nature (London)* **260,** 793–795.

Bury, R. W., and Mashford, M. L. (1974). Biological activity of the C-terminal partial sequence of substance P. *J. Med. Chem.* **19,** 854–856.

Bury, R. W., and Mashford, M. L. (1977). Substance P. Its pharamacological and physiological roles. *Aust. J. Biol. Sci.* **55**, 671–735.

Chang, K-J., and Cuatrecasas, P. (1979). Multiple opiate receptors: Enkephalins and morphine bind to receptors of different specificity. *J. Biol. Chem.* **254**, 2610–2618.

Chang, K-J., Cooper, B. R., Hazum, E., and Cuatrecasas, P. (1979). Multiple opiate receptors: Different regional distribution in the brain and differential binding of opiates and opioid peptides. *Mol. Pharmacol.* **16**, 91–104.

Chang, M. M., and Leeman, S. (1970). Isolation of a sialogogic peptide from bovine hypothalamic tissue and its characterization as substance P. *J. Biol. Chem.* **245**, 4784–4790.

Chang, M. M., Leeman, S. E., and Niall, H. D. (1971). Amino acid sequence of substance P. *Nature (London) (New Biol)* **237**, 86–87.

Chan-Palay, V., Jonsson, G. and Palay, S. L. (1978). On the co-existence of serotonin and substance P in neurons of the rat's central nervous system. *Proc. Natl. Acad. Sci. U.S.A.* **75**, 1582–1586.

Cuello, A. C., and Kanazawa, I. (1978). The distribution of substance P immunoreactivity in the rat central nervous system. *J. Comp. Neurol.* **178**, 129–156.

Cuello, A. C., Del Fiacco, M., and Paxinos, G. (1978). The central and peripheral ends of the substance P-containing sensory neurones in the rat trigeminal system. *Brain Res.* **152**, 499–509.

Dahlstrom, A., and Fuxe, K. (1964). Evidence for the existence of monoamine-containing neurones in the central nervous system. I Demonstration of monoamines in the cell bodies of brain stem neurones. *Acta Physiol. Scand.* **62** (Suppl. 232).

Dale, H. H. (1935). Pharmacology and nerve endings. *Proc. R. Soc. Lond.* **28**, 319–332.

Deschodt-Lanckman, M., Robberecht, P., and Christophe, J. (1977). Characterisation of VIP sensitive adenylate cyclase in guinea pig brain. *FEBS Lett.* **83**, 76–80.

Dickenson, A. H., Oliveras, J-L., and Besson, J-M. (1979). Role of the nucleus raphe magnus in opiate analgesia as studied by the microinjection technique in the rat. *Brain Res.* **170**, 95–111.

Dockray, G. J. (1976). Immunochemical evidence of cholecystokinin-like peptide in brain. *Nature (London)* **264**, 568–570.

Dockray, G. J. (1980). Cholecystokinin in rat cerebral cortex: Identification, purification and characterization by immunochemical methods. *Brain Res.* **188**, 155–165.

Dodd, J., and Kelly, J. S. (1979). Excitation of CA1 pyramidal neurones of the hippocampus by the tetra- and octapeptide C-terminal fragments of cholecystokinin. *J. Physiol.* **295**, 61.

Dodd, J., Kelly, J. S., and Said, S. (1979). Excitation of CA1 neurones of the rat hippocampus by the octasopeptide peptide vasoactive intestinal polypeptide (VIP). *Br. J. Pharmacol.* **66**, 125–126.

Dunlap, K., and Fischback, G. D. (1978). Neurotransmitters decrease the calcium component of sensory neurone action potentials. *Nature (London)* **276**, 837–839.

Emson, P. C. (1979). Peptides as neurotransmitter candidates in the mammalian central nervous system. *Prog. Neurobiol. (Oxford)* **13**, 61–116.

Emson, P. C., and Gilbert, R. F. T. (1980). Time course of degeneration of bulbo-spinal 5-HT/SP neurones after 5,7 dihydroxytryptamine. *Br. J. Pharmacol.,* **69**, 279–280.

Emson, P. C., Fahrenkrug, J., Schaffalitzky de Muckadell, O. B., Jessell, T. M., and Iversen, L. L. (1978). Vasoactive intestinal polypeptide (VIP): Vesicular localization and potassium evoked release from rat hypothalamus. *Brain Res.* **140**, 174–178.

Emson, P. C., Lee, C. M., and Rehfeld, J. (1980a). Cholecystokinin-like peptides: Vesicular localization and calcium-dependent release from rat brain *in vitro. Life Sci.,* **26**, 2157–2163.

Emson, P. C., Arregui, A., Clement-Jones, V., Sandberg, B. E. B., and Rossor, M. (1980b). Regional distribution of methionine-enkephalin and substance P-like immunoreactivity in normal human brain and in Huntington's disease. *Brain Res.,* **199**, 147–160.

Euler, U. S. V., and Gaddum, J. H. (1931). An unidentified depressor substance in certain tissue extracts. *J. Physiol.* **72**, 74–87.

Fahrenkrug, J. (1979). Vasoactive intestinal polypeptide: Radioimmunochemical studies on its distribution and function as a putative neurotransmitter. *Digestion* **19**, 149–169.

Fields, H. L., and Anderson, S. D. (1978). Evidence that raphe-spinal neurons mediate opiate and mid-brain stimulation-produced analgesias. *Pain* **5**, 333–349.

Fields, H. L., Emson, P. C., Leigh, B. K., Gilbert, R. F. T., and Iversen, L. L. (1980). Multiple opiate receptor sites on primary afferent fibres. *Nature (London)* **284**, 351–352.

Fuxe, K., Hökfelt, T., Said, S. I., and Mutt, V. (1977). Vasoactive intestinal polypeptide and the nervous system: Immunohistochemical evidence for localization in central and peripheral neurons, particularly intracortical neurons of the cerebral cortex. *Neurosci. Lett.* **5**, 241–246.

Gamse, R., Molnar, A., and Lembeck, F. (1979a). Substance P release from spinal cord slices by capsaicin. *Life Sci.* **25**, 629–636.

Gamse, R., Holzer, P., and Lembeck, F. (1979b). Decrease of substance P in primary afferent neurones and impairment of neurogenic plasma extravasation by capsaicin. *Br. J. Pharmacol.* **68**, 207–213.

Giachetti, A., Said, S. I., Rolland, C. R., and Koniges, F. C. (1977). Vasoactive intestinal polypeptide in brain: Localization in and release from isolated nerve terminals. *Proc. Natl. Acad. Sci. U.S.A.* **74**, 3424–3427.

Gilbert, R. F. T., Emson, P. C., Bennett, G. W., and Marsden, C. A. (1981). Effects of monoamine depleting drugs on neuropeptides in the ventral horn of the spinal cord of the rat. *Eur. J. Pharmacol.* (in press).

Goldstein, A., Lowney, L. I., and Pal, B. K. (1971). Stereospecific and nonspecific interactions of the morphine congener levorphanol in subcellular fractions of mouse brain. *Proc. Natl. Acad. Sci. U.S.A.* **68**, 1742–1747.

Golterman, N. R., Rehfeld, J. F., and Roigaard-Petersen, H. (1980). In vivo biosynthesis of cholecystokinin in rat cortex. *J. Biol. Chem.* **255**, 6181–6185.

Håkanson, R., Ekman, R., Sundler, F. and Nilsson, R. (1980). A novel fragment of the corticotropin/β-Lipotropin precursor. *Nature (London)* **283**, 789–792.

Hanley, M., and Iversen, L. L. (1980). Substance P receptors. *In* "Receptors and Recognition." Series B. Neurotransmitter Receptors. (H. I. Yamamura and S. J. Enna, eds.), pp. 71–103. Chapman & Hall, London.

Hanley, M., Sandberg, B. E. B., Lee, C. M., Iversen, L. L., Brundish, D. E., and Wade, R. (1980). Specific binding of ^{3}H-substance P to rat brain membranes. *Nature (London)* **286**, 810–812.

Harmar, A., Schofield, J. G., and Keen, P. (1980). Cycloheximidesensitive synthesis of substance P by isolated dorsal root ganglia. *Nature (London)* **283**, 267–269.

Henderson, G., Huhges, J., and Kosterlitz, H. W. (1978). In vitro release of Leu and Met-enkephalin from the corpus striatum. *Nature (London)* **271**, 677–679.

Hiller, J. M., Simon, E. J., Crain, S. M., and Peterson, E. R. (1978). Opiate receptors in cultures of fetal mouse dorsal root ganglia (DRG) and spinal cord: predominance in DRG neurites. *Brain Res.* **145**, 396–400.

Hökfelt, T., Kellerth, J. O., Nilsson, G., and Pernow, B. (1975). Substance P: localization in the central nervous system and in some primary sensory neurones. *Science (Washington, D.C.)* **190**, 889–890.

Hökfelt, T., Elde, T., Johansson, O., Luft, R., Nilsson, G., and Arimura, A. (1976). Immunohistochemical evidence for separate populations of somatostatin-containing and substance P-containing primary afferent neurones in the rat. *Neuroscience* **1**, 131–136.

Hökfelt, T., Ljungdahl, A., Terenius, L., Elde, R., and Nilsson, G. (1977). Immunohistochemical analysis of peptide pathways possibly related to pain and analgesia: Enkephalin and substance P. *Proc. Natl. Sci. U.S.A.* **74**, 3081–3085.

Hökfelt, T., Ljungdahl, A., Steinbusch, H., Verhofstad, A., Nilsson, G., Brodin, E., Pernow, B., and Goldstein, M. (1978). Immunohistochemical evidence of substance P-like immunoreactivity in some 5-hydroxytryptamine-containing neurons in the rat central nervous system. *Neuroscience* **3**, 517–538.

Hökfelt, T., Johansson, O., Ljungdahl, A., Lundberg, J. M., and Schultzberg, M. (1980a). Peptidergic neurones. *Nature (London)* **284**, 515–521.

Hökfelt, T., Lundberg, J., Schultzberg, M., Johansson, O., Skirboll, L., Anggard, A., Fredholm, B., Hamberger, B., Pernow, B., Rehfeld, J., and Goldstein, M. (1980b). Cellular localization of peptides in neural structures. *Proc. R. Soc. London, ser. B* **210**, 91–111.

Hong, J. S., Yang, H-Y.T., and Costa, E. (1978). Substance P content of substantia nigra after chronic treatment with anti-schizophrenic drugs. *Neuropharmacology* **17**, 83–85.

Hughes, J., Smith, T. W., Kosterlitz, H. W., Fothergill, L. A., Morgan, B. A., and Morris, H. R. (1975). Identification of two related penta-peptides from the brain with potent opiate agonist activity. *Nature (London)* **258**, 577–579.

Hunt, S. P., Kelly, J. S., and Emson, P. C. (1980). The electron microscopic localisation of methionine-enkephalin within the superficial layers (I and II) of the spinal cord. *Neuroscience* **5**, 1871–1890.

Iversen, L. L., Iversen, S. D., Bloom, F. E., Vargo, T., and Guillemin, R. (1978). Release of enkephalin from rat globus pallidus in vitro. *Nature (London)* **271**, 679–681.

Iversen, L. L., Iversen, S. D., and Bloom, F. E. (1980). Opiate receptors influence vasopressin release from nerve terminals in the rat neurohypophysis. *Nature (London)* **284**, 350–351.

Jacquet, Y. F., and Marks, N. (1976). The C-fragment of β-lipotropin: an endogenous neuroleptic or antipsychotogen? *Science (Washington, D.C.)* **194**, 632–634.

Jessell, T. M., and Iversen, L. L. (1977). Opiate analgesics inhibit substance P release from rat trigeminal nucleus. *Nature (London)* **268**, 549–551.

Johansson, O., Hökfelt, T., Pernow, B., Jeffcoate, S. L., White, N., Steinbusch, H. W. M., Verhofstad, A. A. J., Emson, P. C., and Spindel, E. (1981). Immunohistochemical support for three putative transmitters in one neuron: Coexistence of 5-hydroxytryptamine, substance P- and TRH-like immunoreactivity in medullary neurons projecting to the spinal cord. *Neuroscience*, in press.

Kanazawa, I., Bird, E., O'Connell, R., and Powell, D. (1977). Evidence for a decrease in substance P content of substantia nigra in Huntington's chorea. *Brain Res.* **120**, 387–392.

Kimura, S., Lewis, R. V., Stern, A. S., Rossier, J., Stein, S., and Uderfriend, S. (1980). Probable precursors of Leu- enkephalin and Met- enkephalin in adrenal medulla: Peptides of 3-5 Kilodaltons. *Proc. Natl. Acad. Sci. U.S.A.* **77**, 1681–1685.

Kobayashi, R. M., Palkovits, M., Miller, J-J, Chang, K. J., and Cuatrecasas, P. (1978). Brain enkephalin distribution is unaltered by hypophysectomy. *Life Sci.* **22**, 527–530.

LaMotte, C., Pert, C. B., and Snyder, S. H. (1976). Opiate receptors binding in primate spinal cord: Distribution and changes after dorsal root section. *Brain Res.* **112**, 407–412.

Larsson, L-I., and Rehfeld, J. F. (1979). Localization and molecular heterogeneity of cholecystokinin in central and peripheral nervous system. *Brain Res.* **165**, 201–218.

Larsson, L-I., Childers, S. R. and Snyder, S. H. (1979). Met- and Leu-enkephalin immunoreactivity in separate neurones. *Nature (London)* **282**, 407–410.

Lawson, S. N., and Nickels, S. M. (1980). The use of morphometric techniques to analyse the effect of neonatal capsaicin treatment on dorsal root ganglia and dorsal roots. *J. Physiol. (London)*, **303**, 12.

Lee, C. M., Emson, P. C., and Iversen, L. L. (1980). The development and application of a novel N-terminal directed substance P antiserum. *Life Sci.* (in press).

Lembeck, F. (1953). Zur Frage der zentralen Ubertagund afferentier Impulse. III Mitteilung das Vorkommen und be deutung der Substance P in der dorsalen worzeln das Ruckemarks. *Naunyn-Schmiedeberg's Arch. Exp. Pathol. Pharmakol.* **219**, 197–213.

Lembeck, F., and Holzer, P. (1980). Sfubstance P as neurogenic mediator of antidromic vasodilation and neurogenic plasma extravasation. *Naunyn-Schmiedeberg's Arch. Pharmacol.* **310**, 175–184.

Ljungdahl, A., Hökfelt, T., and Nilsson, G. (1978a). Distribution of substance P-like immunoreactivity in the central nervous system of the rat. I. Cell bodies and nerve terminals. *Neuroscience* **3**, 861–943.

Ljungdahl, A., Hökfelt, T., Nilsson, G., and Golstein, M. (1978b). Distribution of substance P-like immunoreactivity in the central nervous system of the rat. II. Light microscopical localization in relation to catecholamine-containing neurons. *Neuroscience* **3**, 945–976.

Lord, J. A. H., Waterfield, A. A., Hughes, J., and Kosterlitz, H. (1977). Endogenous opioid peptides: multiple agonists and receptors. *Nature (London)* **267**, 495–500.

Lorèn, I., Alumets, J., Hakanson, R., and Sundler, F. (1979a). Distribution of gastrin and CCK-like peptides in rat brain. *Histochemistry* **59**, 249–257.

Lorèn, I., Emson, P. C., Fahrenkrug, J., Björklund, A., Alumets, J., Hakanson, R., and Sundler, F. (1979b). Distribution of vasoactive intestinal polypeptide (VIP) in the rat and mouse brain. *Neuroscience* **4**, 1153–1976.

Lundberg, J. M., Hökfelt, T., Anggard, A., Pernow, B., and Emson, P. (1979a). Immunohistochemical evidence for substance P immunoreactive nerve fibres in the taste buds of the cat. *Acta Physiol. Scand.* **107**, 389–391.

Lundberg, J. M., Hökfelt, T., Schultzberg, M., Uvnas-Wallersten, K., Kohler, C., and Said, S. (1979b). Occurrence of vasoactive intestinal polypeptide (VIP)-like immunoreactivity in certain cholinergic neurons of the cat: Evidence from combined immunohistochemistry and acetylcholinesterase staining. *Neuroscience* **4**, 1539–1559.

Lundberg, J. M., Anggard, A., Fahrenkrug, J., Hökfelt, T., and Mutt, V. (1980). Vasoactive intestinal polypeptides in cholinergic neurones of exocrine glands: Functional significance of co-existing transmitters for vasodilation and secretion. *Proc. Natl. Acad. Sci. U.S.A.* **77**, 1651–1655.

Lundberg, J. M., Anggard, A., Hökfelt, T., Johansson, O., Fahrenkrug, J., and Emson, P. C. (1981). Co-existence of acetylcholine and vasoactive intestinal polypeptide (VIP) in neurons: Basis for secretion and vasodilation in exocrine glands. *Proc. XXVIII Int. Congr. Physiol. Sci.*, in press.

Mackay, A. V. P. (1979). Psychiatric implications of endorphin research. *Br. J. Psychiatry* **135**, 470–473.

Martin, W. R., Eades, C. G., Thompson, J. A., Huppler, R. E., and Gilbert, D. E. (1976). The effects of morphine and nalorphine-like drugs in the non-dependent, morphine dependent and cyclazocine-dependent chronic spinal dog. *J. Pharmacol. Exp. Ther.* **198**, 66–82.

Muller, J. E., Straus, E., and Yalow, R. S. (1977). Cholecystokinin and its COOH-terminal octapeptide in the pig brain. *Proc. Natl. Acad. Sci. U.S.A.* **74**, 3035–3037.

Mutt, V., and Jorpes, J. E. (1971). Hormonal polypeptides of the upper intestine. *Biochem. J.* **125**, 57–58.

Nagy, J. I., Vincent, S. R., Staines, W. M., and Fibiger, H. C. (1980). Neurotoxic action of capsaicin on spinal substance P neurons. *Brain Res.* **186**, 435–444.

Nakanishi, N., Inoue, A., Kita, T., Nakamura, M., Chang, A. C. Y., Cohen, S. N., and Numa, S. (1979). Nucleotide sequences of cloned ^{c}DNA for bovine corticotropin-β-lipotropin precursor. *Nature (London)* **278**, 423–437.

Nicoll, R. A. (1978). Physiological studies on amino acids and peptides as prospective transmitters in the CNS. *In* "Psychopharmacology: A Generation of Progress" (M. A. Lipton, A. DiMascio, and K. F. Killam, eds.), pp. 103–118. Raven, New York.

Nicoll, R. A., Schenker, C., and Leeman, S. E. (1980). Substance P as a transmitter candidate. *Annu. Rev. Neurosci.* **3**, 227–268.

Olgarth, L., Gazelins, B., Brodin, E., and Nilsson, G. (1977). Release of substance P immunoreactivity from the dental pulp. *Acta. Physiol. Scand.* **101,** 510–512.

Osborne, H., Hollt, V., and Hertz, A. (1978). Subcellular distribution of enkephalins and endogenous opioid activity in rat brain. *Life Sci.* **22,** 611–618.

Pearse, A. G. E. (1979). The endocrine division of the nervous system. A concept and its verification. *In* "Molecular Endocrinology" (I. MacIntyre and M. Szelke, eds.), pp. 3–18. Elsevier, Amsterdam.

Pert, C. B., and Snyder, S. H. (1974). Opiate receptor binding of agonists and antagonists affected differentially by sodium. *Mol. Pharmacol.* **10,** 868–879.

Phillis, J. W., Kirkpatrick, J. R., and Said, S. I. (1978). Vasoactive intestinal polypeptide excitation of central neurons. *Can. J. Physiol. Pharmacol.* **56,** 337–340.

Pinget, M., Straus, E., and Yalow, R. S. (1978). Localization of cholecystokinin-like immunoreactivity in isolated nerve terminals. *Proc. Natl. Acad. Sci. U.S.A.* **75,** 6324–6326.

Quik, M., Iversen, L. L. and Bloom, S. R. (1978). Effect of vasoactive intestinal polypeptide (VIP) and other peptides on cAMP accumulation in rat brain. *Biochem. Pharmacol.* **27,** 2209–2213.

Ramana-Reddy, S. U., and Yaksh, T. L. (1980). Spinal noradrenergic terminal system mediates antinociception. *Brain Res.* **189,** 391–401.

Rehfeld, J. F. (1978). Localization of gastrins to neuro and adenohypophysis. *Nature (London)* **271,** 771–773.

Rossier, J., Vargo, T. M., Minick, S., Ling, N., Bloom, F. E., and Guillemin, R. (1977). Regional dissociation of β-endorphin and enkephalin contents in rat brain and pituitary. *Proc. Natl. Acad. Sci. U.S.A.* **74,** 5162–5165.

Ruda, M., and Gobel, S. (1980). Ultrastructural characterization of axonal endings in the substantia gelatinosa which take up ^{3}H serotonin. *Brain Res.* **184,** 57–83.

Saito, A., Sankaran, H., Goldfine, I. D., and Williams, J. A. (1980). Cholecystokinin receptors in the brain: Characterization and distribution. *Science (Washington, D.C.)* **208,** 1155–1156.

Sar, M., Stumpf, W. E., Miller, R. J., Chang, K-J., and Cuatrecasas, P. (1978). Immunohistochemical localization of enkephalin in rat brain and spinal cord. *J. Comp. Neurol.* **182,** 17–38.

Simon, E. J., Hiller, J. M., and Edelman, I. (1973). Stereospecific binding of the potent narcotic analgesic ^{3}H etophine to rat brain homogenate. *Proc. Natl. Acad. Sci. U.S.A.* **70,** 1947–1949.

Skrabanek, P., and Powell, D. (1977). Substance P. *In* "Annual Research Reviews" (D. F. Horrobin, ed.), Vol. 1. Churchill Livingstone, Edinburgh and London.

Snyder, S. H. (1980). Brain peptides as neurotransmitters. *Science (Washington, D.C.),* in press.

Sternberger, L. A. (1979). Immunocytochemistry. Wiley, New York.

Taylor, D. P., and Pert, C. B. (1979). Vasoactive intestinal polypeptide: specific binding to rat brain membranes. *Proc. Natl. Acad. Sci. U.S.A.* **76,** 660–664.

Terenius, L. (1973). Characteristics of the "receptor" for narcotic analgesics in synaptic plasma membrane fraction from rat brain. *Acta Pharmacol. Toxicol.* **33,** 377–384.

Vanderbaeghen, J. J., Signeau, J. C., and Gepts, W. (1975). New peptide in the vertebrate CNS reacting with antigastrin antibodies. *Nature (London)* **257,** 604–605.

Vanderhaeghen, J. J., Lotstra, F., De Mey, J., and Gilles, G. (1980). Immunohistochemical localization of cholecystokinin- and gastrin-like peptides in the brain and hypophysis of the rat. *Proc. Natl. Acad. Sci. U.S.A.* **77,** 1190–1194.

Viveros, O. H., Dilibertom, E. J., Hazum, E., and Chang, K-J. (1979). Opiate-like material in the adrenal medulla: Evidence for storage and secretion with catecholamines. *Mol. Pharmacol.* **16,** 1101–1108.

Watson, S. J., Akil, H., Richard, C. W., and Barchas, J. D. (1978). Evidence for two separate opiate peptide neuronal systems. *Nature (London)* **275,** 226–228.

Wilkening, D., Sabol, S. L., and Nirenberg, M. (1980). Control of opiate receptor-adenylate cyclase interactions by calcium ions and guanosine-5′-triphosphate. *Brain Res.* **189,** 459–466.

Yaksh, T. L. (1979). Direct evidence that spinal serotonin and noradrenaline terminals mediate the spinal antinociceptive effects of morphine in the periaqueductal grey. *Brain Res.* **160,** 180–185.

Yaksh, T. L., and Wilson, P. R. (1979). Spinal serotonin terminal systems mediates antinociception. *J. Pharmacol. Exp. Ther.* **208,** 446–453.

Young, W. S., and Kuhar, M. J. (1979). A new method for reception autoradiography: [³H]opioid receptors in rat brain. *Brain Res.* **179,** 255–270.

Zakarian, S., and Smyth, D. (1979). Distribution of active and inactive forms of endorphins in rat pituitary and brain. *Proc. Natl. Acad. Sci. U.S.A.* **76,** 5972–5976.

Zieglgansberger, W., and Fry, J. P. (1976). Action of enkephalin on cortical and striatal neurones of naive and morphine tolerant/dependent rats. *In* "Opiates and Endogenous Opioid Peptides" (H. W. Kosterlitz, ed.), pp. 231–238. North-Holland Publ., Amsterdam.

11

MOLECULAR CORRELATES BETWEEN PITUITARY HORMONES AND BEHAVIOR

J. Jolles, V. J. Aloyo, and W. H. Gispen

I. INTRODUCTION

Hormones secreted by the pituitary play an important role in the behavioral adaptation of an organism to its environment. These hormones regulate homeostasis and create the conditions in which the animal can cope optimally with situational demands. Though much research has been directed at elucidating their mechanism of action in peripheral tissues, relatively little attention has been paid to the brain as a target for these hormones. However, it has been shown that pituitary principles are involved in a number of brain functions and that they are important for the maintenance of normal behavioral patterns. It was observed that learning is impaired after removal of the pituitary and that substitution of ad-

Molecular Approaches to Neurobiology

renocorticotropic hormone (ACTH), α-melanocyte-stimulating hormone (α-MSH), or vasopressin restores this behavior (De Wied, 1969). Fragments of these hormones that lack the classic endocrine effects, were able to restore the impaired behavior. Similarly, peptides related to ACTH, MSH, or β-lipotropic hormone (β-LPH) were shown to influence behavior in intact animals (De Wied, 1969; De Wied *et al.*, 1978a). These peptides have not only been found in the pituitary but in many brain structures as well (Rossier *et al.*, 1977; Krieger *et al.*, 1977; Orwall *et al.*, 1979; Watson and Akil, 1980b). It was hypothesized that the pituitary releases peptides that are involved in the formation and maintenance of new behavior patterns. These so-called neuropeptides act directly on the central nervous system (CNS) (De Wied, 1969). In this view, environmental stimuli act to stimulate the release of the neuropeptides from the pituitary or from central cells. They modulate the activity of neuronal systems in the brain, and this altered activity finally results in behavioral adaptation of the animal to its environment (see Wiegant and De Wied, 1980).

This chapter reviews molecular correlates of the pituitary hormones. Relatively little is known about the molecular events that underlie the behavioral effects of the neuropeptides, though some biochemical data have been obtained on the action of peptide hormones in peripheral tissues. Therefore, some attention will be paid to this subject. The wealth of information that has been gathered on behavioral effects of pituitary hormones has forced us to limit ourselves to those behavioral effects that could be related to the available neurochemical data. We have, therefore, focused our attention on ACTH, MSH, β-LPH, and their fragments. The effects on pole-jumping avoidance behavior and grooming behavior are described as well as the opiate-like effects of these hormones.

For a more thorough review of behavioral aspects that are not covered in this chapter, the reader is referred to other papers (grooming behavior: Gispen and Isaacson, 1980; pituitary peptides and behavior: Wiegant and De Wied, 1980, De Wied and Gispen, 1977; lipotropin and CNS: Gispen *et al.*, 1977).

II. BEHAVIORAL EFFECTS OF PITUITARY PEPTIDE HORMONES

A. Acquisition and Retention of New Behavior Patterns

The implication that ACTH has CNS effects was first suggested by observations made on hypophysectomized rats. These animals showed impaired acquisition of shuttlebox avoidance behavior (De Wied, 1964; Applezweig and Baudry, 1955; Applezweig and Moeller, 1959). This behavioral impairment could not only be corrected by treatment with ACTH and α-MSH, but also by fragments of these hormones that are devoid of classical endocrine activity (De Wied, 1969; Bohus *et al.*, 1973). Furthermore, dexamethasone fails to restore shuttlebox avoidance learning in hypophysectomized rats (De Wied, 1971), and avoidance behavior was not impaired in adrenalectomized rats. It was concluded that the behavioral effects are a consequence of a direct action of these peptides on specific areas of the CNS (De Wied, 1977; Greven and De Wied, 1973).

In intact rats, ACTH and congeners have been found to influence the acquisition and retention of new behavior patterns (see Wiegant and De Wied, 1980; De Wied, 1977). These peptides facilitate passive avoidance behavior (Levine and Jones, 1965; Lissák and Bohus, 1972; De Wied, 1974; Kastin *et al.*, 1973; Flood *et al.*, 1976) and delay the extinction of shuttlebox avoidance behavior (Greven and De Wied, 1973), pole-jumping avoidance behavior (De Wied, 1966), food-motivated behavior (Garrud *et al.*, 1974; Guth *et al.*, 1971), conditioned taste aversion (Rigter and Popping, 1976), and sexually motivated approach behavior (Bohus *et al.*, 1975). On the basis of these results it has been suggested that ACTH and related peptides are involved in motivational processes (De Wied, 1977).

These neuropeptides also affect learning and memory; they alleviate the amnesia that was induced by several treatments [inhalation of CO_2, administration of electroconvulsive shock, and treatment with protein synthesis inhibitors (Flexner and Flexner, 1971; Keyes, 1974; Rigter and Van Riezen, 1975; Rigter *et al.*, 1974)]. It has been suggested that the peptide affects memory storage (Gold and Van Buskirk, 1976; Flood *et al.*, 1976) or retrieval (Rigter *et al.* 1974), but motivational effects can not be excluded from most of the experimental paradigms used. So it was hypothesized (De Wied, 1977) that ACTH temporarily increases the motivational value of environmental stimuli, probably by selectively increasing a state of arousal in midbrain limbic structures. The possibility that stimulus-specific behavioral responses occur is thereby increased.

Although under certain conditions ACTH has been shown to improve acquisition of shock-motivated active avoidance behavior in intact rats (Beatty *et al.*, 1970; Guth *et al.*, 1971; Isaacson *et al.*, 1975), extinction of conditioned avoidance behavior seems to be more sensitive to the behavioral effects of the peptides (De Wied and Gispen, 1977). Results obtained on pole-jumping active avoidance behavior show the best dose-response relationship, therefore, structure–activity studies were performed on this paradigm to determine the essential elements required for the behavioral effect of ACTH.

It has been shown that $ACTH_{4-7}$ is the shortest active fragment with essentially the same behavioral potency as ACTH (Greven and De Wied, 1973; De Wied *et al.*, 1975). However, more activity sites are present in ACTH as the fragments $ACTH_{7-10}$ and $ACTH_{11-24}$ also contain some activity (Greven and De Wied, 1977). The residual potency observed for the sequence $ACTH_{7-10}$ could be increased to the same level as that of the reference peptide $ACTH_{4-10}$ by extending the C-terminal sequence to $ACTH_{7-16}$. Thus, the essential elements for avoidance behavior are not exclusively located in the region $ACTH_{4-7}$, but also occur in other areas of the molecule (Greven and De Wied, 1977). Though both MSH and β-LPH contain the sequence $ACTH_{4-10}$ it was concluded that the structural requirements for the effects on pole-jumping avoidance behavior are more related to ACTH than to MSH or β-LPH. First, because of the second affinity site that becomes expressed after chain elongation to $ACTH_{7-16}$. Second, because of the potency of modified ACTH fragments (De Wied et al., 1975; Greven and De Wied, 1977); the analog $[Met(O)^4,D-Lys^8,Phe^9]ACTH_{4-9}$ (Org 2766) is behaviorally 1000 times more active than $ACTH_{4-10}$, but possess 1000 times less MSH activity, and contains no opiate-like activity. Third, omission of either the glycyl residue in position 10 or the lysyl residue in position 16 is accompanied by a drastic decrease in potency (Greven and De Wied, 1977). For behavioral potency a doublet of basic lysine residues apparently is needed at exactly the same distance from the region $ACTH_{7-9}$ as in natural ACTH. This indicates that the structural requirements for behavioral activity in the pole-jumping test are more related to ACTH than to MSH or β-LPH. Extinction of pole-jumping avoidance was also used to assay the behavioral effect of C-terminal fragments of β-LPH (the endorphins). After subcuteneous (sc) injection, α-endorphin (αE; β-LPH_{61-76}) appeared to be the most potent peptide in delaying extinction of pole-jumping avoidance behavior (De Wied et al., 1978a). On a molar basis it was 30 times as active as $ACTH_{4-10}$. After intraventricular administration, however, both peptides were equipotent, indicating that the difference in potencies after systemic administration is related to brain uptake mechanisms rather than to intrinsic behavioral effect.

The relatively weak activity of β-endorphin (βE; β-LPH_{61-91}) is probably the result of metabolic breakdown to fragments with opposite behavioral activity; γ-endorphin (γE; β-LPH_{61-77}) differs from αE by only one extra C-terminal amino acid, but it facilitates rather than delays the extinction of pole-jumping avoidance behavior (De Wied et al., 1978b). Similarly, (des-tyrosine[1])-γ-endorphin (dTγE, β-LPH_{62-77}) was even more potent than γE in affecting avoidance behavior. A number of observations suggest that the influence of endorphins and of ACTH-like peptides on avoidance behavior takes place independently of opiate receptor sites in the brain. First, neither βE- nor ACTH-effects on pole-jumping avoidance behavior could be blocked by specific opiate antagonists (De Wied et al., 1978a). Second, structurally modified peptides— e.g., $[Met(O)^4,D-Lys^8,Phe^9]ACTH_{4-9}$; Org 2766—have increased potency in the avoidance paradigm and have no opiate-like activity (Terenius et al., 1975). Third, removal of the N-terminal tyrosine from the endorphins caused complete loss of opiate-like activity (on the guinea pig ileum) and destroyed the affinity for opiate binding sites (Guillemin et al., 1976; Frederickson, 1977; De Wied et al., 1978b), whereas the activity on pole-jumping active avoidance behavior was preserved (De Wied et al., 1978a).

There is a striking similarity in effect on conditioned avoidance behavior between dTγE and neuroleptic drugs such as haloperidol. The psychophar-macological actions of dTγE such as facilitation of extinction of pole-jumping avoidance behavior, attenuation of passive avoidance behavior (De Wied *et al.*, 1978b), interference with ACTH-induced grooming (Gispen *et al.*, 1980), and its activity in various grip tests (De Wied *et al.*, 1978b) are characteristic of neuroleptic drugs (Kováls and De Wied, 1978). In contrast, αE and related peptides have characteristics that resemble those of psychostimulants like am-phetamine (Kovács and De Wied, 1978).

B. Opiate-Like Activity

The effects of C-terminal β-LPH fragments on extinction and grooming be-havior (Section II,A and C) are obtained with amounts much lower than those needed to induce analgesia. Effects on extinction of pole-jumping avoidance behavior were obtained when amounts less than 1 μg were systemically injected, whereas after intracerebralventricular (icv) administration nanogram quantities are sufficient (De Wied *et al.*, 1978).

Profound analgesia was found after icv injection of higher quantities of βE (Bradbury *et al.*, 1976a), and icv administration of microgram doses of this substance in rats produced a naloxone-reversible catatonia (Bloom *et al.*, 1976). In its antinociceptive effects, βE appeared to be many times more potent than morphine. Similar effects of icv administrered Met-enkephalin (β-LPH$_{61-65}$) have been reported (Belluzzi *et al.*, 1976), but this peptide appears to be less potent than βE. This is probably due to enzymatic degradation *in vivo* (Hambrook *et al.*, 1976). The development of tolerance to βE was similar to that reported for morphine, and cross-tolerance between morphine, βE, or Met-enkephalin (Van Ree and De Wied, 1976; Bläsig and Herz, 1976) could be demonstrated. Morphine and βE also share similar dependence properties as assessed by naloxone-induced withdrawal signs (Wei and Loh, 1976; Loh *et al.*, 1976).

Not only peptides related to β-LPH have opioid activities. Neurophysiological evidence indicated that ACTH and β-MSH counteract the morphine-induced depression of spinal reflex activities *in vivo* (Zimmerman and Krivoy, 1973). $ACTH_{1-24}$ also counteracts morphine effects *in vitro* (Zimmermann and Krivoy, 1974) indicating a morphine–peptide interaction at the level of CNS. Fur-thermore, purified ACTH and $ACTH_{1-24}$ antagonized the analgesic effect of moprhine (Paroli, 1967; Gispen *et al.*, 1976a). The peptides $ACTH_{1-24}$, $ACTH_{1-16}$, $ACTH_{5-16}$, $ACTH_{5-14}$, and [D-Phe[7]]$ACTH_{4-10}$ reduced the analgesic effect of morphine, as measured by the hot-plate test, by 50% (Gispen *et al.*, 1976a). $ACTH_{4-10}$ was less active, whereas $ACTH_{11-16}$, $ACTH_{11-17}$, and $ACTH_{11-24}$ were inactive. These results suggest that the sequence $ACTH_{4-10}$ may contain the active site; a secondary site may provide additional affinity without exerting intrinsic activity.

C. Excessive Grooming Behavior

It has long been known that birds and small mammals display enhanced grooming behavior in situations in which novel or conflicting environmental stimuli are present (Sevenster, 1961; Tinbergen, 1940; Bolles, 1960). Of course, these same stimuli are known to activate the pituitary–adrenal system (Mason, 1968) implying that this system is involved in grooming induction. However, since hypophysectomized rats still show novelty-induced grooming (Jolles *et al.*, 1979a), the pituitary gland can not be directly involved. Two lines of evidence implicate centrally active ACTH as playing a role in the induction of grooming. First, icv administration of antibodies to ACTH reduced novelty-induced grooming (Dunn *et al.*, 1979). Second, intraventricular injection of ACTH or its N-terminal fragments produced an enhanced display of grooming (Ferrari *et al.*, 1963; Izumi *et al.*, 1973; Gispen *et al.*, 1975; Wiegant and Gispen, 1977). In view of the short latency of icv-administered ACTH and since its effects are independent of its endocrine activity (Gispen *et al.*, 1975), this supports a direct CNS effect for ACTH in inducing excessive grooming. Furthermore, systemic administration of the peptide fails to induce grooming, once again implying a direct CNS effect. It has been suggested that grooming dearouses the organism after its activation by ACTH (Jolles *et al.*, 1979a,b; Delius, 1970; Delius *et al.*, 1976).

Structure–activity studies have been performed to determine the elements necessary for ACTH-induced grooming. $ACTH_{1-16}$, α-MSH, and β-MSH were as potent as $ACTH_{1-24}$, but $ACTH_{1-13}NH_2$ was less active (Gispen *et al.*, 1975). As in the avoidance studies (Section II,A), the shortest sequence possessing grooming-inducing activity was $ACTH_{4-7}$ (Wiegant and Gispen, 1977). $ACTH_{4-10}$ and $ACTH_{1-10}$, peptides containing full information for the effect on avoidance behavior, were inactive in the excessive grooming test, but D-Phe^7 substitution rendered them active (Gispen *et al.*, 1975). As the fragments $ACTH_{1-13}NH_2$, $ACTH_{5-14}$, and $ACTH_{5-16}$ also showed grooming-inducing activity, a secondary site may be present beyond the tenth amino acid. This site lacks intrinsic grooming-inducing activity as may be concluded from the inactivity of the fragments $ACTH_{7-16}$ and $ACTH_{11-24}$. In contrast, these fragments are active in the avoidance paradigm (Section II,A), so these data suggest that the peptide has different mechanisms of action in the two behavioral models. In fact, the grooming response may relate more to C-terminal β-LPH than to ACTH. For instance, icv-injected βE is more potent than $ACTH_{1-24}$ in producing excessive grooming (Gispen *et al.*, 1976b); the opioid was able to induce the grooming response in doses as low as 10 ng. The nature of the excessive grooming induced by β-endorphin is somewhat different from the ACTH-induced behavior in that the β-endorphin grooming is frequently interrupted by signs of excitation (quick movements of body and head, jumping, gnawing, and body shakes). Shortening the βE from the C-terminal end resulted in a rapid and progressive loss of activity. Both αE and γE possessed slight grooming-inducing potency, while βLPH$_{61-69}$ was the shortest active sequence. Met-enkephalin was inactive, even if injected in high doses (Gispen *et al.*, 1976b). The des-Tyr fragments, that lack opiate-like activity (Guillemin *et al.*, 1976; Frederickson, 1977; De Wied *et al.*, 1978b) were also inactive, again underscoring the "opiate-like" character of peptide-induced grooming. Similarly, peripheral administration of specific opiate antagonists (naloxone, naltrexone) completely inhibited the grooming induced by ACTH$_{1-24}$ (Gispen and Wiegant, 1976) and by βE (Gispen *et al.*, 1976b). Recently it was found that an "acute tolerance" develops after ACTH$_{1-24}$ (Jolles *et al.*, 1978) and βE administration (Wiegant *et al.*, 1978). After icv injection these peptides were not able to induce grooming when given 1–8 h after a previous injection of the peptide. A cross-tolerance was demonstrated for ACTH and βE, and also for ACTH and morphine, and systemic treatment with naloxone at the time of the first icv injection prevented the development of the single-dose tolerance (Jolles *et al.*, 1978). These results suggest that a close relationship may exist between ACTH- and endorphin-sensitive structures in the CNS.

D. Discussion

ACTH, MSH, β-LPH$_{61-91}$, and their fragments affect avoidance behavior, analgesia, and grooming behavior as described in the previous sections. It was found that the structural requirements for the effects on pole-jumping avoidance behavior were less exacting than those needed for grooming and opiate-like activities (Fig. 1). The fragment ACTH$_{4-10}$, which is present in ACTH, MSH, and β-LPH, contains the prime affinity site; ACTH$_{4-7}$ is the shortest sequence possessing activity in the avoidance paradigm and in groooing behavior. ACTH$_{4-10}$ was fully active on pole-jumping avoidance behavior, but was inactive in grooming test and analgesia. D-Phe[7] substitution rendered this peptide active in the latter test but reversed its effect on pole-jumping avoidance behavior. A second affinity site was found C-terminal to the sequence ACTH$_{4-7}$. This second site exhibited activity in pole-jumping avoidance behavior, but not in grooming and analgesia; ACTH$_{7-16}$ and ACTH$_{11-24}$ were active on avoidance behavior, but not on grooming and analgesia. It was concluded that the effects on avoidance behavior relate more to ACTH than to MSH or β-LPH; in contrast, effects on grooming may relate more to the opiate-like activity and to C-terminal β-LPH.

The metabolic stability of the peptides certainly contributes to the behavioral potency (Greven and De Wied, 1977; Witter *et al.*, 1975). Elongation of the peptide chain and modification of its structure by amino acid substitution can increase the *in vitro* half-life. However, the enhanced effect of [D-Phe[7]]ACTH$_{4-10}$ and [D-Phe[7]]ACTH$_{1-10}$ on grooming behavior can not be explained by an im-

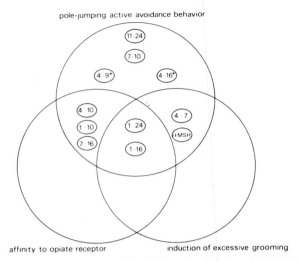

Fig. 1. Venn diagram showing the distribution of a number of ACTH fragments over three overlapping circles. Each circle encloses a collection of ACTH fragments, active on a particular test of CNS activity. This figure suggests that different parts of the ACTH molecule can interact with receptors mediating different functions. The data on "affinity for opiate receptor" are described in Section III,E. The fragments [MET(O)^4, D-Lys^8, Phe^9] ACTH$_{4-9}$, and [Met(O)^4, D-Lys^8, Phe^9, D-Lys^{11}] ACTH$_{4-16}$ are denoted as (4-9*) and (4-16*), respectively.

proved metabolic stability as [D-Arg^8]ACTH$_{4-10}$ and [Met(O)^4, D-Lys^8, Phe^9]-ACTH$_{4-9}$ were inactive in the grooming test (Grispen *et al.*, 1975). So, the strong behavioral potentiation after structural modification may reflect increased activity and/or intrinsic affinity for the receptor sites in the CNS.

In conclusion, there is a redundancy in behavioral information in ACTH, MSH, and β-LPH. These, and possibly other peptides are derived from the same large precursor molecule (Mains *et al.*, 1977; Peng Loh, 1979). The demonstration of specific peptidases in brain membranes (Burbach *et al.*, 1979, 1980; Gráf *et al.*, 1976; Bradbury *et al.*, 1976b) suggest that these peptides in turn may serve as precursor for a series of shorter sequences with a variety of behavioral activities.

III. MOLECULAR CORRELATES OF PITUITARY PEPTIDE HORMONES

From behavioral studies it was concluded that ACTH and related peptides act directly on CNS structures. Indeed, ACTH, α-MSH, and the endorphins have been found in the pituitary as well as in many brain structures (Rossier *et al.*,

1977; Krieger *et al.*, 1977; Orwoll *et al.*, 1979; Watson and Akil, 1980b). A functional connection between ACTH and endorphin systems in the brain is suggested by observations that β-LPH, βE, and ACTH-like peptides can occur in the same neuronal cells (Watson and Akil, 1980a,b; Watson *et al.*, 1978a,b; Pelletier and Leclerc, 1979). Though ACTH-like immunoreactive material has been demonstrated throughout the brain (Krieger *et al.*, 1977), a function for ACTH as a classical neurotransmitter in the CNS lacks experimental support. Currently the influence of ACTH-like neuropeptides on brain neuronal activity is best formulated as neurohormonal or neuromodulatory (Barchas *et al.*, 1978; Gispen *et al.*, 1979). Since ACTH-like peptides have a direct effect on the CNS, many investigators have attempted to relate the behavioral effects of these peptides to a direct effect on neuronal cells. It is expected that the neuropeptide binds to a specific receptor on the cell membrane. Although high affinity binding sites for ACTH and related peptides have as yet not been demonstrated in the brain (Witter, 1980), the multiplicity and specificity of the ACTH–CNS interactions by themselves point to the presence of specific ACTH receptors in the brain. It is conceivable that ACTH interacts with neuronal membranes resulting in events that are similar to those found for other polypeptide hormones or neurotransmitters. The evidence for such a neurochemical action of the neuropeptides will be reviewed in the following sections.

A. Neurotransmitters

The early work of Weiss *et al.* (1970) and Hökfelt and Fuxe (1972) led investigators to assume that ACTH-like peptides would specifically alter the turnover of noradrenaline (NA) in the brain. The prediction was that delay of extinction of a conditioned avoidance response correlates with an increased NA turnover, whereas peptides having the opposite effect would lead to a decreased rate of NA turnover. More recent reports have not been able to substantiate this notion (Versteeg, 1973; Leonard, 1974; Iuvone *et al.*, 1978; Kostrzewa *et al.*, 1975). Similar effects of ACTH on brain serotonin metabolism are by no means convincing (Leonard, 1974; Leonard *et al.*, 1976; Spirtes *et al.*, 1975; Telegdy and Kovács 1979a,b), though Ramaekers *et al.* (1978) reported that the effects of $ACTH_{4-10}$ and [D-Phe[7]]$ACTH_{4-10}$ on passive avoidance behavior correlate to changes in hippocampal serotonin levels. Brain dopamine (DA) has been related to ACTH-induced excessive grooming (Wiegant *et al.*, 1977b; Cools *et al.*, 1978) and acetylcholine (ACh) turnover in the hippocampus may relate to the stretching and yawning syndrome induced by $ACTH_{1-24}$ and α-MSH (Wood *et al.*, 1978, 1979).

A large number of reports have been published concerning the interaction of endorphins, enkephalins, and their analogues with the metabolism of classical neurotransmitters (for a recent review see Versteeg, 1980). Although a majority of these reports indicate that these peptides increase striatal DA metabolism, there are also reports presenting conflicting evidence. One possible explanation of these conflicting reports stems from the work of Van Loon and Kim (1978). They found that the effects of βE on striatal DA are time-dependent resulting in an accelerated and then a decelerated utilization of DA. It has been concluded that part of these effects are mediated via interaction with presynaptic opioid receptors of the nigrostriatal DA terminals (Biggio *et al.*, 1978a,b). However, this accounts for only one-third of the opioid receptors of the striatum (Pollard *et al.*, 1977, 1978; Schwartz *et al.*, 1978, 1979; Carenzi *et al.*, 1978; Trabucchi *et al.*, 1979). It is likely that several mechanisms are involved in the interaction of these peptides with striatal DA activity.

The endorphins and enkephalins also influence brain ACh turnover via opioid receptors. βE modulates hippocampal ACh turnover. 10 μg per rat of βE administered icv decreased ACh turnover in the hippocampus, nucleus accumbens, globus pallidus, and cortex, but not in the caudate nucleus (Moroni *et al.*, 1977). In addition, Botticelli and Wurtman (1979) found that this same amount of βE also given icv elevated hippocampal ACh levels. In both studies opioid antagonists inhibited the effects.

Met-enkephalin also influences brain ACh release. Met-enkephalin reduced the K$^+$-stimulated transmitter release from hippocampal slices (Subramarian *et al.*, 1977) and reduces spontaneous release from cerebral cortex (Jhamandas *et al.*, 1977). However, in striatal slices, Met-enkephalin methylester and two other analogues increase ouabain-induced ACh release (Vizi *et al.*, 1977; Harsing *et al.*, 1978). These effects were all prevented by naloxone pretreatment, indicating the involvement of opioid receptors.

Although enkephalins also have effects on catecholamine metabolism in other brain regions, Versteeg (1980) concludes that our knowledge of these effects is not only fragmentary, but that the significance of the observed changes is obscure.

It is apparent from the previous discussion that ACTH, endorphins, and enkephalins all influence brain neurotransmitter metabolism. In many cases the results are confusing and apparently contradictory perhaps because different species, brain regions, methods as well as peptides (or analogues) were used in the different studies. However, it must also be considered that the effects of these peptides on neurotransmitter metabolism reflect indirect consequences of peptide action. In the following sections we summarize the evidence that these peptides interact directly with CNS neurons and speculate on the possible mechanisms of action of these peptides.

B. Cyclic Nucleotide Metabolism

The mechanism of action of the neuropeptides is not yet understood. However, evidence from studies of neuropeptide action in peripheral tissues suggests that these peptides may act via cAMP. In the second messenger model of Sutherland and Robinson (1966), interaction of the hormone (first messenger) with its membrane receptor activates adenylate cyclase giving rise to an increased intracellular level of cAMP. This cAMP acts as a second messenger by activating various protein kinases.

Cyclic nucleotides are thought to mediate the action of ACTH. Evidence has accumulated that cAMP is involved in the process of steroidogenesis in the adrenal cortex, and of lipolysis in the fat cell (Fain, 1973; Halkerston, 1975). For example, the steroid production that is induced by ACTH is accompanied by a rise in intracellular receptor-bound cAMP (Podesta *et al.*, 1979) and by activation of cAMP-dependent protein kinases (Kudlow *et al.*, 1980).

However, the sole involvement of cAMP in the process of steroidogenesis has been questioned. It has been found that the stimulation of cAMP production in adrenal cortex was dependent on the presence of Ca^{2+} (Bär and Hechter, 1969; Glossman and Gips, 1975, 1976). Bristow *et al.* (1980) found that ACTH may induce steroidogenesis via two different mechanisms: one involves the production of cAMP and the other acts via Ca^{2+} and cGMP. A similar finding was reported by Perchellet and Sharma (1979). They found no steroidogenesis in the absence of Ca^{2+}, even when high concentrations of the hormone were present. Exogenous cGMP could stimulate steroidogenesis.

There have been very few reports concerning the central effect of peptide hormones on brain cyclic nucleotides. It seems likely, however, that there is a parallel between the peripheral and the central mechanism of action. ACTH has been occasionally included in studies on the effect of putative neurotransmitters and hormones on brain adenylate cyclase. Burkard and Gey (1968) and Von Hungen and Roberts (1973) did not detect an effect of ACTH in cell-free membrane preparations. In addition, Forn and Krishna (1971) did not observe an effect of the peptide on cAMP accumulation in slices of cerebral cortex, cerebellum, and hypothalamus. Rudman and Isaacs (1975; Rudman, 1976) were the first to present indirect evidence that ACTH-like peptides might affect brain cyclic nucleotide levels *in vivo*. They showed that intracisternal injection of ACTH or β-MSH in rabbits increased the level of cAMP in the cerebrospinal fluid (CSF). In other studies it was found that chronic treatment of intact as well as hypophysectomized rats with β-MSH resulted in an increased level of cAMP (but not cGMP) in the occipital cortex (Christensen *et al.*, 1976; Spirtes *et al.*, 1978). Direct evidence for the involvement of cAMP in the action of ACTH on the brain was provided by Wiegant *et al.*, (1979). These authors found that the hormone stimulated the production of cAMP in striatal slices *in vitro*. In a broken cell preparation, however, a biphasic dose–response relationship was found. Low concentrations of the peptide (10^{-7}, 10^{-6} M) stimulated the cAMP accumulation, but high concentrations (10^{-5}, 10^{-4} M) inhibited the cAMP accumulation (Wiegant *et al.*, 1979). The effect of Ca^{2+} was similar to that observed in adrenal cortex: EGTA reduced the basal activity to 20% and abolished the effect of the hormone (Fig. 2). The authors showed that the hormone effect is not due to interaction with DA receptors as the DA receptor blocker ergometrine did not counteract the ACTH-inhibited adenylate cyclase activity. Moreover, it was improbable that the observed peptide effect was due to an interaction with the opiate receptor, as the inhibitory effect of ACTH was not counteracted by the specific opiate antagonist naltrexone (Fig. 2; Wiegant *et al.*, 1979). Interestingly, structure activity studies revealed a correspondence between the effects on cAMP production and those on the induction of excessive grooming in rats *in vivo:* $ACTH_{1-16}NH_2$ and $ACTH_{4-7}$ were active, whereas $ACTH_{11-24}$, $ACTH_{1-10}$, $ACTH_{4-10}$, and the D isomers were not (Wiegant *et al.*, 1979). Also, icv injection of $ACTH_{1-16}NH_2$ *in vivo* significantly elevated the cAMP level in the septal area (Wiegant *et al.*, 1979). This area is thought to be an important target for behaviorally active ACTH-like peptides. Such peptides have been shown to specifically accumulate in that region, when administered intracerebroventricularly (Verhoef *et al.*, 1977). Taken together, these studies support the notion that ACTH, in addition to its effects on the production of cyclic nucleotides in peripheral tissue is able to influence their metabolism in the CNS.

C. Protein Phosphorylation

It has been suggested that changes in the state of phosphorylation of membrane proteins may govern the ion permeability of the neuronal membrane (Heald, 1962) and thus may play a key role in determining the functional activity of the neuron (Greengard, 1976, 1978). Alternatively, phosphorylation of membrane enzymes may alter their activity, thereby changing the metabolism of important membrane components (for review on enzyme phosphorylation, see Krebs and Beavo, 1979).

There have been very few reports concerning the effects of peptide hormones on brain protein phosphorylation; enkephalins (Davis and Ehrlich, 1979) and β-endorphin (Ehrlich *et al.*, 1980) have been found to inhibit the phosphorylation of synaptosomal membrane proteins. Also in intact hippocampal slices effects of enkephalins have been noted (Bär *et al.*, 1980). Effects of ACTH-like peptides on brain protein phosphorylation have been described by Zwiers *et al.* (1976). These authors showed that ACTH exerts a dose-dependent inhibitory effect on the phosphorylation of proteins in synaptosomal plasma membrane fractions from rat brain (Zwiers *et al.*, 1976, 1978). The phosphorylation of one of these proteins (B-50 protein) was especially sensitive to ACTH. Structure–activity studies with ACTH-like peptides on the endogenous phosphorylation of B-50 showed that $ACTH_{1-24}$ and $ACTH_{1-16}$ are equally active, and that $ACTH_{1-13}$ and $ACTH_{5-18}$ also possess activity. The sequences 1–10, 11–24, 4–10, 5–16, and 7–16 were inactive (Zwiers *et al.*, 1978). Clearly, this structure–activity relationship is very similar to that found for the induction of excessive grooming

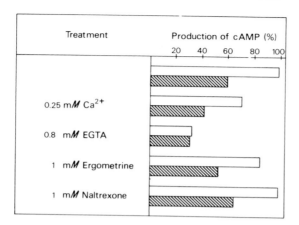

Fig. 2. Effects of various agents on the inhibition of adenylate cyclase in synaptosomal plasma membranes by 100 μM ACTH$_{1\text{-}24}$. These data show that calcium is required for the inhibitory effect of ACTH$_{1\text{-}24}$ and that this effect is not reversed by opiate (naltrexone) or dopamine (ergometrine) antagonists. The data upon which the table is based are from Wiegant *et al.* (1979). A synaptosomal plasma membrane fraction was incubated for 10 min in medium containing [^{3}H]ATP in the presence (shaded bars) or absence (open bars) of 100 μM ACTH$_{1\text{-}24}$. The agents to be tested were dissolved in incubation medium.

(Gispen *et al.*, 1979; Fig. 3). Also when ACTH was administered *in vivo* (icv) dose-dependent effects on the *in vitro* phosphorylation of B-50 were obtained (Zwiers *et al.*, 1977). ACTH specifically inhibited the B-50 kinase and did not affect the phosphoprotein phosphatase. Interestingly, the presence of Ca^{2+} was essential for B-50 kinase activity as no B-50 labeling was obtained after removal of endogenous Ca^{2+} with EGTA (Gispen *et al.*, 1979).

A more general inhibitory effect of ACTH was obtained in Triton-solubilized membrane fractions (Zwiers *et al.*, 1979) and in the presence of cytoplasmic proteins (Jolles *et al.*, 1980c). This suggests that the ACTH-sensitive protein kinase has a broad specificity and is able to phosphorylate other proteins when given access to them (Jolles *et al.*, 1981b).

The B-50 kinase and its protein substrate were isolated in soluble form from the membrane and purified by DEAE-celluose chromatography (Zwiers *et al.*, 1979) and ammonium sulfate precipitation (Zwiers *et al.*, 1980a). The highly purified enzyme complex was still sensitive to ACTH. The B-50 protein (MW 48 K; I.E.P. 4.5) and its kinase (MW 71 K; I.E.P. 5.5) were identified by isoelectric focusing (Zwiers *et al.*, 1980a). The B-50 protein was very susceptible to proteolytic breakdown (Zwiers *et al.*, 1980b). It yielded a large (MW 45 K) and a small (MW 1650) polypeptide. The small basic peptide has 15 amino acids. It was called Phosphorylation Inhibiting Peptide (PIP) due to its inhibitory effect on B-50 phosphorylation. Interestingly, icv injection of PIP into rats elicited a

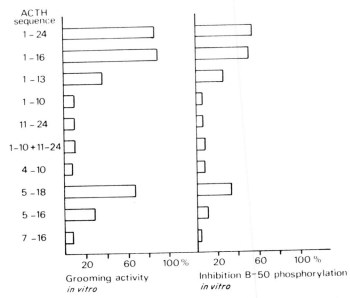

Fig. 3. The effect of ACTH$_{1-24}$ on excessive grooming *in vivo,* and inhibition of membrane protein phosphorylation *in vitro:* structure–activity relationship. This shows the correlation between the effects of ACTH fragments on these two phenomena. The data upon which the table is based are from Gispen *et al.* (1975, 1979) and Zwiers *et al.* (1978). The grooming was elicited by intraventricular administration of doses equimolar to 3 μg ACTH$_{1-24}$. Phosphorylation was studied in a synaptosomal plasma membrane fraction incubated in medium containing [γ-^{32}P]ATP for 20 sec and incorporation into proteins was determined. The results are expressed as percentage of control incubations.

grooming response that was similar to the ACTH-induced behavior (Zwiers *et al.,* 1980b). This again relates the *in vitro* effects of this peptide to its *in vivo* effects on grooming behavior.

Recent evidence suggests that the B-50 kinase/B-50 complex may have a function in the polyphosphoinositide metabolism in the membrane (Section III,D).

D. Polyphosphoinositide Metabolism

A special class of membrane phospholipids, the polyphosphoinositides, has been implicated in the metabolism of Ca^{2+} at the cell membrane (for reviews see Michell, 1975; Hawthrone and Pickard, 1979). Both the binding of Ca^{2+} to the membrane and its permeability are thought to be regulated by the metabolism of phosphatidylinositol (PI) and its phosphorylated derivates phosphatidylinositol 4-phosphate (DPI) and phosphatidylinositol 4,5-diphosphate (TPI; Fig. 4). The relatively high content of DPI and TPI in brain tissue and their rapid metabolism (Hawthorne and Kai, 1970) suggest that these lipids may play an important role in brain cell membrane function. The breakdown of brain PI and its rapid resynthesis via phosphatidic acid (PA) has first been described by Hokin and Hokin (1959). This so-called PI response has been observed in various tissues after receptor activation by hormones and neurotransmitters that utilize Ca^{2+} as their intracellular second messenger (Michell, 1975). An enhanced production of DPI and TPI in adrenal cortex was found after administration of ACTH to rats *in vivo* (Farese *et al.*, 1979). In the authors view, this suggests a direct relation between poly(PI)metabolism and ACTH-induced steroidogenesis, in view of the stimulatory effect of polyphosphorylated lipids (DPI, TPI, cardiolipin) on adrenal pregnenolone synthesis *in vitro* (Farese and Sabir, 1979). The phosphoinositides also mediate the effects of neurohypophyseal hormones.

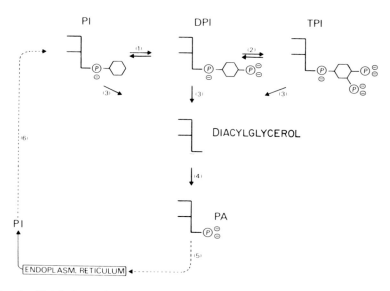

Fig. 4. Metabolism of polyphosphoinositides. Phosphatidyl-*myo*-inositol (PI) and its phosphorylated derivates phosphatidyl-*myo*-inositol 4-phosphate (DPI) and phosphatidyl-*myo*-inositol 4,5-diphosphate (TPI) are rapidly interconverted by lipid kinases and/or phosphomonoesterases (1 and 2). A poly(PI)-specific phosphodiesterase can hydrolyze the phosphodiesterlinkage in a phospholipase C-type manner, thereby yielding 1,2-diacylglycerol (1,2-daG) (3). This substance is rapidly phosphorylated to phosphatidic acid (PA) (4). All these reactions take place at the plasma membrane. Resynthesis of PI from PA takes place in the endoplasmic reticulum, and the transport of the lipid between plasma membrane and intracellular membrane is performed by phospholipid-exchanging proteins (5, 6). The interconversion of these compounds play an important role in calcium metabolism and neuronal cell function.

Effects of pituitary peptide hormones on polyPI metabolism in rat brain have recently been found. A preliminary study showed that both $ACTH_{1-24}$ and lysine-vasopressin (LVP) affected these lipids in a prelabeled synaptosomal fraction (Jolles *et al.*, 1979c). In subsequent studies a lysed synaptosomal fraction was used. It was found that $ACTH_{1-24}$ stimulated the formation of TPI, while at the same time inhibiting the production of PA (Jolles, 1980). These effects were evident after very short time periods (5 sec after the start of the incubation). The peptide was found to act on the lipid kinases, but not on the respective phosphatases (Jolles *et al.*, 1981b). Lipid kinase activity showed a strong calcium dependency (Jolles, 1980): activity was maximal at zero calcium, decreased rapidly as the calcium concentration increased, and was completely inhibited at Ca^{2+} concentrations greater than 1 mM. Similarly, the ACTH stimulation of TPI formation was also very sensitive to Ca^{2+}. The ACTH effect was maximal in the absence of Ca^{2+} resulting in a 3.5-fold stimulation of TPI formation. At 0.1 mM Ca^{2+}, the ACTH effect was reduced such that it gave only a twofold stimulation of TPI labeling. At 1 mM Ca^{2+} the effect of ACTH was completely abolished.

The evidence described so far suggested that a relation might exist between protein phosphorylation and poly(PI) metabolism. Direct evidence in support of this notion was obtained; the ACTH-sensitive B-50 kinase/B-50 protein complex (Section III,C) was solubilized from synaptosomal plasma membranes and purified by DEAE-cellulose chromatography. It appeared that the B-50 kinase/B-50 peak cochromatographed with a peak of DPI kinase activity (Jolles, 1980; Jolles *et al.*, 1980). Moreover, these peak fractions were also able to use DPI as a substrate for the production of PA. This indicated a combined phosphodiesterase/1,2-diacyclglycerol (1,2-daG) kinase activity. Addition of $ACTH_{1-24}$ to the DEAE-cellulose peak fractions inhibited both the phosphorylation of B-50 and the formation of PA, and stimulated the formation of TPI (Jolles *et al.*, 1980). The same peptide effects had been found in intact membranes (Jolles, 1980). Further fractionation of the DEAE peak fractions by ammonium sulfate precipitation isolated the PDE/1,2-daG kinase activity in the ASP 0–55% fraction (Jolles, 1980), but some of the DPI kinase activity was still present in the ASP 55–80% fraction (Jolles, 1980). As had been shown by Zwiers *et al.* (1980a), this fraction is enriched with respect to B-50/B-50 kinase. An inverse relationship was found between B-50 phosphorylation and TPI formation in this purified enzyme fraction; when the B-50 phosphorylation was increased (by prephosphorylation) TPI formation was decreased, but when B-50 phosphorylation was inhibited (by $ACTH_{1-24}$), TPI formation was stimulated (Jolles *et al.*, 1980). We suggested that the B-50 kinase and the DPI kinase originate from one enzyme complex of which B-50 is a regulatory unit. Autophosphorylation of this subunit would regulate the lipid-phosphorylating activity of the enzyme complex (Jolles, 1980; Jolles *et al.*, 1980, 1981b).

In order to further elucidate the mechanism of the effects of ACTH on neuronal membrane lipid phosphorylation, structure activity studies have been performed with ACTH fragments in a lysed synaptosomal fraction (Jolles *et al,* 1981a). A direct correlation was found between the effects of ACTH fragments on PA and TPI phosphorylation and those obtained on B-50 phosphorylation in synaptosomal plasma membranes, again stressing the relationship between B-50 phosphorylation and polyPI metabolism. The effects on TPI formation (stimulation) and PA formation (inhibition) decreased in the order $ACTH_{1-24} > ACTH_{5-18} > ACTH_{1-16} > ACTH_{1-13}$. $ACTH_{1-10}$ was ineffective (Jolles *et al.,* 1981a). Both the sequences $ATCH_{4-7}$ and $ACTH_{11-16}$ may contain information that is critical for the nature of the *in vitro* effects. Loss of the doublet of lysine residues at the position 15 and 16 abolished the effect on polyPI metabolism (Jolles *et al.,* 1981a). Similarly, βE inhibited the formation of PA, whereas γE (β-LPH_{61-77}), αE, the des-Tyr^{61}-derivates, and the enkephalins were not active. It was concluded (Jolles *et al.,* 1981a) that the structure–activity relationship correlates with the *in vivo* activity of the peptides on grooming behavior (Gispen *et al.,* 1975, 1976b), pole-jumping avoidance behavior (Greven and De Wied, 1973, 1977; De Wied *et al.,* 1975), and counteraction of morphine-induced analgesia (Gispen *et al.,* 1976a). We suggested that the correspondence in effects of ACTH and β-endorphin *in vivo* (grooming behavior and opiate-like activity on TPI and PA) might result from the structural similarity of $ACTH_{5-16}$ and β-LPH_{78-91}. Interestingly, the opiate antagonist naloxone had an intrinsic effect on the metabolism of TPI and PA (an inhibition, Jolles 1980) but failed to antagonize the effects of $ACTH_{1-24}$ on TPI and PA formation. It seems that the effects on avoidance behavior *in vivo* correlate with those on DPI metabolism *in vitro* (Section IV) though more research should be directed at elucidating this question (Jolles *et al.,* 1981c).

E. Brain Opiate Receptors

$ACTH_{1-28}$ and $ACTH_{4-10}$ have appreciable affinity for stereospecific opiate binding sites in brain synaptosomal plasma membranes as studied by competition with dihydromorphine (Terenius, 1975). From structure–activity studies it was concluded that $ACTH_{4-10}$ contains the active site. A second affinity site might be present C-terminal to the sequence 4-10 (Gispen *et al.,* 1976a, Terenius *et al.,* 1975). Vasopressin and α-MSH showed no affinity for the opiate receptor (Terenius, 1975; Terenius *et al.,* 1975). Analysis of the binding characteristics of $ACTH_{1-24}$ revealed a relatively low selectivity for agonist and antagonist binding sites, comparable to a partial agonist–antagonist like nalorfine (Terenius, 1976). The magnitude of the affinity constants of ACTH fragments (IC-50) is in the order of 10^{-5}–10^{-6} M, indicating that these peptides may not be powerful endogenous ligands for opiate receptors. Strikingly, the *in vitro* binding characteristics of βE suggest that this peptide also has mixed agonist–antagonist character (Bradbury *et al.*, 1976c). β-Endorphin is also very potent in inducing excessive grooming *in vitro*. In contrast, Met-enkephalin has pronounced agonist properties *in vitro* (Terenius, 1976; Bradbury *et al.*, 1976c) and has no grooming-inducing activity (Gispen *et al.*, 1976b).

These data and the recent findings that ACTH-like peptides have affinity for βE binding sites *in vitro* (Akil *et al.*, 1980) support the notion that ACTH-like peptides interfere with opiods at the level of the CNS (Wiegant *et al.*, 1977a). Similar conclusions have been reached by others. It has recently been proposed (Jacquet, 1978) that the effects of morphine are mediated by two classes of receptor. One, which is stereospecific and naloxone-sensitive (endogenous ligand: β-endorphin) and the other which is nonstereospecific and naloxone-insensitive. ACTH may be the endogenous ligand for the second receptor.

F. Interaction with Membrane Lipids

Membrane lipids may be important for opiate receptor binding. Anionic lipids like cerebroside sulfate are claimed to be part of the opiate receptor (Loh *et al.*, 1978). β-Endorphin specifically interacts with negative lipids (Wu *et al.*, 1979); this peptide formed an α-helix in solutions of cerebroside sulfate, PA, and PS. Ca^{2+} counteracted the helix-forming tendency and the authors suggested that micelles of the amphiphilic lipid cluster around the peptide chain, thus stabilizing the conformation by hydrophobic and hydrophilic interactions (Wu *et al.*, 1979). Likewise, it has been argued that $ACTH_{4-10}$, which is all probability is present in water as a random coil, may assume an α-helix structure in the more hydrophobic environment of the receptor site (Greven and de Wied, 1977). Also phosphatidylserine (Hoss *et al.*, 1977; Abood *et al.*, 1978) and long-chain-polyunsaturated fatty acids (Abood *et al.*, 1978) are claimed to modulate opiate receptor binding, opiate binding was completely inhibited after the release of only 5 nM fatty acid per milligram membrane protein. Similarly, Lin and Simon (1978) showed the importance of long-chain fatty acids for opiate binding. They found that phospholipase A inhibited opiate binding; removal of the released fatty acids with albumin restored the binding.

Long-chain polyunsaturated fatty acids are also implicated in the action of ACTH on the adrenal cortex (Laychock *et al.*, 1978; Schrey and Rubin, 1979). $ACTH_{1-24}$ specifically stimulated the turnover of the pool of arachidonic acid that is bound to PI (Schrey and Rubin, 1979). This effect was dependent on extracellular Ca^{2+} and could be mimicked by the Ca^{2+} ionophore A23187. No effect of cAMP could be found and there seemed to be no relation with the PI response. This increased turnover of arachidonoyl-PI was mediated by a Ca^{2+}-dependent phospholipase A_2 (Rubin *et al.*, 1979) and the authors proposed that this accelerated turnover is an early and specific event in the action of ACTH.

It has been shown (Section III,E) that pituitary peptides can specifically interact with membrane lipids. Schoch *et al.* (1979) exposed $ACTH_{1-24}$ to one side of an artificial lipid bilayer and found that part of the molecule became exposed to the other (trans) side of this artificial membrane. It was presumed that the N-terminal region must be associated with the trans side mainly because the cluster of four positive charges at residues 15–18 would present a formidable barrier to translocation through the hydrophobic interior of the membrane. The peptide–lipid interaction was enhanced in the presence of negative phospholipids. Also Arnaud *et al.* (1980) showed that ACTH is able to interact with these hydrophobic membrane components.

IV. CONCLUDING REMARKS

The pituitary hormones and their fragments influence a variety of biochemical events in membranes (Section III). The question arises are these different molecular events related and do they reveal something about the physiological action of the pituitary hormones *in vivo?*

As a potent tool to bridge the gap between the behavioral action of the neuropeptides and their neurochemical effects, the structure–activity study was used. By the use of this pharmacological technique, it has been possible to discern different regions in the polypeptide structure of ACTH, MSH, and LPH that code for different CNS effects. For instance, both from the effects of ACTH fragments on the behavior (Section II,D) and from the biochemical studies it was concluded that the peptide $ACTH_{1-24}$ contains two affinity sites. Clearly, the *in vitro* effects paralleled those obtained *in vivo;* both sites need to be present simultaneously for effects on grooming behavior (*in vivo*), and adenylcyclase, B-50 phosphorylation, and TPI/PA formation (*in vitro*); the present of either of the sites seems sufficient for effects on avoidance behavior and counteraction of morphine-induced analgesia (*in vivo*), and DPI formation (*in vitro*) (Fig. 5). This correlation may indicate that a change in the neurochemical parameters underlies the effects on the behavior. Finally we will address ourselves to the possible physiological significance of the observed *in vitro* effects.

Among the molecular correlates of pituitary hormone action, two membrane-associated phenomena seem to play a crucial role: the metabolism of calcium and the presence of anionic lipids. The polyanionic lipids DPI and TPI are very potent chelators of Ca^{2+} and Mg^{2+}. By virtue of this aspect these lipids are involved in binding these ions to the membrane (Hawthorne and Kai, 1970; Michell, 1975, 1979), and in regulating the de- or hyperpolarization of the membrane (Torda, 1974). In addition, PI seems to be of crucial importance for both basal and hormone-sensitive adenylate cyclase (Michell, 1975, 1979; Sandermann, 1978) and for ATPase activity (DePont *et al.,* 1978). Furthermore, PI

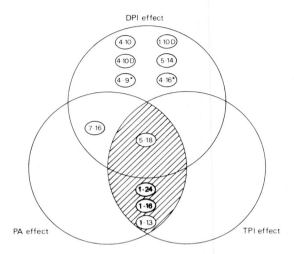

Fig. 5. Venn diagram showing the distribution of a number of ACTH fragments over three overlapping circles. Each circle represents a collection of ACTH fragments active on a particular aspect of polyphosphoinositide metabolism (stimulation of TPI formation; stimulation of DPI formation; inhibition of PA formation). The cross-hatched area represents the collection of ACTH fragments that is active on the phosphorylation of B-50 protein in synaptosomal plasma membranes (inhibition). The grey area represents the ACTH fragments that are active on adenylate cyclase activity in synaptosomal plasma membranes (inhibition). This figure suggests that different parts of the ACTH molecule can interact with different aspects of poly(PI) metabolism, B-50 protein phosphorylation, and adenylate cyclase activity. The sequences $[Met(O)^4$, D-Lys^8, Phe$^9]$ ACTH$_{4-9}$ and $[Met(O)^4$, D-Lys^8, Phe^9, D-Lys$^{11}]$ ACTH$_{4-16}$ are denoted as (4-9*) and (4-16*), respectively.

and its breakdown product 1,2-diacylglycerol activate a cytosolic protein kinase (Kishimoto *et al.*, 1980; Jolles *et al.*, 1980). The breakdown of PI results in the selective release of several membrane enzymes, indicating that PI is also involved in the anchoring of enzymes to the membrane (Arienti and Porcellatti, 1980; Fibean and Shukla, 1980; Kishimoto *et al.*, 1980; Michell, 1975, 1979). This anchoring may occur through the strong electrostatic interactions of the phosphatidylinositides with proteins via Ca^{2+} and Mg^{2+}. Anionic lipids (PI, TPI, and cerebroside sulfate) may also be an integral part of membrane receptors for several hormones such as thyrotropin, cholinerigic drugs, and opiates (Kohn, 1978; Abood *et al.*, 1978, Loh *et al.*, 1978). ACTH and βE seem able to interact with hydrophobic membrane constituents, especially negative lipids and direct effects of the neuropeptides on poly(PI) metabolism have been obtained (Farese *et al.*, 1979; Jolles, 1980; Jolles *et al.*, 1980, 1981a,b; Lo *et al.*, 1979; Shoch *et al.*, 1979).

Taken together, the presence of PI (and its derivative) determines the activity of membrane enzymes, the binding characteristics of membrane receptors, and the binding and permeability of Ca^{2+}. As the poly(PI) are rapidly interconverted (Fig. 4) (the enzymes involved in their metabolism are among the fastest acting known; Hawthorne and Kai, 1970) these lipids play a key role in membrane functioning. Consequently, the dynamic properties of the membrane can be regulated by agents that alter the phosphorylation or breakdown of phosphatidylinositol.

In our view, the experimental evidence that has been provided in Sections II and III supports the notion that neuropeptides like ACTH and βE modulate brain function by virtue of their influence on poly(PI) metabolism. A change in protein phosphorylation may mediate this influence, with a change in cAMP production as a secondary consequence. The experimental support comes from (1) the similar structure–activity relationship on the *in vivo* and *in vitro* parameters, (2) the fact that both $ACTH_{1-24}$ and βE are partial agonist–antagonist for the opiate receptor, (3) the affinity of both peptides for hydrophobic membrane constituents, (4) the proposed lipid–protein character of the opiate receptor, (5) the effect of *in vivo* administered ACTH on phosphorylation of B-50 *in vitro*, (6) the causal relation between B-50 protein phosphorylation and poly(PI) metabolism, (7) the copurification of the ACTH-sensitive B-50/B-50 kinase and the lipid kinases, (8) the sensitivity of the partially purified enzyme complex for βE, (9) the regulation of the dynamic properties of the membrane by Ca^{2+} and anionic lipids, (10) the calcium-sensitive effects of ACTH on adenylate cyclase activity in broken cell preparations.

We propose the following working hypothesis for the action of ACTH-like neuropeptides: the peptide (e.g., $ACTH_{1-24}$) interacts with a mixed protein–lipid receptor. The protein part of the receptor is the regulatory subunit of a DPI kinase, and the lipid part may be a phosphoinositide, or possibly cerebroside sulfate. As a result of the peptide–receptor interaction, the autophosphorylation of its regulatory subunit (the B-50 protein) is inhibited, which results in activation of the DPI kinase (possibly by dissociation of the holoenzyme). TPI is formed and the breakdown of PI is inhibited, resulting in an inhibited Ca^{2+} influx, and inhibition of secondary Ca^{2+}-dependent processes (like cAMP production). The relationship with the inhibited PA formation is not totally clear at present, though it is probable that both processes are manifestations of an inhibited PI breakdown and inhibited Ca^{2+} influx. In the view of Michell (1975) certain extracellular agents (e.g., muscarinic cholinergic or α-adrenergic) act to enhance intracellular Ca^{2+} by the breakdown of PI to 1,2-daG. The consequent membrane depolarization and enhanced phosphorylation of 1,2-daG to PA reflects the breakdown of PI. The influx of Ca^{2+} leads to a rapid breakdown of the polyphosphoinositides (DPI and TPI) at the cytoplasmic side of the membrane (Griffin and Hawthorne, 1978; Abdel-Latif *et al.*, 1978). Consequently, the agonist-induced increase in PA formation and the decrease in TPI are a manifestation of the same phenomenon (see Soukup *et al.*, 1978; Jolles *et al.*, 1981c). Clearly, the effects of the neuropeptides are opposite to those described for the classical neurotransmitters in that $ACTH_{1-24}$ and βE increase TPI and decrease PA. This could indicate that neuropeptides inhibit the Ca^{2+} influx (Guerrero-Munoz *et al.,* 1979), thereby hyperpolarizing the membrane (Torda, 1974) and acting to modulate the effects of the classical neurotransmitters.

It should be stated clearly that this is a working hypothesis; we have tried to fit the pieces of a puzzle together, but more research is needed to substantiate the notion that the neurochemical events are indeed related. For instance, it is not clear whether the modulatory action of the neuropeptides is located pre- or postsynaptically, as the present methods of tissue fractionation do not allow the proper separation of membranes from pre- and postsynaptic origin. Furthermore, though a correlation exists between *in vivo* and *in vitro* effects of the neuropeptides, a causal relation should be established, to determine whether the neurochemical events underlie the physiological action of the behaviorally active neuropeptides.

ACKNOWLEDGMENT

The authors thank Greet Hoekstra, Marleen Schmidt, and Jan Brakkee for their technical assistance, and Elly Looy, Greet Hoekstra, and Annie Endeman for their help in preparing the manuscripts upon which this chapter is based.

REFERENCES

Abdel-Latif, A. A., Green, K., Smith, J. P., McPherson, J. C., and Matheny, J. L. (1978). Norepinephrine-stimulated breakdown of triphosphoinositide of rabbit iris smooth muscle: Effects of surgical sympathetic denervation and in vivo electrical stimulation of the sympathetic nerve of the eye. *J. Neurochem.* **30**, 517–525.

Abood, L. G., Salem, N., MacNeil, M., and Butler, M. (1978). Phospholipid changes in synaptic membranes by lipolytic enzymes and subsequent restoration of opiate binding with phosphatidylserine. *Biochim. Biophys. Acta* **530**, 35–46.

Akil, H., Hewlet, W. A., Barchas, J. D., and Li, C. H. (1980). Binding of ^3H-β-endorphin to rat brain membranes: Characterization of opiate properties and interaction with ACTH. *Eur. J. Pharmacol.* **64**, 69–77.

Applezweig, M. H., and Baudry, F. D. (1955). The pituitary-adrenocortical system in avoidance learning. *Psychol. Rep.* **1**, 417–420.

Applezweig, M. H., and Moeller, G. (1959). The pituitary-adrenocortical system and anxiety in avoidance learning. *Acta Psychol.* **15**, 602–603.

Arienti, G., and Porcellatti, G. (1980). Relationship between phospholipids and receptors. In ''Receptors for Neurotransmitters and Peptide Hormones,''(G. Pepeu, M. J. Kahar, and S. J. Enna, eds.), pp. 43–49. Raven, New York.

Arnaud, J., Nobili, O., and Boyer, J. (1980). In vitro effects of adrenocorticotropic hormone on the pH profile of tri- and diester lipase activities from rat brain. *Biochim. Biophys. Acta* **617**, 524–528.

Bär, H. P., and Hechter, O. (1969). Adenylcyclase and hormone action. III. Calcium requirement for ACTH stimulation of adenylcyclase. *Biochem. Biophys. Res. Commun.* **35**, 681–686.

Bär, P. R., Schotman, P., and Gispen, W. H. (1980). Enkephalins affect hippocampal membrane phosphorylation. *Eur. J. Pharmacol.* **65**, 165–174.

Barchas, J. D., Akil, H., Elliott, G. R., Bruce Holman, R., and Watson, S. J. (1978). Behavioral neurochemistry: Neuroregulators and behavioral states. *Science (Washington, D.C.)* **200**, 964–973.

Beatty, D. A., Beatty, W. A., Bowman, R. E., and Gilchrist, J. C. (1970). The effects of ACTH, adrenalectomy and dexamethasone on the acquisition of an avoidance response in rats. *Physiol. Behav.* **5**, 939–944.

Belluzzi, J. D., Grant, N., Garsky, V., Sarantakis, D., Wise, C. D., and Stein, L. (1976). Analgesia induced in vivo by central administration of enkephalin in rat. *Nature (London)* **260**, 625–626.

Biggio, G., Casu, M., Corda, M. G., Di Bello, C., and Gessa, G. L. (1978a) Stimulation of dopamine synthesis in caudate nucleus by intrastriatal enkephalins and antagonism by naloxone. *Science (Washington, D.C.)* **200**, 552–554.

Biggio, G., Corda, M. G., Casu, M. and Gessa, G. L. (1978b). Striato-cerebellar pathway controlling cyclic GMP content in the cerebellum: Role of dopamine GABA and enkephalins. *Adv. Biochem. Psychopharmacol.* **18**, 227–244.

Bläsig, J., and Herz, A. (1976). Tolerance and dependence induced by morphine-like pituitary peptides in rats. *Naunyn-Schmiedeberg's Arch. Pharmacol.* **294**, 297–300.

Bohus, B., Gispen, W. H., and de Wied, D. (1973). Effects of lysine vasopressin and ACTH$_{4-10}$ on conditioned avoidance behavior of hypophysectomized rats. *Neuroendocrinology* **11**, 137–143.

Bohus, B., Hendrickx, H. H. L., van Kolfschoten, A. A. and Krediet, T. G. (1975). The effect of ACTH$_{4-10}$ on copulatory and sexually motivated approach behavior in the male rat. *In* "Sexual Behavior: Pharmacology and Biochemistry" (M. Sandler and G. L. Gessa, eds.), pp. 269–275. Raven, New York.

Bolles, R. J. (1960). Grooming behavior in the rat. *J. Comp. Physiol. Psychol.* **53**, 306–310.

Bloom, F., Segal, D., Ling, N., and Guillemin, R. (1976). Endorphins: Profound behavioral effects in rats suggest new etiological factors in mental illness. *Science (Washington, D.C.)* **194**, 630–632.

Botticelli, L. J., and Wurtman, R. J. (1979). β-Endorphin administration increases hippocampal acetylcholine levels. *Life Sci.* **24**, 1799–1804.

Bradbury, A. F., Feldberg, W. F., Smyth, D. G., and Snell, C. R. (1976a). Lipotropin C-fragment: An endogenous peptide with potent analgesic activity. *In* "Opiates and Endogenous Opioid Peptides" (H. W. Kosterlitz, ed.), pp. 9–17. Elsevier, Amsterdam.

Bradbury, A. F., Smyth, D. G., and Snell, C. R. (1976b). Lipotropin: Precursor to two biologically active peptides. *Biochem. Biophys. Res. Commun.* **69**, 950–956.

Bradbury, A. F., Smyth, D. G., Snell, C. R., Birdsall, N. J. M., and Hulme, E. G. (1976c). The C-fragment of lipotropin: an endogenous peptide with high affinity for brain opiate receptors. *Nature (London)* **260**, 793–795.

Bristow, A. F., Gleed, C., Fauchère, J.-L., Schwyzer, R., Schulster, D. (1980). Effects of ACTH (corticotropin) analogues on steroidogenesis and cyclic AMP in rat adrenocortical cells. *Biochem. J.* **186**, 599–603.

Burbach, J. P. H., Loeber, J. G., Verhoef, J., de Kloet, E. R., and de Wied, D. (1979). Biotransformation of endorphins by a synaptosomal plasma membrane preparation of rat brain and by human serum. *Biochem. Biophys. Res. Commun.* **86**, 1296–1303.

Burbach, J. P. H., Loeber, J. G., Verhoef, J., Wiegant, V. M., de Kloet, E. R., and de Wied, D. (1980). Selective conversion of β-endorphin into peptides related to γ- and α-endorphin. *Nature (London)* **283**, 96–97.

Burkard, W. P., and Gey, K. F. (1968). Adenylcyclase in Rattenhirn. *Helv. Physiol. Pharmacol. Acta* **26**, 197–198.

Carenzi, A., Frigeni, V., and Della Bella, D. (1978). Synaptic localization of opiate receptors in rat straitum. *Adv. Biochem. Psychopharmacol.* **18**, 265–270.

Christensen, C. W., Harston, C. T., Kastin, K. J., Kostrzewa, R. M., and Spirtes, M. A. (1976). Investigations on alpha-MSH and MIF-I effects on cyclic AMP levels in rat brain. *Pharmacol. Biochem. Behav.* **5**, Suppl. 1, 117–120.

Cools, A. R., Wiegant, V. M., and Gispen, W. H. (1978). Distinct dopaminergic systems in ACTH-induced grooming. *Eur. J. Pharmacol.* **50**, 265–268.

Davis, L. G., and Ehrlich, Y. H. (1979). Opioid peptides and protein phosphorylation. *Adv. Exp. Med. Biol.* **116**, 233–244.

Delius, J. D. (1970). Irrelevant behavior, information processing and arousal homeostatis. *Psychologische Forsch.* **33**, 165–188.

Delius, J. D., Craig, B., and Chaudoir, C. (1976). Adrenocorticotropic hormone, glucose and displacement activities in pigeons. *Z. Tierpsychol.* **40**, 183–193.

De Pont, J. J. H. H. M., van Prooyen-van Eeden, A., and Bonting, S. L. (1978). Role of negatively charged phospholipids in highly purified (Na$^+$-K$^+$) ATPase from rabbit kidney outer medulla. *Biochim. Biophys. Acta* **508**, 464–477.

De Wied, D. (1964). Influence of anterior pituitary on avoidance learning and escape behavior. *Am. J. Physiol.* **207**, 255–259.

De Wied, D. (1966). Inhibitory effect of ACTH and related peptides on extinction of conditioned avoidance behavior in rats. *Proc. Soc. Exp. Biol. Med.* **122**, 28–32.

De Wied, D. (1969). Effects of peptide hormones on behavior. *In* "Frontiers in Neuroendocrinology" (W. F. Ganong and L. Martini, eds.). pp. 97–140. Oxford Univ. Press, London and New York.

DeWied, D. (1971). Pituitary-adrenal hormones and behavior. *In* "Normal and Abrnomal Development of Brain and Behavior" (G. A. A. Stoelinga and J. J. van der Werf ten Bosch, eds.), pp. 315–322. Leiden Univ. Press, Leiden, The Netherlands.

De Wied, D. (1974). Pituitary-adrenal system hormones and behavior. *In* "The Neurosciences: Third Study Program" (F. O. Schmitt and F. G. Worden, eds.), pp. 653–666. MIT Press, Cambridge, Massachusetts.

De Wied, D. (1977). Behavioral effects of neuropeptides related to ACTH, MSH and β-LPH. *Ann. N.Y. Acad. Sci.* **297**, 263–274.

De Wied, D., and Gispen, W. H. (1977). Behavioral effects of peptides. *In* "Peptides in Neurobiology" (H. Gainer, ed.), pp. 397–448. Plenum, New York.

De Wied, D., Witter, A., and Greven, H. M. (1975). Behaviorally active ACTH analogues. *Biochem. Pharmacol.* **24**, 1463–1468.

De Wied, D., Bohus, B., van Ree, J. M., and Urban, I. (1978a). Behavioral and electrophysiological effects of peptides related to lipotropin (β-LPH). *J. Pharmacol. Exp. Ther.* **204**, 570–580.

De Wied, D., Kovács, G. L., Bohus, B., van Ree, J. M. and Greven, H. M. (1978b). Neuroleptic activity of the neuropeptide β-LPH$_{62-77}$ (|Des-tyr|γ-endorphin; dTγE). *Eur. J. Pharmacol.* **49**, 427–436.

Dunn, A. J., Green, E. J., and Isaacson, R. L. (1979). Intracerebral adrenocorticotropic hormone mediates novelty-induced grooming in the rat. *Science (Washington, D.C.)* **203**, 281–283.

Ehrlich, Y. H., Davis, L. G., Keen, P., and Brunngraber, E. G. (1980). Endorphin-regulated protein phosphorylation as a functional entity of the opiate-receptor complex. *In* "Endogenous and Exogenous Opiate Agonists and Antagonists" (E. L. Way, ed.), pp. 229–232. Pergamon, Oxford.

Fain, J. N. (1973). Biochemical aspects of drug and hormone action on adipose tissue. *Pharmacol. Rev.* **25**, 67–118.

Farese, R. V., and Sabir, A. M. (1979). Polyphosphorylated glycerolipids mimic adrenocorticotropin-induced stimulation of mitochondrial pregnenolone synthesis. *Biochim. Biophys. Acta* **575**, 299–304.

Farese, R. V., Sabir, A. M., and Vandor, S. L. (1979). Adrenocorticotropin acutely increases adrenal polyphosphoinositides. *J. Biol. Chem.* **254**, 6842–6844.

Ferrari, W., Gessa, G. L., and Vargiu, L. (1963). Behavioral effects induced by intracisternally injected ACTH and MSH. *Ann. N.Y. Acad. Sci.* **104**, 330–345.

Finean, J. B., and Shukla, S. D. (1980). Enzymes linked to phosphatidylinositol in plasma membranes. *Biochem. Soc. Trans.* **8**, 43.

Flexner, J. B., and Flexner, L. B. (1971). Pituitary peptides and the suppression of memory by puromycin. *Proc. Natl. Acad. Sci. (U.S.A.)* **68**, 2519–2521.

Flood, J. F., Jarvik, M. E., Bennett, E. L., and Orme, A. E. (1976). Effects of ACTH peptide fragments on memory formation. *Pharmacol. Biochem. Behav.* **5**, Suppl. 1, 41–51.

Forn, J., and Krishna, G. (1971). Effect of norepinephrine, histamine and other drugs on cyclic 3′,5′-AMP formation in brain slices of various animal species. *Pharmacology* **5**, 193–204.

Frederickson, R. C. A. (1977). Enkephalin pentapeptides: a review of current evidence for a physiological role in vertebrate neurotransmission. *Life Sci.* **21**, 23–42.

Garrud, P., Gray, J. A., and de Wied, D. (1974). Pituitary-adrenal hormones and extinction of rewarded behaviour in the rat. *Physiol. Behav.* **12**, 109–119.

Gispen, W. H., and Isaacson, R. L. (1980). ACTH-induced excessive grooming in the rat. *Pharmacol. Ther.* **12**, 209–246.

Gispen, W. H., and Wiegant, V. M. (1976). Opiate antagonists suppress $ACTH_{1-24}$-induced excessive grooming in the rat. *Neurosci. Lett.* **2**, 159–162.

Gispen, W. H., Wiegant, V. M., Greven, H. M., and de Wied, D. (1975). The induction of excessive grooming in the rat by intraventricular application of peptides derived from ACTH. Structure-activity studies. *Life Sci.* **17**, 645–652.

Gispen, W. H., Buitelaar, J., Wiegant, V. M., Terenius, L., and de Wied, D. (1976a). Interaction between ACTH fragments, brain opiate receptors and morphine analgesia. *Eur. J. Pharmacol.* **39**, 393–397.

Gispen, W. H., Wiegant, V. M., Bradbury, H. F., Hulme, E. C., Smyth, D. G., Snell, C. R., and de Wied, D. (1976b). Induction of excessive grooming in the rat by fragments of lipotropin. *Nature (London)* **264**, 794–795.

Gispen, W. H., van Ree, J. M., and de Wied, D. (1977). Lipotropin and the central nervous system. *Int. Rev. Neurobiol.* **20**, 209–249.

Gispen, W, H., Zwiers, H., Wiegant, V. M., and Schotman, P., and Wilson, J. E. (1979). The behaviorally active neuropeptide as neurochromone and neuromodulator: The role of cyclic nucleotides and phosphoproteins. *In* "Modulators, Mediators and Specifiers in Brain Function" (Y. H. Ehrlich, J. Volavka, L. G. Davis, and E. G. Brunngraber, eds.), pp. 199–224. Plenum, New York.

Gispen, W. H., Ormond, D., ten Haaf, J., and de Wied, D. (1980). Modulation of ACTH-induced grooming by (des-tyr¹)-γ-endorphin and haloperidol. *Eur. J. Pharmacol.* **63**, 203–207.

Glossman, H., and Gips, H. (1975). Bovine adrenal cortex adenylate cyclase: properties of the particulate enzyme and effects of guanylate cyclase. *Naunyn-Schmiedeberg's Arch. Exp. Path. Pharmakol.* **289**, 77–97.

Glossman, H., and Gips, H. (1976). Adrenal cortex adenylate cyclase. Is Ca^{2+} involved in ACTH stimulation? *Naunyn-Schmiedeberg's Arch. Exp. Path. Pharmakol.* **292**, 199–203.

Gold, P. E., and Van Buskirk, R. (1976). Enhancement and impairment of memory processes with post-trial injections of adrenocorticotropic hormone. *Behav. Biol.* **16**, 387–400.

Gráf, L., Ronai, A. Z., Bajusz, S., Csek, G., and Szekely, J. I. (1976). Opioid agonist activity of β-lipotropin fragments: A possible biological source of morphine-like substances in the pituitary. *FEBS Lett.* **64**, 181–184.

Greengard, P. (1976). Possible role for cyclic nucleotides and phosphorylated membrane proteins in postsynaptic actions of neurotransmitters. *Nature (London)* **260**, 101–108.

Greengard, P. (1978). Phosphorylated proteins as physiological effectors. *Science (Washington, D.C.)* **199**, 146–152.

Greven, H. M., and de Wied, D. (1973). The influence of peptides derived from corticotropin (ACTH) on performance. Structure-activity studies. *In* "Drug Effects on Neuroendocrine Regulation" (E. Zimmermann, W. H. Gispen, B. H. Marks, and D. de Wied, eds), Vol. 39, pp. 429–442, Elsevier, Amsterdam.

Greven, H. M., and de Wied, D. (1977). Influence of peptides structurally related to ACTH and MSH on active avoidance behavior in the rat. A structure relationship. *In* "Frontiers of Hormone Research" (F. Tilders, D. Swaab, and Tj. B. van Wimersma Greidanus, eds.), pp. 140–152. Karger, Basel.

Griffin, H. D., and Hawthorne, J. N. (1978). Calcium-activated hydrolysis of phosphatidyl-myo-inositol 4-phosphate and phosphatidyl-myo-inositol 4,5-diphosphate in guinea-pig syn: ptosomes. *Biochem. J.* **176**, 541–552.

Guerrero-Munoz, F., de Lourdes Guerrero, M., Leong Way, E., and Li, C. H. (1979). Effect of β-endorphin on calcium uptake. *Science (Washington, D.C.)* **206**, 89–91.

Guillemin, R., Ling, N., and Burgus, R. (1976). Endorphins: Hypothalamic and neurohypophysial peptides with morphinomimetic activity. Isolation and primary structure of α-endorphin. *C. R. Acad. Sci. Ser. D.* **282**, 783–785.

Guth, S., Levine, S., and Seward, J. P. (1971). Appetitive acquisition and extinction effects with exogenous ACTH. *Physiol. Behav.* **7**, 195–200.

Halkerston, I. D. K. (1975). Cyclic AMP and adrenocortical function. *Adv. Cyclic Nucleotide Res.* **6**, 99–136.

Hambrook, J. M., Morgan, B. A., Rance, M. J., and Smith, C. F. C. (1976). Mode of deactivation of the enkephalins by rat and human plasma and rat brain homogenates. *Nature (London)* **262**, 782–783.

Harsing, L. G., Vizi, E. S., and Knoll, J. (1978). Increase by enkephalin of acetylcholine release from striatal slices of the rat. *Pol. J. Pharmacol. Pharm.* **30**, 387–395.

Hawthorne, J. N., and Kai, M. (1970). Metabolism of phosphoinositides. *In* "Handbook of Neurochemistry" (A. Lajtha, ed.), pp. 491–508. Plenum, New York.

Hawthorne, J. N., and Pickard, M. R. (1979). Phospholipids in synaptic function. *J. Neurochem.* **32**, 5–14.

Heald, P. J. (1962). Phosphoprotein metabolism and ion transport in nervous tissue: A suggested connexion. *Nature (London)* **193**, 451–454.

Hökfelt, T., and Fuxe, K. (1972). On the morphology and neuroendocrine role of the hypothalamic catecholamine neurons. *In:* "Brain-Endocrine Interaction. Median Emence: Structure and Function" (K. M. Knigge, D. E. Scott, and A. Weindl, eds.), *Int. Symp. München, 1971,* pp. 181–223. Karger, Basel.

Hokin, L. E., and Hokin, M. R. (1959). The mechanism of phosphate exchange in phosphatidic acid in response to acetylcholine. *J. Biol. Chem.* **234**, 1387–1390.

Hoss, W., Abood, L. G., and Smiley, C. (1979). Enhancement of opiate binding to neural membranes with an ethylglycolate ester of phosphatidylserine. *Neurochem. Res.* **2**, 303–304.

Isaacson, R. L., Dunn, A. J., Rees, H. D., and Waldock, B. (1975). $ACTH_{4-10}$ and improved use of information in rats. *Physiol. Psychol.* **4**, 159–162.

Iuvone, P. M., Morasco, J., Delanoy, R. L., and Dunn, A. J. (1978). Peptides and the conversion of (^{3}H) tyrosine to catecholamines: effects of ACTH-analogs, melanocyte-stimulating hormones and lysine-vasopressin. *Brain Res.* **139**, 131–139.

Izumi, K., Donaldson, J., and Barbeau, A. (1973). Yawning and stretching in rats induced by intraventricularly administered zinc. *Life Sci.* **12**, 203–210.

Jacquet, Y. F. (1978). Opiate effects after adrenocorticotropin or β-Endorphin injection in the periaqueductal grey matter of rats. *Science (Washington, D.C.)* **201**, 1032–1034.

Jhamandas, K., Sawynok, J., and Sutak, M. (1977). Enkephalin effects on release of brain acetyl-choline. *Nature (London)* **269**, 433–434.

Jolles, J. (1980). Neuropeptides, brain membrane phosphorylation and grooming behavior. Ph.D. thesis, State University of Utrecht, Utrecht, The Netherlands.

Jolles, J., Wiegant, V. M., and Gispen, W. H. (1978). Reduced behavioral effectiveness of ACTH$_{1-24}$ after a second administration: Interaction with opiates. *Neurosci. Lett.* **9**, 261–266.

Jolles, J., Rompa-Barendregt, J., and Gispen, W. H. (1979a). Novelty and grooming behavior in the rat. *Behav. Neural. Biol.* **25**, 563–572.

Jolles, J., Rompa-Barendregt, J., and Gispen, W. H. (1979b). ACTH-induced excessive grooming in the rat: The influence of environmental and motivational factors. *Horm. Behav.* **12**, 60–72.

Jolles, J., Wirtz, K. W. A., Schotman, P., and Gispen, W. H. (1979c). Pituitary hormones influence polyphosphoinositide metabolism in rat brain. *FEBS Lett.* **105**, 110–114.

Jolles, J., Zwiers, H., van Dongen, C. J., Schotman, P., Wirtz, K. W. A., and Gispen, W. H. (1980). Modulation of brain polyphosphoinositide metabolism by ACTH-sensitive protein phosphorylation. *Nature (London)* **286**, 623–625.

Jolles, J., Bär, P. R., and Gispen, W. H. (1981a). Modulation of brain polyphosphoinositide metabolism by ACTH and β-endorphin: Structure activity studies. *Brain Res.*, in press.

Jolles, J., Zwiers, H., Dekker, A., Wirtz, K. W. A., and Gispen, W. H. (1981b). ACTH$_{1-24}$ affects protein phosphorylation and polyphosphoinositide metabolism in rat brain. *Biochem. J.* **194**, 283–291.

Jolles, J., Schrama, L., and Gispen, W. H. (1981c). Rapid metabolism of polyphosphoinositides *in vitro* after pre-labelling *in vivo*. *Biochim. Biophys. Acta*, in press.

Kastin, A. J., Miller, L. H., Nockton, R., Sandman, C. A., Schally, A. V., and Stratton, L. O. (1973). Behavioral aspects of melanocyte-stimulating hormones (MSH). *In* "Drug Effects on Neuroendocrine Regulation" (E. Zimmermann, W. H. Gispen, B. H. Marks, and D. de Wied, eds.), Vol. 39, pp. 461–470. Elsevier, Amsterdam.

Keyes, J. B. (1974). Effect of ACTH on ECS-produced amnesia of a passive avoidance task. *Physiol. Psychol.* **2**, 307–309.

Kishimoto, A., Takai, Y., Mori, T., Kikkawa, U., and Nishizuka, Y. (1980). Activation of calcium and phospholipid-dependent protein kinase by diacylglycerol, its possible relation to phosphatidylinositol turnover. *J. Biol. Chem.* **255**, 2273–2276.

Kohn, L. D. (1978). Relationships in structure and function of receptors for glycoprotein hormones, bacterial toxins and interferon. *In* "Receptors and Recognition" (P. Cuatrecasas and M. F. Greaves, eds.), Series A, Vol. 5, pp. 134–212, Chapman & Hall, London.

Kostrzewa, R. M., Kastin, A. J., and Spirtes, M. A. (1975). α-MSH and MIF-I effects on catecholamine levels and synthesis in various rat brain areas. *Pharmacol. Biochem. Behav.* **3**, 1017–1023.

Kovács, G. L., and de Wied, D. (1978). Effects of amphetamine and haloperidol on avoidance behavior and exploratory activity. *Eur. J. Pharmacol.* **53**, 103–107.

Krebs, E. G., and Beavo, J. A. A. (1979). Phosphorylation and dephosphorylation of enzymes. *Annu. Rev. Biochem.* **48**, 923–959.

Krieger, D. T., Liotta, A., and Brownstein, M. J. (1977). Presence of corticotropin in brain of normal and hypophysectomized rats. *Proc. Natl. Acad. Sci. (U.S.A.)* **74**, 648–652.

Kudlow, J. E., Rae, P. A., Gutmann, N. S., Schimmer, B. P., and Burrow, G. N. (1980). Regulation of ornithine decarboxylase activity by corticotropin in adrenocortical tumor cell clones: Roles of cyclic AMP and cyclic AMP-dependent protein kinase. *Proc. Natl. Acad. Sci. (U.S.A.)* **77**, 2767–2771.

Laychock, S. G., Shen, J. C., Carmines, E. L., and Rubin, R. P. (1978). The effect of corticotropin on phospholipid metabolism in isolated adrenocortical cells. *Biochem. Biophys. Acta* **528**, 355–363.

Leonard, B. E. (1974). The effect of two synthetic ACTH analogues on the metabolism of biogenic amines in the rat brain. *Arch. Int. Pharmacodyn. Ther.* **207**, 242–253.

Leonard, B. E., Kafoe, W. F., Thody, A. J., and Shuster, S. (1976). The effects of α-melanocyte stimulating hormone (α-MSH) on the metabolism of biogenic amines in the rat brain. *J. Neurosci. Res.* **2**, 39–45.

Levine, S., and Jones, L. E. (1965). Adrenocorticotrophic hormone (ACTH) and passive avoidance learning. *J. Comp. Physiol. Psychol.* **59**, 357–360.

Lin, H.-K., and Simon, E. J. (1978). Phospholipase A inhibition of opiate receptor binding can be reversed by albumin. *Nature (London)* **271**, 383–385.

Lissák, K., and Bohus, B. (1972). Pituitary hormones and avoidance behavior of the rat. *Int. J. Psychobiol.* **2**, 103–115.

Lo, S. J., Chen, T. T., and Taylor, J. D. (1979). ACTH-induced internalisation of plasma membrane by Xantophores of the goldfish *Carassius Auretus L. Biochem. Biophys. Res. Commun.* **86**, 748–754.

Loh, H. H., Tseng, L. F., Wei, E., and Li, C. H. (1976). β-endorphin is a potent analgesic agent. *Proc. Natl. Acad. Sci. (U.S.A.)* **73**, 2895–2898.

Loh, H. H., Law, P. Y., Ostwald, T., Cho, T. M., and Way, E. L. (1978). Possible involvement of cerebroside sulphate in opiate receptor binding. *Fed. Proc., Fed. Am. Soc. Exp. Biol.* **37**, 147–152.

Mains, R. E., Eipper, B. A., and Ling, N. (1977). Common precursor to corticotropins and endorphins. *Proc. Natl. Acad. Sci. (U.S.A.)* **74**, 3014–3018.

Mason, J. W. (1968). A review of psychoneuroendocrine research on the pituitary-adrenal cortical system. *Psychosom. Med.* **30**, 576–607.

Michell, R. H. (1975). Inositol phospholipids and cell surface receptor function. *Biochim. Biophys. Acta* **415**, 81–148.

Michell, R. H. (1979). Inositol lipids in membrane function. *Trends Biochem. Sci. (Pers. Ed.)* **4**, 282–285.

Moroni, F., Cheney, D. L., and Costa, E. (1977). β-endorphin inhibits ACh turnover in nuclei of rat brain. *Nature (London)* **267**, 267–277.

Orwoll, E., Kendall, J. W., Lamorena, L., and McGievra, R. (1979). Adrenocorticotropin and melanocyte-stimulating hormone in the brain. *Endocrinology (Baltimore)* **104**, 1845–1852.

Paroli, E. (1967). Indagini sull'effetto antimorfinico dell'ACTH. I. Relazioni con il corticosurrene ed i livelli ematici degli 11-OH steroidi. *Arch. Ital. Sci. Farmacol.* **13**, 234–237.

Pelletier, G., and Leclerc, R. (1979). Immunohistochemical localization of adrenocorticotropin in the rat brain. *Endocrinology (Baltimore)* **104**, 1426–1433.

Peng Loh, Y. (1979). Immunological evidence for two common precursors to corticotropins, endorphins, and melanotropin in the neurointermediate lobe of the toad pituitary. *Proc. Natl. Acad. Sci. (U.S.A.)* **76**, 796–800.

Perchellet, J.-P., and Sharma, R. K. (1979). Mediatory role of calcium and cGMP in adrenocorticotropin-induced steroidogenesis by adrenal cells. *Science (Washington, D.C.)* **203**, 1259–1261.

Podesta, E. J., Milani, A., Steffen, H., and Neher, R. (1979). Steroidogenesis in isolated adrenocortical cells. *Biochem. J.* **180**, 355–363.

Pollard, H., Llorens-Cortes, C., and Schwartz, J. C. (1977). Enkephalin receptors on dopaminergic neurons in rat striatum. *Nature (London)* **268**, 745–747.

Pollard, H., Llorens, C., Schwartz, J. C., Gros, C., and Dray, F. (1978). Localization of opiate receptors and enkephalins in the rat striatum in relationship with the nigrostriatal dopaminergic system: Lesion studies. *Brain Res.* **151**, 392–398.

Ramaekers, F., Rigter, H., and Leonard, B. E. (1978). Parallel changes in behavior and hippocampal monoamine metabolism in rats after administration of ACTH-analogues. *Pharmacol. Biochem. Behav.* **8**, 547–551.

Rigter, H., and Popping, A. (1976). Hormonal influences of the extinction of conditioned taste aversion. *Psychopharmacology* **46**, 255–261.

Rigter, H., and van Riezen, H. (1975). Anti-amnesic effect of $ACTH_{4-10}$: Its independence of the nature of the amnesic agent and the behavioral test. *Physiol. Behav.* **14**, 563–566.

Rigter, H., van Riezen, H., and de Wied, D. (1974). The effects of ACTH and vasopressin analogues on CO_2-induced retrograde amnesia in rats. *Physiol. Behav.* **13**, 381–388.

Rossier, J., Vargo, T. M., Minick, S., Ling, N., Bloom, F. E., and Guillemin, R. (1977). Regional dissociation of β-endorphin and enkephalin contents of rat brain and pituitary. *Proc. Natl. Acad. Sci. (U.S.A.)* **74**, 5162–5165.

Rubin, R. P., Sink, L. E., Schrey, M. P., Day, A. R., Liao, C. S., and Freer, R. J. (1979). Secretagognes for lysosomal enzyme release as stimulants of arachidonoyl-phosphatidylinositol turnover in rabbit neutrophils. *Biochem. Biophys. Res. Commun.* **90**, 1364–1370.

Rudman, D. (1976). Injection of melatonin into cisterna magna increases concentrations of 3′,5′-cyclic guanosine monophosphate in cerebrospinal fluid. *Neuroendocrinology* **20**, 235–242.

Rudman, D., and Isaacs, J. W. (1975). Effect of intrathecal injection of melanotropic-lipolytic peptides on the concentration of 3′,5′-cyclic adenosine monophosphate in cerebrospinal fluid. *Endocrinology* **97**, 1476–1480.

Sandermann, H., Jr. (1978). Regulation of membrane enzymes by phospholipids. *Biochim. Biophys. Acta* **515**, 209–237.

Schoch, P., Sargent, D. F., and Schwyzer, R. (1979). Hormone-receptor interactions: corticotropin-(1-24)-tetracosapeptide spans artificial lipid-bilayer membranes. *Biochem. Soc. Trans.* **7**, 846–850.

Schrey, M. P., and Rubin, R. P. (1979). Characterization of a calcium-mediated activation of arachidonic acid turnover in adrenal phospholipids by corticotropin. *J. Biol. Chem.* **254**, 11234–11241.

Schwartz, J. C., Pollard, H., Llorens, C., Malfroy, B., Gros, C., Pradelles, Ph., and Dray, F. (1978). Endorphins and endorphin receptors in striatum: Relationship with dopaminergic neurons. *Adv. Biochem. Psychopharmacol.* **18**, 245–264.

Schwartz, J. C., De La Baume, S., Llorens, C., Patey, G., and Pollard, H. (1979). Opiate receptors localized presynaptically on dopaminergic and noradrenergic neurons in brain: Relevance to tolerance and dependence to opioids. *In* "Catecholamines: Basic and Clinical Frontiers" (E. Usdin, I. J. Kopin, and J. Barchas, eds.), pp. 1083–1085. Pergamon, Oxford.

Sevenster, P. (1961). A causal analysis of a displacement activity. *Behaviour Suppl.* **9**, 1–170.

Soukup, J. F., Friedel, R. O., and Shanberg, S. M. (1978). Microwave irradiation fixation for studies of polyphosphoinositide metabolism in brain. *J. Neurochem.* **30**, 635–638.

Spirtes, M. A., Kostrzewa, R. M., and Kastin, A. J. (1975). α-MSH and MIF-I effects on serotonin levels and accumulation in various rat brain areas. *Pharmacol. Biochem. Behav.* **3**, 1011–1015.

Spirtes, M. A., Christensen, C. W., Hartson, C. T., and Kastin, A. J. (1978). α-MSH and MIF-I effects on cGMP levels in various rat brain regions. *Brain Res.* **144**, 189–193.

Subramanian, N., Mitznegg, P., Sprügel, W., Domschke, S., Wünsch, E., and Demling, L. (1977). Influence of enkephalin on K^+-evoked efflux of putative neurotransmitters in rat brain. *Naunyn-Schmiedeberg's Arch. Pharmacol.* **299**, 163–165.

Sutherland, E. W., and Robinson, G. A. (1966). The role of cyclic 3′,5′-AMP in responses to catecholamines and other hormones. *Pharmacol. Rev.* **18**, 145–161.

Telegdy, G., and Kovács, G. L. (1979a). Role of monoamines in mediating the action of ACTH, vasopressin and oxytocin. *In* "Central Nervous System Effects of Hypothalamic Hormones and Other Peptides" (R. Collu, A. Barbeau, J. R. Ducharme, and J. G. Rochefort, eds.), pp. 189–205. Raven, New York.

Telegdy, G., and Kovács, G. L. (1979b). Role of monoamines in mediating the action of hormones on learning and memory. *In* "Brain Mechanisms in Memory and Learning: From Single Neuron to Man" (M. A. B. Brazier, ed.), IBRO Monograph Series, pp. 249–268. Raven, New York.

Terenius, L. (1975). Effect of peptides and amino acids on dihydromorphine binding to the opiate receptor. *J. Pharm. Pharmacol.* **27,** 450–452.

Terenius, L. (1976). Somatostatin and ACTH are peptides with partial agonist-like opiate receptor selectivity. *Eur. J. Pharmacol.* **38,** 211–213.

Terenius, L., Gispen, W. H., and de Wied, D. (1975). ACTH-like peptides and opiate receptors in the rat brain: Structure–activity studies. *Eur. J. Pharmacol.* **33,** 395–399.

Tinbergen, N. (1940). Die Übersprungbewegung. *Z. Tierpsychol.* **4,** 1–40.

Torda, C. (1974). Model of molecular mechanism able to generate a depolarization-hyperpolarization cycle. *Int. Rev. Neurobiol.* **16,** 1–66.

Trabucchi, M., Polo, A., Tonon, G. C., and Spano, P. F. (1979). Interaction among enkephalinergic and dopaminergic systems in striatum and limbic forebrain. *In* "Catecholamines: Basic and Clinical Frontiers" (E. Usdin, I. J. Kopin, and J. Barchas, eds.) pp. 1053–1055. Pergamon, Oxford.

Van Loon, G. R., and Kim, C. (1978). β-endorphin-induced increase in striatal dopamine turnover. *Life Sci.* **23,** 961–970.

Van Ree, J. M., and de Wied, D. (1976). Neurohypophyseal hormones and morphine dependence. *In* "Opiates and Endogenous Opioid Peptides" (H. W. Kosterlitz, ed.), pp. 443–445. North-Holland Publ., Amsterdam.

Verhoef, J., Witter, A., and de Wied, D. (1977). Specific uptake of a behaviorally potent (^{3}H)ACTH$_{4-9}$ analog in the septal area after intraventricular injection in rats. *Brain Res.* **131,** 117–128.

Versteeg, D. H. G. (1973). Effect of two ACTH-analogs on noradrenaline metabolism in rat brain. *Brain Res.* **49,** 483–485.

Versteeg, D. H. G. (1980). Interaction of peptides related to ACTH, MSH and β-LPH with neurotransmitters in the brain. *Pharmacol. Ther.* **11,** 535–557.

Vizi, E. S., Harsing, L. G., and Knoll, J. (1977). Presynaptic inhibition leading to disinhibition of acetylcholine release from interneurons of the caudate nucleus: Effect of dopamine, β-endorphin and D-Ala^2-Pro^5-enkephalinamide. *Neuroscience* **2,** 953–961.

Von Hungen, K., and Roberts, S. (1973). Adenylate cyclases receptors for adrenergic neurotransmitters in rat cerebral cortex. *Eur. J. Biochem.* **36,** 391–401.

Watson, S. J., and Akil, H. (1980a). α-MSH in rat brain: Occurrence within and outside of β-endorphin neurons. *Brain Res.* **182,** 217–223.

Watson, S. J., and Akil, H. (1980b). On the multiplicity of active substances in single neurons: β-endorphin and α-melanocyte stimulating hormone as a model system. *In* "Hormones and the Brain" (D. de Wied and P. A. van Keep, eds.), pp. 73–86. MTP, Lancaster.

Watson, S. J., Akil, H., Richard C. W. III, Barchas, J. D. (1978a). Evidence for two separate opiate peptide neuronal systems. *Nature (London)* **275,** 226–228.

Watson, S. J., Richard, C. W. III, and Barchas, J. D. (1978b). Adrenocorticotropin in rat brain: Immunocytochemical localization in cells and axons. *Science (Washington, D.C.)* **200,** 1180–1182.

Wei, E., and Loh, H. (1976). Physical dependence on opiate-like peptides. *Science (Washington, D.C.)* **193**, 1262–1263.

Weiss, J. M., McEwen, B. S., Silva, M. T., and Kalkut, M. (1970). Pituitary–adrenal alterations and fear responding. *Am. J. Physiol.* **218**, 864–868.

Wiegant, V. M., and de Wied, D. (1981). Behavioral effects of pituitary hormones. *In* "Neuroendocrine Regulation and Altered Behavior" (P. D. Hrdina and R. L. Singhal, eds.), Croon, Helm, Ltd., London, in press.

Wiegant, V. M., and Gispen, W. H. (1977). ACTH-induced excessive grooming behavior in the rat: Latent activity of ACTH$_{4-10}$. *Behav. Biol.* **19**, 554–558.

Wiegant, V. M., Gispen, W. H., Terenius, L., and de Wied, D. (1977a). ACTH-like peptides and morphine: interaction at the level of the CNS. *Psychoneuroendocrinology (Oxford)* **2**, 63–69.

Wiegant, V. M., Cools, A. R., and Gispen, W. H. (1977b). ACTH-induced excessive grooming involves brain dopamine. *Eur. J. Pharmacol.* **41**, 343–345.

Wiegant, V. M., Jolles, J., and Gispen, W. H. (1978). β-Endorphin grooming in the rat: single dose tolerance. *In* "Characteristics and Function of Opioids, Developments in Neuroscience" (J. M. van Ree and L. Terenius, eds.), Vol. 4, pp. 447–450. Elsevier/North-Holland, Amsterdam.

Wiegant, V. M., Dunn, A. J., Schotman, P., and Gispen, W. H. (1979). ACTH-like neurotropic peptides: Possible regulators of rat brain cyclic AMP. *Brain Res.* **168**, 565–584.

Witter, A. (1980). On the presence of receptors for ACTH-neuropeptides in the brain. *In* "Receptors for Neurotransmitters and Peptide Hormones" (G. C. Pepeu, M. Kuhar, and L. Enna, eds.), pp. 407–414. Raven, New York.

Witter, A., Greven, H. M., and de Wied, D. (1975). Correlation between structure behavioral activity and rate of biotransformation of some ACTH$_{4-9}$ analogs. *J. Pharmacol. Exp. Ther.* **193**, 853–860.

Wood, P. L., Malthe-Sørenssen, D., Cheney, D. L., and Costa, E. (1978). Increase of hippocampal acetylcholine turnover rate and the stretching–yawning syndrome elicited by alpha-MSH and ACTH. *Life Sci.* **22**, 673–678.

Wood, P. L., Cheney, D. L., and Costa, E. (1979). Modulation of the turnover rate of hippocampal acetylcholine by neuropeptides: Possible site of action of α-melanocyte-stimulating hormone, adrenocorticotrophic hormone and somatostatin. *J. Pharmacol. Exp. Ther.* **209**, 97–103.

Wu, C. S. C., Lee, N. M., Loh, H. H., Yang, J. T., and Li, C. H. (1979). β-Endorphin: Formation of α-helix in lipid solutions. *Proc. Natl. Acad. Sci. (U.S.A.)* **76**, 3656–3659.

Zimmermann, E., and Krivoy, W. (1973). Antagonism between morphine and the peptides ACTH, ACTH$_{1-24}$ and α-MSH in the nervous system. *In* "Drug Effects on Neuroendocrine Regulation" (E. Zimmermann, W. H. Gispen, B. H. Marks, and D. de Wied, eds.), Vol. 39, pp. 383–394. Elsevier, Amsterdam.

Zimmermann, E., and Krivoy, W. A. (1974). Depression of frog isolated spinal cord by morphine and antagonism by tetracosactin. *Proc. Soc. Exp. Biol. Med.* **146**, 575–579.

Zwiers, H., Veldhuis, D., Schotman, P., and Gispen, W. H. (1976). ACTH, cyclic nucleotides and brain protein phosphorylation in vitro. *Neurochem. Res.* **1**, 669–677.

Zwiers, H., Wiegant, V. M., Schotman, P., and Gispen, W. H. (1977). Intraventricular administered ACTH and changes in rat brain protein phosphorylation: A preliminary report. *In* "Mechanism, Regulation and Special Functions of Protein Synthesis in the Brain (S. Roberts, A. Lajtha, and W. H. Gispen, eds.), pp. 267–273. Elsevier, Amsterdam.

Zwiers, H., Wiegant, V. M., Schotman, P., and Gispen, W. H. (1978). ACTH-induced inhibition of endogenous rat brain protein phosphorylation in vitro: Structure activity. *Neurochem. Res.* **3**, 455–463.

Zwiers, H., Tonnaer, J., Wiegant, V. M., Schotman, P., and Gispen, W. H. (1979). ACTH-sensitive protein kinase from rat brain membranes. *J. Neurochem.* **33**, 247–256.

Zwiers, H., Schotman, P., Gispen, W. H. (1980a). Purification and some characteristics of an ACTH-sensitive protein kinase and its substrate protein in rat brain membranes. *J. Neurochem.* **34**, 1689–1699.

Zwiers, H., Verhoef, J., Schotman, P., and Gispen, W. H. (1980b). A new phosphorylation-inhibiting peptide (PIP) with behavioral activity from rat brain membranes. *FEBS Lett.* **112**, 168–172.

12

MACROMOLECULES AND BEHAVIOR

Adrian J. Dunn

I. INTRODUCTION

There is possibly no greater challenge to neurobiology than to discern specific relationships between molecules in the brain and behavior. Our present understanding of how the brain works and how behavior is generated is rudimentary. Whereas our understanding of biochemical pathways and mechanisms is fairly sophisticated and we have some understanding of how the cell works, our knowledge of the mechanisms of intercellular communication within the brain is very limited. Nevertheless, over the past 40 years a very considerable literature on the associations between behaviors and specific brain molecules has accumulated.

Molecular Approaches to Neurobiology

While my assigned topic is macromolecules and behavior, many of the principles I shall discuss pertain to the relationships between any molecules and behavior. Much of the work on macromolecules has been associated with studies of learning and memory. This is not only a restricted aspect of behavior, but research in this area has had a very chequered history. Despite much effort, I believe we understand very little of the molecular mechanisms of information storage. It may be more useful to study other behaviors, to learn from them and to apply the knowledge gained to the more difficult problems of memory.

The particular studies associating macromolecules with learning and memory have been reviewed relatively frequently and comprehensively, so I shall not attempt that here. The interested reader may consult the reviews of Glassman (1969), Uphouse *et al.* (1974), Rose *et al.* (1976), Dunn (1976b), Agranoff (1976), Rose and Haywood (1977), Agranoff (1980), and Dunn (1980). Instead, I shall analyze approaches to the problems, and illustrate these by specific examples. Thus, I hope that those reading this chapter who have worked in the field will not be offended if I do not cite their work. This chapter is intended for the outsider.

II. GENERAL APPROACHES TO THE RELATIONSHIPS BETWEEN MOLECULES AND BEHAVIOR

As pointed out by so many writers in this field (e.g., Agranoff, 1980), there are only two principal approaches: the correlative and the interventive. This is true whether the physical substrates to be studied are molecules, cells, structures, or pathways. In the correlative approach, the aim is to detect a particular molecular change that occurs during the behavior, and by manipulating a variety of parameters, show that the change and the behavior (or a particular aspect of it) cannot be dissociated. In other words, the presence of the change should always correlate with the behavior. The interventive approach requires that interference with the particular molecular system should predictably affect the behavior. That is to say, inhibition of the process should impair or prevent the behavior; whereas facilitation may increase it. Most researchers have assumed, that at the very least, a combination of these two approaches will be necessary to establish specific associations between molecular changes and behavior.

Some of the difficulties inherent in each approach are obvious, and it is clear that success requires much luck. For correlational studies, it is first necessary to select a molecule whose concentration, disposition, or metabolism changes with a specific behavior. Then it is necessary to manipulate a large number of parameters to establish the correlation, with a relatively high probability that the change will turn out to be correlated with some aspect of the behavior other than the one of major interest. Finally, the establishment of a good correlation is insufficient to prove an association between the molecular event and the behavior.

A major problem with interventive studies is the high probability that a result may be trivial. Clearly, destroying a crucial brain structure or interfering with a metabolic pathway could impair behavior by a mechanism that is not at all specific. Interfering with an animal's ability to see or to move would impair its ability to learn many tasks but would tell us nothing about learning. Although the possibility of getting false positives in impairment studies is the greatest, false negatives are also possible. Thus, when one brain system is blocked, another one may be able to take over. Interventive agents that improve behavior would be much more useful (McGaugh, 1973). However, to elicit an improvement it is necessary that the chosen system be working less than optimally before the intervention (i.e., be rate-limiting). Clearly, this need not always be the case.

The problems of relating brain biochemistry to behavior break down into a number of areas: chemical, anatomical, pharmacological, and behavioral. I shall discuss each of these areas in turn before giving in detail some interesting examples of how progress can, and is, being made.

A. Chemical Problems

Although we know much about the chemistry of the brain, our understanding of its biochemistry still suffers from the limitations of available techniques. We are only able to measure the content of certain chemicals, the activities of certain enzymes, and certain kinetic parameters.

1. Chemical Content

While assays exist for many chemicals found in the brain, an important technical problem is whether the quantities measured in tissue samples accurately reflect *in vivo* concentrations. This problem is most obvious with molecules that have high metabolic activity, such as glucose and ATP, which are degraded very rapidly. For these cases, special techniques such as rapid freezing in liquid nitrogen or microwave fixation have been developed. Even so, the time taken for fixation is finite, and a residual uncertainty exists. Moreover, these techniques generally preclude good regional analysis and eliminate the possibility of cellular or subcellular fractionation. While macromolecules are generally considered to be relatively stable, some very important ones may not be. As will be discussed later, evidence has been presented for an extremely high metabolic turnover of brain nuclear RNA. Protein phosphorylation and dephosphorylation occur extremely rapidly, and it may be impossible to assess the true *in vivo* state (Conway and Routtenberg, 1979; see Chapter 11).

For behavioral studies, the major problem with measures of chemical content is sensitivity. The very best chemical assays would probably not detect a change of less than 5%, and many are less sensitive. Yet how much of a change can be expected in the whole brain content of a chemical in response to a change in one behavior? This is the "Catch 22" of Agranoff *et al.* (1978): if a change is large enough to be detected can it really be specific? The situation can be improved by localizing the change anatomically, but then problems can arise with the sensitivity of the assays. For this reason, relatively few studies have examined the cerebral content of chemicals in relation to behaviors. Notable exceptions to this are behaviorally characteristic genetic disorders, and certain studies of neurotransmitters (see for example, Hornykiewicz, 1974).

2. Enzyme Assays

Assays can or have been developed for most cerebral enzymes. These techniques can be made extremely sensitive especially with the use of radioisotopes or enzymatic recycling (Lowry, 1963), however, they are normally conducted in cell-free preparations. Under these conditions what can be measured at best is the *capacity* of the enzyme to perform a reaction, which may or may not be related to its true *in vivo* activity. The problem is compounded by the natural desire of biochemists to "optimize" activity, altering the incubation conditions to get the greatest possible activity. Often this results in nonphysiological conditions for the assay. For example, tyrosine hydroxylase is frequently assayed at pH 6 with a synthetic cofactor (dimethyltetrahydrobiopterin) and abnormal concentrations of substrate. Unfortunately, it is not always easy to determine the true physiological conditions. The problem is worse for membrane-bound enzymes, whose activity may be influenced by the state of the membrane. Also, membranes *in vivo* have quite different environments on either side, a situation rarely achieved *in vitro*. Thus, most studies of enzyme activity have been limited to long-term adaptive changes. These may be useful for the study of certain behaviors, but useless for others.

3. Kinetic Studies

Kinetic studies attempt to determine the rate of ongoing processes in the brain. The majority of them use a tracer technique in which a small amount of labeled material (normally radioactive) is introduced into the system and its disposition followed. The principle is straightforward; if the tracer is a good one it should be converted into metabolic products at the same rate as the endogenous compound allowing measurements of rate to be made. Much has been written on the problems of interpreting labeling data in brain (see Oja, 1967; Dunlop *et al.*, 1977, Dunn, 1977). Basically the problem is that of knowing the concentration of labeled tracer relative to that of the endogenous compound (i.e., specific activity) in the right place at the right time. Unfortunately, during the experiment the size of the precursor pool and its specific activity may change, and these factors must be taken into account when interpreting the data. Because of the problems of compartmentation (i.e., different specific activities and synthesis rates in different cellular compartments), it is probably not possible to accurately and unequivocally measure metabolic rates in brain cells by this technique (Dunn, 1977). Nevertheless, the technique is extremely sensitive and at the very least can detect metabolic changes although the precise nature of these changes may be hard to discern. Also with suitable choice of precursors, the biochemistry can often be simplified permitting the analysis of large numbers of samples.

The technique has been favored in research on brain macromolecules, particularly because it is important that the metabolic product be stable (otherwise secondary or back reactions must be considered). It is also very sensitive. While it may be unreasonable to expect detectable changes in the brain *content* of a particular protein, at crucial times, synthesis rates may be more dramatically changed, and thus more easily detected.

B. Anatomical Problems

The molecular changes associated with specific behaviors are likely to be associated with specific small populations of cells. Inclusion of unrelated unaffected cells only dilutes the magnitude of the change, and thus renders it less detectable. The neurochemist has frequently been criticized for "grinding up" whole brains or at least large regions, with little or no regard for their intricate anatomy. Ideally, if the expected location of the change is unknown, a large number of small samples should be analyzed. However, not only may the assay not be sensitive enough for this, but processing even 20 samples from the brain of *each* animal can be a formidable problem for many if not most assays. There is also a statistical problem in analyzing the data. If the level of significance is set at 5% ($\alpha = 0.05$), then 5% of the samples should yield significant results by chance alone. Thus if 20 regions are analyzed, on average, one will appear significant falsely (so called type I error).

It should be easiest to determine the crucial anatomical sites associated with behaviors by classical lesioning and stimulation techniques. However, these techniques have their own drawbacks. If lesioning a particular structure disrupts a behavior, it does not necessarily follow that the structure was essential for the behavior. The lesion could have caused secondary trans-synaptic degeneration in other structures, which resulted in the behavioral change. Likewise, the failure of a lesion to affect a behavior does not prove that the lesioned site was not involved. Redundancy in many parts of the brain is such that another intact structure may have assumed the function. It is notable that after more than 50 years, the effects of lesions on learning and memory are still controversial (see Horel, 1979, for example) and we still have no consensus on the structures involved.

Theoretically, biochemical techniques could be used to localize behaviorally relevant sites in the brain. Unfortunately, the biochemical techniques are often too cumbersome to be used as a screening technique. Some recently developed histological techniques may be usable, including the deoxyglucose uptake procedure (see below).

Studies relating neurotransmitters with specific behaviors, have been relatively more successful than those involving macromolecules. This is because the nature of the neurotransmitter often helps to define the anatomical location. Some of the neuropeptides have offered a golden opportunity, because in many cases they are remarkably behaviorally specific, and are localized to small populations of cells.

C. Pharmacological Problems

The pharmacological approach is an important one, not the least because it is extremely easy to use (hence the rapid expansion of behavioral neuropharmacology). It is, therefore, important that its premises be clearly understood. There is an aphorism which runs that there are two actions of every drug: "the one you know about, and the one you don't." This is an oversimplification of the problem of specificity. The specificity of a drug for a metabolic process will never be absolute; good drugs just have less side effects than bad ones. The two most important methods for dealing with the problems of drug specificity are (1) to use a variety of different drugs and (2) to show a correlation between the metabolic and behavioral effects of the drug. By using a variety of different drugs all active on the same metabolic process, it is to be expected that the side effects will be different, at least in degree. Moreover, drugs that are closely structurally related but are inactive on the selected metabolic process should be inactive behaviorally. The correlative technique is most powerful when used with various drugs having different mechanisms of action.

The finest example of such an analysis is that showing the ability of protein synthesis inhibitors to impair the formation of permanent memory. For detailed reviews of this question see Flood and Jarvik (1976), Squire (1976), and Dunn (1980). Cycloheximide, a known inhibitor of cerebral protein synthesis, impairs memory, at least in certain tasks. A side effect of cycloheximide is its complex effect on locomotor activity. However, isocycloheximide has similar effects on activity, but does not affect cerebral protein synthesis or memory (Segal *et al.*, 1971). Moreover, a variety of other protein synthesis inhibitors have similar effects on memory (see above reviews). But what if the side effects are the result of inhibition of cerebral protein synthesis? Protein synthesis inhibitors all inhibit the production of glucocorticoid hormones by the adrenal cortex, so is it possible as suggested by Nakajima (1975) that this adrenocortical effect is really responsible for the amnesia? This possibility can be ruled out, because aminoglutethimide inhibits adrenal steroidogenesis but by a mechanism not involving inhibition of protein synthesis, and it is not amnestic (Squire *et al.*, 1976; Dunn and Leibmann, 1977).

A much more involved problem is whether protein synthesis inhibitors affect memory by interfering with catecholamine metabolism. Protein synthesis inhibitors inhibit cerebral catecholamine synthesis (Flexner *et al.*, 1973). It is still not established whether this is a consequence of inhibited protein synthesis, the elevation of cerebral tyrosine that presumably follows the inhibition of protein synthesis, or an unrelated side effect (or some combination of these). Inhibitors of catecholamine synthesis that do not affect protein synthesis can be amnestic, although it is not clear that the amnestic characteristics of the two classes of drugs are similar (Flood and Jarvik, 1976; Rainbow *et al.*, 1976; Squire, 1976). The problem has been approached by attempting to correlate the biochemical changes with the amnesia. Using cycloheximide, Quinton and Kramarcy (1977) showed that there was a correlation between the dose-dependent inhibition of cerebral protein synthesis and amnesia. Unfortunately, Bloom *et al.* (1977) showed a similar dose-dependent correlation between the inhibition of cerebral catecholamine synthesis and amnesia. These results are not surprising, because it is likely that the three effects would be dose dependent, whether or not they are related. In fact, consideration of the time course of the effects suggests a better correlation between amnesia and inhibition of protein synthesis than between amnesia and inhibition of catecholamine synthesis (see Dunn, 1980). A more powerful approach is to compare the effects of *different* drugs on the biochemical and behavioral processes. This has been done to a limited extent, with the result that the correlation between amnesia and either biochemical measure is not particularly good. The most likely reason is that inhibition of *either* cerebral protein synthesis *or* catecholamine synthesis can cause amnesia (see Dunn, 1980).

An important extension of the pharmacological approach, is the application of drugs in precise cerebral locations. While this approach has been used most extensively with drugs active on neurotransmitter systems, it is also applicable to macromolecules. Several groups have used discrete application of protein synthesis inhibitors to study memory (Berman *et al.*, 1978; Boast and Agranoff, 1978). In these studies, protein synthesis has been monitored by autoradiographic analysis of brain sections following administraion of radioactive amino acid precursors. Thus, the regional and quantitative extent of inhibition of protein synthesis can be related to the amnesia. These studies have implicated the hippocampus, amygdala, caudate, and posterior lateral thalamus in learning.

D. Behavioral Problems

There can be many different problems depending on the behavior to be studied, and a discussion of them all is clearly beyond the scope of this review. One feature common to them all is the problem of timing. Since the first law of neurochemistry is that you can only kill your animal once, when you do it matters very much. With certain ongoing behaviors this does not present a problem, but with others, including learning the memory, it is not at all obvious when to look (see discussion by Entingh *et al.*, 1975; Agranoff, 1976). This is the kind of problem that can best be answered with the interventive approach. It should not be overlooked that behaviors will generally be related to chains of reactions or metabolic sequences, rather than to single reactions or types of molecules. Thus it may be appropriate to study different chemicals at different times during the behavior.

Because so much of the work relating macromolecules to behavior has been related to learning and memory, I shall discuss a few of the problems pertaining to this area here. For a more detailed analysis the reader should refer to Entingh *et al.* (1975), Rose *et al.* (1976), or Agranoff (1976).

Because there is more than one stage of memory (McGaugh, 1966; Gibbs and Ng, 1977), it is necessary to decide first which to study. Most studies have concentrated on the phase shortly after the learning. This is because it seems most likely that this is the time when memory is being formed, but more cogently, because this is the time when memory can be disrupted most readily by a variety of treatments including specific drugs (McGaugh and Dawson, 1971; Flood and Jarvik, 1976). This presents an important problem, because if the experimental subjects are killed and tissues processed at this time the extent of the memory cannot be tested. Thus it is not possible to attempt to correlate the biochemical change with the memory in the same animal. Another problem is that, associated with the learning experience, there will have been exposure to a number of other factors including environmental stimuli and stress, each of which may affect brain chemistry. These factors must thus all be "controlled" for.

A good example of how this may be done is provided by the work of the Glassman-Wilson group. Early studies had indicated a relative increase in the apparent incorporation of labeled uridine into RNA in mice trained in a conditioned avoidance task compared with controls (Zemp *et al.*, 1966). In this task the mice learned that shortly after a light came on and a buzzer was activated (*the conditioning stimuli, CS*), the grid floor of the box was electrified (*unconditioned stimulus, UCS*). To *escape* the footshock the mice could jump to a shelf a few inches above the grid floor. However they could *avoid* it by jumping to the shelf within 3 seconds after presentation of the CS. *Trained* mice clearly differ from *quiet* mice (mice left in their home cage and not exposed to the training apparatus) in a number of respects. One important control used was a *yoked* control. Yoked mice were treated just like trained mice but placed in an adjacent compartment of the "jumpbox" that lacked a shelf. In this way the yoked mouse received all the handling and the conditioning stimuli (including the footshock) that the trained mouse did, but could not learn to avoid because there was no shelf to jump to. Because the footshock was turned off as soon as the trained mouse jumped, the yoked mouse received the same amount of shock as the trained mouse when it escaped, and, like the trained mouse, did not receive footshock when the trained mouse avoided. Trained mice showed increased incorporation of uridine into RNA compared with yoked mice.

Another experiment was to study mice that had previously learned the task and were now *performing* it competently. Mice that were trained for four successive days and their RNA labeling was studied on the fourth day were not different from quiet mice (Adair *et al.*, 1968). A further control was to study *extinction* of the task. Mice were trained in the task for five successive days, but on the sixth day the footshock was no longer presented. The previously trained mouse now had to learn that it need no longer avoid. Mice undergoing extinction training showed increased incorporation of uridine into RNA compared with quiet mice (Coleman *et al.*, 1971). Retraining mice that had learned and extinguished the task, again resulted in increased incorporation of uridine into RNA compared with quiet mice (Coleman *et al.*, 1971). In other experiments, it was shown that *classically conditioned* mice, i.e., mice that received CS-UCS pairings but could not avoid or escape, did not differ from quiet mice in their uridine incorporation into RNA (Adair *et al.*, 1968). Thus, it was apparently necessary for mice to learn to actively avoid the footshock, to observe the neurochemical response. Thus taken at face value, these studies strongly suggested that the changes of brain RNA labeling were specifically related to active avoidance learning.

A similar behavioral analysis has been performed by Rose's group on the chick imprinting model (Horn *et al.*, 1973; Rose, 1980). Various factors are eliminated from the behavioral paradigm one at a time to determine whether each is responsible for the neurochemical response. The problem with this kind of analysis is that the whole may be more than the sum of the parts. Each time the behavioral paradigm is changed the brain will respond differently. It is not necessarily true, for example, that stress will produce one response and light stimuli another. Moreover, memory results from the *combination* of various stimuli, if any one is absent the memory is changed. This point was made earlier by Greenough (1976) and is also pertinent to environmental enrichment paradigms as pointed out in the excellent analysis by Uphouse (1980).

III. DEOXYGLUCOSE PROCEDURE

Changes in neural activity should in theory be associated with changes in the rate of cerebral blood flow or energy utilization (Sokoloff, 1977). Techniques for measurement of cerebral blood flow are not readily adaptable for histological use because the tracers used are by design too readily diffusible. However, a technique to measure cerebral glucose uptake was recently developed for histological analysis (Sokoloff *et al.*, 1977). Since glucose is almost the exclusive energy source for normal brain tissue, glucose uptake is closely related to energy utilization. Extensive evidence suggests that the major use of energy in the brain is to transport ions across membranes, so that changes of glucose uptake will largely reflect changes of neural activity (Sokoloff, 1977). This is why the procedure has been called a "metabolic stain."

Sokoloff's technique uses radioactively labeled 2-deoxyglucose, an analogue of glucose which is taken up by brain cells and phosphorylated by hexokinase. The 2-deoxyglucose 6-phosphate thus formed is not a substrate for any of the enzymes that normally metabolize glucose 6-phosphate, so that the 2-deoxyglucose 6-phosphate accumulates and is trapped forming a measure of past glucose uptake. Using quantitative autoradiography of the trapped [^{14}C]deoxyglucose 6-phosphate, and knowing the specific activity of the plasma deoxyglucose–glucose mixture, it is possible to calculate absolute rates of glucose uptake for particular cerebral structures (Sokoloff *et al.*, 1977). The technique has had important impact on neuroscience research and has even been adapted for human use (Phelps *et al.*, 1979).

Presently, it is an open question whether the deoxyglucose technique is sensitive enough to detect behaviorally significant changes. Results so far have shown a dramatic ability to highlight localized changes (see Sokoloff, 1977), but to date these have been primarily in sensory systems linked directly to receptors for external stimuli. Unfortunately, quantitative autoradiography has a high inherent variance, and in results published so far, only changes that exceed 20% have been found statistically significant. It is not at all clear that changes of this magnitude are likely to be associated with normal changes of ongoing behavior.

For this reason we have recently developed a modification of Sokoloff's procedure, using scintillation counting of free-hand dissected brain structures to determine deoxyglucose uptake (Dunn *et al.*, 1980). A further disadvantage of the classical Sokoloff procedure is that subjects are restrained during the deoxyglucose uptake period because of the intravenous cannula necessary to inject the deoxyglucose and to measure the plasma specific activity of the glucose. This is clearly undesirable for behavioral experiments because of the stress of restraint, and because it is desirable to record behavioral measures during the deoxyglucose uptake, which can later be correlated with the uptake pattern. Therefore, we decided to forego the possibility of measuring absolute glucose uptake rates by neglecting the intravenous cannula, presuming that changes in one cerebral structure relative to another would be sufficient for our purposes. We also showed that we could obtain valid results using subcutaneous injections of label, and economize by substituting [^{3}H]deoxyglucose for the more expensive ^{14}C. Meibach *et al.* (1980) have recently similarly justified the use of [^{3}H]deoxyglucose injected intraperitoneally in rats.

In our first study, we showed that electric footshock of mice caused regionally specific changes in deoxyglucose uptake; decreases in parietal cortex and brain stem, and an increase in hypothalamus (Delanoy and Dunn, 1978). In results, which should be considered preliminary, we have shown that intracerebroventricularly injected $ACTH_{1-24}$, which elicits grooming behavior (Rees *et al.*, 1976, see Chapter 11), and lysine vasopressin, which stimulates a characteristic hyperactivity in mice (Delanoy *et al.*, 1978), caused detectable and highly reproducible changes in the regional deoxyglucose uptake pattern of mouse brain. ACTH consistently decreased deoxyglucose uptake in the pyriform cortex, and vasopressin decreased deoxyglucose uptake in frontal cortex (Dunn *et al.*, 1980). While the technique lacks the anatomical resolution of the autoradiographic procedure, our present data indicate that it is capable of detecting changes of less than 10%.

An obvious and important question is whether the regionally specific changes in deoxyglucose uptake indicate the sites of action of the peptides or are associated with the behavioral changes. Originally, we thought that this might be determined by comparing the results obtained in two different ways: (1) a simple comparison of peptide-injected with saline-injected mice; (2) a correlation of the deoxyglucose uptake changes with the behavioral responses. Theoretically, the former should reveal sites of action of the peptides, and the latter the concomitants of the behavioral changes. In practice, we feel that the variance of the data precludes such clear-cut distinctions, and that the behavioral response often reflects the effectiveness of the injections. Instead we chose a pharmacological approach, using naloxone, which prevents ACTH-induced grooming, and $ACTH_{4-10}$, a behaviorally active analogue of ACTH, which does not elicit grooming (see Chapter 11). Preliminary data suggest that naloxone decreased the change in pyriform cortex, and that this change was not elicited by $ACTH_{4-10}$.

A further variant of the deoxyglucose procedure has been suggested by Altenau and Agranoff (1978). In their method, the tissue is labeled for a long "control" period (1–3 h) with [³H]deoxyglucose, and for a short (30–60 min) "experimental" period with [¹⁴C]deoxyglucose, during which some experimental manipulation is performed. Tissue samples are then "punched" from frozen sections and behavior-associated changes determined by the changes in the ratio of the two isotopes (i.e., ³H:¹⁴C) in the punch. This technique would appear to be very sensitive and applicable to very small pieces of tissue, but few data have yet been published using it.

IV. LEARNING AND MEMORY STUDIES

Learning and memory studies have been both the most exciting and the most frustrating on the chemistry of behavior. They have been focused on macromolecules for the simple reason that researchers have thought that only a macromolecule would be stable enough to retain a memory for many years. This is despite the lack of evidence that any brain molecules, even macromolecules, have half-lives this long. There is no inherent reason why all the molecules in a cell cannot be replaced, and yet the cell retain its original form and function. The same is clearly true for a synapse. Nevertheless, it has been hard to conceive that the images of your best friends' faces are stored in glucose molecules or amino acids. For lack of alternatives, learning has been assumed to occur by changes in "connectivity" between cells, that is, the change of the probability that one cell firing will cause the next in the pathway to do so (Changeux and Danchin, 1976). For obvious reasons, the focus has been on the synapses, the junctions between neurons, but there is no reason whatsoever why changes in dendritic structure, or in the axonal area could not suffice. Indeed, these would be closer to the molecular machinery involved in the production of new proteins and other macromolecules.

It has often been said that a major problem in studying the biochemistry of learning and memory is the respective constraints of experimental design in the two disciplines (Entingh *et al.*, 1975). It is argued that experiments that are well designed behaviorally to measure learning and memory preclude good neurochemical assays, and that good neurochemical assays can only be done under circumstances that make it difficult to measure learning and memory. I believe this analysis to be false. While there is no doubt that there are conflicts between the needs of the two analyses, neither technology is particularly good. We simply do not have good behavioral assays for memory. Nor do current biochemical analyses tell us much about what is really going on in a cell. These criticisms are less cogent for the study of behaviors more simple than learning and memory, but they still persist.

A. Nucleic Acids

Many of the early studies focused on ribonucleic acid (RNA). This was partly because of the finding that amino acid sequences of proteins were coded into base sequences of nucleic acids, suggesting the possibility that cerebral information could be similarly stored. It was also partly because the pioneering work of Hydén (1943) had shown that neuronal RNA metabolism was extraordinarily sensitive to nervous stimulation. Even now there is no adequate explanation for this finding, nor do we have any real understanding of RNA metabolism in mammalian cells, and especially in the brain.

Hydén (1943) showed that stimulation of rats could alter the RNA content of spinal motoneurons measured by cytospectrophotometry, a quantitative light absorption technique used on histological sections. Although the validity of this assay has been questioned (Rose, 1968), there is good evidence that it can be used effectively with appropriate procedures (Pevzner, 1972). Hydén used a variety of stimulants including stress or direct electrical stimulation of spinal roots. He came to the conclusions that "mild" stimulation would increase the neuronal RNA content, and that stronger or excessive stimulation would decrease it. He proposed that stimulation of the neurons somehow increased the use and hence the catabolism of RNA. In the short term, because RNA synthesis was enhanced, the net content increased; however, under excessive stimulation catabolism exceeded synthesis and there was a net loss of RNA. Later he became convinced that the reason for the initial rise was the transfer of RNA from the glial cells to the neurons, and that the eventual decline signaled the exhaustion of the RNA of both the neuron and the associated glia in the neuronal–glial unit. In subsequent studies involving the freehand microdissection of cells and biochemical analysis, he reported not only that the neuronal RNA increased, quantitatively accounting for a glial decrease, but that base composition data were consistent with the transfer of particular classes of RNA from glia to neurons (Hydén and Egyházi, 1963). These studies were later extended to a learning paradigm, transfer of handedness in the rat. The RNA content of cortical neurons was reported to increase, with an altered base composition, consistent with an increased messenger RNA (mRNA) content. Associated glial cells showed a decrease, consistent with glial–neuronal transfer (Hydén and Egyházi, 1964). Unfortunately, no repetition of these experiments outside Hydén's laboratory has ever been published, and they have been severely criticized on a number of grounds (Rose, 1968; Dunn, 1976b).

Evidence for the involvement of RNA in learning from interventive studies is still equivocal. A variety of agents interfering with RNA synthesis including actinomycin D, diaminopurine, and α-amanitin have been shown to impair learning, but others have not (e.g., 6-azaguanine) (see Agranoff, 1976; Dunn, 1980). While a consensus indicates that interference with RNA synthesis probably impairs learning, the drugs are so toxic (actinomycin D and α-amanitin are eventually fatal) that it is hard to be certain that the effect is specific for RNA synthesis.

Most studies other than those of Hydén have attempted to correlate learning with changes in RNA synthesis measured using labeled precursors. Unfortunately, RNA labeling data are almost impossible to interpret. This is over and above the question of accurately determining the specific radioactivity of the precursor in the appropriate "pool" at the appropriate time. Extensive studies have shown that in mammalian cells, the rate of RNA labeling is exceedingly fast, and that the "rapidly labeled" RNA has a half-life of 10 min or less (Harris, 1963; Dunn, 1976a). This means that the newly synthesized RNA has a half-life considerably shorter than the labeling times normally used. Under these circumstances, it is clearly not possible to estimate synthesis rates; the product RNA is simply broken down too fast. Thus, experiments that have used pulses longer than a few minutes have not measured synthesis rates. Furthermore, it is equally well established that most of the RNA labeled in less than 1 h is heterogenous nuclear RNA (hnRNA) and that much of this RNA never enters the cytoplasm (Harris, 1964; Dunn, 1976a). Moreover its cellular function is largely unknown; although a part of it is undoubtedly a precursor to mRNA (see Perry, 1976). Thus, RNA labeling studies in mammalian cells cannot accurately measure synthesis rates, and the function of much of the RNA that is labeled is unknown.

In the studies of the Glassman–Wilson group mentioned in Section II,D, the incorporation of labeled uridine into brain RNA was measured over a 45-min period (Zemp *et al.*, 1966). This is equivalent to several half-lives of the rapidly labeled RNA, but even at this time the RNA labeled is primarily nuclear (Dunn, 1976a). Surprisingly in these studies, the behavioral treatments were not initiated until 30 min after uridine administration. This was to allow time for the mice to recover from the stressful intracranial injections used for uridine administration. In consequence, trained and yoked mice were treated differently only for the last 15 min of labeling. When the RNA was analyzed, the incorporation in this period was averaged in with the previous 30 min of incorporation. Thus, the percentage changes determined need to be inflated to determine the real change in the last 15 min of labeling. Because Zemp *et al.* (1966) reported changes as high as 100% or more and averaging 25%, the behaviorally induced changes would have to be dramatic indeed.

The explanation lies in the correction factors used. To control for the somewhat variable uptake of uridine into the brain following the intracranial injections, the radioactivity in cerebral UMP was determined (UTP was not used because it would not survive the subcellular fractionation procedures used in these experiments). The estimate of RNA synthesis was thus: (radioactivity in RNA)/(radioactivity in UMP). Clearly, this ratio could increase either because the numerator increased or because the denominator decreased. Large changes are more easily derived from the latter. Subsequent experiments in which more quantitative techniques were used, verified that in these experiments the labeling of RNA did not change, whereas that of UMP decreased (Entingh *et al.*, 1974). An adequate biochemical explanation of this change of nucleotide labeling has not been presented.

B. Proteins

Changes in the total content of proteins would only be expected in the long term. Nevertheless, we might expect to observe changes in the rate of protein synthesis associated with specific behaviors. Inhibition of protein synthesis apparently interferes with the formation of memory (see detailed reviews by Squire, 1976; Dunn, 1980), but has little or no effect on other behaviors. This is remarkable considering the high rate of cerebral protein synthesis. Changes of protein synthesis associated with learning have been extensively reviewed previously (Agranoff, 1976; Dunn, 1976b, 1980; Rose and Haywood, 1977). Important questions are those of specificity, both behavioral and molecular.

One series of studies suggests that there is a general increase in the rate of cerebral protein synthesis associated with stress. Cerebral protein synthesis was studied by monitoring the incorporation of radioactive amino acids into protein in the ''jumpbox'' conditioned avoidance task described above. These studies indicated that there was indeed an increase in the incorporation of amino acids into brain protein in the mice that had previously been trained relative to quiet mice (Rees *et al.*, 1974). However, yoked mice also showed a similar effect. This result suggested that the metabolic change might have been associated with one or other of the stimuli used in the conditioned avoidance training (e.g., light, buzzers, shock) or be a nonspecific stress response. Further experiments showed that several of the stimuli (buzzers, shock, etc.) could mimic the biochemical responses to training. Furthermore, repeated training or presentation of the stimuli resulted in a decreased biochemical response (Rees *et al.*, 1974). Thus the changed amino acid incorporation into protein appreared to resemble a stress response. A hormonal factor was suggested both because of the known hormonal responses to stress and because the change of amino acid incorporation into protein was not regionally specific—several gross brain regions showed the change, or even tissue specific—the liver also showed a response (Rees *et al.*, 1974). Subsequently it was shown that adrenalectomized mice showed biochemical responses to footshock stress like intact mice, and that administered corticosterone did not alter amino acid incorporation into brain and liver proteins (Rees and Dunn, 1977). This ruled out adrenal corticosterone (and catecholamines) as mediators of the response. However, the brain response was absent in hypophysectomized mice (see Dunn and Schotman, 1981), and could be mimicked by administration of ACTH (Dunn *et al.*, 1978). These results strongly suggest that ACTH released in response to stress causes the increased labeling of brain protein. It seems likely that this mechanism explains many earlier data indicating increased protein synthesis in response to environmental stimuli (e.g., Appel *et al.*, 1967).

This effect of ACTH is consistent with a variety of other data suggesting an effect of ACTH on brain protein synthesis (Schotman *et al.*, 1976). In particular, cerebral protein synthesis is decreased in hypophysectomized rats, an effect that can be alleviated at least partially by administration of ACTH (Schotman *et al.*, 1972). These and other data suggest that the increased amino acid incorporation into brain protein observed following ACTH administration (or stress) truly reflect an increased rate of cerebral protein synthesis [see Dunn and Schotman (1981) for a detailed review].

It is interesting that studies from three different laboratories have failed to find evidence that this effect of ACTH is specific for any particular protein (Pavlík *et al.*, 1971; Reith *et al.*, 1975; Dunn and Gildersleeve, 1980). Originally, we had speculated that this increased cerebral protein synthesis during stress might facilitate learning the response (Dunn, 1976a). We chose to explain the lack of anatomical specificity by suggesting that the proteins would only be used in locations where the prevailing conditions were right. This explanation avoided the need to duplicate the anatomical specificity already present in the neuronal firing pattern. Clearly, this same rationale could be used to explain the lack of *molecular* specificity of the increased protein synthesis. Increased amounts of all proteins would be available for use either as enzymes or as building blocks for synapses or whatever. Is there any evidence that this is the way the brain works? I believe there is.

Wilson and colleagues have performed a variety of studies in the the proteins synthesized in neurons of sympathetic ganglia and transported axoplasmically. Two-dimensional gel electrophoresis has been used to study over 400 different proteins. Yet under most physiological conditions there do not appear to be detectable changes in the content of any of these proteins. Even after axotomy, the protein pattern was essentially unchanged, although the synthesis rates of several protein spots in the regenerating ganglion was altered (Hall *et al.*, 1978). Even those could be glial. The need to synthesize proteins in the perikaryon and transport them axoplasmically to the nerve terminals, means that a very considerable time can elapse before new proteins can be brought to the terminals (Lasek, 1980). In the case of long-axoned neurons such as are frequently found in the periphery, days, weeks, or even months could be involved. This situation is analogous to getting spare parts for an automobile to a distant part of the globe. The time required for delivery of a new protein from the cell body to a distant terminal could clearly be unacceptable. Under these circumstances, it might be reasonable to synthesize and transport continuously a complete set of all the proteins (spare parts) that could possibly be needed. Proteins that are not needed would simply be degraded at the terminals. This could explain why both the amounts and types of proteins synthesized and transported axoplasmically are so insensitive to environmental stimuli. If this postulate is true, it would have the interesting corollary that the specificity of new proteins incorporated, for example, into new synapses would be determined by degradation not synthesis. Technically specificity determined by degradation is much harder to study, but perhaps could best be done by using very long labeling periods.

C. Protein Derivatives—Glycoproteins and Phosphoproteins

Behavioral neurochemists became interested in protein derivatives because a variety of chemical groups could be added or subtracted very rapidly and thus form the basis of short-term memory (Dunn, 1976b; Agranoff, 1980). Glycoproteins also became of interest because the potential complexity of their structure made them candidates for the determinants of specificity in intercellular interactions (Irwin, 1974). Moreover, the changes of UMP labeling discovered during conditioned avoidance training (Entingh *et al.*, 1974, see above), suggested the possibility of changes in UDP-sugars, substrates for glycoprotein synthesis. Early experiments were largely negative (see Routtenberg, 1979). Then, Damstra-Entingh *et al.* (1974) showed a remarkable behavioral sensitivity of fucose incorporation into brain glycoproteins. Subsequently, Matthies' group (see Matthies, 1979) and Morgan and Routtenberg (1979) discovered changes of the incorporation of [³H]fucose into specific glycoproteins (separated by gel electrophoresis) associated with training in rats. Interestingly, in the latter study these changes were only manifested at long labeling periods (1 and 5 days) but not at 1 or 2 h. This is, of course, inconsistent with a role in short-term memory, but is consistent with our speculations on the determination of specificity by degradation rather than synthesis, since like proteins, most fucosyl glycoproteins appear to be synthesized in the cell body of the neuron and transported axoplasmically to the presynaptic terminals (Zatz and Barondes, 1971). Thus the changes of labeling seen at 1 and 5 days in the Morgan and Routtenberg experiment might reflect those glycoproteins that were selectively degraded or spared. This need not be true for all glycoproteins because there is evidence for the addition and removal of some sugars at synaptic terminals (e.g., glucosamine; Barondes, 1968).

There is also limited evidence for a behavioral role of another glycoconjugate, gangliosides. Like fucosyl glycoproteins, these are largely synthesized in the soma and transported axoplasmically (Forman and Ledeen, 1972). Nevertheless, there is evidence for rapid changes of ganglioside labeling (with glucosamine) associated with avoidance learning (Irwin and Samson, 1971; Dunn and Hogan, 1975).

Phosphate groups on proteins are much more actively metabolized, and this occurs *in situ*. In Chapter 11 evidence is presented for a specific alteration by ACTH of the phosphorylation of a protein in the synaptic plasma membrane. Recently, Routtenberg and Benson (1980) have shown changes in the phosphorylation of synaptic plasma membrane protein F as a result of passive (inhibitory) avoidance training (see also Gispen *et al.*, 1977). However, the phosphorylation state of this protein appears to be extraordinarily sensitive to environmental conditions (see Routtenberg, 1979; Conway and Routtenberg, 1979) and the effect of avoidance training is mimicked by footshock without training (Routtenberg and Benson, 1980). Protein F has been tentatively identified as pyruvate dehydrogenase (Morgan and Routtenberg, 1980). The changes in phosphoproteins may thus not be that behaviorally specific, and as mentioned above it will be difficult to assess their true *in vivo* state.

V. NEUROPEPTIDES AS TOOLS FOR STUDYING THE BIOCHEMISTRY OF BEHAVIOR

The last few years have seen a wealth of data regarding the functional role of neuropeptides in the brain. Whether the neuropeptides act as neurotransmitters or other "neuroregulators" or "neurocommunicators" (see Barchas *et al.*, 1978; Dismukes *et al.*, 1979), they can apparently convey from one cell to another, messages of remarkable behavioral specificity (see Liebeskind *et al.*, 1978). The much-studied endorphins may not be that specific; administered β-endorphin can produce a range of activities including analgesia, hypothermia, hyperthermia, locomotor activation and depression, and a range of endocrinological changes (see Dunn, 1981). However, other peptides injected into the brain in tiny doses can induce specific behaviors, angiotensin will induce drinking (Phillips *et al.*, 1979), and cholecystokinin will cause animals to stop eating (Smith and Gibbs, 1979). This specificity almost rivals the overblown claims of yesteryear for transfer of memory by neuropeptides (Ungar, 1974; see Dunn, 1976b; Rose and Haywood, 1977).

What advantages are there in the use of peptide-induced behaviors for the study of behavioral biochemistry? I believe there are many. First, the behavior can be precisely regulated by altering the dose of the peptide, and by using structural analogues, which can readily be made by substituting amino acids in the sequence. The analogues can also be used to determine specificity. Primary sites of action can be determined by locating the receptors. This can be done by combinations of biochemical, histological, and iontophoretic techniques (cf., the work of Phillips *et al.*, 1979, with angiotensin). But, perhaps most important, the peptides can be used for *in vitro* biochemical studies. This is extremely useful because biochemistry is easiest studied *in vitro,* yet behavior must be studied *in vivo.* The major biochemical problems in studying the biochemistry of behavior derive from the need to use *in vivo* biochemistry, which is relatively poorly developed and subject to error. With peptide-induced behaviors it is possible to use a combination of *in vivo* and *in vitro* approaches to attack the problem. If the behavioral response *in vivo* correlates with the biochemical response *in vitro* using a number of different peptide analogues, then this provides excellent evidence that the two are related. I shall not give any specific examples here, because Chapter 11 indicates the power of the method using ACTH peptides.

VI. RECEPTORS

Important macromolecules that should not be neglected in studies of the biochemistry of behavior are the receptors for various neurotransmitters. While there are a few studies indicating changes in receptor number during learning (see

Rose, 1980), there is an extensive literature detailing changes in receptor sensitivity underlying behavior. Receptor supersensitivity occurs when insufficient ligand for the receptor is available to interact with it; receptor subsensitivity occurs when there is excess ligand. In either case, the change in sensitivity may result from alterations in the number of receptors, their affinity for the ligand, or in the coupling of the receptor to cellular responses.

The classic example of receptor supersensitivity is the *denervation supersensitivity* to acetylcholine that follows denervation of a skeletal muscle (see Changeux and Danchin, 1976). In this case, as in most others studied, the sensitivity change is accomplished by an increase in receptor number. Receptor sensitivity changes are now known to occur in response to a variety of chronic treatments: e.g., chronic neuroleptic treatment induces supersensitivity to dopamine (Burt *et al.*, 1977); chronic tricyclic antidepressant treatment results in subsensitivity to norepinephrine (Crews and Smith, 1978); and chronic footshock stress results in subsensitivity to norepinephrine (Stone, 1979) (for a review see Schwartz *et al.*, 1978). It would appear that changes of receptor sensitivity are a major mechanism of cellular adaptation. In fact changes of this type have been suggested to underly diseases such as depression (see Bunney *et al.*, 1977) and schizophrenia (Owen *et al.*, 1978). Currently, we know little about the mechanism of these changes, except that it is generally a slow process and is often sensitive to inhibitors of protein synthesis (Schwartz *et al.*, 1978). The mechanisms may, however, vary widely, because rapid changes are seen in muscarinic receptors following imprinting (Rose, 1980), and in benzodiazepine receptors following seizures (Paul and Skolnick, 1978) and stress. Given the recent revelations of the complexity of receptor mechanisms, this should prove a fruitful area for future research.

VII. CONCLUDING REMARKS

This chapter has been a personal essay on macromolecules and behavior. I intended an overview, highlighting with specific examples the findings and directions of general interest. Largely it has dwelled on studies associated with learning and memory. This is because, by far the largest literature has accumulated on this topic. Nevertheless, I hope I have included sufficient examples of the study of macromolecules and other behaviors to encourage the budding behavioral neurochemist.

ACKNOWLEDGMENTS

The author's own research reported in this chapter has been supported by the National Institute of Mental Health and the Sloan Foundation. I am grateful to Neal Kramarcy for reading drafts of this chapter.

REFERENCES

Adair, L., Wilson, J. E., and Glassman, E. (1968). Brain function and macromolecules, IV. Uridine incorporation during different behavioral experiences. *Proc. Natl. Acad. Sci., U.S.A.* **61,** 917–922.

Agranoff, B. W. (1976). Learning and memory, approaches to correlating behavioral and biochemical events. *In* "Basic Neurochemistry" (G. J. Siegel, R. W. Albers, R. Katzmann, and B. W. Agranoff, eds.), 2nd ed., pp. 765–784. Little, Brown, Boston, Massachusetts.

Agranoff, B. W. (1980). Biochemical events mediating the formation of short-term and long-term memory. *In* "Neurological Basis of Learning and Memory" (Y. Tsukada and B. W. Agranoff, eds.), pp. 138–147. Wiley, New York.

Agranoff, B. W., Burrell, H. R., Dokas, L. A., and Springer, A. D. (1978). Progress in Biochemical approaches to learning and memory. *In* "Psychopharmacology: A Generation of Progress" (M. A. Lipton, A. DiMiascio, and K. F. Killman, eds.), pp. 623–635. Raven, N.Y.

Altenau, L. L., and Agranoff, B. W. (1978). A sequential double-label 2-deoxyglucose method for measuring regional cerebral metabolism. *Brain Res.,* **153,** 375–381.

Appel, S. H., Davis, W., and Scott, S. (1967). Brain polysomes: Response to environmental stimulation. *Science (Washington, D.C.)* **157,** 836–838.

Barchas, J. D., Akil, H., Elliot, G. R., Holman, R. B., and Watson, S. J. (1978). Behavioral neurochemistry: Neuroregulators and behavioral states. *Science (Washington, D.C.)* **200,** 964–973.

Barondes, S. H. (1968) Incorporation of radioactive glucosamine into macromolecules at nerve endings. *J. Neurochem.,* **15,** 699–706.

Berman, R. F., Kesner, R. P., and Partlow L. M. (1978). Passive avoidance impairment in rats following cycloheximide injection into the amygdala. *Brain Res.,* **158,** 171–188.

Bloom, A. S., Quinton, E. E., and Carr, L. A. (1977). Effects of cycloheximide, diethyldithiocarbamate and D-amphetamine on protein and catecholamine biosynthesis in mouse brain. *Neuropharmacology,* **16,** 411–418.

Boast, C. A., and Agranoff, B. W. (1978). Biochemical and behavioral effects of streptovitacin A in mice. *Neuroscience* **4,** 255 (Abstr.).

Bunney, Jr., W. E., Post, R. M., Anderson, A. E., and Kopanda, R. T. (1977). A neuronal receptor sensitivity mechanism in affective illness (a review of evidence). *Commun. Psychopharmacol.* **1,** 393–405.

Burt, D. R., Creese, I., and Snyder, S. H. (1977). Antischizophrenic drugs: chronic treatment elevates dopamine receptor binding in brain. *Science (Washington, D.C.)* **196,** 326–328.

Changeux, J. P., and Danchin, A. (1976) Selective stabilisation of developing synapses as a mechanism for the specification of neuronal networks. *Nature (London)* **264,** 705–712.

Coleman, M. S., Wilson, J. E., and Glassman, E. (1971). Incorporation of uridine into polysomes of mouse brain during extinction. *Nature (London)* **229,** 54–55.

Conway, R. G., and Routtenberg, A. (1979). Endogenous phosphorylation *in vitro:* differential effects of brain state (anesthesia, post-mortem) on electrophoretically separated brain proteins. *Brain Res.,* **170,** 313–324.

Crews, F. T., and Smith, C. B. (1978). Presynaptic alpha-receptor subsensitivity after long-term antidepressant treatment. *Science (Washington, D.C.)* **202,** 322–324.

Damstra-Entingh, T., Entingh, D., Wilson, J. E., and Glassman, E. (1974). Environmental stimulation and fucose incorporation into brain and liver glycoproteins. *Pharmacol. Biochem. Behav.,* **2,** 73–78.

Delanoy, R. L., and Dunn, A. J. (1978). Mouse brain deoxyglucose uptake after footshock, ACTH analogs, α-MSH, corticosterone or lysine vasopressin. *Pharmacol. Biochem. Behav.,* **9,** 21–26.

Delanoy, R. L., Dunn, A. J., and Tintner, R. (1978). Behavioral responses to intraventricularly administered neurohypophyseal peptides in mice. *Horm. Behav.*, **11**, 348–362.

Dismukes, R. D. *et al.* (1979). New concepts of molecular communication among neurons. *Brain Behav. Sci.*, **2**, 409–448.

Dunlop, D., Lajtha, A., and Toth, J. (1977). Measuring brain protein metabolism in young and adult rats. *In* ''Mechanisms, Regulation and Special Functions of Protein Synthesis in the Brain'' (S. Roberts, A. Lajtha, and W. H. Gispen, eds.), pp. 79–96. Elsevier, Amsterdam.

Dunn, A. J. (1976a). Biochemical correlates of training: a discussion of the evidence. *In* ''Neural Mechanisms of Learning and Memory'' (M. Rosenzweig and E. L. Bennett eds.), pp. 311–320. MIT Press, Cambridge, Massachusetts.

Dunn, A. J. (1976b). The chemistry of learning and the formation of memory. *In* ''Molecular and Functional Neurobiology'' (W. H. Gispen, ed.), pp. 347–387. Elsevier, Amsterdam.

Dunn, A. J. (1977). Measurement of the rate of brain protein synthesis. *In* ''Mechanisms, Regulation and Special Functions of Protein Synthesis in the Brain'' (S. Roberts, A. Lajtha, and W. H. Gispen, eds.), pp. 97–105. Elsevier, Amsterdam.

Dunn, A. J. (1980). Neurochemistry of Learning and Memory: An evaluation of recent data. *Annu. Rev. Psychol.* **31**, 343–390.

Dunn, A. J. (1981). Central nervous system effects of adrenocorticotropin (ACTH), β-lipotropin (β-LPH) and related peptides. *In* ''Molecular and Behavioral Neuroendocrinology'' (A. J. Dunn and C. B. Nemeroff, eds.), Spectrum, New York, in press.

Dunn, A. J., and Gildersleeve, N. B. (1980). Corticotrophin-induced changes in protein labelling: lack of molecular specificity. *Pharmacol. Biochem. Behav.* **13**, 823–827.

Dunn, A. J., and Hogan, E. L. (1975). Brain gangliosides: increased incorporation of (1-³H)glucosamine during training. *Pharmacol. Biochem. Behav.*, **3**, 605–612.

Dunn, A. J., and Leibmann, S. (1977). The amnestic effect of protein synthesis inhibitors is not due to inhibition of adrenal corticosteroidogenesis. *Behav. Biol.*, **9**, 411–416.

Dunn, A. J., and Schotman, P. (1981). Effects of ACTH and related peptides on cerebral RNA and protein synthesis. *Pharmacol. Ther.* **12**, 353–372.

Dunn, A. J., Rees, H. D., and Iuvone, P. M. (1978). ACTH and the stress-induced changes of lysine incorporation into brain and liver protein. *Pharmacol. Biochem. Behav.* **8**, 455–464.

Dunn, A. J., Steelman, S., and Delanoy, R. L. (1980). Intraventricular ACTH and vasopressin cause regionally specific changes in cerebral deoxyglucose uptake. *J. Neurosci. Res.*, **5**, 485–495.

Entigh, D., Damstra-Entigh, T., Dunn, A. J., Wilson, J. E., and Glassman, E. (1974). Brain uridine monophosphate: Reduced incorporation of uridine during avoidance learning. *Brain Res.* **70**, 131–138.

Entigh, D., Dunn, A. J., Wilson, J. E., Glassman, E., and Hogan, E. (1975). Biochemical approaches to the biological basis of memory. *In* ''Handbook of Psychobiology'' (M. S. Gazzaniga and C. Blakemore, eds.), pp. 201–238, Academic Press, New York.

Flexner, L. B., Serota, R. G., and Goodman, R. H. (1973). Cycloheximide and acetoxycy-cloheximide: Inhibition of tyrosine hydroxylase activity and amnestic effects. *Proc. Natl. Acad. Sci. U.S.A.*, **70**, 354–356.

Flood, J. F., and Jarvik, M. E. (1976). Drug influences on learning and memory. *In* ''Neural Mechanisms of Learning and Memory'' (M. R. Rosenzweig and E. L. Bennett, eds.), pp. 483–507. MIT Press, Cambridge, Massachusetts.

Forman, D. W., and Ledeen, R. W. (1972). *Science (Washington, D.C.)* **177**, 630–633.

Gibbs, M. E., and Ng, K. T. (1977). Psychobiology of memory: Towards a model of memory formation. *Biobehav. Rev.* **1**, 113–136.

Gispen, W. H., Perumal, R., Wilson, J. E., and Glassman, E. (1977). Phosphorylation of proteins of synaptosome-enriched fractions of brain during short term training experience: The effects of various behavioral treatments. *Behav. Biol.* **21**, 358–363.

Glassman, E. (1969). The biochemistry of learning: An evaluation of the role of RNA and protein. *Annu. Rev. Biochem.* **38**, 605–646.

Greenough, W. T. (1976). Enduring brain effects of differential experience and training. *In* "Neural Mechanisms of Learning and Memory" (M. Rosenzweig and E. L. Bennett, eds.), pp. 255–278. MIT Press, Cambridge, Massachusetts.

Hall, M. E., Wilson, D. L., and Stone, G. C. (1978). Changes in synthesis of specific proteins following axotomy: Detection with two-dimensional gel electrophoresis. *J. Neurobiol.* **9**, 353–366.

Harris, H. (1963). Nuclear ribonucleic acid. *Prog. Nucleic Acid Res. Mol. Biol.* **2**, 20–59.

Harris, H. (1964). Transfer of radioactivity from nuclear to cytoplasmic ribonucleic acid. *Nature (London)* **202**, 249–250.

Horel, J. A. (1978). The neuroanatomy of amnesia (A critique of the hippocampal memory hypothesis). *Brain* **101**, 403–445.

Horn, C., Rose, S. P. R., and Bateson, P. P. G. (1973). Experience and plasticity in the central nervous system. Is the nervous system modified by experience? Are such modifications involved in learning? *Science (Washington, D.C.)* **181**, 506–514.

Hornykiewicz, O. (1974). The mechanisms of action of L-DOPA in Parkinson's disease. *Life Sci.* **15**, 1249–1259.

Hydén, H. (1943). Protein metabolism in the nerve cell during growth and function. *Acta Physiol. Scand. Suppl.*, **17**, 1–136.

Hydén, H., and Egyházi, E. (1963). Glial RNA changes during a learning experiment in rats. *Proc. Natl. Acad. Sci. U.S.A.*, **49**, 618–624.

Hydén, H., and Egyházi, E. (1964). Changes in RNA content and base composition in cortical neurons of rats in a learning experiment involving transfer of handedness. *Proc. Natl. Acad. Sci. U.S.A.* **52**, 1030–1035.

Irwin, L. N. (1974). Glycolipids and glycoproteins in brain function. *Rev. Neurosci.* **1**, 137–182.

Irwin, L. N., and Samson, F. E. (1971). Content and turnover of gangliosides in rat brain following behavioural stimulation. *J. Neurochem.* **18**, 203–211.

Lasek, R. J. (1980). Axonal transport: A dynamic view of neuronal structures. *Trends Neurosci. Pers. Ed.* **3**, 87–91.

Liebeskind, J. C., Dismukes, R. K., Barker, J. L., Berger, P. A., Creese, I., Dunn, A. J., Segal, D. S., Stein, L., and Vale, W. W. (1978). Peptides and behavior: A critical analysis of research strategies. *Neurosci. Res. Program Bull.* **16**, 490–635.

Lowry, O. H. (1963). The chemical study of single neurons. *Harvey Lect.* **58**, 1–19.

Matthies, J. (1979). Biochemical, electrophysiological, and morphological correlates of brightness discrimination in rats. *In* "Brain Mechanisms in Memory and leaning: From the Single Neuron to Man" (M. A. B. Brazier, ed.), pp. 197–215. Raven, New York.

McGaugh, J. L. (1966). Time dependent processes in memory storage. *Science (Washington, D.C.)* **153**, 1351–1358.

McGaugh, J. L. (1973). Drug facilitation of learning and memory. *Annu. Rev. Pharmacol.* **13**, 229–214.

McGaugh, J. L., and Dawson, R. C. (1971). Modification of memory storage processes. *Behav. Sci.* **16**, 45–63.

Meibach, R. C., Glick, S. D., Ross, D. A., Cox, R. D., and Maayani, S. (1980). Intraperitoneal administration and other modifications of the 2-deoxy-D-glucose technique. *Brain Res.* **195**, 167–176.

Morgan, D. G., and Routtenberg, A. (1979). The incorporation of intrastriatally injected [^{3}H]fucose into electrophoretically separated synaptosomal glycoproteins. II. The influence of passive avoidance training. *Brain Res.* **179**, 343–354.

Morgan, D. G., and Routtenberg, A. (1980). Evidence that a 41,000 Dalton brain phosphoprotein is pyruvate dehydrogenase. *Biochem. Biophys. Res. Commun.* **95**, 569–576.

Nakajima, S. (1975). Amnesic effect of cycloheximide in the mouse mediated by adrenocortical hormones. *J. Comp. Physiol. Psychol.* **88**, 378–385.

Oja, S. S. (1967). Studies on protein metabolism in developing rat brain. (1967). *Ann. Acad. Sci. Fenn. Ser. A5* **131**, 1–81.

Owen, F., Crow, T. J., Poulter, M., Cross, A. J., Longden, A., and Riley, G. J. (1978). Increased dopamine-receptor sensitivity in schizophrenia. *Lancet* **2**, 223–226.

Paul, S. M., and Skolnick, P. (1978). Rapid changes in brain benzodiazepine receptors after experimental seizures. *Science (Washington, D.C.)* **202**, 892–894.

Pavlík, A., Jakoubek, J., Buresova, M., and Hajek, I. (1971). The effect of ACTH on the synthesis of acidic proteins in brain cortical slices. *Physiol. Bohemoslov.* **20**, 399.

Perry, R. P. (1976). Processing of RNA. *Annu. Rev. Biochem.* **45**, 605–630.

Pevzner, L. Z. (1972). Macromolecular changes within neuroneuroglia unit during behavioral events. *In* "Macromolecules and Behavior" (J. Gaito, ed.), pp. 335–358. Appleton, New York.

Phelps, M. E., Huang, S. C., Hoffman, E. J., Selin, C., Sokoloff, L., and Kuhl, D. E. (1979). Tomograhic measurement of local cerebral glucose metabolic rate in humans with (F-18) 2-fluoro-2-deoxy-D-glucose: validation of method. *Ann. Neurol.* **6**, 371–388.

Phillips, M. I., Weyhenmeyer, J., Felix, D., Ganten, D., and Hoffman, W. E. (1979). Evidence of an endogenous brain renin angiotensin system. *Fed. Proc., Fed. Am. Soc. Exp. Biol.* **38**, 260–266.

Quinton, E. E., and Kramarcy, N. R. (1977). Memory impairment correlates clearly with cycloheximide dose and degree of inhibition of protein synthesis. *Brain Res.* **131**, 184–190.

Rainbow, T. C., Adler, J. E., and Flexner, L. B. (1976). Comparison in mice of the amnestic effects of cycloheximide and 6-hydroxydopamine in a one-trial passive avoidance task. *Pharmacol. Biochem. Behav.* **4**, 347–349.

Rees, H. D., and Dunn, A. J. (1977). The role of pituitary-adrenal system in the footshock-induced increase of [³H]lysine incorporation into mouse brain and liver proteins. *Brain Res.* **120**, 317–325.

Rees, H. D., Brogan, L. L., Entingh, D. J., Dunn, A. J., Shinkman, P. G., Damstra-Entingh, T., Wilson, J. E., and Glassman, E. (1974). Effect of sensory stimulation on the uptake and incorporation of radioactive lysine into protein of mouse brain and liver. *Brain Res.* **68**, 143–156.

Rees, H. D., Dunn, A. J., and Iuvone, P. M. (1976). Behavioral and biochemical responses of mice to the intraventricular administration of ACTH analogs and lysine vasopressin. *Life Sci.* **18**, 1333–1340.

Reith, M. E. A., Schotman, P., and Gispen, W. H. (1975). Incorporation of [³H]leucine into brain stem protein fractions: the effect of a behaviorally active, N-terminal fragment of ACTH in hypophysectomized rats. *Neurobiology (Copenhagen)* **5**, 355–368.

Rose, S. P. R. (1968). The biochemistry of neurones and glia. *In* "Applied Neurochemistry" (A. N. Davison and J. Dobbing, eds.), p. 351. Blackwell, Oxford.

Rose, S. P. R. (1980). Neurochemical correlates of early learning in the chick. *In* "Neurobiological Basis of Learning and Memory" (Y. Tsukada and B. W. Agranoff, eds.), pp. 179–191. Wiley, New York.

Rose, S. P. R., and Haywood, J. (1977). Experience, learning and brain metabolism. *In* "Biochemical Correlates of Brain Structure and Function" (A. N. Davison, ed.), pp. 249–292. Academic Press, New York.

Rose, S. P. R., Hambley, J., and Haywood, J. (1976). Neurochemical approaches to developmental plasticity and learning. *In* "Neural Mechanisms of Learning and Memory" (M. R. Rosenzweig and E. L. Bennett, eds.), pp. 293–310. MIT Press, Cambridge, Massachusetts.

Routtenberg, A. (1979). Anatomical localization of phosphoprotein and glycoprotein substrates of memory. *Prog. Neurobiol. (Oxford)* **12**, 85–113.

Routtenberg, A., and Benson, G. E. (1980). In vitro phosphorylation of a 41,000-MW protein band is selectively increased 24 hr after footshock or learning. *Behav. Neural. Biol.* **29**, 168–175.

Schotman, P., Gispen, W. H., Janz, W. S., and de Wied, D. (1972). Effects of ACTH analogues on macromolecule metabolism in the brain stem of hypophysectomized rats. *Brain Res.* **46**, 347–362.

Schotman, P., Reith, M. E. A., van Wimersma Greidanus, Tj. B., Gispen, W. H., and de Wied, D. (1976). Hypothalamic and pituitary peptide hormones and the central nervous system. With special references to the neurochemical effects of ACTH. *In* "Molecular and Functional Neurobiology" (W. H. Gispen, ed.), pp. 310–336. Elsevier, Amsterdam.

Schwartz, J. C., Constentin, J., Martres, M. P., Protais, P., and Baudry, M. (1978). Modulation of receptor mechanisms in the CNS: Hyper and hyposensitivity to catecholamines. *Neuropharmacology* **17**, 665–685.

Segal, D. S., Squire, L. R., and Barondes, S. H. (1971). Cycloheximide: its effects on activity are dissociable from its effects on memory. *Science (Washington, D.C.)* **172**, 82–84.

Smith, G. P., and Gibbs, J. (1979). Postprandial satiety. *In* "Progress in Psychobiology and Physiological Psychology" (J. M. Sprague and A. N. Epstein, eds.), Vol. 8, pp. 223–224. Academic Press, New York.

Sokoloff, L. (1977). Relation between physiological function and energy metabolism in the central nervous system. *J. Neurochem.* **29**, 13–26.

Sokoloff, L., Reivich, M., Kennedy, C., des Rosiers, M. H., Patlak, C. S., Pettigrew, K. D., Sakurada, O., and Shinohara, M. (1977). The [^{14}C]deoxyglucose method for the measurement of local cerebral glucose utilization: Theory, procedure and normal values in the conscious and anesthetized albino rat. *J. Neurochem.* **28**, 897–916.

Squire, L. R. (1976). Amnesia and the biology of memory. *In* "Current Developments in Psychopharmacology" (W. B. Essman and L. Valzelli, eds.), Vol. 3, pp. 1–23. Spectrum Publ., New York.

Squire, L. R., St. John, S., and Davis, H. P. (1976). Inhibitors of protein synthesis and memory: Dissociation of amnesic effects and effects on adrenal steroidogenesis. *Brain Res.* **112**, 200–206.

Stone, E. A. (1979). Subsensitivity to norepinephrine as a link between adaptation to stress and antidepressant therapy: An hypothesis. *Res. Commun. Psychol. Psychiatry Behav.* **4**, 241–255.

Ungar, G. (1974). Molecular coding of memory. *Life Sci.* **14**, 595–604.

Uphouse, L. L. (1980). Reevaluation of mechanisms that mediate brain differences between enriched and impoverished animals. *Psych. Bull.* **88**, 215–232.

Uphouse, L. L., MacInnes, J. W., and Schlesinger, K. (1974). Role of RNA and protein in memory storage: A review. *Behav. Genet.* **4**, 29–81.

Zatz, M., and Barondes, S. H. (1971). Rapid transport of fucosyl glycoproteins to nerve endings in mouse brain. *J. Neurochem.* **18**, 1125–1133.

Zemp, J. W., Wilson, J. E., Schlesinger, K., Boggan, W. O., and Glassman, E. (1966). Incorporation of uridine into RNA of mouse brain during short-term training experience. *Proc. Natl. Acad. Sci. U.S.A.* **55**, 1423–1431.

13

ISOLATION AND CULTURE OF SPECIFIC BRAIN CELLS AND THEIR EXPERIMENTAL USE

H. H. Althaus and V. Neuhoff

Molecular Approaches to Neurobiology

I. INTRODUCTION

It is true to say "if various parts of the nervous system were made up of elements having the same function and chemical composition, the application of cellular analytical methods would not be justified" (Giacobini, 1968). But since the brain is composed of so many cell types such as neurons, glial cells, ependymal cells, endothelial cells, which differ largely in morphological shape, the classical approach to obtain information about the components of this complex network of cells is to disrupt it into individual units. Many approaches have been developed to isolate specific types of brain cells from a wide variety of different animal species. This chapter will focus on studies of mammalian brain tissue. The isolation of neural cells from nonmammalian brain tissue (mainly molluscs and insects, e.g., see Giacobini, 1969a,b; Kostenko *et al.*, 1974; London and Merickel, 1979; Osborne, 1978) will not be considered. Giacobini (1969a,b) has outlined the advantages of working with nonmammalian systems (for example, neuronal cell bodies are generally larger, the same neurons can be easily identified in different preparations, they retain their functional activity after dissection and survive well, and large amounts of glial material can be isolated); however, extrapolation of data from invertebrates to the mammalian system is difficult particularly with glia where great differences occur (Radojcic and Pentreath, 1979). Another potential pool of information (which is not discussed in this chapter) is that originating from tissue culture. That area deals with the enrichment of specific neural cells during, or after, culturing of embryonic or newborn nervous tissue (e.g., Booher and Sensenbrenner, 1972; Labourdette *et al.*, 1979, 1980; McCarthy and DeVellis, 1980; Pettmannn *et al.*, 1979, 1980; Sotelo *et al.*, 1980). The culturing of isolated brain cells from young adult animals will, however, be covered.

Modifications of certain neural cell isolation procedures will not be covered in full, since these have already been extensively reviewed (Nagata and Tsukada, 1978; Pevzner, 1979; Poduslo and Norton, 1975; Sellinger and Azcurra, 1974; Ulas, 1979). A review of methods for the isolation of brain cells reveals that they have been developed for two main objectives (see Nagata and Tsukada, 1978; Poduslo and Norton, 1975; Roots and Johnston, 1972). One is to gain basic information on the chemical makeup of neural cells, and for this purpose it is not absolutely necessary that the isolated cells be viable provided that only protein and lipid composition are determined (e.g., Freysz *et al.*, 1968b; Satake and Abe, 1966). The second objective is to investigate the metabolic activities of various neural cells. In this case isolated cells must retain some activity. Direct comparison of biochemical results obtained from such studies may not be as conclusive as one desires, since the state of viability of different neural types in a preparation can vary (Benjamins *et al.*, 1974; Sinha and Rose, 1971). More recently, some attempts have been made to meet the following statement concerning the viability of bulk-isolated cells: "if alive means suitable for tissue culture, the answer is definitely no" (Hamberger and Sellström, 1975). In this direction a step forward was achieved when more "physiological" conditions were applied to isolation procedures. Neurons and glial cells can now be maintained *in vitro,* permitting studies on their physiological functioning.

All cell isolation techniques in use at present have limitations. It is essential to keep such limitations and problems in mind when working with isolated neural cells.

II. GENERAL APPROACHES TO THE ISOLATION OF BRAIN CELLS

Present methods fall into two categories: (1) microscale, the collecting of single cells by hand and (2) macroscale, isolation of cells by taking advantage of their physical and/or chemical characteristics, which allows their isolation in bulk. Investigators using microscale methods subdivide into those who handle "fixed tissue" (Edström, 1953; Edström and Neuhoff, 1973; Lowry *et al.,* 1956; Lowry and Passonneau, 1972) and those who use fresh tissue (Giacobini, 1956; Hydén, 1959; Roots and Johnston, 1972). Those using the macroscale approach, involving the so-called bulk isolation methods, may be subdivided into researchers who disaggregate the nervous tissue with digestive enzymes (e.g., Althaus *et al.,* 1977; Farooq and Norton, 1978; Hemminki, 1970; Norton and Poduslo, 1970) or without digestive enzymes (e.g., Freysz *et al.,* 1968b; Hamberger *et al.,* 1970; Iqbal and Tellez-Nagel, 1972; Jones *et al.,* 1972; Rappaport and Howze, 1966; Rose, 1965; Satake and Abe, 1966; Sellinger *et al.,* 1971). In the main this chapter will deal with bulk isolation methods, which provide sufficient material for biochemical analysis. Micromanipulation and hand dissection, which yield a limited number of cells from distinct anatomical areas (which may be advantageous) and which require special equipment for microanalysis, will only be discussed briefly.

III. MICROSCALE ISOLATION METHODS

A. Fixed Tissue

The key references for microscale isolation of brain cells from freeze-dried sections are Lowry (1953) and Lowry and Passoneau (1972) and from histological sections Edström (1953) and Edström and Neuhoff (1973).

1. Freeze-Dried Tissue

In an extension of the work by Anfinsen *et al.* (1942) Lowry elaborated a method for dissecting single cell bodies from freeze-dried tissue. Small pieces of nervous tissue are rapidly removed from the animal and frozen by placing them on a hardened filter paper (or a metal microtome chuck) and plunging them into liquid nitrogen or into Freon 12 chilled with liquid nitrogen to $-150°C$. The frozen tissue is allowed to warm to $-10°$ to $-20°C$ (from the time of freezing until the sections are dry the temperature should not rise above $-10°C$ or ice-crystal artifacts may develop). Blocks of tissue, cut with a thin razor-blade, are mounted on a blockholder (a wooden Dowel rod drilled at one end to form a shallow cup). Sections of 5–50 μm thickness (usually at least 20 μm depending on the size of the cell bodies) are cut at a constant low temperature of $-15°$ to $-20°C$. If the sections are too cold they will fragment, and if too warm they will be compressed. Individual sections are transferred to a suitable holder: a drilled aluminium block. One to four section holders are placed in a special drying tube, which is then transferred to a temperature box kept at $-35°$ to $-40°C$. On reaching this temperature the sections are dried by applying a vacuum of 0.01 mm Hg and using a dry ice trap to condense the water removed. This requires 1–6 h. After which time the tube is removed and evacuation continued until the tube and contents reach room temperature. Individual cells are dissected from freeze-dried sections (transferred by hair points) by free-hand teasing with fine needles under suitable magnification. The sections are held down during dissection with hair loops. Novices using this technique should start with dorsal root ganglia, since the cell bodies can be easily recognized in the freeze-dried section, before proceeding to other parts of the nervous system.

2. Histological Sections

Single nerve cells can be isolated from stained or unstained histological sections. A coverslip with a section of tissue is inverted over an oil chamber and the resulting space filled with liquid paraffin. The nerve cells are separated under a phase-contrast microscope with a round tip needle bent upward operated by a micromanipulator. When 20–30 cells are separated (this requires about 1 h) and remain loosely attached to the coverslip they are sucked up by a microconstriction pipette, which has replaced the needle. The cells aggregate in front of the constriction (while the paraffin oil flows by) and can then be ejected into another vial ready for biochemical analysis.

B. Fresh Tissue

The key references are Giacobini (1956) and Hydén (1959) for microscale cell isolation from tissue slices and Roots and Johnston (1964) for the isolation of cells from cell suspensions.

1. Cell Isolation by the Method of Giacobini and Hydén

Giacobini has worked mainly with spinal and autonomic ganglia and motor horn cells of spinal cord of frogs, while Hydén has used mammalian brain. When using this method it is preferable to start with the isolation of large nerve cells (e.g., from the lateral vestibular nucleus of the ox brain) and those which can be easily disentangled (e.g., cells of frog sympathetic ganglion are easier to isolate than those from rat or cat). A slice is taken through the desired locus of the tissue [immersed in isotonic sucrose solution (Hydén), or Ringer's solution (Giacobini)], a fragment of which is placed on a glass slide in an oxygenated moist chamber. The cells are dissected with needles, either free hand or using a micromanipulator, under a stereomicroscope at a suitable magnification. Initially the cells can be stained slightly with methylene blue. When the nerve cells are removed they can be freed from adhering glial tissue, which stick to each other forming glial clumps. Different opinions exist about the proportion of the various components in these clumps, which have also been termed neuropil (Epstein and O'Connor, 1965), for example, do they consist mainly of oligodendroglia or not. A more homogeneous population of glial cells may be dissected from the spinal and sympathetic ganglia.

2. Cell Isolation by the Method of Roots and Johnston

This procedure was developed at a time when other methods, e.g., aqueous two-phase systems, electrophoresis, or even sedimentation procedures, appeared insufficient to obtain a "pure" fraction of individual cells. The method can be applied to any part of the nervous system and the authors claim considerable numbers of neuronal perikarya free from extraneous matter can be obtained. It is once again preferable to start with large cells, e.g., from the lateral vestibular nucleus. The excised tissue is placed in cold (+ 4°C) medium, i.e., 0.25 M sucrose, in which the cells are easily recognizible or later in media such as Ringer's solution with or without additives such as gangliosides (Johnston and Roots, 1965). The tissue is then disrupted by sieving through a series of nylon cloths of different mesh sizes (300–350 μm to 108–110 μm) resulting in a cell-tissue suspension, which is then allowed to settle in a flask for 10 min at 4°C, after which time 10–20 ml are withdrawn from the bottom of the flask. A few drops of this suspension are placed in a cavity slide and observed under a magnification of ×20 with a stereomicroscope fitted with a cold stage. Neuronal perikarya are transferred by means of a nylon loop to another cavity slide containing fresh medium. By this method an experienced worker can collect 70 or more cells per minute. Glial perikarya may also be recognized in the suspension and may be collected in the same way using a smaller nylon loop. In the case of rat cortical cells nylon cloth with mesh sizes of 130 μm and 40 μm should be used.

IV. MACROSCALE ISOLATION METHODS

A. Brief History, General Comments, and Major Objections

Microdissection techniques are restricted to certain anatomical areas and perhaps only highly specified neuronal cells can be isolated, on which only a limited number of biochemical studies can be performed despite the availability of a number of microanalytical techniques (Giacobini, 1978; Neuhoff, 1973). Furthermore, it is not possible to collect glial cell fractions of reasonable degree of purity by these methods. Attempts were, therefore, made to obtain cells of a satisfactory level of purity in bulk.

The genesis of the bulk isolation methods may be attributed to four researchers. McIlwain (1954) who used enzyme (papain) digestion of the tissue prior to disaggregation may well have been the first to prepare a bulk nerve cell fraction, but he did not attempt a further purification. Korey *et al.* (1958) took white matter as starting material and prepared a suspension of glial cells after homogenization in a Waring blender in which the blades had been dulled and flattened. The resulting suspension was passed through siliconized silk grids of two different mesh sizes (450 and 250 μm) and finally centrifuged on a sucrose gradient. However, no claims were made as regards the purity of the fraction. Eight years later Rose (1965) published a method based on the disruption of rat cerebral cortex by simple mechanical sieving, which yielded a "neuron-enriched" and "glial-enriched" fraction after discontinuous density gradient centrifugation (Ficoll 400-sucrose). Another method for isolating rat cortical cells, which had been developed in the interim (Campbell, 1963), was unfortunately never published in a journal (except for a brief note in Rose, 1967), although the principles for two major techniques for cell isolation were described: (1) the perfusion of an animal with an enzyme- (trypsin-hyaluronidase) containing medium and (2) cell sorting by free-flow electrophoresis.

The methodology for the bulk isolation of brain cells has evolved in two different directions: (1) those who disintegrated the tissue mechanically without exogenous enzymes and (2) those who incubated the tissue (*in vitro* or *in situ*) with or without added proteolytic enzymes. A further difference between the various procedures is the separation technique used to acquire individual brain cell fractions, i.e., either continuous or discontinuous density gradients (normally sucrose or Ficoll 400) in combination with either high or low g forces. A compilation of bulk separation methods for brain cells developed up to 1976–1977 is given in the review of Nagata and Tsukada (1978).

A number of objections have been raised as to the validity of biochemical studies performed on specific brain cells isolated by these techniques. The major objections are the following:

1. The isolated cells consist of perikarya largely shorn of processes. The loss of cellular processes leads to a disruption of the plasma membrane, which is

concomitant with a leakage of intracellular components such as amino acids, enzymes, RNA, and mitochondria, in addition to an influx (or absorption) of exogenous compounds.

2. The use of exogenous enzymes, e.g., trypsin (or utilization of endogenous lysosomal enzymes during the incubation) may produce harmful effects on the cell plasma membrane. It has been shown that trypsin affects the cell and its plasma membrane components (Barnard *et al.*, 1969; Cohen and Bernsohn, 1974; Deppert and Walter, 1978; Guarnieri *et al.*, 1975; Hemminki *et al.*, 1970), for example, glycoproteins (Baumann and Doyle, 1979; Vogel, 1978) or receptors (Henn *et al.*, 1980). Furthermore, it has been suggested that trypsin enters the cell (Guarnieri *et al.*, 1976) and, moreover, contamination of trypsin by deoxyribonuclease may alter the levels of DNA (Fewster *et al.*, 1973; Hemminki *et al.*, 1970).

3. Those procedures which result in neurons that retain some of their processes almost exclusively use 10- to 30-day-old rats as the starting material. The morphological appearance of bulk-isolated neurons deteriorates when adult rats are used (Giorgi, 1971; Sellinger and Santiago, 1972).

4. Some procedures are carried out at very nonphysiological pH values, e.g., 4.7 or pH 6.5.

5. As large variations with regard to cell morphology and cell yield occur within the same technique, the reproducibility of the methods has been questioned.

Additional objections are concerned with the purity of the individual cell fractions and whether or not they are truly representative of the individual cell types *in vivo*. Last, the materials used during density gradient centrifugation may cause artifacts, e.g., sucrose can enter the cell and may lead to inhibition of some enzymes (e.g., succinate dehydrogenase or 5' nucleotidase).

Within the last 4 years improvements of the isolation techniques have been published, which meet these objections to some extent. They involve modifications in the procedure used to disaggregate the brain tissue and improvements in the separation of the components of the resulting cell suspension. Since these recent methods result in a better state of preservation with regard to morphological and metabolic criteria, compared to the "classical" methodologies, a detailed description of the "modern" procedures is given in this chapter. In addition, some comments will be made on the alternatives that have been used recently to substitute the density gradient step in the separation of the cell suspensions.

B. Bulk Separation of Neurons

The key references are Althaus *et al.* (1977), Huttner *et al.* (1979), Farooq *et al.* (1977), Farooq and Norton (1978), and Poduslo and McKhann (1977a).

1. Procedure of Althaus and Huttner

Male Sprague-Dawley rats weighing 120–180 g are anesthetized with pentobarbital and perfused for 20 min via the common carotid arteries at a constant flow rate of 4 ml/min at 37°C (the cervical veins are left intact, the submandibular glands are ligated). The perfusion medium consists of 280 mM D-glucose, 280 mM D-fructose, 1% (w/v) bovine serum albumin (Behring-Werke, Marburg) [which can be replaced by 1% (w/v) Ficoll 70—Pharmacia, Uppsala], 0.1% (w/v) collagenase (Worthington type III; less pure collagenase reduces the quality of the nerve cell preparation), 0.1% hyaluronidase (Sigma, type I), 1.5 mM CaCl$_2$, and 10 mM KH$_2$PO$_4$-NaOH pH 7.4. The medium is equilibrated with 100% O$_2$ immediately prior to perfusion. After perfusion the brain is rapidly removed and placed on a cooled (+4°C) plate (as the following steps are performed in the cold). The cerebral cortices are dissected out and chopped with a razor blade (some drops of cold medium without the enzymes and CaCl$_2$ are added). The minced tissue [washed in medium, centrifuged for 5 min at 250 g, and resuspended in 10 ml medium plus 15% Ficoll 400 (w/v)] is gently sieved through nylon bolting cloths of decreasing mesh sizes (down to 150 μm). The resulting cell tissue suspension is layered on a Ficoll 400 (in hexosephosphate buffer)/fluorocarbon (F-decalin, PP9—Kalichemie, Hannover) discontinuous gradient. After centrifugation in a 3 × 30 ml swinging bucket rotor at 8000 g for 20 min the neurons are collected from the PP9, which has served as a cushion.

The procedure can be applied to Sprague-Dawley rats of different weights (30–300 g) by adjusting the flow rate of the perfusion. Application of the procedure to female Sprague-Dawley rats or to other species, e.g., Wistar rats requires direct perfusion into the internal carotid artery. Different animals, e.g., primates may also be used as demonstrated by Althaus *et al.* (1979). Three modifications have been employed:

1. To enhance the percentage of viable neurons (i.e., cells suitable for tissue culture) the perfusion medium is oxygenated in the presence of fluorocarbon vesicles, which serve as a O$_2$ carrier (Althaus *et al.*, in preparation).

2. The nylon sieving is replaced by a homogenization of the cortices with a Dounce homogenizer, which is more convenient when working in sterile conditions.

3. Instead of Ficoll, metrizamide (Nyggard) is now prefered as the density gradient medium, since 30% Ficoll 400 (undialyzed) exhibits a much higher osmolarity as compared to 30% metrizamide. Another compound, Percoll, (Pharmacia) is at present under investigation. The purity of the neuronal cell fraction is satisfactory when a discontinuous Percoll gradient is performed and a swingout rotor used (Althaus *et al.*, 1981).

2. Procedure of Farooq et al.

This method is based on the procedure of Norton and Poduslo (1970), but a novel disaggregation technique is utilized: the concentration of hexoses is raised to 13%, and Ficoll 400 has replaced albumin. Four to six brains from 15- to 30-day-old rats (or in a modified version frozen human or bovine brain) are trimmed of the cerebella and sliced into 8–12 pieces. The tissue is incubated (under oxygen) for 90 min at 37°C in a medium consisting of 8% glucose, 5% fructose, 2% Ficoll 400, and 10 mM KH_2PO_4-NaOH buffer pH 6.0 (enough medium to just cover the slices). When trypsin (type III, two times crystallized, salt-free from bovine pancreas—Sigma) is used (0.1%) the tubes containing brain slices are placed in ice after incubation and the trypsin containing medium is removed. The slices are washed three times with medium containing 0.1% lima bean trypsin inhibitor (Worthington) followed by two washings with plain medium. After incubation (with or without trypsin) the remaining steps are carried out at 0°–4°C. The tissue is sequentially aspirated through nozzles (cut from 9-in. long disposible Pasteur pipettes) of different orifice diameters and lengths. After the first aspiration the suspension is poured through a Nitex nylon screen cloth of mesh size 420 μm and the remaining tissue is aspirated through a nozzle of a smaller orifice diameter. The combined suspension of dissociated tissue is adjusted to 25% sucrose. Mixing is carried out slowly in a cylinder to which a vacuum is applied (about 30 mm Hg). As the solution degasses, bubbles rise which carry to the surface particles of incompletely dissociated tissue. This foamy layer is removed by suction and discarded. The degassed cell suspension (in 25% sucrose plus medium) is layered over a cushion of 70% sucrose in medium and centrifuged (Sorvall HG-4 rotor, 4600 rpm) for 20 min. The cell layer is removed from the interface (25%/70% sucrose) and adjusted with medium and sucrose (30%) to 20 ml/0.75–1.0 g of tissue. After mixing, vacuum is again applied (to remove undisrupted tissue). 20-ml portions of the cell suspension are layered onto a discontinuous sucrose gradient (made up in medium) and centrifuged in a Spinco SW-27 rotor (4000 rpm) for 20 min. The pellet and the layer on 52% sucrose are the neuron-rich fractions.

3. Procedure of Poduslo, Modified

The basic procedure of Norton and Poduslo (1970) has been modified by Poduslo in order to permit tissue culturing of the neurons and to meet criticism regarding the use of trypsin. Brains from rats less than 30 days old (minus cerebellum) are minced and incubated in 0.1% trypsin (type III, 2 times crystallized, salt free—Sigma) containing medium (5% glucose, 5% fructose, 1% albumin, 10 mM KH_2PO_4-NaOH buffer pH 6.0) at 37°C for 90 min (under oxygen). Thereafter the tissue is cooled, a trypsin inhibitor (type II S, from soybean—Sigma) is added, and the tissue is washed three times. The tissue is dissociated by passing it through a series of screens. The cells are separated on a discontinuous sucrose gradient by centrifugation at 3300 g (10 min) and the neurons are collected from the 2 M sucrose layer. After being gently diluted fivefold with medium the cells are collected by centrifugation. The neurons are washed three times in medium containing an equal volume of buffered (pH 6.0) fetal calf serum in order to remove the trypsin activity. The washed neurons are gradually introduced to medium (e.g., Dulbecco's PBS) adjusted to physiological pH.

C. Bulk Separation of Astrocytes

In contrast to the progress made in the isolation of neurons and oligodendrocytes, the number of improvements made in the methodology for the bulk isolation of astrocytes has been limited in the past 4 years. In fact, only one procedure has been published in this time interval for the bulk preparation of astrocytes. The key reference (Farooq and Norton, 1978) is based on a previous procedure (Section IV,B,2; Farooq *et al.*, 1977). Again the hexose concentration is elevated, which may be helpful for obtaining cells which are better preserved with respect to morphological criteria, but for comments on metabolic aspects see Poduslo and Norton (1975).

1. Procedure of Farooq and Norton

Brains from 20- to 60-day-old rats (minus cerebellum and brain stem) are sliced into pieces. The slices are incubated at 37°C with shaking in 9 ml of cell isolation medium (Section IV,B,2) containing 0.1% of acetylated trypsin (type V, from bovine pancreas—Sigma). For 20-day-old brain the slices are incubated for 60 min, for older brains 90 min. After incubation the medium is removed and replaced by an equal amount of medium containing 0.1% soybean trypsin inhibitor. After cooling the slices are washed 4–5 times with ice-cold medium and the following steps are carried out at 0°–4°C. The tissue is disaggregated by successive aspiration steps, through a nozzle (prepared as described in Section IV,B,2) with a diameter at the tip of 2.2.–2.4 mm and a length of 3.4–3.5 cm. The disaggregated tissue is collected in a flask (containing 30 ml of medium) under slight vacuum. This suspension is then filtered through a nylon screen of 420 μm mesh to give both a filtrate and a residue on the screen, which is once again aspirated. This step is repeated twice and the filtrates are combined. During a period of 15–20 min larger particles of the suspension settle and the supernatant is decanted and retained. This step is repeated with the remixed precipitates. The combined supernatants are centrifuged at 1600 rpm (Sorvall HG-4 rotor) for 15 min to obtain a cell-rich pellet (concentration step). The cell-rich pellet is suspended in 64 ml of 7% Ficoll 400 in medium and centrifuged (HG-4 rotor, 1000 rpm, 10 min) to obtain a pellet P1 (neuron rich). The supernatant is centrifuged at a higher speed (1600 rpm, 10 min) to obtain pellet P2 (containing astrocytes and neurons). The supernatant diluted 1:1.125 with medium is then centrifuged at 2000 rpm for 15 min yielding pellet P3 (astrocyte enriched). Each of the pellets are suspended gently in 38 ml of 7% Ficoll 400 in medium and layered on a discontinuous Ficoll 400 gradient. The tubes are centrifuged (Spinco SW 27 rotor) at 8000 rpm for 5 min. The fraction of the pellet P3 banding on 22% Ficoll 400 contains predominantly astrocytes.

D. Bulk Separation of Oligodendrocytes

Since the first reports of bulk isolated oligodendrocytes (Fewster *et al.*, 1967; Poduslo and Norton, 1972a) this cell type has been obtained in a high level of purity (90–95%). However, certain drawbacks are apparent which originate from the starting material, usually white matter dissected from large animals (ox or calf brain) of undetermined ages. Hence, investigations regarding the sequence of oligodendroglia metabolism during myelination were strictly limited (e.g., Benjamins *et al.*, 1974). Furthermore, with exception of one publication that reported an attempt to culture isolated oligodendrocytes (Fewster and Blackstone, 1975), these cells have been subjected only to short-term metabolic studies. Methods recently developed have met these shortcomings to some extent. The key references are Gebicke-Härter *et al.* (1981), Poduslo *et al.* (1978), Snyder *et al.* (1980), and Szuchet *et al.* (1980b).

1. Procedure of Gebicke-Härter and Althaus

Cat brains (8- to 12-week-old) are perfused *in situ* according to the method of Althaus *et al.* (1977) at a constant flow rate of 6 ml/min for 20 min via the common carotid arteries. The brain is cut into slices from which the white matter is carefully prepared avoiding the ventricular layers. After three washings in hexose-containing phosphate-buffered medium pH 7.4, the tissue is chopped and suspended in 20–25 ml of the medium. The suspension is gently squeezed through nylon sieves of decreasing mesh sizes (300–125 μm) using the double syringe technique as described previously (Althaus *et al.*, 1977). The cell suspension is mixed with Percoll (Pharmacia) in a ratio of 3 to 1, and 10 ml of this mixture is placed under a mixture of medium and Percoll (4 parts medium/1 part Percoll) and centrifuged (Swingout rotor SW 30) for 30 min at 10,000 *g*. The cells do not form a distinct layer as in sucrose gradients but float just above the bottom of the tubes. They are collected and processed for culturing.

2. Procedure of Poduslo *et al.*

Corpus callosum and centrum semiovale areas of white matter from the forebrain of one calf or four lamb brains are dissected free of gray matter. The white matter (about 40 g wet weight) is minced and incubated at 37°C for 90 min in a trypsin (type IX—Sigma) solution (0.1% in medium—prepared as in Section IV,B,3). The softened tissue is cooled and equivalent amounts of trypsin inhibitor (soybean, type 2S—Sigma) are added. After 5 min on ice the suspension is centrifuged at low speed (400 rpm, 3 min), followed by two other washings. The tissue is dissociated by sieving through screens of different mesh size (under vacuum to prevent the medium bubbling), and passed through a double layer of nylon and three times through a stainless steel screen. The volume of the cell suspension is adjusted to 280 ml with 0.9 M sucrose and 70 ml aliquots are layered onto each of 4 sucrose gradients and centrifuged (HG-4 rotor, Sorvall RC 3) at 4000 rpm for 15 min at 4°C. The layer on 1.55 M sucrose is removed, gradually diluted fivefold with cell medium containing 5% fetal calf serum pH 6.0, and filtered through (sterile) glass wool to remove capillaries. Thereafter the cells are processed for culturing.

3. Procedure of Snyder *et al.*

In contrast to the above two procedures the whole forebrain of rats is used as starting material in this method instead of isolated white matter. A similar approach has been attempted previously (Banik and Smith, 1976), but the cell fraction obtained was not well characterized by transmission electron microscopy (TEM) or immunocytochemical methods.

The brains of Sprague-Dawley rats (10- to 30- and 60-day-old) are trimmed of olfactory bulbs, cerebellum, and brain stem, and finally minced on a cold glass plate in ice-cold isolation medium [Hanks' BSS and Hepes (25 mM) and 1% (w/v) BSA, pH adjusted with NaOH to pH 7.2]. The minced tissue is transferred to a flask and 3 ml/g tissue trypsinization solution (0.1% trypsin, type III two times crystallized from bovine pancreas, Sigma, 10 μg/ml DNase, 25 mM Hepes in Hanks' BSS pH 7.2) is added. Incubation is performed for 30 min at 37°C. Afterward the suspension is chilled for 5 min on ice and ice-cold undiluted calf serum (1 ml/g) is added. The subsequent steps are carried out at 0°–4°C. Portions of the suspensions (equivalent to five brains) are centrifuged for 5 min at 190 g (HS-4 rotor, Sorvall RC-5). The pellet is resuspended in isolation medium for four other washings. The tissue is disaggregated by forcing through a nylon screen (fitted to a Buchner funnel) of 145 μm mesh size without vacuum and then through a stainless steel screen (74 μm mesh) under slight vacuum. The cell suspension is adjusted with medium to 10 ml/g original tissue and an equal volume of 70% (w/v) sucrose in medium is added. A 15 ml aliquot is placed on a discontinuous sucrose gradient and centrifuged at 3065 g for 15 min (swingout rotor). The oligodendrocytes are harvested by collecting the volume beneath the red-cell layer (below 45–53% sucrose interphase) and the pellet. The cell suspension is diluted fivefold with isolation medium containing 10 μg/ml of DNase and filtered through a nylon screen of 25 μm mesh to trap capillary fragments.

4. Procedure of Szuchet *et al.*

Twelve grams of ovine white matter (4–6 g/brain) are dissected, minced into pieces (on a cold plate), and suspended in 200 mg Hanks' BSS containing 2% (w/v) Ficoll 400 (H–2% F) plus 0.1% trypsin and 0.02% EDTA. The tissue suspension is incubated at 37°C for 3.6 min/g white matter, cooled afterward and mixed with 200 ml of H–2% F containing soybean trypsin inhibitor (type IIS— Sigma). The mixture is centrifuged at low *g* forces (192 *g*) for 3–4 min and the combined pellets (supernatants discarded) washed twice again with H–2% F. The pellets are resuspended in about 50 ml of 0.9 *M* sucrose—Hanks' BSS at half strength and screened through a series of meshes and finally through a 30 μm stainless-steel screen. The final suspension (500–600 ml of solvent has been used for the screening procedure) is centrifuged at 850 *g* for 10 min resulting in a pellet P1 and a floating myelin layer. P1 is resuspended in 2–3 ml of the ''screening'' medium and layered on top of a 1.0–1.2 *M* linear sucrose gradient. Centrifugation is performed at 277 *g* for 40 min using 50 ml centrifuge tubes (42 ml gradient). Bands are recovered by means of a peristaltic pump. Two distinct bands of oligodendrocytes are obtained suggesting that two oligodendroglial subpopulations might be isolated. The cells have been used for culturing.

E. Bulk Separation of Neurons, Astrocytes, and Oligodendrocytes from the Same Brain

As pointed out by Benjamins *et al.* (1974) it may be more conclusive to compare the metabolic activities of different cell types prepared from the same brain. Whereas astrocytes and neurons can be isolated from small laboratory animals, e.g., rats, oligodendrocytes are difficult to obtain in reasonable purity when using gray and white matter as starting material (Banik and Smith, 1976). To dissect sufficient white matter from rat brains is time-consuming and needs an appreciable number of animals (e.g., Deshmukh *et al.*, 1974). The recently published method of Snyder *et al.* (1980) allows the isolation of oligodendro-cytes from the whole rat forebrain. Another method has also been reported by Chao and Rumsby (1977) taking similar advantage of the effect that under special conditions oligodendrocytes retain their integrity while neurons and astrocytes lyse. In their procedure one-half of the total rat brain is incubated (after chopping) in trypsinization medium according to Norton and Poduslo (1970) and Poduslo and Norton (1972b), at pH 6.0 for 90 min (for obtaining neurons and astrocytes). The other one-half is incubated in the same medium, but at pH 7.4 for 60 min (for obtaining oligodendrocytes). Subsequent steps are performed according to the method of Norton and Poduslo (1970) and Poduslo and Norton (1972b), each brain fraction being subjected to the appropriate discontinuous sucrose gradient yielding the respective cell types.

F. Bulk Separation of Cerebellar Cells

Development is an intriguing area of investigation for the biochemist studying cellular "behavior." This is particularly challenging when an appreciable amount of information is available on the morphology and physiology of a system such as the cerebellum (see Balazs *et al.,* 1977; Jacobson, 1978; Llinas, 1969; Palay and Chan-Palay, 1974). Hence attempts have been made to utilize these data by a biochemical characterization of individual cerebellar cells prepared by bulk separation methods. These procedures almost exclusively use immature cerebella (e.g., Balazs *et al.,* 1979; Barkley *et al.,* 1973; Cohen *et al.,* 1978; Sellinger *et al.,* 1974; Wilkin *et al.,* 1976), which facilitates disaggregation of the tissue. Exceptions are the studies of Yanagihara and Hamberger (1973) and Hazama and Uchimura (1974).

Continuous efforts have been made to improve the isolation of individual cerebellar cells by Balazs and co-workers. The key references discussing their latest "approach" are Cohen *et al.* (1978) and Balazs *et al.* (1979).

Two to eight cerebella from 2- to 14-day-old rats (Porton strain) are pooled and sliced at 400 μm intervals by passing through a McIlwain chopper twice. After washing in Krebs-Ringer bicarbonate solution containing 14 mM glucose and 0.2% bovine serum albumin (fraction V—Sigma) (= KRB/BSA) the tissue is resuspended in the same medium containing 0.25 mg/ml trypsin (type III, twice crystallized, from bovine pancreas—Sigma) and incubated for 15 min at 37°C. The incubation is terminated by the addition of 10 ml KRB/BSA containing 4 μg/ml DNase I (Sigma) followed by brief centrifugation (50 g for 5 seconds). The sedimented tissue is again suspended in KBR/BSA containing 0.25 mg soyabean trypsin inhibitor (SBT I—Sigma) and incubated at 37°C for 5 min. After sedimentation the tissue is resuspended in Ca^{2+}- and Mg^{2+}-free KRB/BSA containing 2 mM EDTA. The tissue is washed three times with the latter medium (minus EDTA) and disrupted by passage through a Pasteur pipette. The suspension is centrifuged at 100 g for 5 min to remove most of the debris. The cell pellet is resuspended in KRB/BSA (plus 8 μg/ml DNase) to give a cell suspension with a final concentration of about 3×10^6 cells/ml. The cells are separated by velocity sedimentation under gravity at 4°C for 1.5–2.5 h (chamber device as described by Miller and Phillips, 1969). Ten-milliliter fractions were collected for cell counting and sizing. Granular cells and large neurons (mainly Purkinje cells) can be isolated with a purity of 70–80%. The purity of the Purkinje cell fraction can be enhanced, when the animals were pretreated with 2 mg/g hydroxyurea (ip) to eliminate replicating cells (Woodhams *et al.,* 1980).

G. Isolation of Other Brain Cells

It would go beyond the limits of this chapter to review in detail the attempts that have been undertaken to isolate other types of brain cells. However, for the sake of completeness some recent key references are given. Reports on the successful bulk isolation of microglia or ependymal cells have (to the knowledge of the authors) never been published. Ependymal cells with beating cilia can be easily obtained when rat brains are perfused as described in Section IV,B,1 with the addition of 25 mM tripotassium citrate to the medium. However, no attempts have been made to enrich this fraction (H. H. Althaus, unpublished result). Some effort has been devoted to the isolation of microvessels, as described by Betz and Goldstein (1978); Djuricic and Mrsulja (1977), Goldstein *et al.* (1977); Hjelle *et al.* (1978); Williams *et al.* (1980).

H. Assessment of Bulk-Isolated Neuronal and Glial Cells

Biochemical studies with isolated brain cell populations assume that such preparations have not undergone destructive alterations of their cellular entity. The degree of integrity of the cells with respect to their internal–external morphology and their biochemical compounds (see major objections) obviously cannot be the same as in the *in vivo* situation, since the application of separation methods involves "drastic" alterations in the cellular environment. Processes are shorn off, the cell surface undergoes physical and/or chemical alterations, the intracellular milieu of the cells will reflect the exposure to the media (usually of nonphysiological composition). Undoubtedly, the isolated cells are damaged and this injury may harm neurons differently compared to glial cells. Hence, when one is talking about cell integrity after isolation this term is disputable and may possibly be attributed only to parts of the cell.

Another important parameter to consider is the viability of the separated cells. Viability certainly is not positively indicated by the widely used trypan blue exclusion approach, since trypan blue is the least sensitive test for cell damage (Medzihradsky and Marks, 1975). With few exceptions all bulk-isolated cells are stated as 95% viable by trypan blue exclusion. A better indication might be given with fluorescein diacetate (Hansson *et al.*, 1980) or fluorescein dibutyrate (which does not have such a fast hydrolysis rate, Guilbault and Kramer, 1964), since this technique depends on some metabolic activity of the cell. However, metabolic studies carried out immediately after the isolation procedure reflect the viability of the cells only to some extent. The measured activity may be due to residual activity present in substantially damaged cells. A more conclusive approach is the examination of the cells' resting and action potential, or culturing of the isolated cells which should exhibit either regeneration phenomena or signs of intracellular movement of organelles. A low resting membrane potential (but no action potential) has been recorded from hand-dissected neurons from Deiter's nucleus (Hillman and Hydén, 1965). Membrane potentials of cortical and cerebellar neurons and glia bulk isolated from mature animals have not yet been reported. Neurons bulk isolated by the perfusion method have been found to have

Table I. Criteria by Which Isolated Brain Cells are Characterized[a]

Method	Animal	Cell type isolated	Yield × 10^6 cells/g WW	Percent purity (particle)	Dye exclusion	Metabolic studies	TEM SEM	Processed for tissue culture	Special features
Althaus and Huttner	Rat	N	10–17	90	+	+	+	+	Synaptic complexes retained, maintained in vitro for several weeks
Farooq et al.	Rat	N	37	84	–	–	–	–	
Poduslo and McKhann	Rat	N	Previous data	Previous data	+	+	Previous data	+	Maintained in vitro for 1–2 days
Farooq and Norton	Rat	A N	4–7 10–14	57 90	–	–	–	–	Astrocytes are highly ramified
Gebicke-Haerter	Cat, pig	O	3	90	–	–	+	+	Immunologically characterized,

Author	Species	Cell type					Previous data		Remarks
...and Althaus									outgrowth of cell fibers
Poduslo et al.	Bovine	O	11.4	90	+	+	+	+	Maintained in vitro for some days
Snyder et al.	Rat	O	3.5–7.1	90	+	–	+	–	Isolated from the whole forebrain
Szuchet et al.	Sheep	O	3	Band II 65% Band III 90%	–	+	+	+	Two subpopulations; immunologically characterized; outgrowth of cell fibers
Chao and Rumsby	Rat	N	14.08	95	+	–	–		
		A	6.22	70					
		O	3.64	62					
Cohen et al. and	Rat	G	10/per run	70					
		P	1/per run	38*	+	+	+	+	
Balazs et al.	(A)			57					

[a] N, neuron; A, astroglial cells; O, oligodendroglial cells; G, granular cells; P, Purkinje cells; –, no data given yet; +, performed; *, within a fraction of large neurons; TEM, transmission electron microscopy; SEM, scanning electron microscopy.

357

low membrane potentials of 30–40 mV after several days in culture (Montz, 1981). The culturing of bulk-isolated cells is still in its infancy, however, just to maintain cells for some days does not automatically mean that the cells are fully viable.

Altogether the parameter—cell viability—deserves better documentation using physiological and culturing methods. Otherwise the term viability should be replaced by activity for metabolic studies carried out shortly after the isolation procedure. In addition, certain criteria used to characterize isolated brain cell preparations have to be disputed. For instance, some error may be introduced into percentage purity determinations depending on whether contaminants are estimated by particle or volume count. Furthermore, no quick test exists to discriminate between isolated oligodendrocytes and astrocytes whose processes have been shorn off, or ependymal cells. Table I summarizes the criteria by which brain cells isolated by the above described procedures have (or not) been characterized by the authors.

I. Techniques for Cell Separation Other Than Those Based on Size or Density

The separation technique traditionally applied to bulk isolation of brain cells involves density gradient centrifugation or velocity sedimentation of the cells. In order to improve the purity of the fractions and/or to substitute the gradients, which usually exhibit nonphysiological osmolarity, attempts have been made to find other separation techniques (see Campbell, 1979). These involve free-flow electrophoresis, electronic cell sorting, magnetophoretic cell separation, and affinity column separation. It must be borne in mind that these techniques are still under intensive investigation, restricted to a few laboratories and have not yet all been applied to brain cells.

Free-flow electrophoresis was considered long ago (Campbell, 1963; Johnston and Roots, 1970) as a possible tool for discriminating neural cell types simply on their cell surface charge. However, initial experiments were not successful (Johnston and Roots, 1970). Subsequently, a method has been developed to separate cells magnetophoretically, utilizing cell-specific surface constituents (Kronick *et al.*, 1978; Campbell *et al.*, 1978). Magnetic microspheres are conjugated with a "recognizing" compound such as cholera enterotoxin (affinity for ganglioside GM1) and applied to the separation of C-1300 neuroblastoma cells (Kronick *et al.*, 1978), or goat antirabbit IgG for the detection of rabbit antibovine oligodendrocyte antibodies (Campbell *et al.*, 1978). The cells to which the microspheres adhere are retained in a divergent magnetic field, whereas the remaining cells pass through. Potentially this technique also provides a tool for discriminating cell surface molecules on a quantitative basis, since differential magnetophoretic mobilities can be achieved. The cells separated in this manner are reported to be viable as indicated by regenerating fibers under culture conditions.

Electronic cell sorting provides another approach to the separation of cells using either angle light scatter or differential antiserum binding (fluorescence-activated cell sorting). The principal is that a stream of single cells intersects a laser beam, which is focused on the stream. Light scattering or emitted fluorescence are monitored by detectors. When a cell generates a specific signal of designated value a charging voltage pulse is applied to the stream which breaks into droplets (by low-power ultrasonic vibration). Those droplets containing the cells, which have exhibited the specific signal, remain charged after separation from the stream. These are then deflected by an electric field and collected in a separate vial. The use of large angle light scattering patterns for neural cell separation has been suggested previously (Meyer *et al.,* 1974) and recently forward angle (5°) light scatter has been applied to the separation of cerebellar cell fractions which have been obtained by velocity sedimentation on a BSA discontinuous gradient (Campbell *et al.,* 1977). In the same paper (Campbell *et al.,* 1977) cell analysis using rabbit anticorpus callosum (anti CC) serum was reported, i.e., fluorescein-conjugated IgG fractions of goat antirabbit γ-globulin labeling the anti-CC positive cells, which are sorted out electronically. The *in vitro* growth of the different separate populations was shown (Campbell and Williams, 1978).

Affinity column separation was successfully applied to the separation of tumor cells (e.g., Venter *et al.,* 1976) and lymphocytes (e.g., Truffa-Bachi and Wofsy, 1970). A ligand, which binds to cell surface moieties, is immobilized on a substrate (e.g., Sepharose 6 MB—Pharmacia) that allows a high flow rate of cells through the column. The cells of interest are retained in the column, whereas the remaining cells pass through. The retained cells are then released by washing with an appropriate compound. This method has been successfully applied to the separation for chick embryo sympathetic ganglion neurons using α-bungarotoxin as the affinity ligand (Dvorak *et al.,* 1978) and also for chick embryonal spinal cord cells utilizing immunoglobulins against cord cells (Au and Varon, 1979). The latter report provides additional detailed information on the prerequisites required for this approach. In both papers the separated cells are claimed to be viable by culturing (Au and Varon, 1979) or by electrophysiological measurements (Dvorak *et al.,* 1978).

The potential use of these recent methods in the field of neural cell isolation is undoubtedly great. However, at present their use is restricted, for example, a reduced recovery (affinity column) or a limitation in the cell number to be separated (electronic cell sorting). Furthermore, it may be difficult to provide sterile conditions throughout the procedure. Since the quality of the cell separation is dependent on the specificity of the "recognizing" ligand and on the composition and availability of the target ligand on the cell surface, which may alter during tissue disaggregation, it must be recognized that there are other limitations to this approach.

V. BIOCHEMICAL PROPERTIES OF NEURONAL AND GLIAL CELLS ISOLATED FROM "NORMAL" BRAIN TISSUE

In general three methodological approaches have been employed to obtain information about the biochemical properties of brain cells. Either separated cells are used for experimental studies or initial steps are carried out on brain slices followed by cell isolation. Alternatively, a radioisotope is applied *in vivo* followed by subsequent isolation of the cells. Problems of terminology exist when studies are performed on so-called "glial" fractions, for though these might consist mainly of astrocytes (see Henn, 1980; Henn and Hamberger, 1976; Poduslo and Norton, 1972b), broken-off processes, capillaries, other glial cells, and pinched-off nerve endings are also present as contaminants. This fraction has been referred too by others as "neuropil" (Rose, 1967, 1968a, 1970) and is designated in this chapter as the astroglial fraction.

The attribute "normal" means that the brain tissue from which the cells are prepared is not affected by disease. However, several of the initial steps used in isolation methods, affect brain tissue to some extent causing changes in the *in vivo* situation. For example, such effects can originate from anaesthesia or decapitation of the animal and hypoxic conditions may result. The nonphysiological composition of isolation media will generate reactions in the cells, e.g., swelling of the astrocytes. In addition, rapid shifts of temperature may have an effect on ribosomes or some labile enzymes. Yet another phenomenon which may occur—the reason for it is not yet fully clarified—is the gelling of neuronal cytoplasm on isolation (Johnston and Roots, 1970).

The major goal of biochemical studies on purified neural cells is to characterize the different subcellular compartments in terms of their chemical composition and metabolic properties. Comparison of such features may cast light on their functional interrelationships. As pointed out previously separated brain cells are damaged to various extents in their internal and external morphology although application of recent methods results in less damage. The efflux and influx of compounds, which occurs during isolation will alter the composition of the cytoplasm to some extent and cells will respond to such injury, e.g., chromatolysis in neurons. This response may give erroneous results when, for example, RNA or protein synthesis in neuronal or glial cells is compared. Furthermore, when the biochemical characteristics of neurons and oligodendrocytes are compared it is usually the case that these cell classes have not been prepared from the same animal, nor have they been isolated under identical conditions. Even when the same method is applied it will probably affect neuronal and glial cells differently. An additional problem arises as to what basic denominator will be used when comparing the properties of different cell classes. The usually applied criteria are protein content, cell number, or DNA content. DNA determinations might be preferred, but one has to consider that free DNA from ruptured cells or free nuclei (without cytoplasmic material) might be present. With these problems in mind one must realize that the information acquired from biochemical studies of such preparations is limited and reflects the *in vivo* situation only to some extent.

A. Nucleic Acids

The RNA content of isolated neurons much varies and this is partly attributable to the varying size of the cells. In hand-dissected neurons RNA values for Deiter's nucleus are 700–1500 pg/cell, for hypoglossal neurons 200 pg/cell (Hydén, 1960), and for pyramidal cells of the hippocampus 53–110 pg/cell (Hydén, 1967). RNA values for bulk-isolated cortical neurons range from 14.1 to 25.5 pg/cell (Farooq and Norton, 1978), 24.2 pg/cell (Poduslo and Norton, 1975), 190 pg/cell (Rose, 1965), and up to 2050 pg/cell for pig brain stem neurons (Tamai *et al.,* 1975). The question has been posed as to whether "glial cells" have a lower amount of RNA (Hydén and Pigon, 1960) compared with neurons (see Pevzner, 1979). The RNA content of oligodendroglial cells has been reported as being lower than the above cited values for neurons, i.e., 1.95 pg/cell (Poduslo and Norton, 1972a), 2.1 pg/cell (Iqbal *et al.,* 1977), and 2.5–2.9 pg/cell (Snyder *et al.,* 1980), however, values for astroglial cells range from 8.1 pg/cell (Farooq and Norton, 1978) to 29.1 pg/cell (Norton and Poduslo, 1971). The RNA patterns of neurons and glia, as revealed by acrylamide gel electrophoresis (Jarlstedt and Hamberger, 1971; Satake *et al.,* 1968), seem to be essentially similar, i.e., major band are 28 S, 18 S, 5 S, and 4 S, however, glial cells contain more tRNA (4 S) and less 5 S RNA compared to neurons (Jarlstedt and Hamberger, 1971). These results have been questioned by Azcurra and Carrasco (1978), who additionally found "polydispersed" RNA with labeling characteristics of mRNA in the neural cytoplasm. This latter finding confirms previous results of Løvtrup-Rein and Grahn (1974), who, using the bulk separation method of Freysz *et al.* (1968b), found pronounced effects on polysome profiles due to different cell fractionation procedures. Cell incubation with trypsin resulted in a nearly complete breakdown of the polysomal profile. The RNA base ratios showed a higher proportion of guanine and a lower proportion of adenine in neurons when compared to glia (Hydén and Pigon, 1960). Studies on RNA synthesis revealed a higher specific radioactivity in neurons when compared to glial cells (Flangas and Bowman, 1970) although this depended on several factors, i.e., the concentration of the isotope, the time of the pulse, the age and species of animal, the route of administration of isotope, and the bulk separation method applied. An initial higher activity rate may occur in glial cells (Giuffrida *et al.*, 1976; Jarlstedt and Hamberger, 1971; Volpe and Giuditta, 1967; Yanagihara, 1974). Yanagihara (1979) found a six times higher specific activity in glial cells compared to neurons over a 120-min pulse. As pointed out by the authors the concentration of the isotope might be critical, since with [³H]uridine concentrations below 0.5 mM, an increasingly higher uptake of uridine in astroglial cells compared to neurons was observed. Only minimal differences were observed at [³H]uridine concentrations of 0.5 mM. This might explain why "intact cells" behave differently compared to subfractionated nuclei, where neuronal nuclei are more active in RNA synthesis than glial nuclei. In both neuronal and glial cell fractions, two rapidly labeled RNA species, corresponding to 25 S and 12 S, have been observed (Løvtrup-Rein and Grahn, 1974). However, this preferential labeling at 60 min was not confirmed for neurons by Azcurra and Carrasco (1978).

The DNA content of bulk-isolated neurons and astrocytes has been reported to be in the same range. Values of 8.2 pg/cell for rat neurons are found by Norton and Poduslo (1971), 7.9–8.7 pg/cell by Farooq and Norton (1978) (with some deviation depending on whether trypsin is used or not, see Farooq *et al.* 1977), 6.1–6.8 pg/cell by Satake and Abe (1966) and Freysz *et al.* (1968b). The DNA content of rat astrocytes is 11.2 pg/cell (Norton and Poduslo, 1971), but a somewhat lower figure is given by Farooq and Norton (1978), i.e., 7.7 pg/cell. Some controversy exists with respect to the DNA content of oligodendrocytes, since the value of about 5.2 pg/cell reported by Poduslo and Norton (1972a) and Iqbal *et al.* (1977) was not obtained by Fewster *et al.* (1973, 1974). Her value of 6.7 pg/cell agrees well with the recent estimation of Snyder *et al.* (1980) of 6.2–7.9 pg/cell for oligodendrocytes, isolated from 10- to 60-day-old rats. The possible explanation offered by Fewster *et al.* (1973) is that a partial loss of DNA occurs during the trypsinization treatment due to trace contamination of the trypsin by pancreatic desoxyribonuclease.

Metabolic studies on DNA in separated neuronal and glial cell fractions are rarely reported. Giuffrida *et al.* (1976) incubated slices from different areas of the brains of rats (5- to 30-day-old) with [³H]thymidine and followed by cell separation according to Blomstrand and Hamberger (1970). The results showed that astroglial cells and granule neurons are responsible for the high thymidine incorporation into DNA in the different regions over the developmental period of 5–30 days. (A low rate of incorporation was observed for Purkinje cells.) These results regarding granular and Purkinje cells were confirmed by Cohen *et al.* (1978) who studied the incorporation of [³H]thymidine into cerebellar cells of 9-day-old rats. Fraction C, containing about 70% external and primitive granule cells, incorporated [³H]thymidine several times more actively than fraction E which consists predominantly of Purkinje cells. Investigations of the DNA synthesis in neuronal and glial nuclei isolated from the *adult* guinea pig revealed that the fraction containing 97% glial nuclei synthesized DNA 2.4 less than the neuronal nuclei. DNA polymerase levels were 24.5 units for the glial nuclei and 39.2 for the mixed neuronal–glial nuclei fraction (Inoue *et al.*, 1976).

B. Amino Acids

The free amino acid content of bulk-separated neurons and astroglia is estimated to be one-third to one-fifth of the values reported for fresh whole brain (Nagata *et al.*, 1974). The distribution of free amino acids was similar in both brain cell fractions. Nagata's procedure based on that of Rose (1965) yielded relatively higher concentrations in neurons than in glial cells, particularly GABA and glycine which were twice as high. Rose (1968b, 1970) also found a higher concentration of amino acids in the neuronal fraction compared to the astroglial prior to any incubation. However, Sellström *et al.* (1975), using Hamberger's method, which employs an incubation step, reported considerably lower levels of free amino acids in neurons (4% of the total amino acid concentration present in fresh brain), whereas the level in the astroglial fraction corresponded more to the values given by other authors (20% of the amino acid concentration present in fresh brain). Several possibilities might account for this discrepancy (purity of the fraction, use of incubation or not, different ion content of the medium, etc.), but these are speculations and firm data are missing. Microchromatography reveals that the complete spectrum of amino acids is present in the isolated neurons (H. H. Althaus, unpublished data). An extrapolation back to the *in vivo* pattern of amino acids is problematic since it has been assumed that all amino acids leak to the same extent. This assumption is not true however.

While the data for some amino acids obtained by either method mentioned above are quite controversal, it is noteworthy that the amino acids with proposed transmitter function, having the highest concentration in whole brain tissue, exhibit a similar proportion in the neuronal and glial fraction (Sellström *et al.*, 1975). But again a discrepancy occurs concerning the uptake of amino acids. The fractions of Nagata's preparation take up nearly equal amount of D-glutamate and the final level is about four times above that in the solution (Nagata *et al.*, 1974). The astroglial cell fraction of Hamberger's procedure accumulated amino acids in particular glutamate, aspartic acid (Hamberger, 1971; Weiler *et al.*, 1979a) and threonine, serine, proline, and glycine more actively (Hamberger, 1971). Similar results were found when Norton and Poduslo's isolation method was used by Hamberger. In addition, Hamberger (1971), demonstrated an increase in the accumulation of amino acids into the astroglial fraction when the potassium concentration was increased from 0 to 5 mM, whereas the neuronal fraction was only slightly stimulated. Potassium levels above 10 mM led to a lowered uptake.

Of particular interest is the ability of enriched neuronal and astroglial cell fractions to accumulate norepinephrine, serotonin, dopamine, and GABA, substances believed to act as neurotransmitters. Neurons and astroglial fractions concentrate monoamine transmitters fourfold relative to that present in the medium. Astroglial cells, however, concentrated GABA over 100-fold from the medium in contrast to a neuronal preparation which showed a fourfold concentration (Henn and Hamberger, 1971). The uptake of GABA into astroglial cells which differs from the process present in nerve terminals (Schon and Kelly, 1975), is sodium-, potassium-, and magnesium-dependent. Calcium stimulated the uptake of GABA by synaptosomes at concentrations which inhibited uptake by astroglial cells (Sellström and Hamberger, 1975). Studies on GABA metabolism revealed GABA is mainly formed in the nerve endings, where the specific activity of GAD is several times higher than in astroglial cells (Sellström et al., 1975). On the other hand GABA-T, the catabolic enzyme of GABA, exhibited a higher specific activity in astroglial cells than in neurons or synaptosomes (Sellström et al., 1975). This suggests that the production of GABA takes place mainly at the presynaptic side with GABA destruction in the astroglial cell.

Another enzyme of the glutamate–GABA cycle, glutamine synthetase, which is mainly involved in ammonia detoxification in brain tissue, has been found in higher activity in the neuronal fraction than in the astroglial (Rose, 1968b; Weiler et al., 1979b). However, recent experiments demonstrated the exclusive localization of this enzyme in astrocytes (Norenberg, 1979). Glutamine itself has been shown to be metabolized in astroglial cells to glutamate (Weiler et al., 1979a) which may serve as the small pool for reconstitution. The proposed role for astroglia, functioning as a regulatory unit for certain compounds, was extended by the results of Rose (1968a, 1973). Following the incorporation of [^{14}C]glucose and pyruvate into amino acids in neurons and glia, the absolute incorporation and specific activity of the amino acids were higher in the astroglia than in the neurons by a factor of 3–5. This suggests that astroglial cells, at least in vitro, play an important role in amino acid transport.

C. Proteins

The protein content of isolated brain cells is largely dependent on their mass, on the morphological preservation of the individual cell type, and the purity of the fraction. Hand-dissected nerve cells of the Deiter's nucleus of rabbit have a protein content of 8440–16700 pg/cell (Hydén, 1960) and bulk-isolated pig brain stem neurons 12900 pg/cell (Tamai et al., 1975). Bulk-separated neurons of rat cerebral cortex range in their protein content between 99 pg/cell (Norton and Poduslo, 1971) to 200–230 pg/cell (Farooq et al., 1977; Althaus et al., 1978a). The estimated protein content of astrocytes has been reported to be 307 pg/cell (Norton and Poduslo, 1971), but Farooq and Norton (1978)—with morphologically well-preserved astrocytes of a high degree of purity—found values of 195 pg/cell. This lower value might reflect less contamination of astrocytes by proteins of nonastroglial origin (e.g., albumin is omitted from the isolation medium).

Data for the protein content of oligodendrocytes show little variation between the different separation methods, since this fraction is highly enriched (90–95% purity) in oligodendrocytes with cellular processes torn off. The low protein content of 46 pg/cell to 60 pg/cell in this cell type (Fewster *et al.*, 1973, 1974; Iqbal *et al.*, 1977; Poduslo and Norton, 1972a; Poduslo, 1978; Snyder *et al.*, 1980) of which a considerable proportion has to be attributed to the cell nucleus, is not surprising since mature oligodendrocyte possesses scant cytoplasm. Snyder *et al.* (1980) reported only a slight variation (56–60 pg/cell) in the protein content of oligodendrocytes isolated from rats between 10 and 60 days old. This appears to be questionable since light (and medium) oligodendrocytes, present in rat brain in considerable numbers at 10 days postpartum, exhibit a higher cytoplasm to nucleus ratio (Mori and Leblond, 1970; Privat, 1975).

Bulk-separated brain cells have been subjected to acrylamide gel electrophoresis to reveal their individual protein pattern. Using the detergent free Ornstein and Davis system, Packman *et al.* (1971) found distinct (quantitative and qualitative) differences between neuronal and astroglial fractions (separated according to Hamberger, 1971) with respect to basic and acidic proteins. Some bands were not observed in the neuronal fraction, whereas other bands were not detected in the astroglial fraction. S-100 protein was preferentially localized in astroglial cells and occurred only to small extent in the neurons which agrees well with the finding of Nagata *et al.* (1974). The nerve cell specific protein, 14-3-2, was found to be present equally in both cellular fractions. However, the authors point to the possible contamination of the astroglial fraction by nerve cell processes, whereas the neuronal fraction is negligibly contaminated by glial material. An interesting observation is the comigration of a pronounced acidic protein band in the astroglial fraction with a prominent band present in myelin. This similarity is not due to myelin contamination of the astroglial fraction. It is tempting to speculate (provided that the bands do represent identical proteins) that, whereas these bands correspond to the Wolfgram protein, at least one component among the Wolfgram proteins is more commonly distributed and not particularly localized in the myelin–oligodendroglia unit.

SDS acrylamide gel electrophoresis showed that five major polypeptides (28,000, 44,000, 52,000, 86,000, 95,000) were in common for neuronal and astroglial fractions prepared according to Hamberger (Karlsson *et al.*, 1973). Two fast migrating polypeptides of 14,000 and 16,000 MW were present in the neuronal fraction but were weak in the astroglial fraction. The subcellular fractions of neuronal and astroglial cells revealed only relatively minor differences. However, the astroglial plasma membrane, which has three major proteins of 44,000, 52,000, and 95,000 MW in common with the neuronal plasma membrane, contained a major band, not present in synaptic or neuronal membrane preparations. The electrophoretic pattern of neuronal proteins from 15- to 60-day-old rats revealed mainly quantitative changes of some proteins in the soluble and membrane-bound fraction (Althaus *et al.*, 1978b). Striking differences were observed when protein in the water-soluble and particulate fractions of neurons from primate and rat brains were compared (Althaus *et al.*, 1979). This study indicates that one must be cautious about extrapolating data from mammals of lower order to primates. Electrophoretic protein patterns of isolated oligodendroglia revealed the presence of basic protein (Fewster *et al.*, 1974; Iqbal *et al.*, 1977; Althaus *et al.*, in prep.), the amount of which seems to be dependent on the age of the animal. Immunocytochemical studies have confirmed that basic protein is localized in oligodendrocytes as well as in myelin (Sternberger *et al.*, 1978). S-100 has been detected in oligodendroglia using the method of cell isolation developed in our laboratory (Althaus *et al.*, in prep.), but was not detected by Fewster *et al.* (1973). Support for the presence of S-100 in oligodendroglia comes from the immunochemical studies of Ludwin *et al.* (1976) who demonstrated S-100 in astrocytes and oligodendrocytes. Glycoproteins of isolated brain cells were investigated by Margolis and Margolis (1974) who found a higher glycoprotein content in oligodendrocytes compared with astrocytes and neurons.

Despite the numerous studies on protein synthesis any conclusions as to whether neurons or glial cells are more active in the synthesis of proteins must take into consideration different factors specifically related to this subject. Such factors are

1. The route by which cells are exposed to the labeled amino acids, e.g., whether previously isolated cells or slices are used (Blomstrand *et al.*, 1975) or whether the label is given by *in vivo* intrathecal, intraperitoneal, intracerebral injection, or via arterial perfusion (Yanagihara, 1979).

2. The composition of the incubation medium with respect to its ionic content (Hamberger, 1971; Kohl and Sellinger, 1972; Takahashi *et al.*, 1970) or amino acid concentration (Blomstrand and Hamberger, 1970; Rose, 1973; Haglid and Hamberger, 1973) affects the incorporation rate.

3. The intracellular amino acid pool of the isolated neuronal and glial cells may play a role in the incorporation of a certain labeled amino acid into proteins (Hamberger, 1971; Rose, 1973).

4. As pointed out by Løvtrup-Rein and Grahn (1974) the polysomal profile of the isolated cells is particularly disturbed in those methods which employ an incubation step in their procedure.

5. The denominator usually taken for evaluating specific activity is the protein content. However, the purity of the neuronal and glial fractions that are compared diverges considerably, since the glial fraction particularly contains contaminants which will not participate in an amino acid incorporation. Therefore, any calculation on a protein basis will be erroneous for the glial fraction. Furthermore, an

overall statement about the protein synthesis of brain cells is not very meaningful when cells of different nucleus to cytoplasm ratios are compared over a given time.

Most reports describe a higher activity of protein synthesis in the neuronal fraction (Blomstrand and Hamberger, 1969, 1970; Blomstrand *et al.*, 1975; Haglid and Hamberger, 1973; Hamberger *et al.*, 1971b; Rose, 1973; Yanagihara, 1979). Giorgi (1971), using preparations of neurons and glial cells from brain of young rats, isolated according to Norton and Poduslo (1971), and applying [^{14}C]glycine *in vivo*, found a similar specific activity in astrocytes and large neuron, whereas oligodendrocytes were labeled to a much lower extent. Small neurons possessed a lower specific activity than large neurons, a result which has also been found for large neurons of cerebellar cell fractions prepared from young rats (Cohen *et al.*, 1978). These cells incorporated [^{3}H]glutamate [a strong candidate as neurotransmitter of the granule cells (Hudson *et al.*, 1976)] into protein at higher rate than granular cells. The least active fraction was the one containing predominantly astroglial-like cells. The time course of incorporation of intraperitoneally injected [^{3}H]lysine and [^{14}C]phenylalanine into neuronal and neuropil proteins has been followed up to 8 days (Rose and Sinha, 1974). After intervals longer than 2 h the specific activity of neuronal perikarya falls below that of the neuropil.

Margolis and Margolis (1974) followed the incorporation of intraperitoneally administered [^{3}H]glucosamine into glycoproteins of neuronal perikarya and astroglial cells prepared from 25-day-old rat (using the method of Norton and Poduslo, 1971). Glycoproteins of neurons were slightly more labeled then those of astroglial cells.

The incorporation rates of labeled amino acids into cerebral proteins are known to be faster in young animals than in adults. This is reflected in studies performed by Haglid and Hamberger (1973) and Johnson and Sellinger (1971) who found a significant decrease in the protein-bound radioactivity of neuronal and astroglial-enriched fractions with maturation. But these two reports disagree on (1) the time at which the maximum incorporation and the decline of incorporation occur in the neuronal and glial fractions (for comparison see also Yanagihara, 1979) and (2) the neuron to astroglial ratio of the protein-bound radioactivity. This ratio has been found by Johnson and Sellinger (1971) to reverse in 43-day-old rats. The protein synthesis of isolated neuronal and astroglial cells is sensitive to hypoxic or glucose-deficient conditions (Yanagihara, 1979). Whereas hypoxia resulted in a similar effect in both fractions, the neuronal cells were more sensitive to glucose deficiency. Different results with respect to hypoxia were obtained by Blomstrand and Hamberger (1970) who observed a greater suppression of amino acid incorporation in the neuronal fraction relative to the glial fractions.

Subcellular fractions obtained from neuronal and astroglial fractions prelabeled with [³H]leucine showed a severalfold higher activity in microsomes and soluble proteins of neuronal cells compared to fractions from glial cells, however, the labeling of nuclear and mitochondrial fractions was similar (Hamberger et al., 1971b). In this work the highest activity occurred in the microsomal fraction (Hamberger et al., 1971b). Johnson and Sellinger (1971) although finding initially higher specific activity in the microsomal fraction (after 10 min) noted that by 20 min proteins of the mitochondrial fraction were most highly labeled. The incubation of neuronal, astrocyte, and oligodendrocyte nuclei with ³H- or ¹⁴C-labeled leucine (incubation time up to 60 min) resulted in quite different ratios. Burdman and Journey (1969) observed the highest rate of incorporation in oligodendroglia nuclei followed by neuronal and astroglial nuclei. Løvtrup-Rein (1970) presented results where astrocyte nuclei showed greatest incorporation with lower values for neuronal and oligodendroglial nuclei. Since in these experiments young adult rats were used this discrepancy has to be attributed to other factors besides the age of the animal, i.e., nuclei of adult brain cells exhibit a far lower incorporation rate than nuclei of young animals (Burdman et al., 1970). A few studies on the protein metabolism of bulk-isolated cells are concerned with the effect of changes in the functional state of the nervous system (e.g., Jarlstedt and Hamberger, 1972; Rose and Sinha, 1974; Rose, 1975, 1976). However, more data are necessary before conclusive statements can be made on the field.

D. Lipids

This topic has been comprehensively reviewed several times. The lipid content, and the distribution and proportion of the individual lipids and fatty acids of various brain cells have been analyzed and summarized in tables (see Nagata and Tsukada, 1978; Norton et al., 1975; Pevzner, 1979; Tamai et al., 1975). Hence, the lipid ''section'' of this review contains only some of the basic data on brain cells. This section will outline some discrepancies in the available literature and give newly reported values.

The lipid to protein weight ratio of rat, rabbit, and pig neurons vary very little, i.e., from 0.43 to 0.47 (Hamberger and Svennerholm, 1971; Norton and Poduslo, 1971; Tamai et al., 1971). However, Kohlschütter and Herschkowitz (1973) reported a ratio of about 0.16 for mouse brain neurons. Astroglial fractions exhibit a higher ratio of 0.75 for rat (Norton and Poduslo, 1971) and 0.69 for rabbit astroglial fractions (Hamberger and Svennerholm, 1971) which is somewhat different from the ratio of 0.35 for oligodendrocytes reported by Fewster and Mead (1968a). The estimated value of 0.13 (Poduslo and Norton, 1972a), based on 6.7 pg lipid/cell and an average of 50 pg protein/cell is probably too low.

The lipid composition of brain cells expressed as weight % of total lipids reveals some characteristics for each brain cell type, but also some discrepancies with respect to the individual lipids. The major components are cholesterol and phospholipids, but their percentage of total lipids much varies. For rat neurons Norton and Poduslo (1971) and Freysz *et al.* (1968b) found similar results for cholesterol (10.6 or 10%) and phospholipids (72.3 or 76.4) but rabbit neurons exhibit 21.3% cholesterol and 73.2% phospholipids (Hamberger and Svennerholm, 1971). Tamai *et al.* (1971) reported 9% cholesterol and 86.7% phospholipid for neurons from pig brain stem values, which are more in agreement with the values of Kohlschütter and Herschkowitz (1973) who found 10.4% cholesterol, and 83.8% phospholipids. The data for astroglial fractions are more consistent within different preparations—the percentage of cholesterol is slightly higher than in neurons (14–21.4%), and phospholipids are in the same range (70.9–72.1%) (Freysz *et al.*, 1968b; Hamberger and Svennerholm, 1971; Norton and Poduslo, 1971). Oligodendroglial preparations differ substantially in their cholesterol/phospholipid content: cholesterol contributes between 14.1 (Poduslo and Norton, 1972a) and 30.6% (Fewster and Mead, 1968a) of the total lipids, and phospholipids 62.2 or 48.2%. Nevertheless, oligodendrocytes appear to be the cells with a lower phospholipid concentration compared to others.

Individual phospholipid patterns are rather variable in different preparations, which might be attributed to differences in species, age, and analytical methods. Choline (CPG) and ethanolamine (EPG) phosphoglycerides are the major components in neuronal and glial cells, the level of CPG exceeding that of EPG (Freysz *et al.*, 1968a; Hamberger and Svennerholm, 1971; Kohlschütter and Herschkowitz, 1973; Norton and Poduslo, 1971; Poduslo and Norton, 1972a; Raghavan and Kanfer, 1972; Tamai *et al.*, 1971; Woelk *et al.*, 1974). A reversed CPG/EPG ratio has been reported for glial cells by Davison *et al.* (1966) and for oligodendroglial cells by Fewster and Mead (1968a). Total glycolipids comprise a similar percentage of the total lipids in neuronal (2.1–5.8%) and astroglial fractions (1.8–5.3%) from the same preparations, but differences occur from one method or species to another in the proportions of the total lipids. In contrast to these data Tamai *et al.* (1971) and Davison *et al.* (1966) could not detect any glycolipid (except gangliosides) in neurons. Oligodendroglia seem to contain proportionally about two–fourfold higher glycolipid content (Fewster and Mead, 1968a; Poduslo and Norton, 1972a) compared to neurons and astroglia. Among the glycolipids the ratio of cerebrosides to sulfatides (4.5:1 respectively to 4:1) has been reported to be similar between neurons and astroglia (Hamberger and Svennerholm, 1971). However, Kohlschütter and Herschkowitz (1973) reported the opposite for mice neurons with a ratio of 1 to 3. The corresponding ratio for oligodendroglia varies from 5.5:1 (Poduslo and Norton, 1972a) to 2.7:1 (Fewster and Mead, 1968a).

Gangliosides analyzed as NANA (*N*-acetylneuraminic acid) have been detected in neurons (0.7–1.7 μg/mg dry weight) by Hamberger and Svennerholm (1971), Norton and Poduslo (1971), Tamai *et al.* (1971), but surprisingly the astroglial fraction has been found to contain a 2- to 2.5-fold higher amount of ganglioside NANA (Hamberger and Svennerholm, 1971; Norton and Poduslo, 1971). In contrast Hinzen and Gielen (1973) found a largely similar amount of ganglioside NANA in glia and neurons. Furthermore, hand-dissected neuronal and glial cells from Deiter's nucleus contained more gangliosides in the isolated neurons than in glial cells (Derry and Wolfe, 1967). Oligodendrocytes were in the same range as neurons (Poduslo and Norton, 1972a). Here, however, one has to consider that the isolated neuronal perikarya have lost their processes and the synaptic contacts from other neurons and that in the astroglial fraction synaptosomes are present as contamination to a variable degree (see Hamberger and Sellström, 1975; Hamberger *et al.*, 1975). Nevertheless, gangliosides seem to be present in all brain cell membranes. Ganglioside patterns of neurons, astroglia, and oligodendroglia appear to be rather complex (when compared to cultured cell lines) and similar (predominantly GM_1 and GD_{1a}) when the same animal is used for the cell isolation (Abe and Norton, 1974; Hamberger and Svennerholm, 1971; Norton *et al.*, 1975; Robert *et al.*, 1975; Skrivanek *et al.*, 1978). Oligodendroglia and neurons isolated from human brain differ only slightly in their ganglioside pattern (GM_1, GD_{1a}, in addition GD_{1b} predominant) with the exception of GM_4 (one of the major gangliosides in human myelin), which is present in oligodendrocytes to a higher extent compared to neurons (Yu and Iqbal, 1979).

Among the nonsialic acid containing glycosphingolipids, galactosyl and glucosyl ceramide (the latter not detected by Raghavan and Kanfer, 1972) showed a ratio of about 1 to 1.5 in rat cortical neurons and astroglia and were present in a different amount in tumor cell lines. An important finding was that oligodendroglia (isolated from bovine white matter) exhibit a ratio of about 10 times more galactosyl cerebroside (a main myelin lipid constituent) than glucosyl cerebroside (Abe and Norton, 1974, 1979; Norton *et al.*, 1975). Monogalactosyl diglyceride (MGD), which has been suggested as a "marker" for myelination (Inoue *et al.*, 1971; Pieringer *et al.*, 1973), was detected in rat neuronal and glial cells, but the content in oligodendrocytes is by far the highest and increases during a period from the 16th to the 29th day (81.5 nmoles MGD–136 nmoles/ 100 mg protein) postnatally (Deshmukh *et al.*, 1974).

The fatty acid composition of neurons and astroglia (rabbit brain) revealed no striking differences (Hamberger and Svennerholm, 1971). Fatty acids obtained from different phospholipids—ethanolamine, serine, inositol, and choline phosphoglycerides—are largely similar in neurons and astroglia. Predominantly 16:0 (choline phosphoglyceride), 18:0, 18:1, 20:4, and 22:6 fatty acids are the major species present. Determinations by Tamai *et al.* (1971) for neurons prepared from pig brain stem showed a similar pattern for choline phospho-glyceride, but the values were quite different in the ratios of the predominant fatty acids for serine and ethanolamine phosphoglycerides. Corresponding data for oligodendrocytes (isolated from bovine white matter) by Fewster and Mead (1968b) gave a similar pattern for the choline phosphoglycerides compared to neurons and astroglia, but fatty acids of ethanolamine and serine phospho-glycerides were dissimilar in that 20:4 and 22:6 fatty acids were present only in minor amounts (as for the bovine myelin). The fatty acid composition of choles-terol esters and triglycerides, detected in oligodendrocytes by Fewster and Mead (1968b), resemble those of white matter having mainly saturated and monoenoic fatty acids.

Different species of the glycolipids have been investigated for their fatty acid composition. Assay of galactosyl ceramide (cerebroside), sulfatides, and glucosyl ceramide levels in specific brain cells have been compared and there is some disagreement over the results. Raghavan and Kanfer (1972) and Hinzen and Gielen (1973) reported fatty acid patterns of the cerebroside of neurons and astroglia (rat or rabbit brain) which showed the presence of α-hydroxy fatty acids in astroglia but not in neurons. In addition, Raghavan and Kanfer (1972) have not detected any unsaturated fatty acids in the cerebrosides of either cell type. These data are not in agreement with those of Abe and Norton (1974) and Norton *et al.* (1975), who found in cerebroside of both cell types unsaturated and α-hydroxy fatty acids. Raghavan and Kanfer (1972), using the isolation method of Norton and Poduslo (1970), detected only minor differences between cerebroside fatty acids in neurons and glia. Abe and Norton (1974) found that the differences are more pronounced, i.e., unsubstituted fatty acids in cerebrosides of astroglial had more long chain (> C18) acids than in neurons (62 versus 40%), whereas neuronal sulfatides had shorter chain fatty acids (29% > C18) than those of astroglia (45% > C18). Differences in cerebroside hydroxy fatty acids were less pronounced for both fractions but neurons had more 22 h:0 and less than 24 h:0 than those of astroglia. The fatty acid pattern of glucosyl ceramide was similar in the two cell fractions. Oligodendroglia (bovine) fatty acids (unsubstituted and α-hydroxy) of galactosyl cerebroside and sulfatide consisted of 80–90% long-chain (> C18) fatty acids (Abe and Norton, 1979; Fewster and Mead, 1968b; Norton *et al.*, 1975), and were thus apparently different to those from rat (15- to 20-day-old) neurons and astroglia, resembling closely the pattern from myelin glycosphingolipids. Another difference might be the proportion of monoenoic fatty acids which comprises about 50% in cerebrosides from oligodendrocytes (Fewster and Mead, 1968b), but only 24% and 13%, (Abe and Norton, 1974) in rat neurons and astroglia, respectively. However, Abe and Norton (1979) re-ported only 22% unsaturated fatty acids (of which a high percent are monoenes) in cerebrosides of oligodendrocytes.

When data on lipid metabolism are compared between brain cell fractions obvious differences are noted. The different conditions used, e.g., age of animal, route of application of the label, isolation method, etc. prevent conclusive infor-mation being available. Two examples illustrate the situation.

1. When isolated brain cells are used in labeling experiments *in vitro* effects occur which indicate that the cell is partly damaged. An example of this is the

inability of oligodendrocytes to engage in elongation of [1-^{14}C]linolenate (Cohen and Bernsohn, 1978; Fewster *et al.*, 1975), which occurs *in vivo* rather rapidly. Alterations to metabolic processes in neural cells may occur during isolation. An example of this is shown by alterations in the incorporation of [1-^{14}C]acetate into ethanolamine and choline phosphoglycerides as reported by Cohen and Bernsohn (1974).

2. Experiments of Binaglia *et al.* (1973)—using neuronal and astroglial fractions of rabbit brain, prepared according to Hamberger *et al.* (1971a)—resulted in a higher rate of synthesis of phosphatidylcholine and phosphatidylethanolamine for the neuronal fraction (neuron/glia ratio about 5–9) when labeled cytidine-5'-phosphate choline and cytidine-5'-phosphate ethanolamine were given as precursors. Abdel-Latif *et al.* (1974) found that when brain slices from young rats were incubated with the label (^{32}P$_i$ or [^{14}C]choline) and the tissue then fractionated (according to Norton and Poduslo, 1970; Poduslo and Norton, 1972b) no significant differences occurred in the incorporation of ^{32}P$_i$ or [^{14}C]choline into the total phospholipids of neuronal and astroglial-enriched fractions even though variation occurred within the individual phospholipids. Acetylcholine plus eserine (and norepinephrine) were found to increase the phospholipid labeling similarly for both fractions. When neuronal and astroglial cell fractions were first isolated from the rat cortices then incubated with ^{32}P$_i$ or [^{14}C]choline, labeling of phospholipids in the neuronal fraction was higher than that in the astroglial fraction; acetylcholine plus eserine had no effect on the incorporation of ^{32}P$_i$ into the lipids of either fraction.

The following results may also be viewed in this respect. Incorporation of ^{32}P$_i$ has been reported to be greater into phospholipids of neurons than of astroglia (Freysz *et al.*, 1968a, 1969; Marggraf *et al.*, 1979; Woelk *et al.*, 1974). Individual phospholipids were labeled to different extents in neuronal and astroglial fractions, phosphatidylcholine, and phosphatidylethanolamine more in neuronal, phosphatidylinositol and phosphatidic acid more in glia, and sphingomyelin synthesis seemed to be mainly localized in the neurons (Marggraf *et al.*, 1979). Combined neuronal and astroglial fractions incorporated ^{32}P$_i$ to a much higher rate than the sum of the separate fractions together (Marggraf *et al.*, 1979). This finding and the results of Binaglia *et al.* (1973) and Goracci *et al.* (1975) that the cytidine-dependent enzymatic system for phosphatidylcholine and phosphatidylethanolamine synthesis is concentrated mostly in the neuronal cells, have lead to the suggestion that the phospholipid metabolism of neuronal and glial cells may exhibit some cooperativity or involve the transfer of phospholipids between cells.

Labeled palmitate was incorporated into phospholipids of neuronal and astroglial cells to a nearly equal amount (Marggraf *et al.*, 1979). When ^{14}C-labeled stearic acid, linoleic acid, and linolenic acid were used, astroglial fractions incorporated more than neurons (Cohen and Bernsohn, 1973, 1978), but oligodendrocytes were the most active in incorporating these fatty acids. Kinetic studies on the biosynthesis of phospholipids revealed a faster turnover of neuronal phospholipids than astroglial phospholipids (Francescangeli et al., 1977; Freysz et al., 1969; Goracci et al., 1975). Concerning other lipid classes only a few reports are available. The rate of sterol formation in rat brain (11-day-old) has been studied by Jones et al. (1975) in vitro and in vivo. The in vitro experiments utilized [2-^{14}C]mevalonic acid as precursor. Astroglia-enriched fractions showed a greater ability to incorporate the label into isoprenoid material than neuronal fractions (cell preparation according to Norton and Poduslo, 1970, minus the incubation step). In vivo, sterol biosynthesis was investigated by intraperitoneal injection of sodium [2-^{14}C]acetate and [U-^{14}C]glucose, sacrifice of the animals after 2 or 24 h, and subsequent cell isolation. Astroglial-enriched fractions again contained the bulk of the labeled isoprenoid material. Studies on the biosynthesis of galactosyldiglyceride (rat brain 16- to 29-day-old) revealed that oligodendroglia had the highest and neurons the lowest capacity to enzymatically synthesize and accumulate monogalactosyl diglyceride (Deshmukh et al., 1974). Jones et al. (1972) as well as Maccioni et al. (1978) using different brain cell preparation methods came to the conclusion that the main cellular site of ganglioside biosynthesis is in the neuronal perikarya when compared to glial-enriched or synaptosomal fractions.

E. Enzymes

The reason for determining enzymatic activities in specific neural cells is to characterize metabolic pathways to reveal their metabolic capabilities. A comparison of enzymatic activities in different isolated cell types may reveal the main cellular site of the synthetic or catabolic properties that have been investigated. If the enzymatic activity of one cell type greatly exceeds that of another cell type this enzyme will inevitably be considered to be a "marker." However, great care has to be taken in drawing conclusions in this respect. For example, β-galactosidase (E.C. 3.2.1.23) has been suggested to be a neuronal marker because its specific activity is 10-fold higher in neurons of adult rats compared to the neuropilastroglia fraction (Sinha and Rose, 1972a). However, this result was not confirmed by Radin et al. (1972) and Raghavan et al. (1972) who found no significant variation in the distribution of activities for β-galactosidase between neuronal and glial fractions of 15- to 20-day-old rats. Freysz et al. (1979) reported a 20-fold higher β-galactosidase activity in neurons prepared from adult rabbit brain (about sixfold in those from beef brain) and a 10 times higher activity for acid phosphatase, another lysosomal enzyme. Abe et al. (1979), who measured the β-galactosidase activities toward different substrates (taking into account the fact that different isoenzymes of β-galactosidase exist, see Awasthi and

Srivastava, 1980; Townsend *et al.*, 1979), could not find this difference between either fraction prepared from rat (15- to 20-day-old) brain. In fact, the specific activity for this enzyme was greater in the astroglial than in the neuronal fraction, thus disputing the idea that β-galactosidase is a neuronal marker.

Age differences of the animals have been offered as a possible explanation for discrepancies in enzyme activities by Freysz *et al.* (1979), referring to the work of Arbogast and Arsenis (1974) who investigated specific activities of various lysosomal enzymes during development. At the time of myelination, the period at which the rats were used, the specific activities of the investigated lysosomal enzymes were similar in neurons and glial cells. However, Arbogast and Arsenis (1974) have demonstrated that the greatest differences in the specific activities of lysosomal enzymes occur during development (including the myelination period) of the rat (10–30 days) and that the neuron–glia ratios of acid phosphatase and aryl sulfatase (and other enzymes) do not exceed a "marker level" until the sixtieth day. This is in contrast to the findings of Freysz *et al.* (1979). Further support that β-galactosidase appears unlikely to be a "marker" enzyme can be also drawn from histochemical data of Robins *et al.* (1961) who showed some, but no dramatic differences in the activity of the enzyme in different layers of the cerebellum during development. Even the internal granular layer exhibited an activity during adulthood at best four times greater than the value for white matter. Two points emerge: (1) several factors including the developmental stage of the animal have to be considered before definite conclusions concerning the predominant cellular site of enzyme activity can be deduced and (2) work with bulk isolation of brain cells has given contradictory results leaving the situation undecided.

Another intriguing question is the extent to which the isolation method may produce "false" enzymatic activities. For example, Kohlschütter and Herschkowitz (1973) found that the exposure to trypsin inactivates cerebroside sulfotransferase activity (no details given). On the other hand Benjamins *et al.* (1974) demonstrated that this effect is largely dependent on the trypsin concentration and that sulfotransferase activity was not inactivated when 0.1% trypsin was used during the isolation procedure but that 1% trypsin decreased the activity. Exposure of (rat) brain cells to 1% trypsin had little effect on CNP activity (Benjamins *et al.*, 1974), and while trypsin exposure of bovine neuronal perikarya increased the sialidase activity two- to threefold, acetylcholinesterase activity was completely abolished (Schengrund and Nelson, 1976). It remains an open question whether each brain cell fraction is affected differently by isolation procedures.

In conclusion, our opinion is in agreement with Pevzner (1979) who questions the existence of strictly neuronal or glial enzymatic "markers," particularly when the main metabolic pathways are considered. A predominance of certain aspects of cell metabolism will occur in neurons and glial cells pointing to the idea of cellular interaction. Newly developed bulk isolation methods, which could be helpful in the characterization of cellular enzymatic activities, have been only recently been applied to this subject. Previous methods for the bulk isolation of brain cells can lead, at best, to information on the principal properties of these cells for synthesizing or catabolizing various compounds, rather than allowing a conclusive comparison of activities of different cell types or of stages of development.

Data on enzyme activities of bulk separated and hand-dissected neuronal and glial cells have been almost completely tabulated by Pevzner (1979), who cites the literature up to 1976. For the *oxidoreductases,* lactate dehydrogenase (E.C. 1.1.1.27) and malate dehydrogenase (E.C. 1.1.1.37) have been investigated for their isoenzyme patterns. No differences were reported for malate dehydrogenase in neuronal and astroglial fractions (rabbit) by Packman *et al.* (1971), but the pattern of LDH was found to be different. Isoenzymes 1 (H_4)–4 (HM_3) (5 was not separated) appeared to be of similar magnitude in the astroglial fraction, whereas isoenzymes 2 (H_3M) and 4 (HM_3) showed a somewhat lower activity in the neuronal fraction. Nagata *et al.* (1974) found in contrast that at any particular age of the rat brain the ratios of the isoenzymes LDH-4 and LDH-1 were the same in both fractions; these ratios changed during development. Cytospectrophotometric data of Brumberg and Pevzner (1975) again confirmed the similarity of the LDH isoenzymes between neurons and glia. The activities of the mitochondrial glycerol-3-phosphate dehydrogenase (E.C. 1.1.99.5) and succinate dehydrogenase (E.C. 1.3.99.1) were similar (some change within the postnatal period) in neuronal and astroglial fractions from rat brain (Svoboda and Lodin, 1972), but Sellinger and Santiago (1972) and Sellinger *et al.* (1973) found that the activity of α-glycerolphosphate dehydrogenase in glial fractions was sixfold higher than neuronal fractions, throughout the first month postpartum (rat).

The hydrocortisone induction of the cytoplasmic glycerol-3-phosphate dehydrogenase (NAD^+) (E.C. 1.1.1.8) activity has been claimed to take place exclusively in oligodendrocytes (DeVellis *et al.*, 1977; McCarthy and DeVellis, 1980). The induction of this enzyme in subclones of C_6 astrocytoma cell lines (up to 20-fold increase of specific activity and less inducible in rapidly dividing or in old cells) has also been reported (DeVellis and Inglish, 1973). Support for the view that cytoplasmic glycerol phosphate dehydrogenase activity may be oligodendroglial in origin came from results of Hirsch *et al.* (1980) who observed a reduced activity in multiple sclerosis plaques (where oligodendrocytes are nearly absent). Recently glycerol phosphate dehydrogenase has been immunocytochemically localized in oligodendrocytes (Leveille *et al.*, 1980). The rate-limiting enzyme in cholesterol synthesis—hydroxymethylglutaryl-Co-A-reductase—(E.C. 1.1.1.34) was found to be enriched in oligodendroglial fraction compared to white matter (Pleasure *et al.*, 1977). Monoamine oxidase (MAO) (E.C. 1.4.3.4) activity varies very little between neuronal and astroglial fractions (Hamberger *et al.*, 1970; Hemminki *et al.*, 1973; Sinha and Rose, 1972b). Hazama *et al.* (1976) investigated six regions of the rat brain for their MAO activity and found that the neuronal fractions (except the caudate nucleus) exhibited 1.5- to 3-fold greater activity than the glial fraction. This ratio, has also been reported by Sinha and Rose (1972b). In the cerebellum MAO activity was distributed between the fractions as follows: Purkinje cells > granular cells > glial cells. Cytochrome oxidase (E.C. 1.9.3.1) activity was similar in isolated neurons and astroglia (Arbogast and Arsenis, 1974) and their respective mitochondria (Hamberger *et al.*, 1970).

Elevated activity in neurons compared to astroglia has been found for the following *transferases:* choline acetyltransferase (E.C. 2.3.1.6) (Hemminki *et al.*, 1973; Nagata *et al.*, 1976), sialytransferase (E.C. 2.4.99.1), acylneuraminatecytidyltransferase (E.C. 2.7.7.43) (Gielen and Hinzen, 1974), and ethanolaminephosphotransferase (E.C. 2.7.8.1) (Roberti *et al.*, 1975). The investigation of the pyruvate kinase isoenzymes (E.C. 2.7.1.40) revealed a similar pattern in neuronal and astroglial fractions from rat and mouse brain (Tolle *et al.*, 1976). Ceramide glucosyltransferase (E.C. 2.4.1.80) activity, not detected in glial fractions (Radin *et al.*, 1972), has been suggested as a possible marker enzyme for neurons. In another report this enzyme has been found to have similar activities in neuronal and glial fractions (Deshmukh *et al.*, 1974). Both authors agree on the timing of galactocerebroside formation and the activity of ceramide galactosyltransferase (E.C. 2.4.1.45) appears to be two- to five times higher in oligodendrocytes than in neurons or astroglia (Deshmukh *et al.*, 1974; Radin *et al.*, 1972). Cerebroside sulfotransferase (E.C. 2.8.2.11) activity has been attributed to oligodendrocytes (Pleasure *et al.*, 1977; Szuchet and Stefansson, 1980; Tennekoon *et al.*, 1980), although some activity has also been detected in neurons from mouse brain (Kohlschütter and Herschkowitz, 1973) and in neurons and astroglia from rat brain (Benjamins *et al.*, 1974). Its activity in oligodendroglia from calf brain is eightfold greater per cell than in calf neurons (Benjamins *et al.*, 1974), but this ratio is markedly less when compared to that of rat neurons and astroglia. Neuronal sulfotransferase activity increases substantially from the seventh to the sixteenth day of development in the rat. The distribution of *hydrolases* in neuronal and glial cells has been studied by several laboratories resulting in a volume of controversial data (see Freysz *et al.*, 1979; Raghavan *et al.*, 1972; Sinha and Rose, 1972a). Butyrylcholinesterase (E.C. 3.1.1.8) has been mainly attributed to glial cells (Koelle, 1951; Giacobini, 1964). Studies with bulk-isolated rat brain cells have revealed that astroglial cell fractions had a higher activity than in neuronal cells; maximum specific activity was found between the seventeenth and twenty-sixth day postnatally (Arbogast and Arsenis, 1974). However, Sinha and Rose (1972b) reported the opposite, whereas Hemminki *et al.* (1973) found a more even distribution of the same enzyme.

Surprisingly, biochemical measurements of Na^+,K^+-ATPase (E.C. 3.6.1.3) on bulk-isolated brain cells contradict the histochemical view that there is a higher activity in neurons than in astroglial cells (Stahl and Broderson, 1975). Repeatedly (except for a report by Hamberger *et al.*, 1970) higher values for the activity of the Na^+,K^+-ATPase have been reported by several groups for astroglial-enriched fractions, using different preparation methods and animals of different species and age (Grisar *et al.*, 1979; Hamberger and Henn, 1973; Kimelberg *et al.*, 1978; Medzihradsky *et al.*, 1972; Nagata *et al.*, 1974). A sharp increase of the ATPase activity was observed between the sixth and twelfth day postnatally in rat astroglial fractions, whereas the neuronal fraction remained at nearly the same level (Medzihradsky *et al.*, 1972). These authors emphasize that the measurements of ATPase made on neuronal perikaryal preparations exclude ATPase associated with the neuronal processes and with the synaptic membranes. The K_m values for K^+, Na^+, and ATP were not markedly different in the neuronal and astroglial fractions (Kimelberg *et al.*, 1978).

Some agreement between different authors is observed for the activity of the myelin "marker" enzyme 2',3'-cyclic nucleotide 3'-phosphodiesterase (CNP) (E.C. 3.1.4.37). The predominant cellular site of activity has been observed to be in the oligodendroglial fraction (Deshmukh *et al.*, 1974; Pleasure *et al.*, 1977; Poduslo and Norton, 1972a; Poduslo and McKhann, 1975; Szuchet and Stefansson, 1980). Isolated plasma membranes of oligodendrocytes (a new method has been developed recently by Szuchet and Polak, 1980) showed a particular enrichment of this enzyme (Poduslo, 1975). Nagata *et al.* (1974) reported a higher activity of CNP in a glial preparation, a somewhat conflicting result. This preparation usually designated as a more astroglial-like fraction may be contaminated with either myelin or oligodendroglia since other authors have found that levels of CNP in astroglia are of the same low magnitude as for neurons (Deshmukh *et al.*, 1974; Poduslo and Norton, 1972a).

Among the *lyases* one enzyme that has been studied very intensively is carbonic anhydrase (CA) (E.C. 4.2.1.1). Astroglial-enriched cell fractions exhibited a severalfold higher CA activity than the neuronal fraction (Kimelberg *et al.*, 1978; Nagata *et al.*, 1974; Rose, 1967). The two molecular forms of CA (CAI, II) have been studied immunohistochemically during development of the rat cerebellum. Isoenzyme CAII appeared to be an oligodendrocytic (myelin related) marker (Ghandour *et al.*, 1980; Mandel *et al.*, 1978). Discrete areas of the rat and rabbit brain have been investigated for their adenylate cyclase (E.C. 4.6.1.1) activity in neuronal and glial fractions. The higher activity found for the neuronal fraction could be partly enhanced by norepinephrine or dopamine depending on the area studied (Palmer, 1973; Palmer and Manian, 1976). Glial cell membrane fractions prepared by using glial cells isolated from horse brain striatum contained an adenylate cyclase activated by 5-HT which is different compared to dopamine-stimulated adenylate cyclase present in the same preparation (Fillion *et al.*, 1980). Enolase, the 14-3-2 protein, (E.C. 4.2.1.11) which has been considered to be a neuron-specific marker in one isoenzyme form (Bock and Dissing,

1975; Schmechel *et al.*, 1978) although it is dependent on the developmental stage and not expressed by all neurons (Cimino *et al.*, 1977), has recently been localized also in the endocrine cells of the pancrease and thyroid (Polak *et al.*, 1980).

Passing over the *isomerases* the main interest among the *ligases* has been concerned with the glutamine synthetase (E.C. 6.3.1.2). Rose (1968b) and Weiler *et al.* (1979b) found an even distribution of glutamine synthetase activity in bulk-isolated neurons and astroglia. These results do not agree with immunohistochemical findings which have localized glutamine synthetase specifically in astroglial cells (Martinez-Hernandez *et al.*, 1977; Norenberg, 1979).

F. "Markers" for Neural Cells

It is appropriate to outline criteria for cell viability and integrity before discussing specific cell characteristics, particularly when enzyme activities are compared and defined as being "cell-specific markers." A conclusive and rapid test to determine the viability of neural cells is not available and might be difficult to establish, since it has to comprise the cell in all its complexity. Up to now any statement concerning cell viability has to be substantiated by physiological and culturing means. Hence, judging the viability of the isolated cells has to be a prerequisite, but even then some problems are inherent in the use of a "marker," a term which "carries the unfortunate implication of an operational tool rather than an informational feature" (Varon, 1978). Varon and Somjen (1979) have pointed out that the "ideal" marker should fulfill the following criteria

1. All cells of a particular category should have it. 2. It should be displayed in all circumstances, *ex situ* as well as *in situ*. 3. It should be quantitatively absent in all other neural cell types. Obviously no current marker meets such rigorous criteria.

Varon (1977) states that

One needs to distinguish carefully between the intrinsic capabilities of a cell (differentiated programm) and the extent to which they become expressed in a given cellular and humoral environment (extrinsic regulation). Thus a cell-specific marker can be best defined as a property that is expressed uniquely or prevalently in a given cell class when correspondingly optimal conditions prevail.

Recently developed bulk isolation methods allow the preparation of brain cells that can be successfully subjected to culture conditions (Althaus *et al.*, 1978a, 1980; Gebicke-Härter *et al.*, 1981; Poduslo and McKhann, 1976, 1977b; Shahar *et al.*, in preparation; Szuchet *et al.*, 1980a,b) thus providing a better basis for the analysis of neural cell characteristics. However, one has to be aware that under culture conditions some of the cell properties might be lost, for example, CNP activity (Szuchet and Stefansson, 1980) or the antigenic site for antigalactocerebroside (Mirsky *et al.*, 1980). Previous bulk isolation methods are substantially limited in their information regarding cellular markers. Some of the conflicting biochemical results concerning whether a compound was a "marker" or not, appeared because of the rigorous alteration of the environment of the brain cells during the preparation method (e.g., isolated neurons lacked synaptic sites) or because of insufficient documentation of the purity of the individual cell fractions (e.g., see Cremer *et al.*, 1968). Other problems which have inhibited clear-cut decisions (e.g., β-galactosidase, see above) arose because of differences in animal species and age. While neuronal and astroglia-enriched fractions were particularly involved in these controversies, isolated oligodendrocytes—obtained in a purity of 90–95%—seem to be less affected. Studies with bulk-isolated cells have also revealed that previously proposed neuronal specific properties, e.g., neurotransmitter–receptor binding, are not unique features (see Henn *et al.*, 1980; Henn and Henn, 1980). Putative biochemical markers that have been or remain to be examined on bulk-isolated brain cells are compiled in Table II. Immunological markers that have been developed for recognizing brain cells are also listed.

VI. BIOCHEMICAL PROPERTIES OF NEURAL CELLS ISOLATED FROM PATHOLOGICALLY ALTERED BRAIN TISSUE

Brain cells isolated from pathological tissue have been used for investigating biochemical disorders; however, the literature available on this subject is scarce. The inherent problem in this approach is to ensure that the enriched specific cell type prepared from pathologically altered brain tissue is identical in origin with the population isolated from normal brain tissue. The first investigators to apply bulk isolation methods in this respect seems to have been McKhann *et al.* (1969). They examined postmortem brain (stored frozen) from patients with a storage disease. They obtained a neuron-enriched fraction, which was subsequently subjected to tissue culture. Outgrowth of fibers and a continued accumulation of PAS and oil-red O positive material was observed. Iqbal and Tellez Nagel (1972) adapted the method of Norton and Poduslo (1970) without trypsinization for the isolation of neurons and glial cells from normal human brain and from brains affected by Alzheimer's dementia and Huntington's chorea. A biochemical analysis was not included in their paper. This procedure was also used for the isolation of brain cells from human patients suffering with various types of mucopolysaccharidoses (Constantopoulos *et al.*, 1980). The content of glucosaminoglycans in cell fractions from normal brains was 2.2 μg/mg protein in neuronal perikarya, 2.0 μg/mg protein in astroglia, and 2.1 μg/mg protein in oligodendroglia. Chondroitin sulfates constituted more

Table II. A Compilation of Putative Neural Cell Markers Which Have Been or Remain to Be Examined on (Bulk) Isolated Brain Cells

Putative biochemical/ immunological marker	Neurons	Astroglia	Oligodendroglia	Reference
1. Cerebroside sulfotransferase	+	+	+++	Benjamins et al., 1974; Pleasure et al., 1977
2. CNP	+	+	+++	Deshmukh et al., 1974; Poduslo and McKhann, 1975; Poduslo and Norton, 1972a
3. HC-inducible GPDH	+	+	O	McCarthy and DeVellis, 1980; Leveille et al., 1980[b]
4. Carbonic anhydrase (CA)	+	+++	+++ (CAII)	Kimelberg et al., 1978; Nagata et al., 1974; Rose, 1967; Ghandour et al., 1980[b]; Roussel et al., 1979[b]
5. Hydroxymethylglutaryl-CoA-reductase	ni	ni	++	Pleasure et al., 1977
6. Ceramide galactosyl-transferase	+	+	++	Radin et al., 1972; Deshmukh et al., 1974
7. Glutamine synthetase	NL +	NL +	ni	Rose, 1968[b]; Weiler et al., 1979b; Norenberg, 1979
8. Enolase (NSE) (14-3-2) NSE/NNE	ni O +	ni +	ni	Bock and Dissing, 1975[b]; Schmechel et al., 1978[b] Bock and Hamberger, 1976
9. β-galactosidase	NL +++	NL +	NL +	Abe and Norton, 1979; Freysz et al., 1979;

				References
10. Antigalactosylcerebroside	nd	nd	+++	Radin et al., 1972; Raghavan et al., 1972; Sinha and Rose, 1972a
11. Antioligodendrocyte antiserum	nd	nd	++	Gebicke-Härter et al., 1981; Szuchet and Stefansson, 1980; Raff et al., 1978b
12. NS-1 antigen	ni	ni	ni O	Abramsky et al., 1978; Pleasure et al., 1977; Poduslo et al., 1977; Traugott et al., 1978
13. (a) Anti-basic protein	ni	ni	+++	Schachner, 1974
(b) Anti-Wolfgram protein	nd	nd	++	Szuchet and Stefansson, 1980
14. Corpus callosum antiserum	nd	++	+++	Labourdette et al., 1980; Campbell et al., 1977; Schachner et al., 1977
15. Antineuron antiserum	++	nd	nd	Poduslo et al., 1977
16. GFA-P	NL	++	ni	Bock and Hamberger, 1976; Bignami et al., 1972b
17. S-100	+	++	+	Bock and Hamberger, 1976; Haglid et al., 1976
18. GM₄ ganglioside	nd	ni	+	Yu and Iqbal, 1979
19. Tetanus toxin	ni O	ni	ni	Dimpfel et al., 1977b; Raff et al., 1979b
20. Cholera toxin (GM₁)	ni O	ni	ni	Hansson et al., 1977b; Willinger and Schachner, 1978b

[a] +++, Strong or clear expression of activity or staining; ++, moderate expression of activity or staining or present when other cell types are not compared; +, weak expression of activity or staining; ni, not investigated; nd, not detected; NL, conflicting results; O, attributed cell type.
[b] Basic reports, not dealing with (bulk) isolated cells.

than one-half of the total glycosaminoglycan and hyaluronic acid, heparin sulfate, and dermatan sulfate were also present in all the cell types. Cell fractions from the brain of patients with mucopolysaccharidoses exhibited a four- to sixfold increase in the concentration of total glycosaminoglycan; the concentration of the gangliosides GM2, GM3, GD3, and ceramide dihexoside were also increased.

Kohlschütter and Herschkowitz (1973) isolated neuronal perikarya according to Sellinger *et al.* (1971) from the brain of Jimpy mice where the main morphological feature in the brain is a myelin deficiency, (the basic disease is a genetic leukodystrophy). It was thought that the Jimpy mutation affects primarily the metabolism of oligodendrocytes (Neskovic *et al.*, 1970), but the reduced capacity of isolated neurons to synthesize sulfatide indicate that other cell types may also be involved.

Brain cells have also been isolated and analyzed from animals with experimentally induced diseases. The protein pattern in particulate and water-soluble fractions from isolated neuron preparations obtained from α-methylphenylalanine-treated rats (which induces hyperphenylalaninemia) and normal controls revealed a substantial difference, which suggests that the high level of phenylalanine in the blood may affect neuronal protein metabolism (Althaus *et al.*, 1978b; Lane and Neuhoff, 1980). *In vitro* incubation of brain slices from guinea pigs with [^{14}C]leucine, in which an experimental allergic encephalomyelitis (EAE) was induced, resulted in an initial (10 days postinduction) increase of [^{14}C]leucine incorporation into protein of the astroglial-enriched fraction. This was followed (17–18 days postinduction) by an increase in the neuronal fraction. This finding suggests a more generalized alteration of the brain tissue (Babitch *et al.*, 1975).

Purified human and bovine oligodendroglia have been used to induce EAE disease in different animals (Raine *et al.*, 1977). Guinea pigs were the most susceptible, whereas rabbits were less affected. A subsequent study (Raine *et al.*, 1978) revealed that isolated oligodendrocytes which were morphologically less well preserved were the only cells that induced the disease. It was concluded that some myelin basic protein (MBP) is "artificially" adsorbed to these cells during the isolation procedure and, therefore, MBP is responsible for the induction of EAE.

VII. CULTURING OF ISOLATED NEURONS AND GLIA FROM BRAINS OF YOUNG ADULT ANIMALS—A PRELIMINARY SURVEY

Experiments on culturing isolated brain cells, particularly bulk-isolated brain cells, have two objectives: one is simply to look at the viability of isolated brain cells since this still remained to be documented, whereas the other objective is to fill in gaps in our information about the biochemical and physiological potentialities of fully differentiated nontransformed neural cells. The first point demands a further comment. A simple transfer of isolated cells to culture conditions and the observation that they do not disintegrate for several hours reveals nothing more than that the isolated cells have retained some functions. The state of vitality has still to be documented, since cell death may not be immediate even though irreversible damage has occurred in the cell. Signs (positive or negative in the outcome) for viable bulk-isolated cells which survive in culture for 4–5 days may be the following: (1) immediate cell death (cell disintegration) after blocking respiration, (2) functioning of the main cellular metabolic pathways, (3) the presence of a membrane resting or action potential, (4) movement of organelles within the cell or uptake and transport of vesicles by the cell as revealed by microcinematography or observation under the microscope, and (5) regeneration (outgrowth of cell processes).

Culturing of individual nontransformed brain cell classes has been reported to be successful by following the "traditional" procedures for culturing embryonic or newborn brain tissues. Changing culture conditions (e.g., species or percentage of the serum) has resulted in oligodendroglia, astroglia, and neuron-enriched cultures (e.g., Labourdette *et al.,* 1979, 1980; Pettmann *et al.,* 1979, 1980; Sensenbrenner, 1977; Yavin and Yavin, 1979). Addition of specific inhibitors of DNA synthesis (e.g., arabinosylcytosine) have yielded neuron-enriched cultures (e.g., Sotelo *et al.,* 1980). Harvesting of specific glial cell types by physical means after culturing of the brain tissue for a certain time has been reported by, for example, McCarthy and DeVellis (1980), Pfeiffer *et al.* (1980), and Pruss *et al.* (1979). Reports on the culturing of bulk-isolated neurons or glial cells prepared from young–adult animals have appeared recently. Oligodendrocytes appear to be less seriously damaged, or their feeding requirements are less sensitive to changes than astrocytes and neurons. These cells have been shown to regenerate their cell processes, when plated cultures are prepared. An extensive study of culture conditions for isolated oligodendroglia was first carried out by Fewster and Blackstone (1975); however, the regenerating cells were not characterized by morphological or immunological means as being oligodendrocytes. Bulk-isolated oligodendrocytes, prepared from lamb white matter and suitable for maintenance (2–7 days) in suspension tissue culture, were obtained by Poduslo and McKhann (1977b) and Poduslo (1978). These cells could actively synthesize cholesterol, cerebrosides, sulfatides, phospholipids, and some proteins of predominantly higher molecular weight. Specific myelin proteins, for example, proteolipid protein and basic protein, as well as gangliosides were not synthesized. These cells have active machinery for the synthesis of some myelin components (Poduslo, 1978; Poduslo *et al.,* 1978), however, the statement that these oligodendrocytes are producing myelin (Poduslo, 1978) must be rejected. An outgrowth of fibers was not observed, and the glial whorls seen may represent instead focal cell lysis. Studies by Pleasure *et al.* (1977, 1979) confirmed that oligodendroglia from calf white matter in suspension culture retain some enzymatic activity (e.g., CNP, cerebroside sulfotransferase, 3-methylglutaryl-CoA reductase, acetoacetyl-CoA synthetase, and thiolase) and can synthesize sterol and sulfatide.

Plated cultures, either on plastic or glass, ($\pm$) L-polylysine or fetuin, of oligodendrocytes separated either by using magnetic microspheres (Campbell *et al.*, 1978) or by using density gradients (Gebicke-Härter *et al.*, 1981; Szuchet *et al.*, 1978, 1980a,b) have revealed the outgrowth of newly formed fibers (Fig. 1). These cells were characterized by immunological (antigalactocerebroside, anti-MBP, anti-oligodendrocyte serum) and by electron microscope means (TEM, SEM), and could be maintained in culture for several months (Gebicke-Härter *et al.*, 1981; Szuchet *et al.*, 1980a). Szuchet *et al.* (1980a) provided evidence that the cells prepared from ovine white matter remain differentiated as indicated by their ability to synthesize cerebrosides, gangliosides, and basic protein. However, CNP, which was present in the cells after their isolation, was not detected after several days in culture (Szuchet and Stefansson, 1980).

Bulk-isolated astrocytes have only been successfully cultured from the developing cerebellum (Cohen *et al.*, 1979). Six-day-old Porton rats received 2 mg/g of hydroxyurea ip to inhibit cell replication and then, 2 days later, the cerebellar tissue was enzymatically dispersed into suspensions of single cells and then fractionated as described in Section IV,F. The cell fraction enriched in astroglia as revealed by GFA-P antibody binding, was grown in monolayer cultures on polylysine-coated glass coverslips for periods of up to 3 days. The cells formed a network of GFA-P positive fibers. As yet no biochemical data are available.

Reports on neurons isolated by hand or in bulk which have then been subjected to tissue culture are rare. In a paper by Hillman (1966), the growth of processes from single hand-isolated dorsal root ganglion cells of young rats was reported. Rose (1968a) has presented information on bulk-isolated neurons growing in slide culture from 10-day-old rat. No processes were visible in cells from adult neurons. However, dissociated neurons from adult human trigeminal and superior cervical ganglia (plated on polylysine or gelatin-coated coverslips) have been shown to regenerate their processes *in vitro* (Kim *et al.*, 1979). They do this slowly (in contrast to embryonic tissue) not usually extending an axonal process before 7–10 days after dissociation. A similar observation has been made on the regeneration of neurons in cell suspensions obtained after perfusing the brain of 2- to 3-month-old rats. After a "quiescence" phase lasting usually 8 days neurons commenced the outgrowth of cellular processes (Althaus *et al.*, 1978c). When the cell suspension was further processed for the separation of a neuronal fraction and the purified neurons were subjected to culture conditions, the result differed somewhat. Even after 6–8 weeks in culture, only veil-like

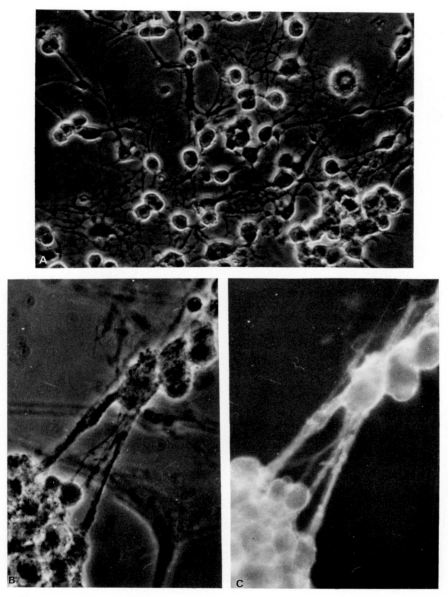

Fig. 1.   Oligodendrocytes, isolated from cat brain (12-week-old), maintained *in vitro* for 2 weeks; phase contrast pictures, (A) × 600; (B) and (C) × 960. C shows GC⁺ cells, which are brightly fluorescent, labeled with an anti-GC-antiserum produced in rabbits.

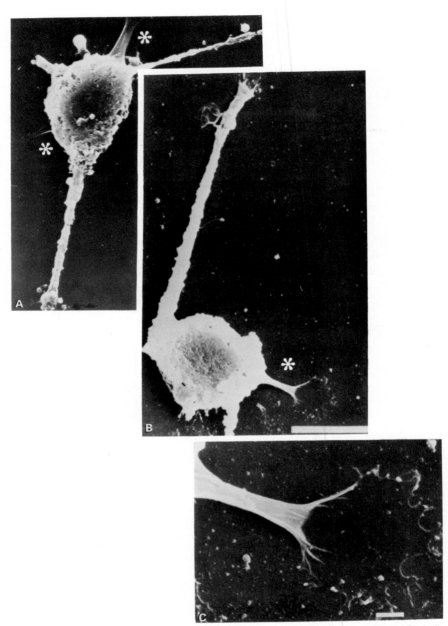

Fig. 2.   Veil-like structures (asterisk) are protruded from cortical neurons (A and B), bulk isolated from rat brain (ca. 2-month-old), maintained *in vitro* for 12 days. C shows the newly formed fiber of B, in higher magnification; bar represents 10 μm (B) or 1 μm (C); SEM pictures from Shahar *et al.* (in preparation).

structures protruded from the neurons (Fig. 2), which were obviously in need of some essential factors (Althaus *et al.*, 1980, 1981; Shahar *et al.*, 1981; in preparation). Further support that this type of regeneration is not due to the "slow death" of the neuron may be the observation of an increasing number of such structures over a culture of up to 3 weeks; furthermore, at different time points in the culture period histochemical staining of several enzymes was positive, uptake of [^{14}C]leucine was noted and a relatively low membrane resting potential (30–40 mV) was recorded (Shahar *et al.*, in preparation).

Granular neuron-enriched fractions, prepared from mouse cerebellum (10-day-old) have been plated on polylysine (succinylated) coated Falcon Petri dishes or on ACLAR coverslips (Allied Chemicals) (Campbell and Williams, 1978). After seeding (within 1 h) the cells quickly started to extend processes. Suspension cultures of bulk-isolated neurons from rat brain (5- to 15-day-old) have been reported by Poduslo and McKhann (1976, 1977a). The neurons were maintained for 18–24 h, during which the cells remained morphologically intact, but showed naturally no sign of regenerating processes. The neurons were capable of incorporating different labeled substrates, [^{3}H]leucine, [^{14}C]uridine, during this period. Maintained neurons prepared from 5-day-old animals incorporated [^{3}H]leucine to a higher degree compared to freshly isolated cells. Cells prepared from 11- to 12-day-old animals showed a decreasing rate of incorporation after maintenance. The incorporation of labeled leucine into proteins was inhibited by puromycin and cyclohexamide, but not by chloramphenicol indicating that the incorporation measured was not due to isolated mitochondria or bacterial contamination.

VIII. SUMMARY AND OUTLOOK

Isolation of specific brain cells has been used as a tool to gain more information on the biochemical and functional properties of neural cells. The field of bulk isolation of brain cells has expanded greatly since its beginning. Neurons and glial cells have been isolated from different animals (e.g., mouse, rat, cat, ox, horse, primate) and their biochemical properties have been investigated. Much of the information in this field has been comprehensively reviewed in this chapter. Certain areas such as studies on carbohydrate metabolism, ion flux, or respiration have been omitted.

Given the large volume of literature available, the proportion of data providing conclusive information on bulk-isolated neural cells leaves a great deal to be desired. This can be attributed to variations in the methods employed and also to the attitude of the investigators themselves. All workers have agreed that many of the objections (Section IV) against most of the bulk separation methods are valid, but regardless of the inherent problems which might inhibit meaningful biochemical results, much data have been published leading to a rather confusing situation. A careful examination of the individual cell preparations (e.g., contaminants, cell damage) has often been neglected. This has been noted by Rose (1976), who comments "it is regrettable how often one still reads papers reporting results for which these precautions have been omitted." Furthermore, in reviewing the present literature on isolated brain cells it is remarkable to notice how often conflicting results are presented by different authors. These originate in part from simple facts such as the use of animals of different ages and species and modification of methods without explanations. Nevertheless certain advances must be credited to the approach of bulk isolation of brain cells. Data obtained by other techniques have been confirmed and new aspects have been revealed. The following "points" underline this: the chemical make up of neurons and glial cells has been characterized in great detail, the presence of major metabolic pathways has been demonstrated, the developmental profile of protein synthesis has been studied, interactions of neurons and glia with respect to their amino acid pool sizes and effect of transmitters (e.g., GABA) have been established, the capability of isolated oligodendrocytes to synthesize myelin components has been examined, some specificities of lipid metabolism in neuronal and glial cells have been elucidated, and the major site of some enzymatic activities (e.g., carboanhydrase, CNP) in neural cells has been determined. Some progress has undoubtedly been made recently in achieving viable and pure isolated cell fractions from nervous tissue. Neurons and oligodendrocytes can be subjected to tissue culture conditions and can be maintained for several weeks. Signs of regeneration (neurons) and newly formed network of fibers (oligodendrocytes) have been observed. Astrocytes have been obtained in improved morphological shape and purity.

What are the tasks which remain to be tackled? First, all biochemical data which still lack validity or which are contradictory should be replaced by conclusive results. This concerns uptake and content of amino acids in neural cells, the synthesis of nucleic acids, and the distribution of particular enzymes. The isolation of astrocytes has to be further improved with respect to their viability, since reliable data are not available.

What are the goals for future work? Isolated neurons that retain a considerable percentage of synaptic boutons (Huttner *et al.*, 1979) may be useful in the separation of synaptosomes, which still are contaminated by glial material to a certain extent (Delaunoy *et al.*, 1979). Preliminary experiments have shown that a synaptosomal fraction is obtained in a reasonable amount from one nerve cell fraction (Althaus and Campbell, unpublished). A very promising approach is opened up by affinity and cell sorting methods for separating neural cells. Subpopulations of neurons and oligodendrocytes may be achieved using this method. Already different neurons of the cerebellum have been isolated. A further evaluation of plasma membrane characteristics, including binding sites for certain chemical compounds using differential centrifugation technique is needed to furnish information on cell-specific ligands. Further knowledge of the biochemical and functional behavior of bulk-isolated brain cells grown under culture conditions may permit a better understanding of the cellular potentialities and reactions of neural cells under various conditions. Major progress can be expected in the next years on how myelination proceeds and how certain viruses interact with their target cells; perhaps this latter aspect will cast some light on the pathogenesis of multiple sclerosis.

REFERENCES

Abdel-Latif, A. A., Yan, S.-J., and Smith, J. P. (1974). Effect of neurotransmitters on phospholipid metabolism in rat cerebral—cortex slices—cellular and subcellular distribution. *J. Neurochem.* **22**, 383–393.

Abe, T., and Norton, W. T. (1974). The characterization of sphingolipids from neurons and astroglia of immature rat brain. *J. Neurochem.* **23**, 1025–1036.

Abe, T., and Norton, W. T. (1979). The characterization of sphingolipids of oligodendroglia from calf brain. *J. Neurochem.* **32**, 828–832.

Abe, T., Miyarake, T., Norton, W. T., and Suzuki, K. (1979). Activities of glycolipid hydrolases in neurons and astroglia from rat and calf brains and in oligodendroglia from calf brain. *Brain Res.* **161**, 179–182.

Abramsky, O., Lisak, R. P., Pleasure, D., Gilden, D. H., and Silberberg, D. H. (1978). Immunologic characterization of oligodendroglia. *Neurosci. Lett.* **8**, 311–316.

Althaus, H.-H., Huttner, W. B., and Neuhoff, V. (1977). Neurochemical and morphological studies of bulk-isolated rat brain cells. I. A new approach to the separation of cerebral neurones. *Hoppe-Seyler's Z. Physiol. Chem.* **358**, 1155–1159.

Althaus, H.-H., Gebicke, P. J., Meyermann, R., Huttner, W. B., and Neuhoff, V. (1978a). Some biochemical and morphological characteristics of bulk isolated neurones from adult rat brain. (V. Neuhoff, ed.), Vol. 1, p. 557. Verlag Chemie, Weinheim.

Althaus, H.-H., Huttner, W. B., Gebicke, P. J., Neuhoff, V., and Lane, J. D. (1978b). Biochemical characterization and in vitro maintenance of neurons from adult rat brain isolated by the perfusion method. *Abstr. Soc. Neurosci.* **4**, 311.

Althaus, H.-H., Neuhoff, V., Huttner, W. B., Monzain, R., and Shahar, A. (1978c). Maintenance and regeneration *in vitro* of bulk-isolated cortical neurones from adult rats. *Hoppe–Seyler's Z. Physiol. Chem.* **359**, 773–775.

Althaus, H.-H. Gebicke, P. J., and Neuhoff, V. (1979). Primate brain nerve cells bulk isolated by the perfusion method. *Naturwissenschaften* **66**, 117.

Althaus, H.-H., Montz, H., Schwartz, P., Rittner, I., and Neuhoff, V. (1980). Kultivierung präparativ gewonnener Hirn-Neurone der Ratte. *Hoppe-Seyler's Z. Physiol. Chem.* **361**, 213.

Althaus, H.-H., Stoykova, A., Schwartz, P., and Neuhoff, V. (1981). Culturing of bulk isolated neurons and oligodendrocytes from rat brain. *Int. Symp. Nerv. Syst. Regeneration, Catania,* p. 35 (Abstr.).

Althaus, H.-H., Montz, H., and Neuhoff, V. Improved state of bulk isolated brain neurones by perfusion with a fluorocarbon containing medium, in preparation.

Althaus, H.-H., Gebicke-Härter, P. J., Benatar, N., and Neuhoff, V. Some biochemical characteristics of oligodendrocytes isolated from pigs during postnatal development. In preparation.

Anfinsen, C. B., Lowry, O. H., and Hastings, A. B. (1942). The application of the freezing-drying technique to retinal histochemistry. *J. Cell Comp. Physiol.* **20**, 231–237.

Arbogast, B. W., and Arsenis, C. (1974). The enzymatic ontogeny of neurons and glial cells isolated from postnatal rat cerebral gray matter. *Neurobiology* **4**, 21–37.

Au, A. M-J., and Varon, S. (1979). Neural cell sequestration on immunoaffinity columns. *Exp. Cell Res.* **120**, 269–276.

Awasthi, Y. C., and Srivastava, S. K. (1980). Structure, function and metabolism of glycosphingolipids. *In* "Biochemistry of Brain" (S. Kumar, ed.), pp. 1–20. Pergamon, Oxford.

Azcurra, J. M., and Carrasco, A. E. (1978). Identification of polydisperse RNA in the cytoplasm of neurons isolated from the immature brain cortex. *Neurochem. Res.* **3**, 521–534.

Babitch, J. A., Blomstrand, Chr., and Hamberger, A. (1975). Amino acid incorporation into neurons and glia of guinea pigs with experimental allergic encephalomyelitis. *Brain Res.* **86**, 459–467.

Balazs, R., Patel, A. J., and Lewis, P. D. (1977). Metabolic influences on cell proliferation in the brain. *In* "Biochemical Correlates of Brain Structure and Function" (A. N. Davison, ed.), pp. 43–83. Academic Press, New York.

Balazs, R., Cohen, J., Woodhams, P. L., Patel, N. J., and Garthwaite, J. (1979). Isolation and characterisation of cell types from the developing cerebellum. *In* "Neural Growth and Differentiation" (E. Meisami and M. A. B. Brazier, eds.), pp. 113–132. Raven, New York.

Banik, N. L., and Smith, M. E. (1976). *In vitro* protein synthesis by oligodendroglial cells. *Neurosci. Lett.* **2**, 235–238.

Barkley, D. S., Rakic, L. L., Chaffee, J. K., and Wong, D. L. (1973). Cell separation by velocity sedimentation of postnatal mouse cerebellum. *J. Cell Physiol.* **81**, 271–280.

Barnard, P. J., Weiss, L., and Ratcliffe, T. (1969). Changes in the surface properties of embryonic chick neural retina cells after dissociation. *Exp. Cell Res.* **54**, 293–301.

Baumann, H., and Doyle, D. (1979). Effect of trypsin on the cell surface proteins of hepatoma tissue culture cells. *J. Biol. Chem.* **254**, 3935–3946.

Benjamins, J. A., Guarnieri, M., Miller, K., Sonneborn, M., and McKhann, G. M. (1974). Sulphatide synthesis in isolated oligodendroglial and neuronal cells. *J. Neurochem.* **23**, 751–757.

Betz, A. L., and Goldstein, G. W. (1978). Polarity of the blood–brain barrier: Neutral amino acid transport into isolated brain capillaries. *Science (Washington, D.C.)* **202**, 225–226.

Bignami, A., Eng, L. F., Dahl, D., and Uyeda, C. T. (1972). Localization of the glial fibrillary acidic protein in astrocytes by immunofluorescence. *Brain Res.* **43**, 429–435.

Binaglia, L., Goracci, G., Porcellati, G., Roberti, R., and Woelk, H. (1973). The synthesis of choline and ethanolamine phosphoglycerides in neuronal and glial cells of rabbit *in vitro*. *J. Neurochem.* **21**, 1067–1082.

Blomstrand, C., and Hamberger, A. (1969). Protein turnover in cell-enriched fractions from rabbit brain. *J. Neurochem.* **16**, 1401–1407.

Blomstrand, C., and Hamberger, A. (1970). Amino acid incorporation *in vitro* into protein of neuronal and glial cell-enriched fractions. *J. Neurochem.* **17**, 1187–1195.

Blomstrand, C., Hamberger, A., Sellström, Å., and Steinwall, O. (1975). Protein bound radioactivity in neuronal and glial fractions following intra-carotid ^{3}H-leucine perfusion. *Neurobiology (Copenhagen)* **5**, 178–187.

Bock, E., and Dissing, J. (1975). Demonstration of enolase activity connected to the brain-specific protein 14.3.2. *Scand. J. Immunol. Suppl.* **4**, 2, 31–36.

Bock, E., and Hamberger, A. (1976). Immunoelectrophoretic determination of brain specific antigens in bulk-prepared neuronal and glial cells. *Brain Res.* **112**, 329–335.

Booher, J., and Sensenbrenner, M. (1972). Growth and cultivation of dissociated neurons and glial cells from embryonic chick, rat and human brain in flask cultures. *Neurobiology (Copenhagen)* **2**, 97–105.

Brumberg, V. A., and Pevzner, L. Z. (1975). Cytospectrophotometric studies on the lactate dehydrogenase isoenzymes in functionally different neuron-neuroglia units. *Acta Histochem.* **54**, 218–229.

Burdman, J. A., Haglid, K., and Dravid, A. R. (1970). Protein synthesis in fractions from isolated brain cell nuclei. *J. Neurochem.* **17**, 669–676.

Burdman, J. A., and Journey, L. J. (1969). Protein synthesis in isolated nuclei from adult rat brain. *J. Neurochem.* **16**, 493–500.

Campbell, G.-L. (1963). Some characteristics of rat cerebral cortical cells prepared and separated by a new technique. Ph.D. thesis, Rutgers Univ., New Brunswick, New Jersey (Univ. Microfilms Inc., Michigan).

Campbell, G.-L. (1979). An immunological approach to cell separation of the central nervous system. *Fed. Proc., Fed. Am. Soc. Exp. Biol.* **38**, 2386–2390.

Campbell, G.-LeM., and Williams, M. P. (1978). *In vitro* growth of glial cell-enriched and depleted populations from mouse cerebellum. *Brain Res.* **156**, 227–239.

Campbell, G.-LeM., Schachner, M., and Sharrow, S. O. (1977). Isolation of glial cell-enriched and -depleted populations from mouse cerebellum by density gradient centrifugation and electronic cell sorting. *Brain Res.* **127**, 69–86.

Campbell, G.-LeM., Abramsky, O., and Silberberg, D. H. (1978). Isolation of oligodendrocytes from mouse cerebellum using magnetic microspheres. *Abstr. Soc. Neurosci.* **4**, 64.

Chao, S.-W., and Rumsby, M. G. (1977). Preparation of astrocytes, neurones and oligodendrocytes from the same rat brain. *Brain Res.* **124**, 347–351.

Cimino, M., Hartman, B. K., and Moore, B. W. (1977). Immunohistochemical localization of 14-3-2 protein during development in chick and rat. *Proc. Int. Soc. Neurochem.* **6**, 304.

Cohen, J., Balazs, R., Hajos, F., Currie, D. N., and Dutton, G. R. (1978). Separation of cell types from the developing cerebellum. *Brain Res.* **148**, 313–331.

Cohen, J., Woodhams, P. L., and Balazs, R. (1979). Preparation of viable astrocytes from the developing cerebellum. *Brain Res.* **161**, 503–514.

Cohen, S. R., and Bernsohn, J. (1973). Incorporation of 1-^{14}C labelled fatty acids into isolated neuronal soma, astroglia and oligodendroglia from calf brain. *Brain Res.* **60**, 521–525.

Cohen, S. R., and Bernsohn, J. (1974). Differential trypsin effect upon (1-^{14}C) acetate incorporation into choline and ethanolamine glycerophosphatides of rat brain and liver. *Lipids* **9**, 72–76.

Cohen, S. R., and Bernsohn, J. (1978). The *in vivo* incorporation of linolenic acid into neuronal and glial cells and myelin. *J. Neurochem.* **30**, 661–669.

Constantopoulos, G., Iqbal, K., and Dekaban, A. S. (1980). Mucopolysaccharidosis Types 14, 35, II, and III A: Glycosaminoglycans and lipids of isolated brain cells and other fractions from autopsied tissues. *J. Neurochem.* **34**, 1399–1411.

Cremer, J. E., Johnston, P. V., Roots, B. J., and Trevor, A. J. (1968). Heterogeneity of brain fractions containing neuronal and glial cells. *J. Neurochem.* **15**, 1361–1370.

Davison, A. N., Cuzner, M. L., Banik, N. L., and Oxberry, J. (1966). Myelinogenesis in the rat brain. *Nature (London)* **212**, 1373–1374.

Delaunoy, J.-P. Hog, F., DeFeudis, F. V., and Mandel, P. (1979). Estimation of glial contamination of synaptosomal-mitochondrial fractions of rat brain by radioimmunassay of carbonic anhydrase (CA II). *J. Neurochem.* **33**, 611–612.

Deppert, W., and Walter, G. (1978). Cell surface glycosyltransferases—Do they exist? *J. Supramol. Struct.* **8**, 19–37.

Derry, D. M., and Wolfe, L. S. (1967). Gangliosides in isolated neurons and glial cells. *Science (Washington, D.C.)* **158**, 1450–1452.

Deshmukh, D. S., Flynn, T. J., and Pieringer R. A. (1974). The biosynthesis and concentration of galactosyl diglyceride in glial and neuronal enriched fractions of actively myelinating rat brain. *J. Neurochem.* **22**, 479–485.

DeVellis, J., and Inglish, D. (1973). Age-dependent changes in the regulation of glycerolphosphate dehydrogenase in the rat brain and in a glial cell line. In "Neurobiological Aspects of Maturation and Aging" (D. H. Ford, ed.). *Prog. Brain Res.* **40**, 321–330.

DeVellis, J., McGinnis, J. F., Breen, G. A. M., Leveille, P., Bennett, K., and McCarthy, K. (1977). Hormonal effects on differentiation in neural cultures. *In* "Cell, Tissues, and Neurobiology" (S. Fedoroff and L. Hertz, eds.), pp. 485–511, Academic Press, New York.

Dimpfel, W., Huang, R. T. C., and Habermann, E. (1977). Ganglioside in nervous tissue cultures and binding of ^{125}J-labelled tetanus toxin, a neuronal marker. *J. Neurochem.* **29**, 329–334.

Djuricic, B. M., and Mrsulja, B. B. (1977). Enzymic activity of the brain: Microvessels vs. total forebrain homogenate. *Brain Res.* **138**, 561–564.

Dvorak, D. J., Gipps, E., and Kidson, C. (1978). Isolation of specific neurones by affinity methods. *Nature (London)* **271**, 564–566.

Edström, J. E. (1953). Ribonucleic acid mass and concentration in individual nerve cells. A new method for quantitative determinations. *Biochim. Biophys. Acta* **12**, 361–386.

Edström, J. E., and Neuhoff, V. (1973). Micro-Electrophoresis for RNA and DNA base analysis in V. Neuhoff "Micromethods in Molecular Biology." *In* "Molecular Biology, Biochemistry and Biophysics" (A. Kleinzeller, G. F. Springer, and H. G. Wittmann, eds.), pp. 215–238. Springer-Verlag, Berlin and New York.

Epstein, M. H., and O'Connor, J. S. (1965). Respiration of single cortical neurones and of surrounding neuropile. *J. Neurochem.* **12**, 389–395.

Farooq, M. and Norton, W. T. (1978). A modified procedure for isolation of astrocyte- and neuron-enriched fractions from rat brain. *J. Neurochem.* **31**, 887–894.

Farooq, M., Ferszt, R., Moore, C. L., and Norton, W. T. (1977). The isolation of cerebral neurons with partial retention of processes. *Brain Res.* **124**, 69–81.

Fewster, M. E., and Blackstone, S. (1975). *In vitro* study of bovine oligodendroglia. *Neurobiology (Copenhagen)* **5**, 316–328.

Fewster, M. E., and Mead, J. F. (1968a). Lipid composition of glial cells isolated from bovine white matter. *J. Neurochem.* **15**, 1041–1052.

Fewster, M. E., and Mead, J. F. (1968b). Fatty acid and fatty aldehyde composition of glial cell lipids isolated from bovine white matter. *J. Neurochem.* **15**, 1303–1312.

Fewster, M. E., Scheibel, A. B., and Mead, J. F. (1967). The preparation of isolated glial cells from rat and bovine white matter. *Brain Res.* **6**, 401–408.

Fewster, M. E., Blackstone, S. C., and Ihrig, T. J. (1973). The preparation and characterization of isolated oligodendroglia from bovine white matter. *Brain Res.* **63**, 263–271.

Fewster, M. E., Einstein, E. R., Csejtey, J., and Blackstone, S. C. (1974). Proteins in the bovine oligodendroglia cells at various stages of brain development. *Neurobiology (Copenhagen)* **4**, 388–401.

Fewster, M. E., Ihrig, T. J., and Mead, J. F. (1975). Biosynthesis of long chain fatty acids by oligodendroglia isolated from bovine white matter. *J. Neurochem.* **25**, 207–213.

Fillion, G., Beaudoin, D., Rousselle, J. C., and Jacob, J. (1980). [^{3}H]5-HT binding sites and 5-HT-sensitive adenylate cyclase in glial cell membrane fraction. *Brain Res.* **198**, 361–374.

Flangas, A. L., and Bowman, R. E. (1970). Differential metabolism of RNA in neuronal-enriched and glial-enriched fractions of rat cerebellum. *J. Neurochem.* **17**, 1237–1245.

Francescangeli, E., Goracci, G., Piccinin, L., Mozzi, R., Woelk, H., and Porcellati, G. (1977). The metabolism of labelled choline in neuronal and glial cells of the rabbit *in vivo*. *J. Neurochem.* **28**, 171–176.

Freysz, L., Bieth, R., and Mandel, P. (1968a). Einbau von ^{32}P in die verschiedenen Phosphatide der Neuronen und der Gliazellen des Cortex der Ratte. *Hoppe-Seyler's Z. Physiol. Chem.* **349**, 6.

Freysz, L., Bieth, R., Judas, C., Sensenbrenner, M., Jacob, M., and Mandel, P. (1968b). Distribution quantitative des divers phospholipides dans les neurones et les cellules gliales isoles du cortex cerebral de rat adulte. *J. Neurochem.* **15**, 307–313.

Freysz, L., Bieth, R., and Mandel, P. (1969). Kinetics of the biosynthesis of phospholipids in neurons and glial cells isolated from rat brain cortex. *J. Neurochem.* **16**, 1417–1424.

Freysz, L., Farooqui, A. A., Adamezewsky-Goncerzewicz, Z., and Mandel, P. (1979). Lysosomal hydrolases in neuronal, astroglial, and oligodendroglial enriched fractions of rabbit and beef brain. *J. Lipid Res.* **20,** 503–508.

Gebicke-Härter, P.-J., Althaus, H.-H., and Neuhoff, V. (1981). Oligodendrocytes from postnatal cat brain in cell culture. I. Regeneration and maintenance. *Develop. Brain Res.* **1,** 497–518.

Ghandour, M. S., Vincendon, G., Gombos, G., Limozin, N., Filippi, D., Dalmasso, C., and Laurent, G. (1980). Carbonic anhydrase and oligodendroglia in developing rat cerebellum: A biochemical and immunohistological study. *Dev. Biol.* **77,** 73–83.

Giacobini, E. (1956). Histochemical demonstration of AChE activity in isolated nerve cells. *Acta Physiol. Scand.* **36,** 276–290.

Giacobini, E. (1964). Metabolic relations between glia and neurons studied in single cells. *In* "Morphological and Biochemical Correlates of Neural Activity" (M. M. Cohen and R. S. Snider, eds.), pp 15–38. Harper, New York.

Giacobini, E. (1968). Chemical studies on individual neurons. Part I. Vertebrate nerves. *In* "Neuroscience Research" (S. Ehrenpreis and O. C. Solnitzky, eds.), Vol. I, pp. 1–71. Academic Press, New York.

Giacobini, E. (1969a). Chemistry of isolated invertebrate neurons. *In* "Handbook of Neurochemistry," Vol. II, Structural Neurochemistry (A. Lajtha, ed.), pp. 195–239. Plenum, New York.

Giacobini, E. (1969b). Chemical studies on individual neurons. Part II. Invertebrate nerve cells. *In* "Neuroscience Research" (S. Ehrenpreis and O. C. Solnitzky, eds.), Vol. II, pp. 111–202. Academic Press, New York.

Giacobini, E. (1978). Validity of single neuron chemical analysis. *In* "Biochemistry of Characterised Neurons" (N. N. Osborne, ed.), pp. 3–17. Pergamon, Oxford.

Gielen, W., and Hinzen, D. H. (1974). Acetylneuraminat-cytidylyltransferase und Sialyltransferase in isolierten Neuronal und Gliazellen des Rattengehirns. *Hoppe-Seyler's Z. Physiol. Chem.* **355,** 895–901.

Giorgi, P. P. (1971). Preparation of neurons and glial cells from rat brain by zonal centrifugation. *Exp. Cell Res.* **68,** 273–282.

Giuffrida, A. M., Hamberger, A., and Serra, I. (1976). Biosynthesis of DNA and RNA in neuronal and glial cells from various regions of developing rat brain. *J. Neurosci. Res.* **2,** 203–215.

Goldstein, G. W., Csejtey, J., and Diamond, I. (1977). Carrier mediated glucose transport in capillaries isolated from rat brain. *J. Neurochem.* **28,** 725–728.

Goracci, G., Francescangeli, E., Piccinin, G. L., Binaglia, L., Woelk, H., and Porcellati, G. (1975). The metabolism of labelled ethanolamine in neuronal and glial cells of the rabbit. *J. Neurochem.* **24,** 1181–1184.

Grisar, T., Frere, J.-M., and Franck, G. (1979). Effect of K$^+$ ions on kinetic properties of the (Na$^+$,K$^+$)-ATPase (E.C. 3.6.1.3) of bulk isolated glial cells, perikarya and synaptosomes from rabbit brain cortex. *Brain Res.* **165,** 87–103.

Guarnieri, M., Krell, L. S., McKhann, G. M., Pasternak, G. W., and Yamamura, H. I. (1975). The effects of cell isolation techniques on neuronal membrane receptors. *Brain Res.* **93,** 337–342.

Guarnieri, M., Cohen, S. R., and Ginns, E. (1976). Cells isolated from trypsin-treated brain contain trypsin. *J. Neurochem.* **26,** 41–44.

Guilbault, G. G., and Kramer, D. N. (1964). Fluorometric determination of lipase, acylase, alpha- and gamma-chymotrypsin and inhibitors of these enzymes. *Anal. Chem.* **36,** 409–412.

Haglid, K. G., and Hamberger, A. (1973). Cellular and subcellular distribution of protein-bound radioactivity in the rat brain during maturation after incorporation of ^{3}H leucine *in vitro*. *Brain Res.* **52,** 277–287.

Haglid, K. G., Hamberger, A., Hansson, H.-A., Hydén, H., Persson, L., and Rönnbäck, L. (1976). Cellular and subcellular distribution of the S-100 protein in rabbit and rat central nervous system. *J. Neurosci. Res.* **2,** 175–191.

Hamberger, A. (1971). Amino acid uptake in neuronal and glial cell fractions from rabbit cerebral cortex. *Brain Res.* **31**, 169–178.

Hamberger, A., and Henn, F. A. (1973). Some aspects of the differential biochemistry and functional relationships between neurons and glia. *In* "Metabolic Compartmentation in the Brain" (R. Balazs and J. E. Cramer, eds.), pp. 305–318. Macmillan, New York.

Hamberger, A., and Sellström, Å. (1975). Techniques for separation of neurons and glia and their application to metabolic studies. *In* "Metabolic Compartmentation and Neurotransmission: Relation to Brain Structure and Function" (S. Berl, D. D. Clarke, and D. Schneider, eds.), pp. 145–166. Plenum, New York.

Hamberger, A., and Svennerholm, L. (1971). Composition of gangliosides and phospholipids of neuronal and glial cell enriched fractions. *J. Neurochem.* **18**, 1821–1829.

Hamberger, A., Blomstrand, C., and Lehninger, A. L. (1970). Comparative studies on mitochondria isolated from neuron-enriched and glia-enriched fractions of rabbit and beef brain. *J. Cell Biol.* **45**, 221–234.

Hamberger, A., Eriksson, O., and Norrby, K. (1971a). Cell size distribution in brain suspensions and in fractions enriched with neuronal and glial cells. *Exp. Cell Res.* **67**, 380–388.

Hamberger, A., Blomstrand, C., and Yanagihara, T. (1971b). Subcellular distribution of radioactivity in neuronal and glial-enriched fractions after incorporation of [^{3}H]leucine *in vivo* and in vitro. *J. Neurochem.* **18**, 1469–1478.

Hamberger, A., Hansson, H.-A., and Sellström, Å. (1975). Scanning and transmission electron microscopy on bulk prepared neuronal and glial cells. *Exp. Cell Res.* **92**, 1–10.

Hansson, H.-A., Holmgren, J, and Svennerholm, L. (1977). Ultrastructural localization of cell membrane GM_1 ganglioside by cholera toxin. *Proc. Natl. Acad. Sci. U.S.A.* **74**, 3782–3786.

Hansson, E., Sellström Å, Persson, L. I., and Rönnbäck, L. (1980). Brain primary culture—A characterization. *Brain Res.* **188**, 233–246.

Hazama, H., and Uchimura, H. (1974). Bulk separation of Purkinje cells and granular cells from small amounts of cerebellar tissue. *Exp. Cell Res.* **87**, 412–415.

Hazama, H., Ito, M., Hirano, M., and Uchimura, H. (1976). Monoamine oxidase activities in neuronal and glial fractions from regional areas of rat brain. *J. Neurochem.* **26**, 417–419.

Hemminki, K. (1970). An improved method for preparing brain cell suspensions. *FEBS Lett.* **9**, 290–292.

Hemminki, K., Huttunen, M. O., and Järnefelt, J. (1970). Some properties of brain cell suspensions prepared by a mechanical enzymic method. *Brain Res.* **23**, 23–34.

Hemminki, K., Hemminki, E., and Giacobini, E. (1973). Activity of enzymes related to neurotransmission in neuronal and glial fractions. *Int. J. Neurosci.* **5**, 87–90.

Henn, F. A. (1980). Separation of neuronal and glial cells and subcellular constituents. *In* "Advances in Cellular Neurobiology" (S. Fedoroff and L. Hertz, eds.), Vol. 1, pp. 373–403. Academic Press, New York.

Henn, F. A., and Hamberger, A. (1971). Glial cell function: Uptake of transmitter substances. *Proc. Natl. Acad. Sci. U.S.A.* **68**, 2686–2690.

Henn, F. A., and Hamberger, A. (1976). Preparation of glial plasma membrane from a cell fraction enriched in astrocytes. *Neurochem. Res.* **1**, 261–273.

Henn, F. A., and Henn, S. W. (1980). The psychopharmacology of astroglial cells. *Prog. Neurobiol. (Oxford)* **15**, 1–17.

Henn, F. A., Deering, J., and Anderson, D. (1980). Receptor studies on isolated astroglial cell fractions prepared with and without trypsin. *Neurochem. Res.* **5**, 459–464.

Hillman, H. (1966). Growth of processes from single isolated dorsal root ganglion cells of young rats. *Nature (London)* **209**, 102–103.

Hillman, H., and Hydén, H. (1965). Membrane potentials in isolated neurones *in vitro* from Deiter's nucleus of rabbit. *J. Physiol.* **177**, 398–410.

Hinzen, D. H., and Gielen, W. (1973). Distribution and composition of gangliosides and galac-tocerebrosides in neuronal and glial cells isolated from rabbit cerebral cortex. *ISN Tokyo Meeting*, p. 200.

Hirsch, H. E., Blanco, C. E., and Parks, M. E. (1980). Glycerol phosphate dehydrogenase: Reduced activity in multiple sclerosis plaques confirms localization in oligodendrocytes. *J. Neurochem.* **34**, 760–762.

Hjelle, J. T., Baird-Lambert, J., Cardinale, G., Spector, S., and Udenfriend, S. (1978). Isolated microvessels: The blood–brain barrier *in vitro*. *Proc. Natl. Acad. Sci. U.S.A.* **75**, 4544–4548.

Hudson, D. B., Valcana, T., Bean, G., and Timiras, P. S. (1976). Glutamic acid: A strong candidate as the neurotransmitter of the cerebellar granule cell. *Neurochem. Res.* **1**, 73–81.

Huttner, W. B., Meyermann, R., Neuhoff, V., and Althaus, H.-H. (1979). Neurochemical and morphological studies of bulk isolated rat brain cells. II. Preparation of viable cerebral neurons which retain synaptic complexes. *Brain Res.* **171**, 225–237.

Hydén, H. (1959). Quantitative assay of compounds in isolated fresh nerve cells and glial cells from control and stimulated animals. *Nature (London)* **184**, 433–435.

Hydén, H. (1960). The neuron. *In* "The Cell" Biochemistry, Physiology, Morphology" (J. Brachet and A. E. Mirsky, eds.), Vol. IV, Part 1, pp. 215–323. Academic Press, New York.

Hydén H. (1967). RNA in brain cells. *In* "The Neurosciences" (G. C. Quarton, Th. Melnechuk, and F. O. Schmitt, eds.), pp. 248–266. Rockefeller Univ. Press, New York.

Hydén, H., and Pigon, A. (1960). A cytophysiological study of the functional relationship between oligodendroglial cells and nerve cells of Deiters' nucleus. *J. Neurochem.* **6**, 57–72.

Inoue, T., Deshmukh, D. S., and Pieringer, R. A. (1971). The association of the galactosyl dig-lycerides of brain with myelination. *J. Biol. Chem.* **246**, 5688–5694.

Inoue, N., Suzuki, O., and Kato, T. (1976). DNA synthesis in neuronal, Glial and liver nuclei isolated from the adult guinea pig. *J. Neurochem.* **27**, 113–119.

Iqbal, K., and Tellez-Nagel. J. (1972). Isolation of neurons and glial cells from normal and patholog-ical human brains. *Brain Res.* **45**, 296–301.

Iqbal, K., Grundke-Iqbal, J., and Wisniewski, H. M. (1977). Oligodendroglia from human autop-sied brain: Bulk isolation and some chemical properties. *J. Neurochem.* **28**, 707–716.

Jacobson, M. (1978). "Developmental Neurobiology." 2nd. ed., pp. 76–92. Plenum, New York.

Jarlstedt, J., and Hamberger, A. (1971). Patterns and labelling characteristics in neuronal and glial RNA. *J. Neurochem.* **18**, 921–930.

Jarlstedt, J., and Hamberger, A. (1972). Experimental alcoholism in rats. Effects of acute ethanol intoxication on the *in vitro* incorporation of ^{3}H-leucine into neuronal and glial proteins. *J. Neurochem.* **19**, 2299–2306.

Johnson, D. E., and Sellinger O. Z. (1971). Protein synthesis in neurons and glial cells of the developing rat brain: An in vivo study. *J. Neurochem.* **18**, 1445–1460.

Johnston, P. V., and Roots, B. I. (1965). The neurone surface. *Nature (London)* **205**, 778–780.

Johnston, P. V., and Roots, B. I. (1970). Neuronal and glial perikarya preparations: An appraisal of present methods. *Int. Rev. Cytol.* **29**, 265–280.

Jones, J. P., Ramsey, R. B., Aexel, R. T., and Nicholas, H. J. (1972). Lipid biosynthesis in neuron-enriched and glial-enriched fractions of rat brain: Ganglioside biosynthesis. *Life Sci.* **11**, 309–315.

Jones, J. P., Nicholas, H. J., and Ramsey R. B. (1975). Rate of sterol formation by rat brain glia and neurons *in vitro* and *in vivo*. *J. Neurochem.* **24**, 123–126.

Karlsson, J.-O., Hamberger, A., and Henn, F. A. (1973). Polypeptide composition of membranes derived from neuronal and glial cells. *Biochim. Biophys. Acta* **298**, 219–229.

Kim, S. U., Warren, K. G., and Kalia, M. (1979). Tissue culture of adult human neurons. *Neurosci. Lett.* **11**, 137–141.

Kimelberg, H. K., Biddlecome, S., Narumis, S., and Bourke, R. S. (1978). ATPase and carbonic anhydrase activities of bulk isolated neuron, glia and synaptosome fractions from rat brain. *Brain Res.* **141**, 305–323.

Koelle, G. B. (1951). Elimination of enzymatic diffusion artefacts in the histochemical localization of cholinesterases and a survey of their cellular distribution. *J. Pharmacol. Exp. Ther.* **103**, 153–171.

Kohl, H. H., and Sellinger, O. Z. (1972). Protein synthesis in neuronal perikarya isolated from cerebral cortex of the immature rat. *J. Neurochem.* **19**, 699–711.

Kohlschütter, A., and Herschkowitz, N. N. (1973). Sulfatide synthesis in neurons: A defect in mice with a hereditary myelination disorder. *Brain Res.* **50**, 379–385.

Korey, S. R., Orchen, M., and Brotz M. (1958). Studies of white matter I. Chemical constitution and respiration of neuroglial and myelin enriched fractions of white matter. *J. Neuropath. Exp. Neurol.* **17**, 430–438.

Kostenko, M. A., Geletyuk, V. I., and Veprintsev, B. N. (1974). Completely isolated neurons in the mollusc, Lymnaea stagnalis. A new objective for nerve cell biology investigation. *Comp. Biochem. Physiol.* **49A**, 89–100.

Kronick, P. L., Campbell, G.-LeM., and Joseph, K. (1978). Magnetic ;microspheres prepared by redox polymerization used in a cell separation based on gangliosides. *Science (Washington, D.C.)* **200**, 1074–1076.

Labourdette, G., Roussel, G., Ghandour, M. S., and Nussbaum, J. L. (1979). Cultures from rat brain hemispheres enriched in oligodendrocyte-like cells. *Brain Res.* **179**, 199–203.

Labourdette, G., Roussel, G., and Nussbaum, J. L. (1980). Oligodendroglia content of glial cell primary cultures, from newborn rat brain hemispheres, depends on the initial plating density. *Neurosci. Lett.* **18**, 203–209.

Lane, J. D., and Neuhoff, V. (1980). Phenylketonuria—Clinical and experimental considerations revealed by the use of animal models. *Naturwissenschaften* **67**, 227–233.

Leveille, P. J., McGinnis, J. F., Maxwell, D. S., and DeVellis, J. (1980). Immunocytochemical localization of glycerol-3-phosphate dehydrogenase in rat oligodendrocytes. *Brain Res.* **196**, 287–305.

Llinas, R. (1969). Neurobiology of cerebellar evolution and development. *Proc. I. Int. Symp. Inst. Biomed. Res.*

London, J. A., and Merickel, M. (1979). Single cell isolation: A way to examine network interactions. *Brain Res.* **179**, 219–230.

Løvtrup-Rein, H. (1970). Protein synthesis in isolated nuclei of nerve and glial cells from rat brain. *Brain Res.* **19**, 433–444.

Løvtrup-Rein, H., and Grahn, B. (1974). Polysomes and polysomal RNA from nerve and glial cell fractions. *Brain Res.* **72**, 123–136.

Lowry, O. H. (1953). The quantitative histochemistry of the brain—Histological sampling. *J. Histochem. Cytochem.* **1**, 420–428.

Lowry, O. H., Roberts, N. R., and Chang, M.-L. W. (1956). The analysis of single cells. *J. Biol. Chem.* **222**, 97–107.

Lowry, O. H., and Passonneau, J. V. (1972). "A Flexible System of Enzymatic Analysis," pp. 223–235. Academic Press, New York.

Ludwin, S. K., Kosek, J. C., and Eng, L. F. (1976). The topographical distribution of S-100 and GFA proteins in the adult rat brain: An immunohistochemical study using horseradish peroxidase labelled antibodies. *J. Comp. Neurol.* **165**, 197–208.

Maccioni, H. J. F., Defilpo, S. S., Lauda, C. A., and Caputto, R. (1978). The biosynthesis of brain gangliosides. Ganglioside glycosylating activity in rat brain neuronal perikarya fraction. *Biochem. J.* **174**, 673–680.

Mandel, P., Roussel, G. Delaunoy, J.-P., and Nussbaum, J.-L. (1978). Wolfgram proteins, oligodendroglial cell and myelin markers, carbonic anhydrase C, a glial marker. *In* "Dynamic

Properties of Glia Cells'' (E. Schoffeniels, G. Franck, L. Hertz, and D. B. Tower, eds.), pp. 267–274. Pergamon, New York.

Marggraf, W. D., Wang, H., and Kanfer, J. N. (1979). Phospholipid metabolism in isolated neuronal and glial cells from brains of 17-day old rats. *J. Neurochem.* **32,** 353–361.

Margolis, R. U., and Margolis R. K. (1974). Distribution and metabolism of mucopolysaccharides and glycoproteins in neuronal perikarya, astrocytes and oligodendroglia. *Biochemistry* **13,** 2849–2852.

Martinez-Hernandez, A., Bell, K. P., and Norenberg, M. D. (1977) Glutamine synthetase: Glial localization in brain. *Science (Washington, D.C.)* **195,** 1356–1358.

McCarthy, K. D., and DeVellis, J. (1980). Preparation of separate astroglial and oligodendroglial cell cultures from rat cerebral tissue. *J. Cell Biol.* **85,** 890–902.

McIlwain. H. (1954). *Proc. Univ. Otago Medical Sch.* **32,** 17, cited *in* McIlwain, H., and Rodnight, R. (1972). ''Practical Neurochemistry,'' pp. 132–133. Livingston, Edinburgh.

McKhann, G. M., Ho, W., Raiborn, Ch., and Varon, S. (1969). The isolation of neurons from normal and abnormal human cerebral cortex. *Arch. Neurol. (Chicago)* **20,** 542–547.

Medzihradsky, F., and Marks, M. J. (1975). Measures of viability in isolated cells. *Biochem. Med.* **13,** 164–177.

Medzihradsky, F., Sellinger, O. Z., Nandhasri, P. S., and Santiago, J. C. (1972). ATPase activity in glial cells and in neuronal perikarya of rat cerebral cortex during early postnatal development. *J. Neurochem.* **19,** 543–545.

Meyer, R. A., Haase, S. F., Poduslo, S. E., and McKhann, G. M. (1974). Light-scattering patterns of isolated oligodendroglia. *J. Histochem. Cytochem.* **22,** 594–597.

Miller, R. G., and Philipps, R. A. (1969), Separation of cells by velocity sedimentation. *J. Cell. Physiol.* **73,** 191–201.

Mirsky, R., Winter, J., Abney, E. R., Pruss, R. M., Gavrilovic, J., and Raff, M. C. (1980). Myelin-specific proteins and glycolipids in rat Schwann cells and oligodendrocytes in culture. *J. Cell Biol.* **84,** 483–494.

Montz, H. (1981). Isolierung und Kultivierung von Nervenzellen aus dem Hirn erwachsener Ratten. M. D. Thesis, Univ. of Göttingen.

Mori, S., and Leblond, C. P. (1970). Electron microscopic identification of three classes of oligodendrocytes and a preliminary study of their proliferative activity in the corpus callosum of young rats. *J. Comp. Neurol.* **139,** 1–30.

Nagata, Y., and Tsukada, Y. (1978). Bulk separation of neuronal cell bodies and glial cells from mammalian brain and some of their biochemical properties. *In* ''Reviews of Neuroscience 3'' (S. Ehrenpreis and I. J. Kopin, eds.), pp. 195–221. Raven, New York.

Nagata, Y., Mikoshiba, K., and Tsukada, Y. (1974). Neuronal cell body enriched and glial cell enriched fractions from young and adult rat brains; preparative and morphological and biochemical properties. *J. Neurochem.* **22,** 493–503.

Nagata, Y., Nanba, T., and Ando, M. (1976). Changes in some enzymic activities of separated neuronal and glial cell-enriched fractions from rat brains during development. *Neurochem. Res.* **1,** 299–312

Neskovic, N., Nussbaum, J. L., and Mandel, P. (1970). A study of glycolipid metabolism in myelination disorder of Jimpy and Quaking mice. *Brain Res.* **21,** 39–53.

Neuhoff, V. (1973). Micromethods in molecular biology. *In* ''Molecular Biology, Biochemistry and Biophysics'' (A. Kleinzeller, G. F. Springer, H. G. Wittmann, eds.), Vol. 14. Springer-Verlag, Berlin and New York.

Norenberg, M. D. (1979). The distribution of glutamine synthetase in the rat central nervous system. *J. Histochem. Cytochem.* **27,** 756–762.

Norton, W. T., and Poduslo, S. E. (1970). Neuronal soma and whole neuroglia of rat brain: A new isolation technique. *Science (Washington, D.C.)* **167,** 1144–1145.

Norton, W. T., and Poduslo, S. E. (1971). Neuronal perikarya and astroglia of rat brain: Chemical composition during myelination. *J. Lipid Res.* **12**, 84–90.

Norton, W. T., Abe, T., Poduslo, S. E., and DeVries, G. H. (1975). The lipid composition of isolated brain cells and axons. *J. Neurosci. Res.* **1**, 57–75.

Osborne, N. N. (1978). "Biochemistry of Characterised Neurons." Pergamon, Oxford.

Packman, P. M., Blomstrand, C., and Hamberger, A. (1971). Disc electrophoretic separation of proteins in neuronal, glial and subcellular fractions from cerebral cortex. *J. Neurochem.* **18**, 479–487.

Palay, S. L., and Chan-Palay, V. (1974). "Cerebellar Cortex. Cytology and Organization." Springer-Verlag, Berlin and New York.

Palmer, G. C. (1973) Adenyl cyclase in neuronal and glial enriched fractions from rat and rabbit brain. *Res. Commun. Chem. Pathol. Pharmacol.* **5**, 603–613.

Palmer, G. C., and Manian, A. A. (1976). Actions of phenothiazine analogues on dopamine-sensitive adenylate cyclase in neuronal and glial enriched fractions from rat brain. *Biochem. Pharmacol.* **25**, 63–71.

Pettmann, B., Louis, J. C., and Sensenbrenner, M. (1979). Morphological and biochemical maturation of neurones cultured in the absence of glial cells. *Nature (London)* **281**, 378–380.

Pettmann, B., Delaunoy, J.-P., Courageot, J., Devilliers, G., and Sensenbrenner, M. (1980). Rat brain glial cells in culture: effects of brain extracts on the development of oligodendroglia-like cells. *Dev. Biol.* **75**, 278–287.

Pevzner, L. Z. (1979). "Functional Biochemistry of the Neuroglia," pp. 15–64. Consultants Bureau, New York.

Pfeiffer, S. E., Barbarese, E., and Bhat, S. (1980). Myelinogenic properties expressed in mixed primary cultures of rat brain and in long-term cultures of isolated oligodendrocytes. *Abstr. I Meet. Int. Soc. Develop. Neurosci., Straßbourg*, p. 62.

Pieringer, R. A., Deshmukh, D. S., and Flynn, T. J. (1973). The association of the galactosyl diglycerides of nerve tissue with myelination. *Progr. Brain Res.* **40**, 397–405.

Pleasure, D., Abramsky, O., Silberberg, D., Quinn, B., Parris, J. and Saida, T. (1977). Lipid synthesis by an oligodendroglial fraction in suspension culture. *Brain Res.* **134**, 377–382.

Pleasure, D., Lichtman, C., Eastman, S., Lieb, M., Abramsky, O., and Silberberg, D. (1979). Acetoacetate and D-(-)-Beta-Hydroxybutyrate as precursors for sterol synthesis by calf oligodendrocytes in suspension culture: Extramitochondrial pathway for acetoacetate metabolism. *J. Neurochem.* **32**, 1447–1450.

Poduslo, S. E. (1975). The isolation and characterization of a plasma membrane and a myelin fraction derived from oligodendroglia of calf brain. *J. Neurochem.* **24**, 647–654.

Poduslo, S. E. (1978). Studies of isolated, maintained oligodendroglia; biochemistry, metabolism, and *in vitro* myelin synthesis. *In* "Myelination and Demyelination" (J. Palo, ed.), Vol. 100, pp. 71–94. Plenum, New York.

Poduslo, S. E., and McKhann, G. M. (1975). The isolation of specific cell types from mammalian brain. *In* "The Nervous System," Vol. 1: The Basic Neurosciences (D. B. Tower, ed.), pp. 437–441. Raven, New York.

Poduslo, S. E., and McKhann, G. M. (1976). Maintenance of bulk isolated neurons. *Neurosci. Lett.* **2**, 267–271.

Poduslo, S. E., and McKhann, G. M. (1977a). Maintenance of neurons isolated in bulk from rat brain: Incorporation of radiolabelled substrates. *Brain Res.* **132**, 107–120.

Poduslo, S. E., and McKhann, G. M. (1977b). Synthesis of cerebrosides by intact oligodendroglia maintained in culture. *Neurosci. Lett.* **5**, 159–163.

Poduslo, S. E., and Norton, W. T. (1972a). Isolation and some chemical properties of oligodendroglia from calf brain. *J. Neurochem.* **19**, 727–736.

Poduslo, S. E., and Norton, W. T. (1972b). The bulk separation of neuroglia and neuron perikarya. *In* "Research Methods in Neurochemistry" (N. Marks and R. Rodnight, eds.), Vol. 1, pp. 19–32. Plenum, New York.

Poduslo, S. E., and Norton, W. T. (1975). Isolation of specific brain cells. *Methods Enzymol.* **35**, 561–579.

Poduslo, S. E., McFarland, H. F., and McKhann, G. M. (1977). Antiserums to neurons and to oligodendroglia from mammalian brain. *Science (Washington, D.C.)* **197**, 270–272.

Poduslo, S. E., Miller, K., and McKhann, G. M. (1978). Metabolic properties of maintained oligodendroglia purified from brain. *J. Biol. Chem.* **252**, 1592–1597.

Polak, J. M., Facer, P., Marangos, P. J., and Age, P. (1980). Immunohistochemical localization of neurone specific enolase (NSE) in the gastrointestinal tract. *J. Histochem. Cytochem.* **28**, 618.

Privat, A. (1975). Postnatal gliogenesis in the mammalian brain. *Int. Rev. Cytol.* **40**, 281–323.

Pruss, R. M., Lawson, D., and Bartlett, P. (1979). Oligodendrocytes can be isolated from cultures of rat central nervous system. *In* "Biochemical Development of Nervous Tissue in Culture," p. 67. BAT-SHEVA Seminar (Abstr.).

Radin, N. S., Brenkert, A., Arora, R. C., Sellinger, O. Z., and Flangas, A. L. (1972). Glial and neuronal localization of cerebroside-metabolizing enzymes. *Brain Res.* **39**, 163–169.

Radojcic, T., and Pentreath, V. W. (1979). Invertebrate glia. *Prog. Neurobiol. (Oxford)* **12**, 115–179.

Raff, M. C., Mirsky, R., Fields, K. L., Lisak, R. P., Dorfman, S. H., Silberberg, D. H., Gregson, N. A., Leibowitz, S., and Kennedy M. C. (1978). Galactocerebroside is a specific cell-surface antigenic marker for oligodendrocytes in culture. *Nature (London)* **274**, 813–816.

Raff, M. C., Fields, K. L., Hakomori, S.-I., Mirsky, R., Pruss, R. M., and Winter, J. (1979). Cell-type-specific markers for distinguishing and studying neurons and the major classes of glial cells in culture. *Brain Res.* **174**, 283–308.

Raghavan, S., and Kanfer, J. N. (1972). Ceramide galactoside of enriched neuronal and glial fractions from rat brain. *J. Biol. Chem.* **247**, 1055–1056.

Raghavan, S., Rhoads, D. B., and Kanfer, J. N. (1972). Acid hydrolases in neuronal and glial enriched fractions of rat brain. *Biochim. Biophys. Acta* **268**, 755–760.

Raine, C. S., Wisniewski, H. M., Iqbal, K., Grundke-Iqbal, S., and Norton, W. T. (1977). Studies on the encephalitogenic effects of purified preparations of *human* and *bovine* oligodendrocytes. *Brain Res.* **120**, 269–286.

Raine, C. S., Traugott, U., Iqbal, K., Snyder, D. S., Cohen, S. R., Farooq, M., and Norton, W. T. (1978). Encephalitogenic properties of purified preparations of bovine oligodendrocytes tested in guinea pigs. *Brain Res.* **148**, 85–96.

Rappaport, C., and Howze, G. B. (1966). Further studies on the dissociation of adult mouse tissue. *Proc. Soc. Exp. Biol. Med.* **121**, 1016–1021.

Robert, J., Freysz, L., Sensenbrenner, M., Mandel, P., and Rebel, G. (1975). Gangliosides of glial cells: A comparative study of normal astroblasts in tissue culture and glial cells isolated on sucrose-ficoll gradients. *FEBS Lett.* **50**, 144–146.

Roberti, R., Binaglia, L., Francescangeli, E., Goracci, G., and Porcellati, G. (1975). Enzymic synthesis of 1-alkyl-2-acyl-sn-glycero-3-phosphorylethanolamine through ethanolamine-phosphotransferase activity in the neuronal and glial cells of rabbit *in vitro. Lipids* **10**, 121–127.

Robins, E., Fisher, H. K., and Lowe, I. P. (1961). Quantitative histochemical studies of the morphogenesis of the cerebellum. II. Two β-galactosidases. *J. Neurochem.* **8**, 96–104.

Roots, B. I., and Johnston, P. V. (1964). Neurons of ox brain nuclei: their isolation and appearance by light and electron microscopy. *J. Ultrastruct. Res.* **10**, 350–361.

Roots, B. I., and Johnston, P. V. (1972). Nervous system cell preparations: microdissection and micromanipulation. *In* "Research Methods in Neurochemistry" (N. Marks and R. Rodnight, eds.), Vol. 1, pp. 3–17. Plenum, New York.

Rose, S. P. R. (1965). Preparation of enriched fractions from cerebral cortex containing isolated, metabolically active neuronal cells. *Nature (London)* **206,** 621–622.

Rose, S. P. R. (1967). Preparation of enriched fractions from cerebral cortex containing isolated, metabolically active neuronal and glial cells. *Biochem. J.* **102,** 33–43.

Rose, S. P. R. (1968a). The biochemistry of neurones and glia. *In* "Applied Neurochemistry" (A. N. Davison and J. Dobbing eds.), pp. 332–355. Blackwell, Oxford.

Rose, S. P. R. (1968b). Glucose and amino acid metabolism in isolated neuronal and glial cell fractions in vitro. *J. Neurochem.* **15,** 1415–1429.

Rose, S. P. R. (1970). The compartmentation of glutamate and its metabolites in fractions of neuron cell bodies and neuropil; studied by intraventricular injection of U-^{14}C glutamate. *J. Neurochem.* **17,** 809–816.

Rose, S. P. R. (1973). Cellular compartmentation of metabolism in the brain. *In* "Metabolic Compartmentation in the Brain" (R. Balazs and J. E. Cramer, eds.), pp. 287–304. Macmillan, New York.

Rose, S. P. R. (1975). Cellular compartmentation of brain metabolism and its functional significance. *J. Neurosci. Res.* **1,** 19–30.

Rose, S. P. R. (1976). Functional biochemistry of neurons and glial cells. *In* "Progress in Brain Research" (M. A. Corner and D. F. Swaab, eds.), Vol. 45, pp. 67–82. Elsevier, Amsterdam.

Rose, S. P. R., and Sinha, A. K. (1974). Incorporation of amino acids into proteins in neuronal and neuropil fractions of rat cerebral cortex: Presence of a rapidly labelling neuronal fraction. *J. Neurochem.* **23,** 1065–1076.

Roussel, G., Delaunoy, J.-P., Nussbaum, J.-L, and Mandel, P. (1979). Demonstration of a specific localization of carbonic anhydrase C in the glial cells of rat CNS by an immunohistochemical method. *Brain Res.* **160,** 47–55.

Satake, M., and Abe, S. (1966). Preparation and characterization of nerve cell perikarya from rat cerebral cortex. *J. Biochem.* **59,** 72–75.

Satake, M., Hasegawa, S.-J., Abe, S., and Tanaka, R. (1968). Preparation and characterization of nerve cell perikaryon from pig brain stem. *Brain Res.* **11,** 246–250.

Schachner, M. (1974). NS-1 (nervous system antigen-1), a glial-cell-specific antigenic component of the surface membrane. *Proc. Natl. Acad. Sci. U.S.A.* **71,** 1795–1799.

Schachner, M., Wortham, K. A., Ruberg, M. Z., Dorfman, S., and Campbell, G. LeM. (1977). Brain cell surface antigens detected by anti-corpus callosum antiserum. *Brain Res.* **127,** 87–97.

Schengrund, C.-L., and Nelson, J. T. (1976). Ganglioside sialidase activity in bovine neuronal perikarya. *Neurochem. Res.* **1,** 171–180.

Schmechel, D., Marangos, P. J., Zis, A. P., Brightman, M., and Goodwin, F. K. (1978). Brain enolases as specific markers of neuronal and glial cells. *Science (Washington, D.C.)* **199,** 313–315.

Schon, F., and Kelly, J. S. (1975). Selective uptake of ^{3}H β-alanine by glia: Association with the glial uptake system for GABA. *Brain Res.* **86,** 243–257.

Sellinger, O. Z., and Azcurra, J. M. (1974). Bulk separation of neuronal cell bodies and glial cells in the absence of added digestive enzymes. *In* "Research Methods in Neurochemistry" (N. Marks and R. Rodnight, eds.), Vol. 2, pp. 3–38. Plenum, New York.

Sellinger, O. Z., and Santiago, J. C. (1972). Unequal development of two mitochondrial enzymes in neuronal cell bodies and glial cells of rat cerebral cortex. *Neurobiology (Copenhagen)* **2,** 133–146.

Sellinger, O. Z., Azcurra, J. M., Johnson, D. E., Ohlsson, W. G., and Lodin, Z. (1971). Independence of protein synthesis and drug uptake in nerve cell bodies and glial cells isolated by a new technique. *Nature (London) New Biol.* **230,** 253–256.

Sellinger, O. Z., Johnson, D. E., Santiago, J. C., and Idyaga-Vargas, V. (1973). A study of the biochemical differentiation of neurons and glia in the rat cerebral cortex. *Prog. Brain Res.* **40,** 331–347.

Sellinger, O. Z., Legrand, J., Clos, J., and Ohlsson, W. G. (1974). Unequal patterns of development of succinate-dehydrogenases and acetylcholinesterase in Purkinje cell bodies and granule cells isolated in bulk from the cerebellar cortex of the immature rat. *J. Neurochem.* **23,** 1137–1144.

Sellström, Å., and Hamberger, A. (1975). Neuronal and glial systems for γ-aminobutyric acid transport. *J. Neurochem.* **24,** 847–852.

Sellström, Å, Sjöberg, L.-B., and Hamberger, A. (1975). Neuronal and glial systems for γ-aminobutyric acid metabolism. *J. Neurochem.* **25,** 393–398.

Sensenbrenner, M. (1977). Dissociated brain cells in primary cultures. *In* "Cell, Tissue, and Organ Cultures in Neurobiology" (S. Fedoroff and L. Hertz, eds.), pp. 191–213. Academic Press, New York.

Shahar, A., Schupper, H., Kimhi, Y., Mizrachi, Y., and Schwartz, M. (1981). Regeneration of adult rat brain neurons in culture is enhanced by glial factor(s). *Int. Symp. Nerv. Syst. Regeneration, Catania,* p. 40 (Abstr.).

Shahar, A., Althaus, H.-H., Montz, H., Schwartz, P., Schupper, H., and Neuhoff, V. Scanning microscopy of bulk isolated rat brain cells in culture, in preparation.

Sinha, A. K., and Rose, S. P. R. (1971). Bulk separation of neurones and glia: A comparison of techniques. *Brain Res.* **33,** 205–217.

Sinha, A. K., and Rose, S. P. R. (1972a). Compartmentation of lysosomes in neurones and neuropil and a new neuronal marker. *Brain Res.* **39,** 181–196.

Sinha, A. K., and Rose, S. P. R. (1972b). Monoamineoxidase and cholesterase activity in neuronal and neuropil from the rat cerebral cortex. *J. Neurochem.* **19,** 1607–1610.

Skrivanek, J., Ledeen, R., Norton, W., and Farooq, M. (1978). Ganglioside distribution in rat cortex. *Trans. Am. Soc. Neurochem.* **9,** 185.

Snyder, D. S., Raine, C. S., Farooq, M., and Norton, W. T. (1980). The bulk isolation of oligodendroglia from whole rat forebrain: A new procedure using physiologic media. *J. Neurochem.* **34,** 1614–1621.

Sotelo, J., Gibbs, C. J., Gajdusek, C., Toh, B. H., and Wurth, M. (1980). Method for preparing cultures of central neurons: cytochemical and immunochemical studies. *Proc. Natl. Acad. Sci. U.S.A.* **77,** 653–657.

Stahl, W. L., and Broderson, S. H. (1975). Localization of Na$^+$,K$^+$-ATPase in brain. *Fed. Proc., Fed. Am. Soc. Exp. Biol.* **35,** 1260–1265.

Sternberger, N. H., Itoyama, Y., Kies, M. W., and Webster, H. deF. (1978). Myelin basic protein demonstrated immunocytochemically in oligodendroglia prior to myelin sheath formation. *Proc. Natl. Acad. Sci. U.S.A.* **75,** 2521–2524.

Svoboda, P., and Lodin, Z. (1972). Postnatal development of some mitochondrial enzyme activities of cortical neurons and glial cells. *Physiol. Bohemoslov.* **21,** 457–465.

Szuchet, S., and Polak, P. E. (1980). Isolation of the plasma membrane from oligodendrocytes. *Fed. Proc., Fed. Am. Soc. Exp. Biol.* **39,** 1995.

Szuchet, S., and Stefansson, K. (1980). *In vitro* behaviour of isolated oligodendrocytes. *In* "Advances in Cellular Neurobiology" (S. Fedoroff and L. Hertz, eds.), Vol. 1, pp. 314–346. Academic Press, New York.

Szuchet, S., Arnason, B. G. W., and Polak, P. E. (1980b). Separation of ovine oligodendrocytes isolation. *Biophys. J.* **2,** 51a.

Szuchet, S., Stefansson, K., Wollmann, R. L., Dawson, B., and Arnason, B. G. W. (1980a). Maintenance of isolated oligodendrocytes in long-term culture. *Brain Res.* **200,** 151–164.

Szuchet, S., Arnason, B. G. W., and Polak, P. E. (1980b). Separation of bovine oligodendrocytes into two distinct bands on a linear sucrose grandient. *J. Neurosci. Methods* **3,** 7–19.

Takahashi, Y., Hsü, C. S., and Honma, S. (1970). Potassium and glutamate effects on protein synthesis in isolated neuroglial cells. *Brain Res.* **23**, 284–287.

Tamai, Y., Matsukawa, S., and Satake, M. (1971). Lipid composition of nerve cell perikarya. *Brain Res.* **26**, 149–157.

Tamai, Y., Araki, S., Komai, Y., and Satake, M. (1975). Studies on neuronal lipid. *J. Neurosci. Res.* **1**, 161–182.

Tennekoon, G. I., Kishimoto, Y., Singh, J., Nonaka, G., and Bourre, J.-M. (1980). The differentiation of oligodendrocytes in the rat optic nerve. *Dev. Biol.* **79**, 149–158.

Tolle, S. W., Dyson, R. D., Newburgh, R. W., and Cardenas, J. M. (1976). Pyruvate kinase isozymes in neurons, glia, neuroblastoma, and glioblastoma. *J. Neurochem.* **27**, 1355–1360.

Townsend, R. R., Li, Y.-T., Li, S.-C. (1979). Brain glycosidases. *In* "Complex Carbohydrates of Nervous Tissue" (R. Margolis and R. K. Margolis, eds.), pp. 127–137. Plenum, New York.

Traugott, U., Snyder, D. S., Norton, W. T., and Raine, C. S. (1978). Characterization of antioligodendrocyte serum. *Ann. Neurol.* **4**, 431–439.

Truffa-Bachi, P., and Wofsy, L. (1970). Specific separation of cells on affinity columns. *Proc. Natl. Acad. Sci. U.S.A.* **66**, 685–692.

Ulas, J. (1979). Methods of separation of neuronal and glial cells: Their application for biochemical investigation (poln.). *Postepi. Biochem.* **25**, 417–436.

Varon, S. (1977). Neural cell isolation and identification. *In* "Cell, Tissue and Organ Cultures in Neurobiology" (S. Fedoroff and L. Hertz, eds.), pp. 237–261. Academic Press, New York.

Varon, S. (1978). Macromolecular glial cell markers. *In* "Dynamic Properties of Glial Cells" (E. Schoffeniels, G. Franck, L. Hertz, and D. B. Tower eds.), pp. 93–103. Pergamon, New York.

Varon, S., and Somjen, G. G. (1979). Neuron-Glia interactions. *Neurosci. Res. Program Bull.* **17**, 42–65.

Venter, B. R., Venter, J. C., and Kaplan, N. O. (1976). Affinity isolation of cultured tumor cells by means of drugs and hormones covalently bound to glass and sepharose beads. *Proc. Natl. Acad. Sci. U.S.A.* **73**, 2013–2017.

Vogel, K. G. (1978). Effects of hyaluronidase, trypsin, and EDTA on surface composition and topography during detachment of cells in culture. *Exp. Cell Res.* **113**, 345–357.

Volpe, P., and Giuditta, A. (1967). Biosynthesis of RNA in neuron- and glia-enriched fractions. *Brain Res.* **6**, 228–240.

Weiler, C. T., Nyström, B., and Hamberger, A. (1979a). Characteristics of glutamine vs glutamate transport in isolated glia and synaptosomes. *J. Neurochem.* **32**, 559–565.

Weiler, C. T., Nyström, B., and Hamberger, A. (1979b). Glutaminase and glutamine synthetase activity in synaptosomes, bulk isolated glia and neurons. *Brain Res.* **160**, 539–543.

Wilkin, G. P., Balazs, R., Wilson, J. E., Cohen, J., and Dutton, G. R. (1976). Preparation of cell bodies from the developing cerebellum: Structural and metabolic integrity of the isolated cells. *Brain Res.* **115**, 181–199.

Williams, S. K., Gillis, J. F., Matthews, M. A., Wagner, R. C., and Bitensky, M. W. (1980). Isolation and characterization of brain endothelial cells: Morphology and enzyme activity. *J. Neurochem.* **35**, 374–381.

Willinger, M., and Schachner, M. (1978). Distribution of GM_1 ganglioside on cerebellar cells in tissue culture. *J. Supramolek. Struct. Suppl.* **2**, 128.

Woelk, H., Kanig, K., and Peiler-Ischikawa, K. (1974). Incorporation of ^{32}P into the phospholipids of neuronal and glial cell enriched fractions isolated from rabbit cerebral cortex: effect of norepinephrine. *J. Neurochem.* **23**, 1057–1063.

Woodhams, P. L., Cohen, J., Mallet, J., and Balazs, R. (1980). A preparation enriched in Purkinje cells identified by morphological and immunocytochemical criteria. *Brain Res.* **199**, 435–442.

Yanagihara, T. (1974). RNA metabolism in rabbit brain: Study with neuron-glia and subcellular fractions. *J. Neurochem.* **23**, 833–837.

Yanagihara, T. (1979). Protein and RNA syntheses and precursor uptake with isolated nerve and glial cells. *J. Neurochem.* **32,** 169–177.

Yanagihara, T., and Hamberger, A. (1973). A method for separation of Purkinje cell- and granular cell-enriched fractions from rabbit cerebellum. *Brain Res.* **59,** 445–448.

Yavin, Z., and Yavin, E. (1979). Survival and maturation of cerebral neurons on poly-L-lysine surfaces in the absence of serum. *In* "Biochemical Development of Nervous Tissue in Culture," p. 76. BAT-SHEVA Seminar.

Yu, R. K., and Iqbal, K. (1979). Sialosylgalactosyl ceramide as a specific marker for human myelin and oligodendroglial perikarya: Gangliosides of human myelin, oligodendroglia and neurons. *J. Neurochem.,* 293–300.

INDEX

A

Acetylation, of brain histones, 53, 54–55
Acetylcholine
  axonal transport of, 139
  coexistence with ACh, 274, 276
  supersensitivity to, 335
Acetylcholine receptor
  different forms of, genes and, 3
  immediate translation product, 9
  synaptic junction and, 111–112
Acetylcholinesterase
  axonal transport of, 141
  in isolated brain cells, 374
  NFG and, 167, 172–173
Acidic brain nuclear proteins, gene derepression
  and, 56
Acid phosphatase, in isolated brain cells, 373,
  374
ACTH
  active avoidance behavior and, 286–288
  anionic lipids and, 302–303
  counteraction of morphine analgesia, 289
  cyclic nucleotide metabolism and, 295–296
  deoxyglucose uptake and, 327
  gene for, 4–5
  grooming behavior and, 290–291, 296, 298,
    301, 303
  importance of Ca^{2+}, 295, 296
  labeling of brain protein and, 331–332, 333
  localization of, 257, 292–293
  membrane effects of, 302–303
  neurotransmitters and, 293, 294
  opiate receptors and, 289, 301–302
  polyphosphoinositide metabolism and, 298–
    301

protein phosphorylation and, 296–298
  receptors, 292, 293
  single dose tolerance to, 291
  steroidogenesis and, 295, 299
  working model of neurochemical action,
    303–306
Actin
  axonal transport of, 136
  brain, DNA probes for, 9
  in synaptic junctions and postsynaptic de-
    nsities, 107–108, 112, 113
α-Actinin, in postsynaptic densities, 108
Actinomycin D
  AChE induction and, 167
  learning and, 329
  neurite outgrowth and, 162, 165
  NGF and, 160
Adenosine monophosphate, cyclic, *see* Cyclic
  adenosine monophosphate
Adenylate cyclase, 30
  calmodulin and, 117
  in isolated brain cells, 377
  mutants and, 8
Adrenal medulla, enkephalins in, 262
Adrenal steroids, learning and, 322–323
β-Adrenergic receptor, 19
  adenylate cyclase and, 8, 28
  in fibroblasts, 28
Adrenocorticotropic hormone, *see* ACTH
Affinity columns, for separation of brain cell
  types, 359
Aging
  effect on brain RNA complexity, 90–92
  effect on cell RNA content, 91
Allergic encephalomyelitis, protein synthesis in
  brain and, 382

CELL BIOLOGY: A Series of Monographs

EDITORS

D. E. BUETOW

*Department of Physiology
and Biophysics
University of Illinois
Urbana, Illinois*

I. L. CAMERON

*Department of Anatomy
University of Texas
Health Science Center at San Antonio
San Antonio, Texas*

G. M. PADILLA

*Department of Physiology
Duke University Medical Center
Durham, North Carolina*

A. M. ZIMMERMAN

*Department of Zoology
University of Toronto
Toronto, Ontario, Canada*

G. M. Padilla, G. L. Whitson, and I. L. Cameron (editors). THE CELL CYCLE: Gene-Enzyme Interactions, 1969

A. M. Zimmerman (editor). HIGH PRESSURE EFFECTS ON CELLULAR PROCESSES, 1970

I. L. Cameron and J. D. Thrasher (editors). CELLULAR AND MOLECULAR RENEWAL IN THE MAMMALIAN BODY, 1971

I. L. Cameron, G. M. Padilla, and A. M. Zimmerman (editors). DEVELOPMENTAL ASPECTS OF THE CELL CYCLE, 1971

P. F. Smith. The BIOLOGY OF MYCOPLASMAS, 1971

Gary L. Whitson (editor). CONCEPTS IN RADIATION CELL BIOLOGY, 1972

Donald L. Hill. THE BIOCHEMISTRY AND PHYSIOLOGY OF *TETRAHYMENA*, 1972

Kwang W. Jeon (editor). THE BIOLOGY OF AMOEBA, 1973

Dean F. Martin and George M. Padilla (editors). MARINE PHARMACOGNOSY: Action of Marine Biotoxins at the Cellular Level, 1973

Joseph A. Erwin (editor). LIPIDS AND BIOMEMBRANES OF EUKARYOTIC MICROORGANISMS, 1973

A. M. Zimmerman, G. M. Padilla, and I. L. Cameron (editors). DRUGS AND THE CELL CYCLE, 1973

Stuart Coward (editor). DEVELOPMENTAL REGULATION: Aspects of Cell Differentiation, 1973

I. L. Cameron and J. R. Jeter, Jr. (editors). ACIDIC PROTEINS OF THE NUCLEUS, 1974

Govindjee (editor). BIOENERGETICS OF PHOTOSYNTHESIS, 1975

James R. Jeter, Jr., Ivan L. Cameron, George M. Padilla, and Arthur M. Zimmerman (editors). CELL CYCLE REGULATION, 1978

Gary L. Whitson (editor). NUCLEAR–CYTOPLASMIC INTERACTIONS IN THE CELL CYCLE, 1980

Danton H. O'Day and Paul A. Horgen (editors). SEXUAL INTERACTIONS IN EUKARYOTIC MICROBES, 1981

Ivan L. Cameron and Thomas B. Pool (editors). THE TRANSFORMED CELL, 1981

Arthur M. Zimmerman and Arthur Forer (editors). MITOSIS/CYTOKINESIS, 1981

Ian R. Brown (editor). MOLECULAR APPROACHES TO NEUROBIOLOGY, 1982

In preparation

Henry C. Aldrich and John W. Daniel (editors). CELL BIOLOGY OF *PHYSARUM* AND *DIDYMIUM*, Volume 1: Organisms, Nucleus, and Cell Cycle, 1982; Volume 2: Differentiation, Metabolism, and Methodology, 1982

John A. Heddle (editor). MUTAGENICITY: New Horizons in Genetic Toxicology, 1982

Potu N, Rao, Robert T. Johnson, and Karl Sperling (editors). PREMATURE CHROMOSOME CONDENSATION: Application in Basic, Clinical, and Mutation Research, 1982

George M. Padilla and Kenneth S. McCarty, Sr. (editors). GENETIC EXPRESSION IN THE CELL CYCLE, 1982

David S. McDevitt (editor). CELL BIOLOGY OF THE EYE, 1982

Govindjee (editor). PHOTOSYNTHESIS, Volume I: Energy Conversion by Plants and Bacteria; Volume II: Development, Carbon Metabolism, and Plant Productivity, 1982